LIFE SCIENCES RESEARCH AND DEVELOPMENT

INNOVATIONS IN LIFE SCIENCE RESEARCH

Life Sciences Research and Development

Additional books and ebooks in this series can be found on Nova's website under the Series tab.

LIFE SCIENCES RESEARCH AND DEVELOPMENT

INNOVATIONS IN LIFE SCIENCE RESEARCH

RAJESHWAR P. SINHA
SHASHI PANDEY-RAI
AND
NANDITA GHOSHAL
EDITORS

nova
science publishers
New York

Library of Congress Cataloging-in-Publication Data

ISBN: 978-1-53615-868-7
Library of Congress Control Number:2019944561

Published by Nova Science Publishers, Inc. † New York

Dedicated to Prof. em. Dr. Dr. h.c. Donat-P. Häder, Department of Biology, Emeritus from Friedrich-Alexander University (Erlangen), Neue Str. 9, Möhrendorf- 91096, Germany

On the occasion of his 75th birthday

Contents

PREFACE

This edited book titled "*Innovations in Life Science Research*" published by "*Nova Science Publishers*" attempts to present a wide spectrum of the state-of-the-art research in various fields of life science and their applications in the real-world challenges and complexities. This includes major efforts to develop effective techniques and tools for emergent areas and domains, including identification, mechanism and application of natural resources for the benefit of human being in the era of global climate change. This book includes 15 chapters. Chapter one deals with the sources and mode of action of natural sunscreens whereas chapters two and three describe about the biomedical and pharmaceutical potentials of secondary metabolites and genomics and proteomics of mycosporine-like amino acids respectively, involving cyanobacteria. Chapters four and five deal with targeted genome editing and molecular markers respectively. Chapters six, seven and eight describes genetics and epigenetic research and their role in crop improvement and bioremediation. Chapter nine explains the abiotic stress tolerance and chapter ten about disease management. Chapter eleven, twelve and thirteen deal with therapeutic potential of various plants. Chapter fourteen and fifteen deals with causes and consequences of land use change and multi-functionality of tropical forests. We hope this edited book will provide a better understanding of the current scientific problems and help in designing strategies to solve them.

We present this book with great enthusiasm and are pleased to express our sincere thanks to all the authors who have contributed chapters to this book, without their great efforts it would not have been possible to bring this book in its present form. In addition, we appreciate all reviewers and assistance provided by Dr. Jainendra Pathak, Dr. Rajneesh, Ms. Vidya Singh, Mr. Deepak Kumar, Mr. Haseen Ahmed, Mr. Deepak K. Singh, Ms. Abha Pandey, Ms. Sonal Mishra and Ms. Neha Kumari in formatting the book.

We are also extremely thankful to Nova Science Publishers, Inc., for their continued support and timely attention given in publishing this book.

Rajeshwar P. Sinha
Shashi Pandey-Rai
Nandita Ghoshal

March, 2019

In: Innovations in Life Science Research ISBN: 978-1-53615-868-7
Editors: R. P. Sinha, S. Pandey-Rai and N. Ghoshal ©2019 Nova Science Publishers, Inc.

Chapter 1

NATURAL SUNSCREENS: SOURCES AND MODE OF ACTION

Jainendra Pathak, Deepak K. Singh, Abha Pandey, Vidya Singh, Deepak Kumar and Rajeshwar P. Sinha*

Laboratory of Photobiology and Molecular Microbiology,
Centre of Advanced Study in Botany, Institute of Science,
Banaras Hindu University, Varanasi, India

ABSTRACT

Climate change and continuous enhancement of ultraviolet radiation (UVR; 280-400 nm), strongly affects terrestrial and aquatic ecosystems. Solar UVR are detrimental for most of the sun-exposed organisms. Natural sunscreens have emerged as important photo protectants in taxonomical diverse organisms ranging from prokaryotes to eukaryotes. Important 'natural sunscreen' compounds include phenylpropanoids, flavonoids, melanins, mycosporines, scytonemin, mycosporine-like amino acids (MAAs) and other chemical compounds distributed in microorganisms, plants and animals. Vitamins can be useful by themselves or as addition to sunscreens preparation. Carotenoids are being explored as UV-protective additives to limit photochemical damage. In many ways natural sunscreens are more effective over synthetic chemicals. Study of these sunscreens has led to the listing of new classes of photo protectants and their distribution among diverse biota and has enhanced understanding of their possible mode of action. Such studies are helpful in the area of photobiology for potential dermatological/biomedical applications. Natural sunscreens have a crucial role in different life forms such as scytonemin and MAAs act as a third line of defense mechanism in cyanobacteria and mycosporines are important for UV-induced sporulation in fungi which also dissipate absorbed radiation as heat, without generation of reactive oxygen species (ROS). Besides Photo protection, many of UV-absorbing chemicals are also known for their multipurpose role for example flavonoids have antioxidative and antimicrobial properties whereas MAAs and scytonemin are

* Corresponding Author's Email: rpsinhabhu@gmail.com.

efficient antioxidant, and can scavenge ROS produced during UV exposure. Present chapter deals with diverse sunscreens of biological origin and their possible modes of actions.

Keywords: melanin, carotenoids, mycosporine-like amino acids (MAAs), scytonemin, flavonoids, ultraviolet radiation (UVR)

INTRODUCTION

It is now a well-established fact that certain chemical compounds synthesized by diverse groups of living organisms act as sunscreens. Sunscreens are cosmetic products to protect skin from damage mediated by solar radiation (Fageon et al. 2009). These are sole chemical components that provide Photo protection against the direct effects of short wavelength solar radiation, enabling living organisms to survive and maintain the energy dynamics of an ecosystem. The natural shield of Earth, the ozone layer is being depleted and leaving "holes" above large tropical zones that have resulted in progressive risk of overexposure to harmful shortwave ultraviolet radiations (UVR) (Manney et al. 2011; Singh et al. 2017). Anthropogenically, as well as naturally released ozone depleting substances (ODS), are the key factors for negative impacts of UVR on the living biota. UVR absorbed by the major biomolecules such as nucleic acids, proteins and lipids ultimately result in lethal symptoms in the biological systems (Richa et al. 2016). Solar UV-B and UV-A radiation are detrimental for most of the sun-exposed organisms. To counteract the photo damage various organisms including microorganisms, plants and animals have developed a number of photo protective compounds functioning as sunscreens.

While investigating, sunscreen producing taxonomical groups, melanins are one of the best known sunscreens, which intercepts UV photons in the skin of humans and other animals (Scotto and Fraummeni 1982; Kollias et al. 1991). Apart from humans, melanins are also reported in fungi and some bacteria (Butler and Day 1998; Plonka and Grabacka 2006). Melanins are dark, generally black, biological macromolecules, composed of various types of phenolic or indolic monomer, usually complexed with protein and carbohydrates (Butler and Day 1998; Plonka and Grabacka 2006). Melanins along with their three main types, eumelanins, pheomelanins and allomelanins are distributed in prokaryotes and eukaryotes (Gao and Garcia-Pichel 2011a). Fungal melanins are usually located in the cell wall, either enmeshed within the structure of the wall, or as its outermost layer (Durrell 1964; Borisyuk et al. 1991). Melanins localization occurs in the mammalian melanosome, where the melanin is located internally in specialized vacuoles.

In higher plants, phenylpropanoids and flavonoids are the important photo protectants that play crucial role against UVR damage (Sinha et al. 2007; Gutiérrez-Hernández et al. 2016). Excessive exposure to UVR triggers production of non-photosynthetic pigments and flavonoids which are responsible for blocking UVR in addition to their antioxidant activity (Cho et al. 2007). There are enough evidences for a protective role of flavonoids in defence against UV-B in plants. However, confirmatory proof was obtained after mutant studies (chalcone synthase (CHS) and chalcone flavanone isomerase (CHI)) in *Arabidopsis thaliana* that showed hypersensitivity towards UV-B (Li et al. 1993).

Lingin is synthesized in plants by phenylpropanoid pathway. Lignin derived complex, lignin-carbohydrate complexes (LCCs) show excellent anti-UV activity (Nanbu et al. 2011, 2013). However, synthesized lignin nanoparticles (lignin NPs) showed more efficacies towards UVR absorption (Gutiérrez-Hernández et al. 2016). Besides, microorganisms possess diverse range of other compounds such as scytonemins, mycosporines, mycosporine-like amino acids (MAAs) and several other substances of unknown chemical structures that absorb UVR (Sinha et al. 1998; Cockell and Knowland 1999; Pathak et al. 2015a; 2015b; 2017a; 2017b; 2017c; Rajneesh et al. 2017a; 2019; Richa et al. 2018; Pathak et al. 2019) and have been proposed to have a sunscreen role.

Scytonemin is a yellow to brown, lipid-soluble, non-fluorescent, very stable pigment that is excreted and deposited in the extracellular polysaccharide sheaths of some cyanobacteria (Rastogi et al. 2015; Pathak et al. 2017a). Role of scytonemin as photo protective compound has been reported in several cyanobacteria inhabiting diverse habitats (Garcia-Pichel and Castenholz 1991; Sinha et al. 1999; Rastogi et al. 2013; Pathak et al. 2017a; 2017b; 2017c). Scytonemin absorbs and dissipates energy through harmless thermal de-excitation and also act as a potent antioxidant, preventing cellular damage resulting from UV-induced production of reactive oxygen species (ROS) (Takamatsu et al. 2003; Matsui et al. 2012; Rastogi and Incharoensakdi 2014). Scytonemin showed photo protective function that can be measured in terms of *in vivo* ROS and thymine dimer formation (Rastogi et al. 2013).

Mycosporines were discovered and named because it was found to be responsible for light-induced sporulation in fungi (Trione et al. 1966; Favre-Bonvin et al. 1976; Dehorter et al. 1980). MAAs are colourless compounds found intracellularly in many marine and freshwater organisms (Sinha et al. 2007; Klisch and Hader 2008; Llewellyn and Airs 2010; Richa and Sinha 2015; Richa et al. 2018; Pathak et al. 2019). MAAs have been reported in different taxonomic groups, including micro/macroalgae, lichen symbionts and several aquatic animals (Sinha et al. 2007; Pathak et al. 2017b). MAAs and synthetic derivatives have inspired applications as skin care products (Bandaranayake 1998). MAAs enable the absorbed energy to dissipate as heat into their surroundings, without producing ROS (Conde et al. 2000) and play an important role in photo protection. Figure 1 shows important natural sunscreens produced by various groups of organisms for their protection from solar radiation.

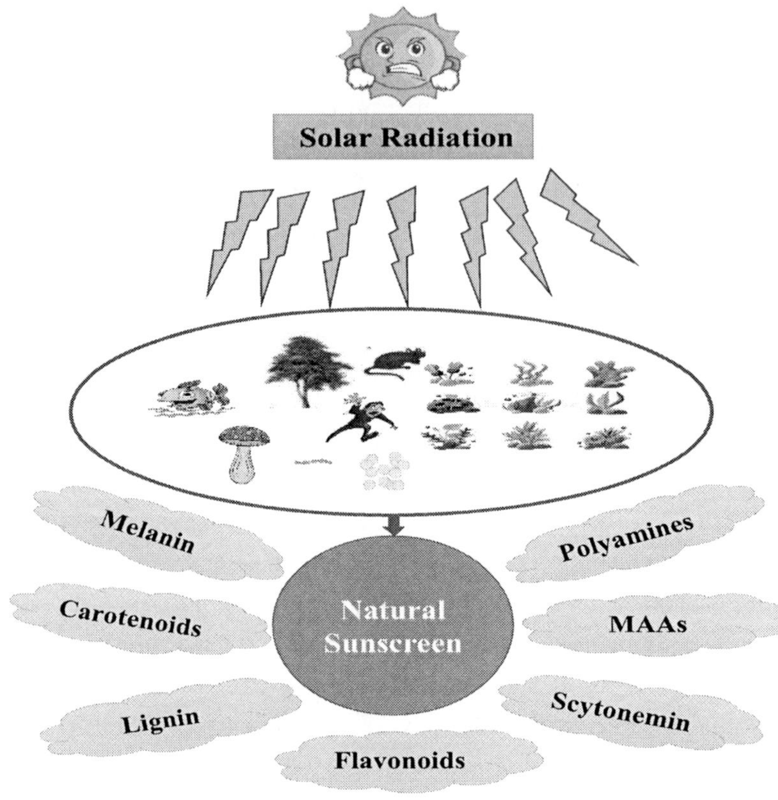

Figure 1. Important natural sunscreens synthesized by different organisms for their protection from high intensity solar radiation/UV radiation.

Apart from these, several antioxidants and other natural products such as carotenoids are currently being explored as UV-protective additives to limit photochemical damage (Morabito et al. 2011). Antioxidants are found naturally in plants. Earlier studies indicated that "natural" antioxidants (vitamin C and E) can be useful by themselves or as addition to sunscreens preparation (Darr et al. 1996). Carotenoids are important pigment because they play a role in protecting plants from UV damage. Carotenoids are produced in plants and other photosynthetic organisms. Humans cannot self-produce carotenoids and depend on a dietary supplement (Britton 1995; Olson and Krinsky 1995; Demmig-Adams and Adams 2002; Stahl and Sies 2005). They are effective natural antioxidants because of their singlet oxygen quenching ability and scavenging of peroxyl radicals (Burton 1984; Conn et al. 1991; Stahl and Sies 2003). Because of their antioxidant function and blue light filtering, carotenoids are plausible components to be used in a sunscreen formulation. Carotenoids have potentials in skin protection applications. Dietary mixture of the β-carotene, lutein and lycopene protects against UV induced erythema (Heinrich et al. 2003). Considering these information about the importance of natural sunscreens, this chapter bioprospects "natural sunscreens" in different groups of organisms and their possible mode of action has also been discussed.

Biology of Natural Sunscreens

Plant kingdom produces secondary metabolites that are synthesized to develop protective functions against hostile environments, such as oxidation by solar radiation. Plants secondary metabolites such as flavonoids, tannins, alkaloids, anthraquinones and polyphenols are well known for their photo protection and are considered as natural sunscreens, since they have the property of absorbing light in the UV region (Bizari 2013; Saewan and Jimtaisong 2013). Carotenoids are important photosynthetic pigment that influences biological activities including energy transfer, Photo protection and quenching of damaging singlet oxygen in plants and also bioactivities in humans and other organisms. In animals including human and fungi, melanin is dominating pigment produced as physiological response under high intensity light/UVR exposure. However, great diversity in photo protective/UV-absorbing compounds have been reported in microalgae including cyanobacteria such as scytonemins, mycosporine-like amino acids (MAAs) and several other substances of unknown chemical structures (Sinha et al. 1998; Cockell and Knowland 1999). These chemical substances have been proposed to have a sunscreen role. A list of natural sunscreens with their diverse sources has been shown in Table 1.

Melanin

Melanin is the generic name used to refer the most ubiquitous, resistant, heterogeneous, and ancient pigments found in nature. Melanins appeared very early in most living kingdoms on the Earth. Thus, melanins have been found in very old fossils from dinosaurs, early birds (Zhang et al. 2010; Wogelius et al. 2011), and primitive cephalopods (Glass et al. 2012). These findings will probably make melanin a new biomarker in life evolution. Melanins can be found in all living kingdoms prokaryotes as well as in eukaryotes (Gao and Garcia-Pichel 2011a).

A widespread and simple definition to include all types of melanins would be "heterogeneous polymer derived by the oxidation of phenols and subsequent polymerization of intermediate phenols and their resulting quinones". Melanins have been classified in at least five main types according to their source: animal melanins, plant melanins, fungal melanins, bacterial melanins, and synthetic melanins. The phenolic units of the polymer and the properties of the oxidases needed for the polymer formation show significant differences. On the basis of extraction of different melanins from different sources, melanin are named as, eumelanin, pheomelanin, neuromelanin, allomelanin, and pyomelanin. Melanin is the main pigment responsible for the various pigmentations found in animal and human skin, hair and eyes. Melanins are mainly synthesized from the amino acid L-tyrosine with the participation of tyrosinase, the key enzyme of melanogenesis in animals and many microorganisms. Thus, L-tyrosine should be considered the main monophenol precursor of melanin. Amino acid L-tyrosine is not used for the synthesis of plant melanin. In plants, the most common precursor is just catechol, so the melanin formed is also named catechol-melanin (Ghiani et al. 2008) and the enzymatic system involved in the synthesis is named catechol oxidase (Mayer 1986). Tyrosine is a poor substrate of these catechol oxidases, but it should be clear that catechol-melanin is a subtype of allomelanin, and catechol oxidase which is the most frequent type of plant phenol oxidases (Mason 1956), related to animal tyrosinases. Fungal melanin is quite abundant and appears in the cell wall rather than in specialized subcellular organelles such as the animal melanosomes. Usually, in fungi and yeast, melanin precursors are secreted and then oxidized outside the cell wall (Bell and Wheeler 1986). One of the main reasons for importance of fungal melanin is it plays an essential role in the virulence of plant and human pathogenic fungi (Taylor et al. 1987; Langfelder 2003) In fact, melanin formation was proposed as an appropriate technique for the classification of *Streptomyces* (Arai and Mikami 1972).

These species had been good models for the study of the enzyme tyrosinase, its structure, mechanism of catalysis, and regulation of its expression (Lerch and Ettlinger 1972; Jackman et al. 1991). Melanin is produced in living organisms as physiological responses under UVR exposure. Melanin pigments are one of the main determinants of skin and hair sensitivity to UV light and susceptibility to sun damage.

Table 1. Organisms belonging to different groups produce variety of sunscreens

Group of organism	Name	Sunscreening compound	Reference
		Mycosporine-like amino acids (MAAs)	
Rhodophyta	*Chondrus yendoi*	Shinorine	Tsujino et al. (1980)
	Palmaria palmate	Usujirine	
	Porphyra tenera	Porphyra	Takano et al. (1979)
Cyanobacteria	*Trichodesmium* sp.	Asterina-330, Palythene	Subramaniam et al. (1999)
	Euhalothece sp.	Euhalothece	Volkmann et al. (2006)
	Synechocystis PCC 6803	Mycosporine-taurine, M-343, Dehydroxylusujirene	Zhang et al. (2007)
	Anabaena doliolum	Mycosporine-glycine, Shinorine Porphyra-334	Singh et al. (2008)
	Anabaena variabilis PCC 7937	Palythine-serine, Shinorine	Singh et al. (2010)
	Nostoc commune	Palythine-threonine, Porphyra	Nazifi et al. (2013)

Group of organism	Name	Sunscreening compound	Reference
Sea anemone	*Alvania elegantissima*	Mycosporine-taurine	Stochaj et al. (1994)
Scleractinian coral	*Pocillopora damicornis*	Mycosporine methylamine-threonine	Wu Won et al. (1995)
	Stylophora pistillata	Palythine, threonine-sulphate	Wu Won et al. (1997)
	Pocillopora eydouxii	Mycosporine methylamine - serine	Teai et al. (1997)
Starfish	*Asterina pectinifera*	Asterina	Nakamura et al. (1981)
Zoanthid	*Palythoa tuberculosa*	Palithinol, Palythene	Takano et al. (1978)
Ascidian	*Halocyanthia roretizi*	(z)-palythenic acid	Kobayashi et al. (1981)
Ciliate	*Bursaridium*	Porphyra, Palythine, Asterina, Mycosporine-glycine, Usujerine	Sonntag et al. (2017)
Bacteria	*Micrococcus* sp. AK-334	Shinorine	Arai et al. (1992)
Cyanobacteria	*Nostoc punctiforme*, *Chroococcidiopsis* sp., *Lyngbya aestuarii*	Scytonemin	Fleming and Castenholz 2008; Rath et al. (2012)
	Nostoc muscorum, *Phormidium foveolarum*, *and Spirulina platensis*	Carotenoids, β-Carotene, Lutein, Zeaxanthin, Cryptoxanthin, α-Carotene, Lycopene	Cardozo et al. (2007) Prasanna et al. (2010) Kumar et al. (2016)
	Oscillatoria	Biopterin glucoside (BG)	Matsunaga et al. (1993)
	Nostoc sp. (TAU strain IL-235)	Prenostocdione	Ploutno and Carmeli (2001)
	Synechocystis sp. PCC 6803	Polyamines	Jantaro et al. (2003)
Bacteria	*Streptomyces* sp., *Azotobacter*, *Mycobacterium*, *Micrococcus*, *Bacillus*, *Streptomyces*, *Rhizobium*, *Vibrio*, *Proteus*, *Azospirillum*, *Pseudomonas aeruginosa*, *Hypomonas* sp., *Burkholderia cepacia*, *Escherichia coli*	Melanin	Coyne and Al-Harthi (1992); Geng et al. (2010); Li et al. (2018)
Fungi	*Cryptococcus neoformans*, *Aspergillus niger*, *Penicillium marneffei*, *Paracoccidioides brasiliensis*, *Cryptococcus neoforman*		Youngchim et al. (2004)
Plants	*Cinnamomum burmannii*, *Osmanthus fragrans*		Huang et al. (2011)
Animals	Mammals		Slominski et al. (2004)

FLAVONOIDS AND PHENYLPROPANOIDS

The term "flavonoid" is generally used to describe a broad collection of natural products that include a C6-C3-C6 carbon framework, or more specifically a phenyl benzopyran functionality. Depending on the position of the linkage of the aromatic ring to the benzopyrano (chromano) moiety, this group of natural products may be divided into three classes: the flavonoids (2-phenylbenzopyrans), iso flavonoids (3-benzopyrans) and the neoflavonoids (4-benzopyrans). These groups usually share a common chalcone precursor, and therefore are biogenetically and structurally related. Flavonoids are produced from phenylalanine *via* 4-coumaroyl-CoA. It has ability to absorb UVR due to the presence of chromophores in their structure. Rutin (3-*O* rutinoside-quercetin), can then be used as sunscreen in cosmetic formulations (Lopes et al. 2000; Violante et al. 2008; Bizari 2013; Panche et al. 2016). Chalcone synthase (CHS) catalyzes the first step reaction of the flavonoid biosynthesis; it may be possible to increase the production of UV-B-protective flavonoids by genetically improving the expression of CHS. Flavodiiron proteins (FDPs) also called A-type flavoproteins or Flvs, play important role in the photo protection of the photosynthetic machinery of cyanobacteria (Wasserfallen et al. 1998; Zhang et al. 2009, 2012; Allahverdiyeva et al. 2013; Ermakova et al. 2013; Pathak et al. 2019).

Anthocyanins are the most abundant flavonoid pigments that get sequestered in vacuoles and are largely responsible for diverse pigmentation from orange to red, purple and blue in flowers and fruits. Often, these compounds also occur in leaves, stems, seeds, and other tissues. Flavonoids play an important role in photosynthetic resistance against the UV-B component of sun radiation (Solovchenko and Schmitz-Eiberger 2003). Plants respond to abiotic UV-B stress with enhanced expression of genes for flavonoid production, especially the key-enzyme chalcone synthase (CHS). Most flavonoid pathway genes (FPGs) are highly UV-B inducible (Schenke et al. 2011) and plants lacking flavonoids are hypersensitive to UV-B stress (Li et al. 1993; Landry et al. 1995). Apart, flavonoids also play key role in plant defenses towards both abiotic and biotic stress. Propolis is a complex mixture, which includes the presence of flavonoids, aromatic acids and esters, aldehydes, ketones, terpenes, steroids, amino acids, phenyl propanoid, fatty acids, hydrocarbons and many other compounds in smaller proportions. Its chemical composition varies according to the collection site, due to the vegetation. Propolis has numerous pharmacological properties; among them is the ability to absorb UV rays. Various aspects of the gene structure and regulation, genetics, epigenetics, biochemistry and regulation of flavonoid biosynthesis have been reviewed (Springob et al. 2003; Schwinn and Davies 2004; Tanner 2004; Koes et al. 2005; Tanaka et al. 2005). Phenyl propanoids pathway associated with the first committed steps of the flavonoid biosynthesis (Heller and Forkmann 1988). Phenyl propanoid-derived compounds lingin, coumarins, etc. were synthesized in plants by using phenyl propanoid pathway that have crucial role as sunscreen (Heller and Forkmann 1988). Lignin is a very good candidate for the development of new materials due to the phenolic and aliphatic hydroxyl groups in its structure. Lignins have pharmacological applications playing a protective role against the development of different diseases (Martinez et al. 2012). In many cases, the effects of lignins could be explained by their antioxidant capacity too. These confer considerable potential for chemical modification (Kai et al. 2016) on the macromolecule. Another phenyl propanoid derivative, sinapic acid esters, also partially replaced flavonoids as UV-B screeners (Li et al. 1993).

Carotenoids

A major role of carotenoids in plants is the contribution to light capture in photosynthesis and photo protection under high light conditions (Polívka and Frank 2010; Dall'Osto et al. 2012; Qin et al. 2015; Hashimoto et al. 2016; Leonelli et al. 2017; Emiliani et al. 2018; Rajneesh et al. 2019). Without carotenoids in photosynthetic tissue, plants do not survive outside of the laboratory (Robertson 1955; Anderson and Robertson 1960; Matthews et al. 2003). Typical carotenoid structures are forty carbons in length with a central conjugated double bond system (Britton et al. 2004). The conjugated double bonds absorb light and are responsible for the unique colors, spectral properties, and biochemical activities. The plant kingdom is known to produce hundreds of different carotenoid structures including geometrical isomers (cis/trans or Z/E), carotenoids with different end-groups and modifications, and altered positions of double bonds. Astaxanthin is a keto carotenoid with stronger biological antioxidant activities compared with other compounds such as β-carotene. β-carotene ketolase (BKT) and β-carotene hydroxylase (CrtR-B) are two key enzymes of in astaxanthin biosynthesis (Maheswaran et al. 2007; Jia et al. 2019). Recently, these two genes have been successfully integrated in 'Brookfield Gala' apple, obtaining transgenic plantlets with pink color and higher antioxidant activity (Jia et al. 2019). In such way development of such transgenic plants could enhance photo oxidation resistance and prevent photo-oxidative injury during photosynthesis. Carotenoids have diverse bioactivities and potential for retooling to meet emerging needs of agriculture. There has been great progress made in understanding mechanisms controlling carotenoid biosynthesis, modification, degradation, and properties related to energy transfer (Wurtzel 2018). An extensive collection of genes and enzymes are available and already much progress has been made in metabolic engineering of carotenoids in multiple species. The study of carotenoids from chemical structure elucidation to molecular biology has been ongoing for over 120 years (Britton et al. 2017). The chemistry and biochemistry of carotenoids and carotenoid enzymes has been recently reviewed (Moise et al. 2014; Rodriguez-Concepcion et al. 2018). Chemical structural variation is the foundation for carotenoid functional diversity in plants and in other species. Specific apocarotenoids (enzymatic cleaved products of carotenoids) serve as signaling molecules (Hou et al. 2016). These include internal signals that mediate stress responses and developmental programs and external signals for communication with other organisms. Carotenoid chemical structures differ widely in terms of shape, biophysical properties, and how the molecules also interact and alter the physical properties of plant membranes where hydrophobic carotenoids are localized. For example, various carotenoid structures uniquely impact membrane permeability, membrane fluidity, and diffusion of small molecules across artificial membranes (Gruszecki and Strzałka 2005). Variation in carotenoid structures not only influences their spectral properties or color but can also influence biological activities including energy transfer, photo protection and quenching of damaging singlet oxygen in plants and bioactivities in humans and other organisms.

The carotenoid biosynthetic pathway in plants is mediated by nuclear-encoded enzymes which are imported into plastids of varying types (Moise et al. 2014; Rodriguez-Concepcion et al. 2018; Sun et al. 2018). The structural genes of the biosynthetic pathway are under transcriptional control as evidenced by the correlation of transcripts with carotenoid content or composition, whether by natural genetic variation (Harjes et al. 2008; Vallabhaneni et al. 2009; Vallabhaneni and Wurtzel 2009), or by transgenic overexpression (Burkhardt et al. 1997; Ye et al. 2000; Giuliano 2017). Thus, the pathway is amenable to metabolic engineering by overexpression of genes controlling flux into the pathway (Giuliano 2017) or by controlling the transcript levels of other genes that lead to specific pathway intermediates (Harjes et al. 2008; Vallabhaneni et al. 2009). Binding of a carotenoid to a protein also alters the carotenoid spectral properties and bioactivity. A classic example is the astaxanthin-binding carotenoprotein, alpha-crustacyanin, which imparts a blue color to lobster (*Homarus gammarus*). A boiled lobster turns red as the red carotenoid is released upon denaturation of the carotenoprotein (Weesie et al. 1995). The cyanobacterial orange carotenoid protein mediates Photo protection by relying on protein-carotenoid interactions that exhibit unique spectral states associated with the photo protection mechanism (Maksimov et al. 2017). Depending upon light intensity, photosynthetic bacteria are known to control the carotenoid pigment-protein composition to alter the spectral characteristics and efficiency of energy transfer (Magdaong et al. 2014). These observations point out the prospects of improving energy transfer or photo protection in plants by engineering bioenergetics changes in carotenoids and caroteno-protein complexes. Simply changing the suite of photo protective carotenoids can improve plant responses to high light stress, as seen when the xanthophyll cycle pigment composition was altered (Leonelli et al. 2017). The unusual features of carotenoids represent an opportunity for creating synthetic systems based on novel carotenoids to further improve photo protection and efficiency of energy capture. Molecular dynamics simulations currently being applied to investigate protein-carotenoid interactions (Daskalakis 2018) will be valuable in designing optimal interactions between photosystem proteins and non-native carotenoids. Photosynthetic bacteria are good test beds to build and test features of this system (Nagashima et al. 2017; Yukihira et al. 2017). Carotenogenic bacteria and marine algae have also been used to engineer carotenoid biosynthesis (Wang et al. 2014) and there is also the option of utilizing plant tissue culture systems (Oleszkiewicz et al. 2018; Schaub et al. 2018).

Scytonemin and Mycosporine-Like Amino Acids

Scytonemin is predominantly produced by cyanobacteria and algal symbionts of some lichens. It is highly stable, lipid-soluble, pale yellow to brown, dimeric phenolic pigment, accumulated in the extracellular polysaccharide sheath of some cyanobacteria and it acts as a passive UV sunscreen (Garcia-Pichel and Castenholz 1991; Rastogi and Incharoensakdi 2013; 2014; Rastogi et al. 2013; Pathak et al. 2015a; 2015b; 2017a; 2017b; 2017c; Richa et al. 2018; Pathak et al. 2019). Scytonemin is highly stable against different stressors and performs its UV-absorbing/screening activity without any further metabolic investment (Brenowitz and Castenholz 1997). *In vivo,* scytonemin shows absorption maximum at 370 nm and purified scytonemin shows maximum absorption at 386 nm, but it also absorbs significantly at 252, 278, and 300 nm and provides photo protection to the organisms inhabiting under UV-A/B exposed environments (Pathak et al. 2017b; Rajneesh et al. 2017b). Scytonemin reduces ROS generation in cyanobacteria that can be detect by using oxidant sensing probe 2',7'-dichlorodihydrofluorescein diacetate (DCFH-DA) (Rajneesh et al. 2017c).

Different derivatives of scytonemin, such as dimethoxy scytonemin, tetramethoxy scytonemin, scytonin, and scytonemin-3a-imine, have been reported from different cyanobacteria (Bultel-Poncé et al. 2004; Grant and Louda 2013; Rastogi et al. 2013). Scytonemin exists in both a reduced (Mw 546 Da) and an oxidized (Mw 544 Da) forms. Genes responsible for scytonemin biosynthesis have been investigated in several cyanobacterial genomes (Sorrels et al. 2009; Soule et al. 2007; 2009). In *Nostoc punctiforme* ATCC 29133, a gene cluster consisting of 18 unidirectionally transcribed ORFs (NpR1276-NpR1259) was identified (Soule et al. 2007). Tyrosine and Trptophan derivatives serve as key precursors for scytonemin biosynthesis. A probable route for scytonemin biosynthesis was presented by Balskus and Walsh (2009). They recognized the acycloin reaction as a key step in assembling the carbon framework of scytonemin. Role of scytonemin in survival and protection of cyanobacterial communities exposed to intense solar radiation is now well established (Pathak et al. 2017b).

MAAs are small, hydrophilic, colorless, compounds composed of a cyclohexenimine or cyclohexenone chromophore, conjugated to the nitrogen substituent of an amino acid or its imino alcohol. High molar extinction coefficients (ε = 28,100 - 50,000 M^{-1} cm^{-1}) and strong UV absorption (λmax: 310-362 nm) suggest their role in UV-screening. Different MAAs have been reported and characterized from cyanobacteria and other eukaryotic microalgae (Xiong et al. 1999; Sinha et al. 2003; Llewellyn and Airs 2010; Rastogi et al. 2012; Carignan and Carreto 2013; Rastogi and Incharoensakdi 2014; Richa et al. 2016; Pathak et al. 2017b). Different cyanobacteria also synthesize some novel glycosylated MAAs (Böhm et al. 1995; Matsui et al. 2011; Nazifi et al. 2014).

They are found in various organisms from tropical to polar region (Sinha et al. 2007). The ring system generally contains a glycine subunit at the third carbon atom. Some MAAs also contain sulphate esters or glycosidic linkages through the imine substituents (Böhm et al. 1995; Wu Won et al. 1997). Differences among the absorption spectra of MAAs are due to the variations in the side groups and nitrogen substituents. The biosynthesis of MAAs is presumed to occur *via* the first part of shikimate pathway, but conclusive evidence is lacking to a great extent. 3-Dehydroquinate, which is formed in the shikimate pathway, acts as a precursor for the synthesis of fungal mycosporines and MAAs *via* gadusols (Bandaranayake 1998; Shick and Dunlap 2002). The primary MAA mycosporine-glycine synthesized by the shikimate pathway is then transformed by chemical and/or biochemical conversions into other MAAs (Callone et al. 2006; Conde et al. 2007). MAAs are synthesized in bacteria, fungi, cyanobacteria, phytoplankton and macroalgae but not in animals, which lack the shikimate pathway. MAAs provides protection from UVR not only in their producers but also to primary and secondary consumers through the food chain (Whitehead et al. 2001). MAAs are highly resistant against abiotic stressors such as temperature, UVR, various solvents and pH (Gröniger and Häder 2000; Sinha et al. 2000). Genomic mining of these four cyanobacteria reveals the presence of a unique combination of two genes (DHQS; YP_324358 and O-methyltransferase; YP_324357) in *A. variabilis* PCC 7937 responsible for MAAs synthesis which are missing in other non-MAA-synthesizing cyanobacteria (Singh et al. 2010). Studies of *in vivo* heterologous gene expression in *E. coli* and *in vitro* enzyme reconstitution experiments proved that in *A. variabilis* ATCC 29413 and *N. punctiforme* ATCC 29133, MAA biosynthesis proceeds from the intermediate sedoheptulose 7-phosphate of the pentose phosphate pathway (Gao and Garcia-Pichel 2011b). Furthermore, it can be also concluded that MAA biosynthesis occurs *via* pentose phosphate as well as shikimate pathways (Balskus and Walsh 2010; Gao and Garcia-Pichel 2011b). Deletion of O-methyltransferase (*OMT*) gene (*Ava-3857*) in *A. variabilis* ATCC 29413 revealed that this gene product is essential to both the pentose phosphate and shikimate pathways for MAA biosynthesis (Pope et al. 2015).

OTHER CHEMICALS, ANTIOXIDANTS AS SUNSCREEN

Besides scytonemin and MAAs, some other UV-absorbing biomolecules, such as pteridines, prenostodione (λmax: 318 nm) and biopterin glucoside (λmax: 362 nm) have also been reported in some cyanobacteria (Matsunaga et al. 1993; Ploutno and Carmeli 2001). Some other compounds such as polyamines (PAs) are found in microalgae which aids in photo protection. It was found that polyamines affect non-photochemical quenching in higher plants (Ioannidis and Kotzabasis 2007). The three main PAs, putrescine (Put), spermidine (Spd) and spermine (Spm), are compounds implicated in many biological processes such as cell division, growth and senescence (Igarashi and Kashiwagi 2000; Paschalidis and Roubelakis-Angelakis 2005). In cyanobacteria spermidine is the major PA found (Jantaro et al. 2003; Incharoensakdi et al. 2010). Reversible changes were observed in xanthophylls along with PAs content at the LHCII level under UV-B stress (Sfichi et al. 2004; Sfichi-Duke et al. 2008). The changes in the thylakoid-associated PAs along with xanthophylls might be involved in dissipation of excess excitation energy during UV-B stress. PAs may also be involved in the regulation of antenna size. Phycobiliproteins (PBPs) are water-soluble, brilliantly colored and multi-chain proteins assembled into supra molecular complexes on stromal surfaces of the thylakoid membranes (Singh and Montgomery 2011; Kannaujiya and Sinha 2014, 2016, 2017). PBPs are characterized into three groups such as phycocyanin (PC, max 610-620 nm), phycoerythrin (PE, max 540–570 nm) and allophycocyanin (APC, max 650–655 nm) (Grossman et al. 1993; Kannaujiya and Sinha 2017; Kumar et al. 2018). Apart from above known chemical compounds, there are number of sunscreens of unknown chemical structures frequently functioning as photo protectants.

Mode of Action

Many defence mechanisms as post-UVR damage processes such as DNA check pointing and repair processes can identify and rectify much of the damage caused by UVR (Sancar et al. 2004; Mason and Reichrath 2013; Rajneesh et al. 2019). Apart from that, different antioxidative enzymes counteract the damage caused by UVR stress and thus contribute to the survival of cyanobacteria (Zeeshan and Prasad 2009; Singh et al. 2013; Rajneesh et al. 2019) and other living organism under UVR. Roles of different enzymatic antioxidants such as superoxide dismutase (SOD), catalase (CAT), glutathione peroxidase (GPX) are well documented (He and Häder 2002; Pathak et al. 2018). While revealing the mode of action of natural sunscreen, most of the natural sunscreens reduces free radical generation, thereby reducing skin damages along with blocking of UV-rays . It is vital to protect skin and eyes from damaging effects of UVR as the skin is an important part of our immune system.

The shielding effect of melanins, especially eumelanin, is achieved by their ability to serve as a physical barrier which scatters UVR, and act as an absorbent filter that reduces the penetration of UVR through the epidermis (Kaidbey et al. 1979). The efficacy of melanin as a sunscreen was assumed to be about 1.5 - 2.0 sun protective factors (SPF); possibly as high as 4 SFP, implying that melanin absorbs 50% to 75% of UVR. An SPF of 2 means the doubling of protection of the skin against sunburn. Dark skin, which contains more eumelanin than fair skin is better protected against UV-induced damage, and eumelanin is thought to be superior to pheomelanin in its photo protective properties. Besides this melanin also has an important role as electron acceptors in dissimilating bacteria and as a factor of virulence in pathogenic bacteria. Epidemiological data strongly support the photo protective role of melanin as there exists an inverse correlation between skin pigmentation and the incidence of sun-induced skin cancers (Gilchrest et al. 1999) and subjects with white skin are ~ 70 times more likely to develop skin cancer than subjects with black skin (Halder and Bang 1988). The primary photo protective mechanism is skin pigmentation with melanin synthesis which, among its many physiological roles, reduces UVR damage occurring in the first place (Slominski et al. 2004; Fedorow et al. 2005).

Flavonoids have three different actions: capability of absorbing UVR, antioxidant properties, and modulating several signaling pathways. The UV absorption spectrum of a flavonoid generally shows two maximum peaks of absorption, one between 240 and 280 nm and the other at 300-550 nm, which shows the possibility of using these extracts as sunscreens in both UV-B and UV-A photo protection preparations (Violante et al. 2008; Kale et al. 2010).

The penetration and permeation of flavonoids has been studied by *in vitro* diffusion cell methodologies (Bobin et al. 1995; Lopes et al. 2000; Filho et al. 2016). Bobin et al. (1995) evaluated the permeation of rutin, catechin and quercetin in the presence of adjuvants hydrophilic (wetting) and their influence on the permeation of the flavonoids through an artificial membrane. According to the experimental results obtained, these acted as inhibitors of the permeation of flavonoids under study, however, at different levels of inhibition (Baby et al. 2008). Carotenoid influences the biological activities including energy transfer, photo protection and quenching of damaging singlet oxygen in plants and bioactivities in humans and other organisms. Some carotenoids are interconverted through epoxidase/de-epoxidase activities, such as activities which control the formation of energetically optimal pigments for photo protection (Leonelli et al. 2017; Leuenberger et al. 2017). Carotenoid epoxides also form non-enzymatically, as a consequence of saturating light intensities during photosynthesis. For example, excess light at the reaction center leads to non-enzymatic formation of β-carotene epoxides and consequential harmful ROS. The repair of β-carotene epoxide has been postulated to be controlled by CruP, an enzyme that likely evolved from a carotenoid biosynthetic enzyme, lycopene cyclase (Bradbury et al. 2012; Li et al. 2016).

Synthesis of UV-absorbing compounds (scytonemin and MAAs) act as a third line of the defense mechanism in several cyanobacteria to overcome the damaging effects of UVR under prolonged UVR exposure (Cockell and Knowland 1999; Pathak et al. 2017b). MAAs provide passive defense mechanisms which allow cyanobacteria to capture photons and prevent their interaction with proteins and DNA. In cyanobacteria, MAAs can prevent three of every 10 photons from striking the cytoplasmic targets (Garcia-Pichel et al. 1993). The photo protective role of MAAs has already been well established in different microalgae including cyanobacteria (Garcia-Pichel et al. 1993; Xiong et al. 1997; Klisch and Häder 2000; Rastogi et al. 2015; Richa and Sinha 2015; Richa et al. 2018). MAAs effectively dissipates absorbed radiation as heat, without generation of ROS (Conde et al. 2000, 2007).

Table 2. Structures and mode of actions of different sunscreens produced by organisms

Natural Sunscreens	Chemical Structures	Mode of Action	References
Melanin	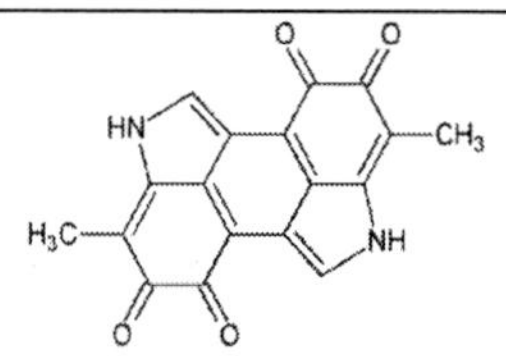	Functions as a physical barrier, scatters UVR, protects against UV-induced damages. Photo protection occurs by skin pigmentation with melanin synthesis	Kaidbey et al. (1979); Slominski et al. (2004); Fedorow et al. (2005)
Flavonoids	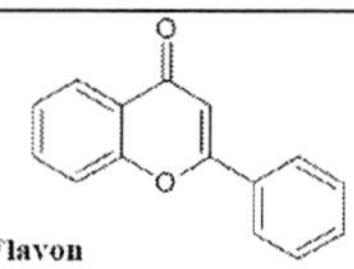 Flavon 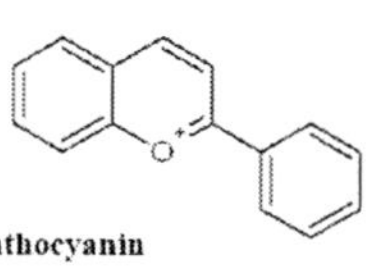 Anthocyanin	Absorbs harmful UVR, have antioxidant properties and modulate several signaling pathways	Lopes et al. (2000); Cho et al. (2007); Sinha et al. (2007); Violante et al. (2008); Allahverdiyeva et al. (2013); Bizari (2013); Gutiérrez-Hernández et al. (2016); Panche et al. (2016)
Phenyl propanoid	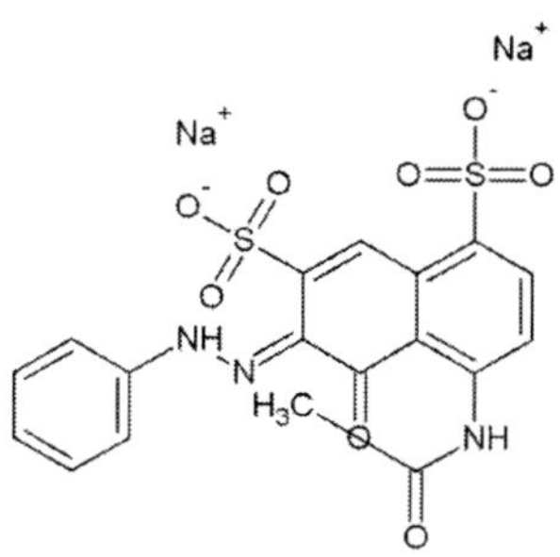 **Lignin**	Shows antioxidant capacity, Lignin-carbohydrate complexes (LCCs) show excellent anti-UV activity	Sinha et al. (2007); Nanbu et al. (2011), (2013); Gutiérrez-Hernández et al. (2016)
Carotenoid	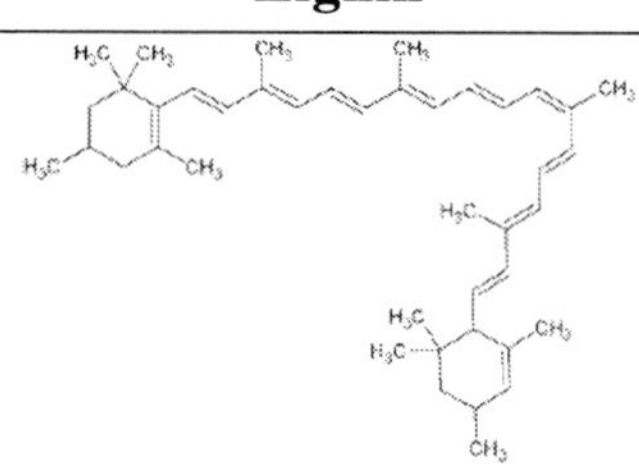 **Lutein**	Plays role in energy transfer, as antioxidants, in photo protection, quenching of damaging singlet oxygen developed after UVR exposure and serve as signalling molecules	Hou et al. (2016); Kaur et al. (2017)

Table 2. (Continued)

Natural Sunscreens	Chemical Structures	Mode of Action	References
Scytonemin	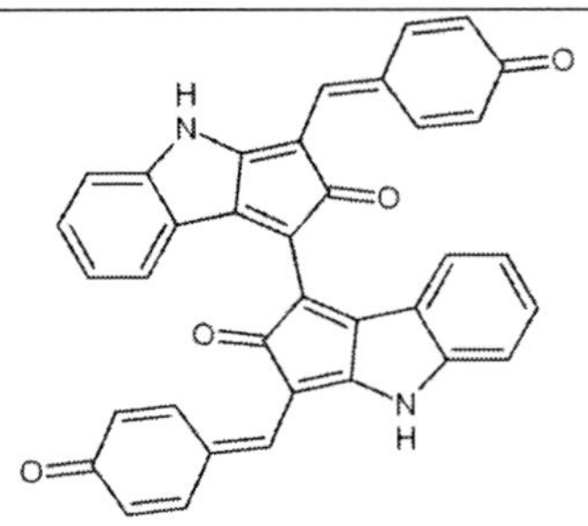	Photo protective function by reducing generation of ROS and thymine dimer formation, efficient antioxidant, absorbs and dissipates energy	Takamatsu et al. (2003); Matsui et al. (2012); Rastogi et al. (2013); Rajneesh et al. (2017a); Pathak et al. (2017b)
MAAs	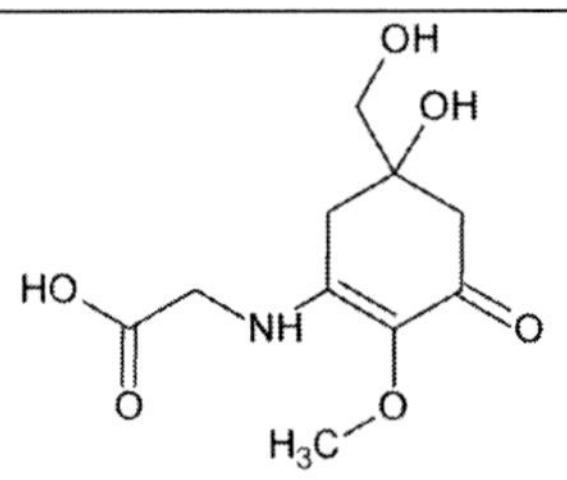 **Mycosporine-glycine** 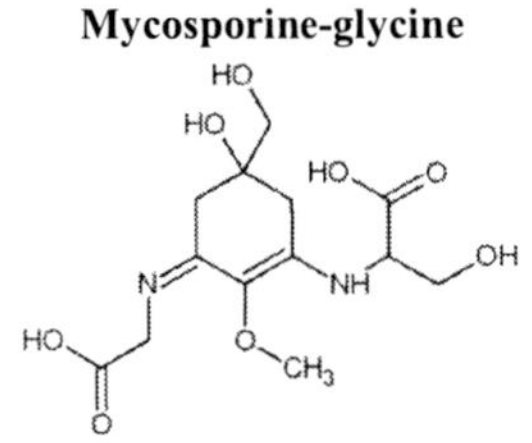 **Shinorine**	Capture photons and prevent their interaction with proteins and DNA, effectively dissipates absorbed radiation as heat, without generation of ROS	Conde et al. (2000, 2007); Oren and Gunde-Cimerman (2007); Pathak et al. (2017b)
Polyamines	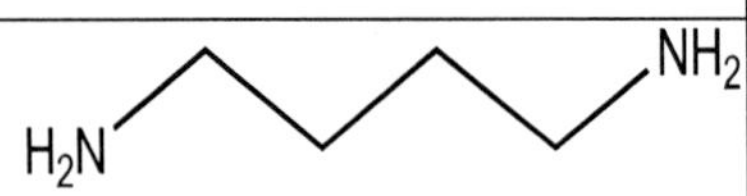 **Putrescine** 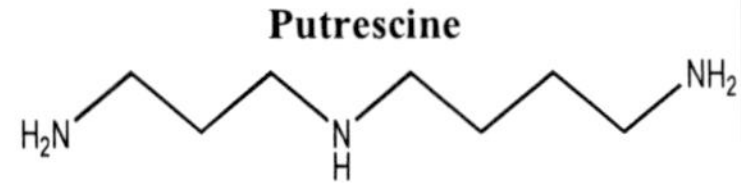 **Spermidine**	Accumulate under UV-B induced condition, affect non-photochemical quenching	Jantaro et al. (2003); Ioannidis and Kotzabasis (2007)

In cyanobacteria light harvesting complexes arrange themselves in a specific manner known as phycobilisomes. The accessory light-harvesting antenna complex (PBPs) is well described as a primary receptor site for effects of light and dark period of the solar radiation (Kannaujiya and Sinha 2017). However, up to now the use of natural secondary metabolites for human skin photo protection is limited by their ability to diminish free radical and singlet oxygen mediated damage to cells and extracellular matrix related to photochemical reactions (Table 2).

Future Perspectives

There is a huge diversity of natural chemicals like polyphenols (flavonoids, tannins), carotenoids, anthocyanidins, few vitamins, fixed oils, volatile oils, scytonemins and MAAs from plant parts, algae and lichens (Ashawat et al. 2009; Nisakorn and Ampa 2015; Saewan and Jimtaisong 2015), and other natural occurring pigments are more effective sunscreen over synthetic chemicals. Plants and marine organisms are considered as valuable sources of antiphotoageing and photo protective compounds for the development of cosmeceutical and topical pharmaceutical products (Chanchal and Swarnlata 2009; Pallela et al. 2010; Saewan and Jimtaisong 2015; Korkina et al. 2017). Nowadays *Spirulina* sp. and *Chlorella* sp. are sold in the open market at a price of about 40 to 50 US-$/kg. ß-carotene derived from carotenoids act as antioxidants and suppresses the adverse effect of free radicals in the human digestion up-to multiple folds (Kaur et al. 2017). This is one of the main reasons for using carotenoids as "functional food" supplement. Different types of carotenoids like lutein and zeaxanthin have commercial value in nutraceuticals. It has importance in eye health because it is the crucial pigment in the macula. Lutein has been clinically confirmed to inhibit cataract (Bernstein et al. 2016). Besides, fucoxanthin is another carotenoid which is produced by cyanobacteria and is popular for its anti-cancer, anti-inflammatory and anti-obesity properties (Sharma et al. 2017). Certain carotenoids are enzymatically converted into visual pigments and signals needed to control development while other carotenoids and their derivatives (apocarotenoids) contribute to eye health, cardiovascular health, function of the immune system, cognitive function, antioxidant activity, general fitness and interspecies communication (Rivera-Madrid et al. 2016; de Carvalho and Caramujo 2017; Rubin et al. 2017; Toews et al. 2017; Zhao and Nabity 2017; Eggersdorfer and Wyss 2018; Harrison and Quadro 2018; Saint et al. 2018).

Skin cancer is a major health problem worldwide. It is a disease caused by the abnormal and uncontrolled growth of the skin cells. Enhanced production of melanins is one of the defence mechanisms against consequences of UVR. Melanin pigments are very common in fungi, although melanogenesis is restricted to certain developmental stages on mycelium, sporulation, or defensive reactions to wounding. Bacterial melanins were first characterized in *Streptomyces*, as these species have been actively pursued in the search of antibiotics (Mencher and Heim 1962). Besides photo protective role, scytonemin may also act as an antioxidant, preventing cellular damage resulting from UV-induced production of ROS (Takamatsu et al. 2003; Matsui et al. 2012). Moreover, due to its potential UV-absorbing/screening capacity as well as non-toxic and antiproliferative properties, scytonemin may be biotechnologically exploited by pharmaceutical and cosmetic industries (Pathak et al. 2017a; Rajneesh et al. 2017a).

Moreover, in addition to UV-absorbing properties, evidence is accumulating that MAAs possess other physiological functions, such as antioxidative, scavengers of toxic oxygen radicals (ROS), and as an osmoprotectant (Oren and Gunde-Cimerman 2007).

These characteristics indicate MAAs to be a promising candidate for natural sunscreen(s) and with pharmaceutical and cosmetic applications.

Conclusion

Natural sunscreens are widely distributed among the living organisms from prokaryotes to eukaryotes and are useful in survival during different life stages of living organisms. Most of them are well known for their multiple roles after photo protection that are useful in existence of organisms. Most of natural sunscreens share common basic mode of action against UVR which is well documented. However, extensive work is needed to explore the different photo protective compounds and their derivatives that play crucial role against upcoming future environment.

Acknowledgments

Jainendra Pathak, Deepak K. Singh, Vidya Singh and Abha Pandey are thankful to Council of Scientific and Industrial Research (CSIR), New Delhi, India, for the financial support in the form of senior research fellowships. Deepak Kumar is grateful to DST-INSPIRE for providing the grant in the form of senior research fellowship.

References

Allahverdiyeva, Y., Mustila, H., Ermakova, M., Bersanini, L., Richaud, P., Ajlani, G., Battchikova, N., Cournac, L. & Aro, E. M. (2013). Flavodiiron proteins *Flv1* and *Flv3* enable cyanobacterial growth and photosynthesis under fluctuating light. *Proc Natl Acad Sci USA*, *110*, 4111-4116.

Anderson, I. C. & Robertson, D. S. (1960). Role of carotenoids in protecting chlorophyll from photodestruction. *Plant Physiol*, *35*, 531-534.

Arai, T. & Mikami, Y. (1972). Chromogenicity of streptomyces. *Appl Microbiol*, *23*, 402-406.

Arai, T., Nishijima, M., Adachi, K. & Sano, H. (1992). Isolation and structure of a UV absorbing substance from the marine bacterium *Micrococcus* sp. AK-334. *MBI Report*. Marine Biotechnology Institute, 2-35-10 Hongo, Bunkyoku, Tokyo, *113*, Japan, 88-94.

Ashawat, M. S., Madhuri, B. S. S. & Swarnlata, S. (2009). Herbal cosmetics: trends in skin care formulation. *Pharmacognosy Rev*, *3*, 72-79.

Baby, A. R., Haroutiounian-Filho, C. A., Sarruf, F. D., Tavante-Júnior, C. R., Pinto, C. A. S. D. O., Zague, V., Arêas, E. P. G., Kaneko, T. M. & Velasco, M. V. R. (2008). Stability and *in vitro* skin penetration study of the routine conveyed in a cosmetic emulsion through an alternative biomembrane model. *Brazilian J Pharm Sci*, *44*, 233-248.

Balskus, E. P. & Walsh, C. T. (2009). An enzymatic cyclopentyl [b] indole formation involved in scytonemin biosynthesis. *J Am Chem Soc*, *131*, 14648-14649.

Balskus, E. P. & Walsh, C. T. (2010). The genetic and molecular basis for sunscreen biosynthesis in cyanobacteria. *Science*, *329*, 1653-1656.

Bandaranayake, W. M. (1998). Mycosporines are they nature's sunscreens? *Nat Prod Rep*, *15*, 159-172.

Bell, A. A. & Wheeler, M. H. (1986). Biosynthesis and functions of fungal melanins. *Ann Rev Phytopathol*, *24*, 411-451.

Bernstein, P. S., Li, B., Vachali, P. P., Gorusupudi, A., Shyam, R., Henriksen, B. S. & Nolan, J. M. (2016). Lutein, zeaxanthin, and meso-zeaxanthin: the basic and clinical science underlying carotenoid-based nutritional interventions against ocular disease. *Prog Retin Eye Res*, *50*, 34-66.

Bizari, M. (2013). *Development of solar filters from Zn^{2+} and/or Ce^{3+} metal complexes with quercetin for sunscreens*, IDE Chemistry, Dissertation, Universidade Estadual Paulista, Campus de Araraquara, São Paulo, Brazil.

Bobin, M. F., Raymond, M. & Martini, M. C. (1995). UVA/UVB absorption properties of natural products. *Cosmet Toil*, *7*, 44-50.

Böhm, G. A., Pfleiderer, W., Böger, P. & Scherer, S. (1995). Structure of a novel oligosaccharide mycosporine-amino acid ultraviolet A/B sunscreen pigment from the terrestrial cyanobacterium *Nostoc commune*. *J Biol Chem*, *270*, 8536-8539.

Borisyuk, L. G., Kharat'yan, E. F. & Zhdanova, N. N. (1991). Localization and dynamics of melanin accumulation in cells of *Cladosporium cladosporiodies* (Fresen) de Vries. *Mikrobiol Zh* (Kiev), *53*, 10-16.

Bradbury, L. M. T., Shumskaya, M., Tzfadia, O., Wu, S. B., Kennelly, E. J. & Wurtzel, E. T. (2012). Lycopene cyclase paralog CruP protects against reactive oxygen species in oxygenic photosynthetic organisms. *Proc Nat Acad Sci USA*, *109*, 1888-1897.

Brenowitz, S. & Castenholz, R. W. (1997). Long-term effects of UV and visible irradiance on natural populations of a scytonemin-containing cyanobacterium (*Calothrix* sp.). *FEMS Microbiol Ecol*, *24*, 343-352.

Britton, G. (1995). Structure and properties of carotenoids in relation to function. *FASEB J*, *9*, 1551-1558.

Britton, G., Liaaen-Jensen, S. & Pfander, H. (2004). *Carotenoids Handbook*. Birkhäuser Verlag, Basel.

Britton, G., Liaaen-Jensen, S. & Pfander, H. (2017). *Carotenoids: A Colourful History*. Fagtrykk Trondheim AS, Norway Carote Nature GmbH, Erlenauweg, Münsingen, Switzerland.

Bultel-Poncé, V., Felix-Theodore, F., Sarthon, C., Ponge, J. F. & Bodo, B. (2004). New pigments from the terrestrial cyanobacterium *Scytonema* sp. collected on the Mitaraka Inselberg, French Guyana. *J Nat Prod*, *67*, 678-681.

Burkhardt, P. K., Beyer, P., Wünn, J., Klöti, A., Armstrong, G. A., Schledz, M., von Lintig, J. & Potrykus, I. (1997). Transgenic rice (*Oryza sativa*) endosperm expressing daffodil (*Narcissus pseudonarcissus*) phytoene synthase accumulates phytoene, a key intermediate of provitamin A biosynthesis. *Plant J*, *11*, 1071-1078.

Burton, G. W. & Ingold, K. U. (1984). beta-Carotene: an unusual type of lipid antioxidant. *Science*, *224*, 569-573.

Butler, M. J. & Day, A. W. (1998). Fungal melanins: a review. *Can J Microbiol*, *44*, 1115-1136.

Callone, A. I., Carignan, M., Montoya, N. G. & Carreto, J. I. (2006). Biotransformation of mycosporine like amino acids (MAAs) in the toxic dinoflagellates *Alexandrium tamarense*. *J Photochem Photobiol B Biol*, *84*, 204-212.

Cardozo, K. H., Guaratini, T., Barros, M. P., Falcão, V. R., Tonon, A. P., Lopes, N. P., Campos, S., Torres, M. A., Souza, A. O., Colepicolo, P. & Pinto, E. (2007). Metabolites from algae with economical impact. *Comp Biochem Physiol* Part C. *Toxicol Pharmacol*, *146*, 60-78.

Carignan, M. O. & Carreto, J. I. (2013). Characterization of mycosporine-serine-glycine methyl ester, a major mycosporine-like amino acid from dinoflagellates: a mass spectrometry study. *J Phycol*, *49*, 680-688.

Chanchal, D. & Swarnlata, S. (2009). Herbal photo protective formulations and their evaluation. *Open Nat Prod J*, *2*, 71-76.

Cho, H. S., Lee, M. H., Lee, J. W., No, K. O., Park, S. K., Lee, H. S., Kang, S., Cho, W. G., Park, H. J., Oh, K. W. & Hong, J. T. (2007). Antiwrinkling effects of the mixture of vitamin C, vitamin E, pycnogenol and evening primrose oil, and molecular mechanisms on hairless mouse skin caused by chronic ultraviolet B irradiation. *Photodermatol Photoimmunol Photomed*, *23*, 155-162.

Cockell, C. S. & Knowland, J. (1999). Ultraviolet radiation screening compounds. *Biol Rev*, *74*, 311-345.

Conde, F. R., Churio, M. S. & Previtali, C. M. (2000). The photoprotector mechanism of mycosporine-like amino acids. Excited-state properties and photostability of porphyra-334 in aqueous solution. *J Photochem Photobiol B Biol*, *56*, 139-144.

Conde, F. R., Churio, M. S. & Previtali, C. M. (2007). Experimental study of the excited-state properties and photostability of the mycosporine-like amino acid palythine in aqueous solution. *Photochem Photobiol Sci*, *6*, 669-674.

Conn, P. F., Schalch, W. & Truscott, T. G. (1991). The singlet oxygen and carotenoid interaction. *J Photochem Photobiol B Biol*, *11*, 41-47.

Coyne, V. E. & Al-Harthi, L. (1992). Induction of melanin biosynthesis in *Vibrio cholera*. *Appl Environ Microbiol*, *58*, 2861-2865.

Dall'Osto, L., Holt, N. E., Kaligotla, S., Fuciman, M., Cazzaniga, S., Carbonera, D., Frank, H. A., Alric, J. & Bassi, R. (2012). Zeaxanthin protects plant photosynthesis by modulating chlorophyll triplet yield in specific light-harvesting antenna subunits. *J Biol Chem*, *287*, 41820-41834.

Darr, D., Dunston, S., Faust, H. & Pinnell, S. (1996). Effectiveness of antioxidants (vitamin C and E) with and without sunscreens as topical photo protectants. *Acta Derm Vcnereol*, *76*, 264-268.

Daskalakis, V. (2018). Protein–protein interactions within photosystem II under Photo protection: the synergy between CP29 minor antenna, subunit S (PsbS) and zeaxanthin at all-atom resolution. *Phys Chem Chem Phys*, *20*, 11843-11855.

De Alencar Filho, J. E. M. T., Sampaio, P. A., Pereira, E. C. V., de Oliveira Junior, R. G., Silva, F. I. S., da Silva Almeida, J. R. G., Araújo Rolim, L., Nunes, X. P. & da Cruz Araújo, E. C. (2016). Flavonoids as photo protective agents: A systematic review. *J Med Plants Res*, *10*, 848-864.

De Carvalho, C. C. C. R. & Caramujo, M. J. (2017). Carotenoids in aquatic ecosystems and aquaculture: a colorful business with implications for human health. *Front Mar Sci*, *4*, 93.

Dehorter, B., Brocquet, M. F., Lacoste, L. & Alais, J. (1980). Influence of light and mycosporin on the sterol composition of *Nectria galligena* during sexual morphogenesis *Phytochem*, *19*, 2311-2315.

Demmig-Adams, B. & Adams, III. W. W. (2002). Antioxidants in photosynthesis and human nutrition. *Science*, *298*, 2149-2153.

Durrell, L. W. (1964). The composition and structure of walls of dark fungus spores. *Mycopathol Mycol Appl*, *23*, 339-345.

Eggersdorfer, M. & Wyss, A. (2018). Carotenoids in human nutrition and health. *Arch Biochem Biophys*, *652*, 18-26.

Emiliani, J., D'Andrea, L., Lorena Falcone Ferreyra, M., Maulión, E., Rodriguez, E., Rodriguez-Concepción, M. & Casati, P. (2018). A role for β,β-xanthophylls in *Arabidopsis* UV-B Photo protection. *J Exp Bot*, *69*, 4921-4933.

Ermakova, M., Battchikova, N., Allahverdiyeva, Y. & Aro, E. M. (2013). Novel heterocyst-specific flavodiiron proteins in *Anabaena* sp. PCC 7120. *FEBS Lett*, *587*, 82-87.

Fageon, L., Moyal, D., Coutet, J. & Candau, D. (2009). Importance of sunscreen products spreading protocol and substrate roughness for *in vitro* sun protection factor assessment. *Int J Cosmet Sci*, *31*, 405-418.

Favre-Bonvin, J., Arpin, N. & Brevard, C. (1976). Structure of mycosporine (P-310). *Can J Chem*, *54*, 1105-1113.

Fedorow, H., Tribl, F., Halliday, G., Gerlach, M., Riederer, P. & Double, K. L. (2005). Neuromelanin in human dopamine neurons: comparison with peripheral melanins and relevance to Parkinson's disease. *Prog Neurobiol*, *75*, 109-124.

Fleming, E. D. & Castenholz, R. W. (2008). Effects of nitrogen source on the synthesis of the UV-screening compound, scytonemin, in the cyanobacterium *Nostoc punctiforme* PCC 73102. *FEMS Microbiol Ecol*, *63*, 301-308.

Gao, Q. J. & Garcia-Pichel, F. (2011a). Microbial ultraviolet sunscreens. *Nat Rev Microbiol*, *9*, 791.

Gao, Q. J. & Garcia-Pichel, F. (2011b). An ATP-Grasp ligase involved in the last biosynthetic step of the iminomycosporine shinorine in *Nostoc punctiforme* ATCC 29133. *J Bacteriol*, *193*, 5923-5928.

Garcia-Pichel, F. & Castenholz, R. W. (1991). Characterization and biological implications of scytonemin, a cyanobacterial sheath pigment. *J Phycol*, *27*, 395-409.

Garcia-Pichel, F., Wingard, C. E. & Castenholz, R. W. (1993). Evidence regarding the UV sunscreen role of a mycosporine like compound in the cyanobacterium *Gloeocapsa* sp. *Appl Environ Microbiol*, *59*, 170-176.

Geng, J., Yuan, P., Shao, C., Yu, S. B., Zhou, B., Zhou, P. & Chen, X. D. (2010). Bacterial melanin interacts with double stranded DNA with high affinity and may inhibit cell metabolism *in vivo*. *Arch Microbiol*, *192*, 321-329.

Ghiani, S., Baroni, S., Burgio, D., Digilio, G., Fukuhara, M., Martino, P., Monda, K., Nervi, C., Kiyomine, A. & Aime, S. (2008). Characterization of human hair melanin and its degradation products by means of magnetic resonance techniques. *Mag Reso Chem*, *46*, 471-479.

Gilchrest, B. A., Eller, M. S., Geller, A. C. & Yaar, M. (1999). The pathogenesis of melanoma induced by ultraviolet radiation. *N Engl J Med*, *340*, 1341-1348.

Giuliano, G. (2017). Provitamin A biofortification of crop plants: a gold rush with many miners. *Curr Opin Biotechnol*, *44*, 169-180.

Glass, K., Ito, S., Wilby, P. R., Sota, T., Nakamura, A., Bowers, C. R., Vinther, J., Dutta, S., Summons, R., Briggs, D. E. & Wakamatsu, K. (2012). Direct chemical evidence for eumelanin pigment from the jurassic period. *Proc Natl Acad Sci USA*, *109*, 10218-10223.

Grant, C. S. & Louda, J. W. (2013). Scytonemin-imine, a mahogany-colored UV/Vis sunscreen of cyanobacteria exposed to intense solar radiation. *Org Geochem*, *65*, 29-36.

Gröniger, A. & Häder, D. P. (2000). Stability of mycosporine-like amino acids. *Recent Res Devel Photochem Photobiol*, *4*, 247-252.

Grossman, A. R., Schaefer, M., Chiang, G. G. & Collier, J. L. (1993). The phycobilisome, a light harvesting complex responsive to environmental conditions. *Microbiol Rev*, *57*, 725-749.

Gruszecki, W. I. & Strzałka, K. (2005). Carotenoids as modulators of lipid membrane physical properties. *BBA Mol Basis Dis*, *1740*, 108-115.

Gutiérrez-Hernández, J. M., Escalante, A., Murillo-Vázquez, R. N., Delgado, E., González, F. J. & Toríz, G. (2016). Use of Agave tequilana-lignin and zinc oxide nanoparticles for skin Photo protection. *J Photochem Photobiol B Biol*, *163*, 156-161.

Halder, R. M. & Bang, K. M. (1988). Skin cancer in blacks in the United States. *Dermatol Clin*, *6*, 397-405.

Harjes, C. E., Rocheford, T. R., Bai, L., Brutnell, T. P., Kandianis, C. B., Sowinski, S. G., Stapleton, A. E., Vallabhaneni, R., Williams, M., Wurtzel, E. T., Yan, J. & Buckler, E. S. (2008). Natural genetic variation in lycopene epsilon cyclase tapped for maize biofortification. *Science*, *319*, 330-333.

Harrison, E. H. & Quadro, L. (2018). Apocarotenoids: emerging roles in mammals. *Annu Rev Nutr*, *38*, 153-172.

Hashimoto, H., Uragami, C. & Cogdell, R. J. (2016). Carotenoids and photosynthesis. *Subcell Biochem*, *79*, 111-139.

He, Y. Y. & Häder, D. P. (2002). UV-B-induced formation of reactive oxygen species and oxidative damage of the cyanobacterium *Anabaena* sp.: protective effects of ascorbic acid and *N*-acetyl-L-cysteine. *J Photochem Photobiol B*, *66*, 115-124.

Heinrich, U., Gärtner, C., Wiebusch, M., Eichler, O., Sies, H., Tronnier, H. & Stahl, W. (2003). Supplementation with beta-carotene or a similar amount of mixed carotenoids protects humans from UV-induced erythema. *J Nutr*, *133*, 98-101.

Heller, W. & Forkmann, G. (1988). Biosynthesis. In: *The Flavonoids. Advances in Research since 1980* (Ed. Harborne JB), Chapman and Hall Ltd., Springer, pp. 399-425.

Hou, X., Rivers, J., Leon, P., McQuinn, R. P. & Pogson, B. J. (2016). Synthesis and function of apocarotenoid signals in plants. *Trends Plant Sci*, *21*, 792-803.

Huang, S., Pan, Y., Gan, D., Ouyang, X., Tang, S., Ekunwe, S. I. N. & Wang, H. (2011). Antioxidant activities and UV-protective properties of melanin from the berry of *Cinnamomum burmannii* and *Osmanthus fragrans*. *Med Chem Res*, *20*, 475-481.

Igarashi, K. & Kashiwagi, K. (2000). Polyamines: mysterious modulators of cellular functions. *Biochem. Biophys. Res Commun*, *271*, 559-564.

Incharoensakdi, A., Jantaro, S., Raksajit, W. & Maenpaa, P. (2010). Polyamines in cyanobacteria: biosynthesis, transport and abiotic stress response. In: *Current Research: Technology and Education: Topics in Applied Microbiology and Microbial Biotechnology* (Ed. Mendez-Vilas A), Microbiology Book Formatex, Badajoz, pp. 23-32.

Ioannidis, N. E. & Kotzabasis, K. (2007). Effects of polyamines on the functionality of photosynthetic membrane *in vivo* and *in vitro*. *Biochim Biophys Acta*, *1767*, 1372-1382.

Jackman, M. P., Hajnal, A. & Lerch, K. (1991). Albino mutants of *Streptomyces glaucescens* tyrosinase. *Biochem J*, *274*, 707-713.

Jantaro, S., Mäenpää, P., Mulo, P. & Incharoensakdi, A. (2003). Content and biosynthesis of polyamines in salt and osmotically stressed cells of *Synechocystis* sp. PCC 6803. *FEMS Microbiol Lett*, *228*, 129-135.

Jia, D., Fana, L., Shena, J., Qin, S., Lib, F. & Yuana, Y. (2019). Genetic transformation of the astaxanthin biosynthetic genes bkt and crtR-B into apple tree to increase photooxidation resistance. *Sci Hortic*, *243*, 428-433.

Kai, D., Tan, M. J., Chee, P. L., Chua, Y. K., Yap, Y. L. & Loh, X. J. (2016). Towards lignin-based functional materials in a sustainable world. *Green Chem*, *18*, 1175-1200.

Kaidbey, K. H., Agin, P. P., Sayre, R. M. & Kligman, A. M. (1979). Photo protection by melanin-a comparison of black and Caucasian skin. *J Am Acad Dermatol*, *1*, 249-260.

Kale, S., Ghoge, P., Ansari, A., Waje, A. & Sonawane, A. (2010). Formulation and in-vitro determination of sun protection factor of *Nigella sativa* Linn. seed oil sunscreen cream. *Int J Pharm Tech Res*, *4*, 2194-2197.

Kannaujiya, V. K., Rastogi, R. P. & Sinha, R. P. (2014). GC constituents and relative codon expressed amino acid composition in cyanobacterial phycobiliproteins. *Gene*, *546*, 162-171.

Kannaujiya, V. K. & Sinha, R. P. (2016). Thermokinetic stability of phycocyanin and phycoerythrin in food-grade preservatives. *J Appl Phycol*, *28*, 1063-1070.

Kannaujiya, V. K. & Sinha, R. P. (2017). Impacts of diurnal variation of ultraviolet-B and photosynthetically active radiation on phycobiliproteins of the hot-spring cyanobacterium *Nostoc* sp. strain HKAR-2. *Protoplasma*, *254*, 423-433.

Kaur, G., Sandal, A. & Dhillon, N. S. (2017). Lycopene and human health-A review. *Agri. Rev.*, *38*(4). DOI: 10.18805/ag.R-1741.

Klisch, M. & Häder, D. (2008). Mycosporine-like amino acids and marine toxins-The common and the different. *Mar drugs*, *6*, 147-163.

Klisch, M. & Häder, D. P. (2000). Mycosporine-like amino acids in the marine dinoflagellate *Gyrodiniumdorsum*: induction by ultraviolet irradiation. *J. Photochem. Photobiol. B Biol.*, *55*, 178-182.

Kobayashi, J., Nakamura, N. & Hirata, Y. (1981). Isolation and structure of a UV absorbing substance 337 from the ascidian *Halocynthia roretzi*. *Tetrahedron Lett*, *22*, 3001-3002.

Koes, R., Verweij, W. & Quattrocchio, F. (2005). Flavonoids: a colorful model for the regulation and evolution of biochemical pathways. *Trends Plant Sci*, *10*, 236-242.

Kollias, N., Sayre, R. M., Zeise, L. & Chedekel, M. R. (1991). Photo protection by melanin. *J Photochem Photobiol B Biol*, *9*, 135-160.

Korkina, L., Mayer, W., Talib, N. & De Luca, C. (2017). Meristem plant cells as a sustainable source of redox actives for skin rejuvenation. *Biomolecules*, *7*, 40.

Kumar, D., Kannaujiya, V. K., Richa, Pathak J., Sundaram, S. & Sinha, R. P. (2018). Composition and functional property of photosynthetic pigments under circadian rhythm in the cyanobacterium *Spirulina platensis. Protoplasma*, *255*, 885-898.

Kumar, J., Parihar, P., Singh, R., Singh, V. P. & Prasad, S. M. (2016). UVB induces biomass production and nonenzymatic antioxidant compounds in three cyanobacteria. *J Appl Phycol*, *28*, 131-140.

Landry, L. G., Chapple, C. C. & Last, R. L. (1995). *Arabidopsis* mutants lacking phenolic sunscreens exhibit enhanced ultraviolet-B injury and oxidative damage. *Plant Physiol*, *109*, 1159-1166.

Langfelder, K., Streibel, M., Jahn, B., Haase, G. & Brakhage, A. A. (2003). Biosynthesis of fungal melanins and their importance for human pathogenic fungi. *Fungal Genet Biol*, *38*, 143-158.

Leonelli, L., Brooks, M. D. & Niyogi, K. K. (2017). Engineering the lutein epoxide cycle into *Arabidopsis thaliana*. *Proc Natl Acad Sci USA*, *114*, 7002-7008.

Lerch, K. & Ettlinger, L. (1972). Purification and characterization of a tyrosinase from *Streptomyces glaucescens*. *Eur J Biochem*, *31*, 427-437.

Leuenberger, M., Morris, J. M., Chan, A. M., Leonelli, L., Niyogi, K. K. & Fleming, G. R. (2017). Dissecting and modeling zeaxanthin and lutein dependent non-photochemical quenching in *Arabidopsis thaliana*. *Proc Natl Acad Sci USA*, *114*, 7009-7017.

Li, C., Ji, C. & Tang, B. (2018). Purification, characterization and biological activity of melanin from *Streptomyces* sp. *FEMS Microbiol Lett*, *365*, 1-8.

Li, J., Ou-Lee, T. M., Raba, R., Amundson, R. G. & Last, R. L. (1993). *Arabidopsis* flavonoid mutants are hypersensitive to UV-B irradiation. *Plant Cell*, *5*, 171-179.

Li, Z., Peers, G., Dent, R. M., Bai, Y., Yang, S. Y., Apel, W., Leonelli, L. & Niyogi, K. K. (2016). Evolution of an atypical de-epoxidase for Photo protection in the green lineage. *Nat Plants*, *2*, 16140.

Llewellyn, C. A. & Airs, R. L. (2010). Distribution and abundance of MAAs in 33 species of microalgae across 13 classes reported in several microalgae. *Mar Drugs*, *8*, 1273-1291.

Lopes, R., Oliveira, T. T., Nagem, T. J. & Pinto, A. D. S. (2000). Flavonoids: pharmacology of flavonoids in the hyperlipidemic control in experimental animals. *Biotechnology One Hundred Development*, *3*, 18-22.

Magdaong, N. M., LaFountain, A. M., Greco, J. A., Gardiner, A. T., Carey, A. M., Cogdell, R. J., Gibson, G. N., Birge, R. R. & Frank, H. A. (2014). High efficiency light harvesting by carotenoids in the LH2 complex from photosynthetic bacteria: unique adaptation to growth under low-light conditions. *J Phys Chem B*, *118*, 11172-11189.

Maheswaran, G., Pridmore, L., Franz, P. & Anderson, M. A. (2007). A proteinase inhibitor from *Nicotiana alata* inhibits the normal development of light-brown apple moth, *Epiphyas postvittana* in transgenic apple plants. *Plant Cell Rep*, *26*, 773-782.

Maksimov, E. G., Sluchanko N. N., Slonimskiy, Y. B., Slutskaya, E. A., Stepanov, A. V., Argentova-Stevens, A. M., Shirshin, E. A., Tsoraev, G. V., Klementiev, K. E., Slatinskaya, O. V., Lukashev, E. P., Friedrich T., Paschenko, V. Z. & Rubin, A. B. (2017). The photocycle of orange carotenoid protein conceals distinct intermediates and asynchronous changes in the carotenoid and protein components. *Sci Rep*, *7*, 15548.

Manney, G. L., Santee, M. L., Rex, M., Livesey, N. J., Pitts, M. C., Veefkind, P., Nash, E. R., Wohltmann, I., Lehmann, R., Froidevaux, L., Poole, L. R., Schoeberl, M. R., Haffner, D. P., Davies, J., Dorokhov, V., Gernandt, H., Johnson, B., Kivi, R., Kyrö, E., Larsen, N., Levelt, P. F., Makshtas, A., McElroy, C. T., Nakajima, H., Parrondo, M. C., Tarasick, D. W., von der Gathen, P., Walker, K. A. & Zinoviev, N. S. (2011). Unprecedented Arctic ozone loss in 2011. *Nature*, *478*, 469-475.

Martinez, V., Mitjans, M. & Vinardell, M. P. (2012). Pharmacological applications of lignins and lignins related compounds: An overview. *Curr Organ Chem*, *16*, 1863-1870.

Mason, H. S. (1956). Structures and functions of the phenolase complex. *Nature*, *177*, 79-81.

Mason, R. S. & Reichrath, J. (2013). Sunlight vitamin D and skin cancer. *Adv. Anticancer Agents Med Chem*, *13*, 83-97.

Matsui, K., Nazifi, E., Hirai, Y., Wada, N., Matsugo, S. & Sakamoto, T. (2012). The cyanobacterial UV absorbing pigment scytonemin displays radical scavenging activity. *J Gen Appl Microbiol*, *58*, 137-144.

Matsui, K., Nazifi, E., Kunita, S., Wada, N., Matsugo, S. & Sakamoto, T. (2011). Novel glycosylated mycosporine-like amino acids with radical scavenging activity from the cyanobacterium *Nostoc commune*. *J Photochem Photobiol B Biol*, *105*, 81-89.

Matsunaga, T., Burgess, J. G., Yamada, N., Komatsu, K., Yoshida, S. & Wachi, Y. (1993). An ultraviolet (UV-A) absorbing biopterin glucoside from the marine planktonic cyanobacterium *Oscillatoria* sp. *Appl Microbiol Biotechnol*, *39*, 250-253.

Matthews, P. D., Luo, R. & Wurtzel, E. T. (2003). Maize phytoene desaturase and zeta-carotene desaturase catalyse a poly-Z desaturation pathway: implications for genetic engineering of carotenoid content among cereal crops. *J Exp Bot*, *54*, 2215-2230.

Mayer, A. M. (1986). Polyphenol oxidases in plants-recent progress. *Phytochemistry*, *26*, 11-20.

Mencher, J. R. & Heim, A. H. (1962). Melanin biosynthesis by *Streptomyces lavendulae*. *Microbiology*, *28*, 665-670.

Moise, A. R., Al-Babili, S. & Wurtzel, E. T. (2014). Mechanistic aspects of carotenoid biosynthesis. *Chem Rev*, *114*, 164-193.

Morabito, K., Shapley, N. C., Steele, K. G. & Tripathi, A. (2011). Review of sunscreen and the emergence of non-conventional absorbers and their applications in ultraviolet protection. *Int J Cosmet Sci*, *33*, 385-390.

Nagashima, K. V. P., Sasaki, M., Hashimoto, K., Takaichi, S., Nagashima, S., Yu, L. J., Abe, Y., Gotou, K., Kawakami, T., Takcnouchi, M., Shibuya, Y., Yamaguchi, A., Ohno, T., Shen, J. R., Inoue, K., Madigan, M. T., Kimura, Y. & Wang-Otomo, Z. Y. (2017). Probing structure-function relationships in early events in photosynthesis using a chimeric photocomplex. *Proc Natl Acad Sci USA*, *114*, 10906-10911.

Nakamura, H., Kobayashi, J. & Hirata, Y. (1981). Isolation and structure of a 330 nm UV absorbing substance, asterina-330 from the starfish *Asterina pectinifera*. *Chem Lett*, *28*, 1413-1414.

Nanbu, T., Matsuta, T., Sakagami, H., Shimada, J., Maki, J. & Makino, T. (2011). Anti-UV activity of *Lentinus edodes* mycelia extract (LEM). *In Vivo*, *25*, 733-740.

Nanbu, T., Shimada, J., Kobayashi, M., Hirano, K., Koh, T., Machino, M., Ohno, H., Yamamoto, M. & Sakagami, H. (2013). Anti-UV activity of lignin-carbohydrate complex and related compounds. *In Vivo*, *27*, 133-139.

Nazifi, E., Wada, N., Asano, T., Nishiuchi, T., Iwamuro, Y., Chinaka, S., Matsugo, S. & Sakamoto, T. (2014). Characterization of the chemical diversity of glycosylated mycosporine-like amino acids in the terrestrial cyanobacterium *Nostoc commune*. *J Photochem Photobiol B Biol*, *142*, 154-168.

Nazifi, E., Wada, N., Yamaba, M., Asano, T., Nishiuchi, T., Matsugo, S. & Sakamoto, T. (2013). Glycosylated porphyra-334 and palythine-threonine from the terrestrial cyanobacterium *Nostoc commune*. *Mar Drugs*, *11*, 3124-54.

Nisakorn, S. & Ampa, J. (2015). Natural products as Photo protection. *J Cosmet Dermatol*, *14*, 47-63.

Oleszkiewicz, T., Klimek-Chodacka, M., Milewska-Hendel, A., Zubko, M., Stroz, D., Kurczynska, E., Boba, A., Szopa, J. & Baranski, R. (2018). Unique chromoplast organisation and carotenoid gene expression in carotenoid-rich carrot callus. *Planta*, *248*, 1455-1471.

Olson, J. A. & Krinsky, N. I. (1995). Introduction: the colorful, fascinating world of the carotenoids: important physiologic modulators. *FASEB J*, *9*, 1547-1550.

Oren, A. & Gunde-Cimerman, N. (2007). Mycosporines and mycosporine-like amino acids: UV protectants or multipurpose secondary metabolites? *FEMS Microbiol Lett*, *269*, 1-10.

Pallela, R., Young, Y. N. & Kim, S. W. (2010). Anti-photoaging and photo protective compounds derived from marine organisms. *Mar. Drugs*, *8*, 1189-1202.

Panche, A. N., Diwan, A. D. & Chandra, S. R. (2016). Flavonoids: an overview. *J Nutr Sci*, *47*, 10-19.

Paschalidis, K. & Roubelakis-Angelakis, K. A. (2005). Spatial and temporal distribution of polyamine levels and polyamine anabolism in different organs/tissues of the tobacco plant. Correlations with age, cell division/expansion and differentiation. *Plant Physiol, 138*, 142-152.

Pathak, J., Richa, Rajneesh., Sonker, A. S., Kannaujiya, V. K. & Sinha, R. P. (2015a). Isolation and partial purification of scytonemin and mycosporine-like amino acids from biological crusts. *J Chem Pharma Res*, *7*, 362-371.

Pathak, J., Rajneesh, Richa., Kannaujiya, V. K., Sonker, A. S. & Sinha, R. P. (2015b). Diverse functions and applications of novel and unique cyanobacterial sheath pigment, scytonemin. In: *Biological Sciences: Innovations and Dynamics* (Eds. Sinha RP, Richa, Rastogi RP), New India Publishing Agency, New Delhi, pp. 237-262.

Pathak, J., Rajneesh, Richa., Sonker, A. S., Kannaujiya, V. K. & Sinha, R. P. (2017a). Cyanobacterial extracellular polysaccharide sheath pigment, scytonemin: a novel multipurpose pharmacophore. In: *Marine Glycobiology: Principles and Applications* (Ed. Se-Kwon K), Taylor and Francis Group, CRC Press, Boca Raton, Florida, pp. 323-338.

Pathak, J., Rajneesh, Ahmed H. & Richa, Sinha R. P. (2017b). Metabolomic profiling of cyanobacterial UV-protective compounds. *Curr Metabolomics*, *5*, 138-163.

Pathak, J., Sonker, A. S., Richa, Rajneesh., Kannaujiya, V. K., Singh, V., Ahmed, H. & Sinha, R. P. (2017c). Screening and partial purification of photo protective pigment, scytonemin from cyanobacteria dominated biological crusts dwelling on the historical monuments in and around Varanasi, India. *Microbiol Res (Pavia)*, *8*, 6559.

Pathak, J., Rajneesh Kannaujiya, V. K., Richa, Singh, V., Kumar, D., Singh, S. P. & Sinha, R. P. (2018). Enzymatic and non-enzymatic protection strategies of cyanobacteria against ultraviolet radiation. In: *Trends in Life Sci* (Eds. Sinha RP, Srivastava UP), Nova Publishers, New York NY, pp. 1-28.

Pathak, J., Ahmed, H., Singh, P. R., Singh, S. P., Häder, D. P. & Sinha, R. P. (2019). Mechanisms of Photo protection in Cyanobacteria. In: *Cyanobacteria* (Eds. Mishra AK, Tiwari DN, AN Rai), Elsevier Inc., Academic Press, pp. 145-171.

Plonka, P. M. & Grabacka, M. (2006). Melanin synthesis in microorganisms-biotechnological and medical aspects. *Acta Biochim Pol*, *53*, 429-443.

Ploutno, A. & Carmeli, S. (2001). Prenostodione, a novel UV-absorbing metabolite from a natural bloom of the cyanobacterium *Nostoc* sp. (TAU strain IL-235). *J Nat Prod*, *64*, 544-549.

Polívka, T. & Frank, H. A. (2010). Molecular factors controlling photosynthetic light harvesting by carotenoids. *Acc Chem Res*, *43*, 1125-1134.

Pope, M. A., Spence, E., Seralvo, V., Gacesa, R., Heidelberger, S., Weston, A. J., Dunlap, W. C., Shick, J. M. & Long, P. F. (2015). O-methyltransferase is shared between the pentose phosphate and shikimate pathways and is essential for mycosporine-like amino acid biosynthesis in *Anabaena variabilis* ATCC 29413. *Chembiochem*, *16*, 320-327.

Prasanna, R. A., Sood, A., Jaiswal, S., Nayak, S., Gupta, V. & Chaudhary, V. (2010). Rediscovering cyanobacteria as valuable sources of bioactive compounds (review). *Appl Biochem Microb*, *46*, 119-134.

Qin, X., Suga, M., Kuang, T. & Shen, J. R. (2015). Photosynthesis. Structural basis for energy transfer pathways in the plant PSI-LHCI super complex. *Science*, *348*, 989-995.

Rajneesh, Singh S. P., Pathak, J. & Sinha, R. P. (2017a). Cyanobacterial factories for the production of green energy and value-added products: An integrated approach for economic viability. *Renew Sust Energ Rev*, *69*, 578-595.

Rajneesh, Pathak J., Srivastava, D., Ahmed, H., Singh, V., Singh, S. P. & Sinha, R. P. (2017b). Role of photosynthetically active radiation (PAR) and dark conditions in the repair of ultraviolet-B radiation (UV-B) induced damages in the cyanobacterium *Nostoc* sp. strain HKAR-2. *Focus Sci*, *3*, 9-14.

Rajneesh, Pathak J., Chatterjee, A., Singh, S. P. & Sinha, R. P. (2017c). Detection of reactive oxygen species (ROS) in cyanobacteria using the oxidant sensing probe 2',7'-Dichlorodihydrofluorescein Diacetate (DCFH-DA). *Bio-Protoc*, *7*, 2545.

Rajneesh, Pathak J., Richa, Hader D. P. & Sinha, R. P. (2019). Impacts of ultraviolet radiation on certain physiological and biochemical processes in cyanobacteria inhabiting diverse habitats, *Environ Exper Bot*, *161*, 375-387.

Rastogi, R. P. & Incharoensakdi, A. (2013). UV radiation-induced accumulation of photo protective compounds in the green alga *Tetraspora* sp. CU2551. *Plant Physiol Biochem*, *70*, 7-13.

Rastogi, R. P. & Incharoensakdi, A. (2014). Characterization of UV-screening compounds, mycosporine-like amino acids, and scytonemin in the cyanobacterium *Lyngbya* sp. CU2555. *FEMS Microbiol Ecol*, *87*, 244-256.

Rastogi, R. P., Kumari, S., Richa, Han T. & Sinha, R. P. (2012). Molecular characterization of hot spring cyanobacteria and evaluation of their photo protective compounds. *Can J Microbiol*, *58*, 719-727.

Rastogi, R. P., Madamwar, D. & Incharoensakdi, A. (2015). Sun-screening bioactive compounds mycosporine-like amino acids in naturally occurring cyanobacterial biofilms: role in Photo protection. *J Appl Microbiol*, *119*, 753-762.

Rastogi, R. P., Sinha, R. P. & Incharoensakdi, A. (2013). Partial characterization, UV-induction and photo protective function of sunscreen pigment, scytonemin from *Rivularia* sp. HKAR-4. *Chemosphere*, *93*, 1874-1878.

Rath, J., Mandal, S. & Adhikary, S. P. (2012). Salinity induced synthesis of UV-screening compound scytonemin in the cyanobacterium *Lyngbya aestuarii*. *J Photochem Photobiol*, *115*, 5-8.

Richa, Pathak J., Sonker, A. S., Singh, V. & Sinha, R. P. (2018). Potential applications of natural bioactive cyanobacterial UV protective compounds. In: *Blue technologies: production and uses of marine molecules* (Eds. La Barre S, Bates S), Wiley VCH, Germany, pp. 693-717.

Richa, Sinha R. P. (2015). Biochemical characterization of sunscreening mycosporine-like amino acids from two *Nostoc* species inhabiting diverse habitats. *Protoplasma*, *252*, 199-208.

Richa, Sinha R. P. & Häder, D. P. (2016). Effects of global change, including UV and UV-screening compounds. In: *The Physiology of Microalgae* (Eds. Borowitzka MA, Beardall J, Raven JA), Springer International Publishing, pp. 373-409.

Rivera-Madrid, R., Aguilar-Espinosa, M., Cardenas-Conejo, Y. & Garza-Caligaris, L. E. (2016). Carotenoid derivates in achiote (*Bixa orellana*) seeds: synthesis and health promoting properties. *Front Plant Sci*, *7*, 1406.

Robertson, D. S. (1955). The genetics of vivipary in maize. *Genetics*, *40*, 745-760.

Rodriguez-Concepcion, M., Avalos, J., Bonet, M. L., Boronat, A., Gomez-Gomez, L., Hornero-Mendez, D., Limon, M. C., Meléndez-Martínez, A. J., Olmedilla-Alonso, B., Palou, A., Ribot, J., Rodrigo, M. J., Zacarias, L. & Zhu, C. (2018). A global perspective on carotenoids: Metabolism, biotechnology, and benefits for nutrition and health. *Prog Lipid Res*, *70*, 62-93.

Rubin, L. P., Ross, A. C., Stephensen, C. B., Bohn, T. & Tanumihardjo, S. A. (2017). Metabolic effects of inflammation on vitamin a and carotenoids in humans and animal models. *Adv Nutr*, *8*, 197-212.

Saewan, N. & Jimtaisong, A. (2013). Photo protection of natural flavonoids. *J Appl Pharm Sci*, *9*, 129-141.

Saewan, N. & Jimtaisong, A. (2015). Natural products as Photo protection. *J Cosmet Dermatol*, *14*, 47-63.

Saint, S. E., Renzi-Hammond, L. M., Khan, N. A., Hillman, C. H., Frick, J. E. & Hammond, B. R. (2018). The macular carotenoids are associated with cognitive function in preadolescent children. *Nutrients*, *10*, 193.

Sancar, A., Lindsey-Boltz, L. A., Unsal-Kaçmaz, K. & Linn, S. (2004). Molecular mechanisms of mammalian DNA repair and the DNA damage checkpoints. *Annu Rev Biochem*, *73*, 39-85.

Schaub, P., Al-Babili, S., Drake, R. & Beyer, P. (2005). Why is golden rice golden (yellow) instead of red? *Plant Physiol*, *138*, 441-450.

Schenke, D., Böttcher, C. & Scheel, D. (2011). Crosstalk between abiotic UV-B stress and biotic stress signaling in *Arabidopsis* prevents flavonol accumulation in favour of pathogen defense compound production. *Plant Cell Environ*, *34*, 1849-1864.

Schwinn, K. E. & Davies, K. M. (2004). Flavonoids. In: *Plant pigments and their manipulation* (Ed. Davies K), Oxford: Blackwell Publishing Ltd, pp. 92-149.

Scotto, J. & Fraumeni, J. F. (1982). Skin (other than melanoma). In: *Cancer Epidemiology and Prevention* (Eds. Schottenfeld D, Fraumeni JF), Philadelphia: W.B. Saunders Co., pp. 996-1011.

Sfichi, L., Ioannidis, N. & Kotzabasis, K. (2004). Thylakoid-associated polyamines adjust the UV-B sensitivity of the photosynthetic apparatus by means of light-harvesting complex II changes. *Photochem Photobiol*, *80*, 499-506.

Sfichi-Duke, L., Ioannidis, N. E. & Kotzabasis, K. (2008). Fast and reversible response of thylakoid-associated polyamines during and after UV-B stress: a comparative study of the wild type and a mutant lacking chlorophyll b of unicellular green alga *Scenedesmus obliquus*. *Planta*, *228*, 341-53.

Sharma, Y. C. & Singh, V. (2017). Microalgal biodiesel: A possible solution for India's energy security. *Renew Sust Eneg Rev*, *67*, 72-88.

Shick, J. M. & Dunlap, W. C. (2002). Mycosporine-like amino acids and related gadusols: biosynthesis, accumulation, and UV-protective functions in aquatic organisms. *Ann Rev Physiol*, *64*, 223-262.

Singh, D. K., Richa, Kumar D., Chatterjee, A., Rajneesh, Pathak J. & Sinha, R. P. (2017). Response of the cyanobacterium *Fischerella* sp. strain HKAR-5 against combined stress of UV-B radiation, PAR and pyrogallic acid. *JSM Environ Sci Ecol*, *5*, 1049.

Singh, G., Babele, P. K., Sinha, R. P., Tyagi, M. B. & Kumar, A. (2013). Enzymatic and non-enzymatic defense mechanisms against ultraviolet-B radiation in two *Anabaena* species. *Process Biochem*, *48*, 796-802.

Singh, S. P., Kilsch, M., Sinha, R. P. & Häder, D. P. (2010). Sulfur deficiency changes mycosporine-like amino acid (MAA) composition of *Anabaena variabilis* PCC 7937: a possible role of sulphur in MAA bioconversion. *Photochem Photobiol*, *86*, 862-870.

Singh, S. P., Klisch, M., Sinha, R. P. & Häder, D. P. (2010). Genome mining of mycosporine-like amino acid (MAA) synthesizing and non-synthesizing cyanobacteria: A bioinformatics study. *Genomics*, *95*, 120-128.

Singh, S. P. & Montgomery, B. L. (2011). Determining cell shape: adaptive regulation of cyanobacterial cellular differentiation and morphology. *Trends Microbiol*, *19*, 278-285.

Singh, S. P., Sinha, R. P., Kilsch, M. & Häder, D. P. (2008). Mycosporine-like amino acids (MAAs) profile of a rice-field cyanobacterium *Anabaena doliolum* as influenced by PAR and UVR. *Planta*, *229*, 225-233.

Sinha, R. P., Ambasht, N. K., Sinha, J. P., Klisch, M. & Häder, D. P. (2003). UVB-induced synthesis of mycosporine-like amino acids in three strains of *Nodularia* (cyanobacteria). *J Photochem Photobiol B Biol*, *71*, 51-58.

Sinha, R. P., Klisch, M., Gröniger, A. & Häder, D. P. (1998). Ultraviolet absorbing/screening substances in cyanobacteria, phytoplankton and macroalgae. *J Photochem Photobiol B Biol*, *47*, 83-94.

Sinha, R. P., Klisch, M., Gröniger, A. & Häder, D. P. (2000). Mycosporine-like amino acids in the marine red alga *Gracilaria* cornea-effects of UV and heat. *Environ Exp Bot*, *43*, 33-43.

Sinha, R. P., Klisch, M., Vaishampayan, A. & Häder, D. P. (1999). Biochemical and spectroscopic characterization of the cyanobacterium *Lyngbya* sp. inhabiting Mango (*Mangifera indica*) trees: Presence of an ultraviolet absorbing pigment, Scytonemin. *Acta Protozool*, *38*, 291-298.

Sinha, R. P., Singh, S. P. & Häder, D. P. (2007). Database on mycosporines and mycosporine-like amino acids (MAAs) in fungi, cyanobacteria, macroalgae, phytoplankton and animals. *J Photochem Photobiol B Biol*, *89*, 29-35.

Slominski, A., Tobin, D. J., Shibahara, S. & Wortsman, J. (2004). Melanin pigmentation in mammalian skin and its hormonal regulation. *Physiol Rev*, *84*, 1155-1228.

Solovchenko, A. & Schmitz-Eiberger, M. (2003). Significance of skin flavonoids for UV-B protection in apple fruits. *J Exp Bot*, *54*, 1977-1984.

Sonntag, B., Kammerlander, B. & Summerer, M. (2017). Bioaccumulation of ultraviolet sunscreen compounds (mycosporine-like amino acids) by the heterotrophic freshwater ciliate *Bursaridium* living in alpine lakes. *Inland Waters*, *7*, 55-64.

Sorrels, C. M., Proteau, P. J. & Gerwick, W. H. (2009). Organization, evolution, and expression analysis of the biosynthetic gene cluster for scytonemin, a cyanobacterial UV absorbing pigment. *Appl Environ Microbiol*, *75*, 4861-4869.

Soule, T., Palmer, K., Gao, Q., Potrafka, R. M., Stout, V. & Garcia-Pichel, F. (2009). A comparative genomics approach to understanding the biosynthesis of the sunscreen scytonemin in cyanobacteria. *BMC Genomics*, *10*, 336-345.

Soule, T., Stout, V., Swingley, W. D., Meeks, J. C. & Garcia-Pichel, F. (2007). Molecular genetics and genomic analysis of scytonemin biosynthesis in *Nostoc punctiforme* ATCC 29133. *J Bacteriol*, *189*, 4465-4472.

Springob, K., Nakajima, J., Yamazaki, M. & Saito, K. (2003). Recent advances in the biosynthesis and accumulation of anthocyanins. *Nat Prod Rep*, *20*, 288-303.

Stahl, W. & Sies, H. (2003). Antioxidant activity of carotenoids. *Mol Aspects Med*, *24*, 345-351.

Stahl, W. & Sies, H. (2005). Bioactivity and protective effects of natural carotenoids. *Biochim Biophys Acta*, *1740*, 101-107.

Stochaj, W. R., Dunlap, W. C. & Shick, J. M. (1994). Two new UV absorbing mycosporine-like amino acids from the sea anemone *Anthopleura elegantissima* and the effects of zooxanthellae and spectral irradiance on chemical composition and content. *Mar Biol*, *118*, 149-156.

Subramaniam, A., Carpenter, E. J., Karentz, D. & Falkowski, P. G. (1999). Bio-optical properties of the marine diazotrophic cyanobacteria *Trichodesmium* sp. I. Absorption and photosynthetic action spectra. *Limnol Oceanogr* 44: 608-617.

Sun, T., Yuan, H., Cao, H., Yazdani, M., Tadmor, Y. & Li, L. (2018). Carotenoid metabolism in plants: the role of plastids. *Mol Plant*, *11*, 58-74.

Takamatsu, S., Hodges, T. W., Rajbhandari, I., Gerwick, W. H., Hamann, M. T., Nagle, D. G. (2003). Marine natural products as novel antioxidant prototypes. *J Nat Prod*, *66*, 605-608.

Takano, S., Nakanishi, A., Uemura, D. & Hirata, Y. (1979). Isolation and structure of a 334 nm UV-absorbing substance, Porphyra-334 from the red algae *Porphyra tenera* Kjellman. *Chem Lett*, *26*, 419-420.

Takano, S., Uemura, D. & Hirata, Y. (1978). Isolation and structure of two new amino acids, palythinol and palythene, from the zoanthid *Palythoa tuberculosa*. *Tetrahedron Lett*, *49*, 4909-4912.

Tanaka, Y., Katsumoto, Y., Brugliera, F. & Mason, J. (2005). Genetic engineering in floriculture. *Plant Cell Tiss Org Cult*, *80*, 1-24.

Tanner, G. J. (2004). Condensed tannins. In: *Plant pigments and their Manipulation* (Ed. Davies K), Oxford: Blackwell Publishing Ltd., pp. 150-184.

Taylor, B. E., Wheeler, M. H. & Szaniszlo, P. J. (1987). Evidence for pentaketide melanin biosynthesis in dematiaceous human pathogenic fungi. *Mycologia*, *79*, 320-322.

Teai, T. T., Raharivelomanana, P., Bianchini, J. P., Faure, R., Martin, P. M. V. & Cambon, A. (1997). Structure de deux nouvelles iminomycosporines isolates de *Pocillopora eydouxi*. *Tetrahedron Lett*, *38*, 5799-5800.

Toews, D. P. L., Hofmeister, N. R. & Taylor, S. A. (2017). The evolution and genetics of carotenoid processing in animals. *Trends in Genet*, *33*, 171-182.

Trione, E. J., Leach, C. M. & Mutch, J. T. (1966). Sporogenic substances isolated from fungi. *Nature*, *212*, 163-164.

Tsujino, I., Yabe, K. & Sekekawa, I. (1980). Isolation and structure of a new amino acid, shinorine, from the red algae, *Chondrus yendoi* Yarnada et Mikami. *Bot Mar*, *23*, 65-66.

Vallabhaneni, R., Gallagher, C. E., Licciardello, N., Cuttriss, A. J., Quinlan, R. F. & Wurtzel, E. T. (2009). Metabolite sorting of a germplasm collection reveals the hydroxylase-3 locus as a new target for maize provitamin A biofortification. *Plant Physiol*, *151*, 1635-1645.

Vallabhaneni, R. & Wurtzel, E. T. (2009). Timing and biosynthetic potential for carotenoid accumulation in genetically diverse germplasm of maize. *Plant Physiol*, *150*, 562-572.

Violante, I. M. P., Souza, I. M., Venturini, C., Ramalho, A. F. S., Santos, R. A. N. & Ferrari, M. (2008). Preliminary study of the *in vitro* photo protective activity of plant extracts from. Mato Grosso ccv. *Revista Brasileira de Farmäcia*, *89*, 175-179.

Volkmann, M., Gorbushina, A. A., Kedar, L. & Oren, A. (2006). Structure of euhalothece-362, a novel red-shifted mycosporine-like amino acid, from a halophilic cyanobacterium (*Euhalothece* sp.). *FEMS Microbiol Lett*, *258*, 50-54.

Wang, C., Kim, J. H. & Kim, S. W. (2014). Synthetic biology and metabolic engineering for marine carotenoids: new opportunities and future prospects. *Mar Drugs*, *12*, 4810-4832.

Wasserfallen, A., Ragettli, S., Jouanneau, Y. & Leisinger, T. (1998). A family of flavoproteins in the domains archaea and bacteria. *Eur J Biochem*, *254*, 325-332.

Weesie, R. J., Askin, D., Jansen, F. J., de Groot, H. J., Lugtenburg, J. & Britton, G. (1995). Protein-chromophore interactions in alpha-crustacyanin, the major blue carotenoprotein from the carapace of the lobster, *Homarus gammarus*. A study by ^{13}C magic angle spinning NMR. *FEBS Lett*, *362*, 34-38.

Whitehead, K., Karentz, D. & Hedges, J. I. (2001). Mycosporine-like amino acids (MAAs) in phytoplankton, a herbivorous pteropod (*Limacina helicina*), and its pteropod predator (*Clione antarctica*) in McMurdo Bay, Antarctica. *Mar Biol*, *139*. 1013-1019.

Wogelius, R. A., Manning, P. L., Barden, H. E., Edwards, N. P., Webb, S. M., Sellers, W. I., Taylor, K. G., Larson, P. L., Dodson, P., You, H. & Da-Qing, L. (2011). Trace metals as biomarkers for eumelanin pigment in the fossil record. *Science*, *333*, 1622-1626.

Wu Won, J. J., Chalker, B. E. & Rideout, J. A. (1997). Two new UV-absorbing compounds from *Stylophora pistillata*: sulfate esters of mycosporine-like amino acids. *Tetrahedron Lett*, *38*, 2525-2526.

Wu Won, J. J., Rideout, J. A. & Chalker, B. E. (1995). Isolation and structure of a novel mycosporine-like amino acid from the reef building corals *Pocillopora damicornis* and *Stylophora pistillata*. *Tetrahedron Lett*, *36*, 5255-5256.

Wurtzel, E. T. (2018). Changing form and function through carotenoids and synthetic biology. *Plant Physiol* DOI:10.1104/pp.18.01122.

Xiong, F., Kopecky, J., Kopecky, J. & Nedbal, L. (1997). Strategies of ultraviolet-B protection in microscopic algae. *Physiol Plant*, *100*, 378-388.

Xiong, F., Kopecky, J. & Nedbal, L. (1999). The occurrence of UV-B absorbing mycosporine-like amino acids in freshwater and terrestrial microalgae (chlorophyta). *Aquat Bot*, *63*, 37-49.

Ye, X., Al-Babili, S., Kloti, A., Zhang, J., Lucca, P., Beyer, P. & Potrykus, I. (2000). Engineering the provitamin A (beta-carotene) biosynthetic pathway into (carotenoid-free) rice endosperm. *Science*, *287*, 303-305.

Youngchim, S., Morris-Jones, R., Hay, R. J. & Hamilton, A. J. (2004). Production of melanin by *Aspergillus fumigates*. *J Med Microbiol*, *53*, 175-181.

Yukihira, N., Sugai, Y., Fujiwara, M., Kosumi, D., Iha, M., Sakaguchi, K., Katsumura, S., Gardiner, A. T., Cogdell, R. J. & Hashimoto, H. (2017). Strategies to enhance the excitation energy-transfer efficiency in a light-harvesting system using the intra-molecular charge transfer character of carotenoids. *Faraday Discuss*, *198*, 59-71.

Zeeshan, M. & Prasad, S. M. (2009). Differential response of growth, photosynthesis, antioxidant enzymes and lipid peroxidation to UV-B radiation in three cyanobacteria. *S Afr J Bot*, *75*, 466-474.

Zhang, F., Kearns, S. L., Orr, P. J., Benton, M. J., Zhou, Z., Johnson, D., Xu, X. & Wang, X. (2010). Fossilized melanosomes and the colour of cretaceous dinosaurs and birds. *Nature*, *463*, 1075-1078.

Zhang, L., Li, L. & Wu, Q. (2007). Protective effects of mycosporine-like amino acids of *Synechocystis* sp. PCC 6803 and their partial characterization. *J Photochem Photobiol*, *86*, 240-245.

Zhang, P., Allahverdiyeva, Y., Eisenhut, M. & Aro, E. M. (2009). Flavodiiron proteins in oxygenic photosynthetic organisms: Photo protection of photosystem II by *Flv2* and *Flv4* in *Synechocystis* sp. PCC 6803. *PLoS ONE*, *4*, 5331.

Zhang, P., Eisenhut, M., Brandt, A. M., Carmel, D., Silén, H. M., Vass, I., Allahverdiyeva, Y., Salminen, T. A. & Aro, E. M. (2012). Operon *flv4-flv2* provides cyanobacterial photosystem II with flexibility of electron transfer. *Plant Cell*, *24*, 1952-1971.

Zhao, C. & Nabity, P. D. (2017). Phylloxerids share ancestral carotenoid biosynthesis genes of fungal origin with aphids and adelgids. *PLoS One*, *12*, 185-484.

In: Innovations in Life Science Research ISBN: 978-1-53615-868-7
Editors: R. P. Sinha, S. Pandey-Rai and N. Ghoshal ©2019 Nova Science Publishers, Inc.

Chapter 2

BIOMEDICAL AND PHARMACEUTICAL POTENTIALS OF CYANOBACTERIAL SECONDARY METABOLITES

Jainendra Pathak, Neha Kumari, Sonal Mishra, Haseen Ahmed and Rajeshwar P. Sinha*

Laboratory of Photobiology and Molecular Microbiology, Centre of Advanced Study in Botany, Institute of Science, Banaras Hindu University, Varanasi, India

ABSTRACT

The Gram-negative photosynthetic prokaryotes, cyanobacteria, produce variety of secondary metabolites having biomedical and biotechnological importance. These cyanobacterial secondary metabolites range from pharmacologically important compounds showing biological activities such as antibiotics, antiviral, antifungal, immunosuppressant, anti-inflammatory agents to the toxins having various ecological roles such as allelochemicals and show properties such as herbicides, algicides and insecticides. Production of such a wide diversity of secondary metabolites by cyanobacteria is due to their ability to integrate non-ribosomal peptide synthethases with polyketide synthases. Cyanobacteria also synthesize novel category of secondary metabolites such as mycosporine-like amino acids (MAAs) and scytonemin which provide protection to them against various stresses, primarily from ultraviolet radiation and brightly lit environments. Hence, these secondary metabolites hold a promising future in biomedical research and have tremendous applications in fields such as biotechnology, agriculture, cosmetic industries and most importantly, in new drugs discovery.

Keywords: cyanobacteria, secondary metabolites, cyanotoxins, antibacterial, antiviral, antifungal activities

---

* Corresponding Author's Email: rpsinhabhu@gmail.com.

INTRODUCTION

Cyanobacteria are an ancient group of photosynthetic prokaryotic organisms that are thought to be the first organisms to carry out oxygenic photosynthesis and were responsible for the creation of an oxygenic environment (Sergeev et al. 2002). The most common cyanobacterial structures in the fossil records include stromatolites and oncolites (Flores and Herrero 2008). The stromatolites of fossilized oxygen-producing cyanobacteria have been reported from 2.8 billion years ago (Olson 2008). Cyanobacteria are ubiquitous in nature and are widely distributed in aquatic and terrestrial environments. Cyanobacteria represent important components of terrestrial as well as aquatic ecosystems and possess a central position in the nutrient cycling due to their unique capacity to fix atmospheric nitrogen, thereby enhancing productivity in a variety of environments. N_2-fixing cyanobacteria form a prominent component of microbial populations in wetland soils, especially in rice-fields, where they significantly contribute as natural biofertilizers (Singh1961; Roger and Watanabe 1986; Sinha et al. 2008; Rajneesh et al. 2017; Pathak et al. 2018a). Some other important habitats include polluted lakes, ponds, water reservoirs, hypersaline lakes, tree barks, rocks and stones, deserts, ice-shelves, thermal springs, estuaries etc., as well as in the form of endosymbionts in plants, lichens and several protists (Car and Whitton1982; Vincet and Quasada 1994; Quasada and Vincet 1997; Sanchez-Baracaldo et al. 2005; Thajuddin and Subramanian 2005; Pathak et al. 2017a). In warm, nutrient-rich water bodies, planktonic cyanobacteria grow quickly and cover the water surface, resulting in the formation of water blooms. Due to their persistent survival in various habitats, some authors have attributed this adaptability, in part, to the ability of cyanobacteria to produce a wide range of secondary metabolites, each with specialized functions to compete successfully on the planet (Gademann and Portmann 2008; Rastogi and Sinha 2009; Yadav et al. 2011; Rajneesh et al. 2017). Several cyanobacterial species produce certain photo protective compounds such as scytonemin and mycosporine-like amino acids (MAAs) as their secondary metabolites that play a significant role in screening of ultraviolet radiation (UVR) (Sinha et al. 1998; Sinha and Häder 2008; Pathak et al. 2015a, 2015b, 2017b, 2018b, 2019). The production of antioxidant enzymes such as superoxide dismutase, catalase and peroxidases and non-enzymatic antioxidants such as vitamins B, C and E as well as cysteine and glutathione can quench or scavenge UV-induced excited states and reactive oxygen species (ROS) (Vincent and Quesada 1994; Richa and Sinha 2015; Rajneesh et al. 2019; Pathak et al. 2019). The diverse array of cyanobacterial secondary metabolites can cause undesirable tastes and odors (toxic or odorous metabolites) or otherwise biochemically active (bioactive) metabolites, which have been studied in marine and freshwater as well as in extensive and intensive aquaculture systems. (Smith et al. 2008). The odorous metabolites produced by certain cyanobacteria from marine and freshwater habitats are detrimental for several organisms including humans as they alter the quality of drinking water and recreational activities

(Zurawell et al. 2005; Dittmann and Wiegand 2006). Cyanobacteria are known to produce various toxic metabolites (cyanotoxins), such as microcystins, nodularins, anatoxins, cylindrospermopsins and aplysiatoxins which have haepatotoxic, neurotoxic and cytotoxic potentials (Tyagi et al. 2006; Karl and Cyril 2008; Yadav and Sinha 2015). The allelochemical roles of cyanobacterial toxins and their applications as algicides, herbicides and insecticides have been investigated by several workers (Gleason and Case 1986; Vepritskii et al. 1991; Jang et al. 2007; Berry et al. 2008).

The ability of cyanobacteria to synthesize numerous complex secondary metabolites such as peptides, depsipeptides, polyketides and alkaloids etc. has attracted the attention of the researchers for their pharmaceutical and biotechnological exploitations (Thajuddin and Subramanian 2005; Sielaff et al. 2006; Spolaore 2006; Rastogi and Sinha 2009). These compounds are mainly formed through large multi modular non-ribosomal peptide synthetase (NRPS), polyketide synthase (PKS) and mixed NRPS-PKS enzymatic systems might be exploited as drug leads (Wase and Wright 2008). Several indole alkaloids from simple carbolines (bauerines and nostocarboline) as well as complex polycyclic structures (hapalindole, welwitindolinone and ambiguine) have been reported in various cyanobacteria (Van et al. 2007). The production of various iron chelators (siderophores) such as schizokinen, synechobactin and anachelin has also been reported in cyanobacteria. Protease inhibitors such as cyanopeptolins, micropeptin and oscillapeptin from certain cyanobacteria and their selectivity for trypsin/chymotrypsin have also been described. In recent years cyanobacteria have been recognized as a potential source of toxins as well as novel bioactive compounds with pharmaceutical applications (Raja et al. 2008; Abed et al. 2009; Rajneesh et al. 2017). Cyanobacterial metabolites show various biological activities such as immunosuppressant, anticancer, anti-HIV, antibacterial, anticoagulant, antifungal, anti-inflammatory, antimalarial, algicide, cytotoxic activities, antiprotozoal, antituberculosis, antiviral, and antitumor activities (Rao 1994; Schlegel et al. 1998; Issa 1999; Schaeffer et al. 2000; Gademann et al. 2008; Wase and Wright 2008; Mayer et al. 2009). Cyanobacteria are promising but still underexplored natural resources, which offer a plethora of chemicals for the discovery of new drugs. This chapter presents an overview of various secondary metabolites synthesized by cyanobacteria and their biomedical and biotechnological importance.

ANTI-INFLAMMATORY METABOLITES

Several malyngamides were quite potent inhibitors, especially those in the F series (Figure 1) (Villa et al. 2010). Mechanistic investigations revealed that while Interleukins 1 and 6 (IL1 and 6) were depressed, Tumor Necrosis Factor a (TNFa) was enhanced, and subsequently, that the agent was acting by inhibition of the MyD88 pathway versus the

TRIF pathway (MyD88-independent pathway) of inflammation. Additional malyngamides such as malyngamide 2 also show anti-inflammatory activity (Malloy et al. 2010).

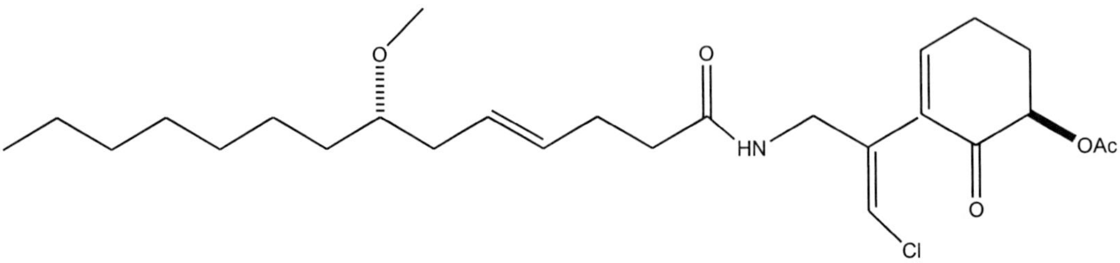

(A)

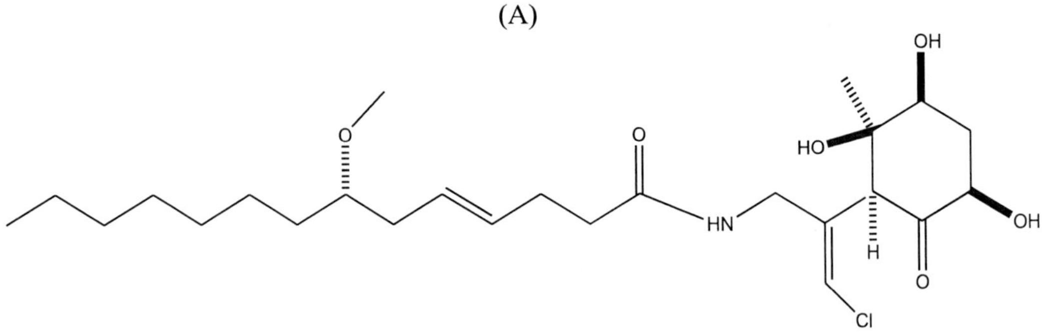

(B)

Figure 1. Structure of Malyngamide F acetate (A) and Malyngamide 2 (B).

Anti-HIV Compounds

Lyngbyatoxin

Targeted lyngbyatoxin A and debromo aplysiatoxin are structurally different and highly inflammatory metabolites isolated from the toxic strains of *Lyngbya majuscula* (Koehn et al. 1992) and anatoxin a from *Anabaena ciecinalis* (Gerwick et al. 1994) (Figure 2). Compounds extracted from *Lyngbya lagerhaimanii* and *Phormidium tenue* showed some anti-HIV activities (Carte 1996; Luescher-Mattli 2003).

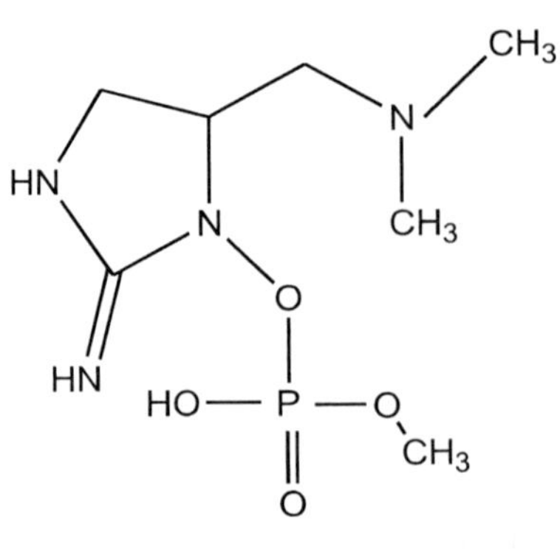

Figure 2. Structure of Anatoxin-a(s).

Spirulan and Ca-Spirulan

Spirulan and Ca-spirulan (Ca-SP) isolated from *Spirulina* sp. show potent *in vitro* antiviral activity. These compounds showed potent and broad-spectrum activities against HIV-1, HIV-2, H, influenza, and a series of other enveloped viruses. Reverse transcriptase activity of HIV-1 (like azidothymidine) was inhibited by these compounds. These sulfated polysaccharides prevent virus-cell attachment and fusion with host cells. They also inhibit the fusion between HIV-infected and not HIV infected $CD4^+$ lymphocytes, a mechanism that greatly enhances viral infectivity. These compounds have advantages as antiviral agents over other sulfated polysaccharides because of reduced anticoagulant properties (Sielaff et al. 2006). Ca-SP, the sulfated polysaccharide is produced by *Arthrospira platensis* (*Spirulina platensis*) (Hayashi et al. 1996a). Ca-SP is composed of mannose, ribose, fructose, glucose, xylose, galactose, rhamnose, galacturonic acid, glucuronic acid, Ca and sulfate has been shown to inhibit the replication of various enveloped and non-enveloped viruses, for example, HIV-1, HSV-1, Human CytoMegalo Virus (HCMV), influenza A, coxsackievirus, measles, polio and mumps, possibly by preventing virus entry into host cells (Hayashi et al. 1996b; Ahmadi et al. 2015). Ca-SP, exhibit the potent activity against HIV (Hayashi et al. 1996a; Santoyo et al. 2006). Ca-SP shows activity against HIV-1, HIV-2, Influenza virus and other enveloped viruses. Ca-SP inhibits the activity of reverse transcriptase of HIV-1 and also inhibits the attachment of virus to the host. Ca-SP prevent the fusion of HIV infected and uninfected CD4 lymphocytes (Feldmann et al. 1999).

Cyanovirin-N

It is a cyanobacterial lectin from *Nostoc ellipsosporum* considered to be a broad-spectrum virucidal agent against enveloped viruses, such as HIV and ebola virus, through the binding to N-linked high-mannose glycans on the viral envelope (Barrientos and Gronenborn 2005). Cyanovirin-N, is a 101-amino-acid-long, 11 kDa polypeptide which shows potent *in vitro* and *in vivo* activities against HIV and other lentiviruses at nanomolar concentrations. It interferes with the binding of HIV gp120 proteins with $CD4^+$ receptors and the chemokine CCR5 or CXCR4 core receptors of target cells, and thus inhibits fusion of the HIV virus with the CD4 cell membrane. It also inhibits herpes simplex virus-6 and measles virus *in vitro* (Singh et al. 2005).

Scytovirin (SVN)

It is another cyanobacterial lectin isolated from *Scytonema varium* that binds with high affinity to mannose residues on the envelope glycoproteins of viruses and inhibits the virus replication, as observed with the Zaire *Ebola* virus (Garrison et al. 2014). It is a 95-amino-acid-long, 9.7 kDa polypeptide containing five intrachain disulfide bonds. It binds to the envelope glycoprotein of HIV (gp120, gp160 and gp41) and inactivates the virus at low nanomolar concentrations. A nuclear magnetic resonance analysis has revealed that scytovirin has two domains. The first domain (48 amino acids) has similar anti-HIV activity to that of the full-length Scytovirin (Karl and Cyri 2008). In addition, two cyclic depsipeptides, ichthyopeptins A and B, were isolated from *Microcystis ichthyoblabe*. They have shown antiviral activity against the influenza A virus. SVN also demonstrated *in vitro* activity against genotypes I and II hepatitis C virus through the binding to the envelope glycoproteins E1 and E2, blocking the viral penetration in human hepatocytes (Takebe et al. 2013). It is a 95-amino-acid-long, 9.7 kDa polypeptide containing five intrachain disulfide bonds. It was first isolated from the aqueous extract of *Scytonema varium*. It binds to the envelope glycoprotein of HIV (gp120, gp160 and gp41) and inactivates the virus at low nanomolar concentrations. A nuclear magnetic resonance analysis has revealed that scytovirin has two domains. The first domain (48 amino acids) has similar anti-HIV activity to that of the full-length scytovirin (Arment and Carmichael 1996). In addition, two cyclic depsipeptides, ichthyopeptins A and B, were isolated from *Microcystis ichthyoblabe*. They have shown antiviral activity against the influenza A virus.

Lectins

Lectins are natural proteins and some are potent antimicrobials through binding to carbohydrates on microbial surfaces. Lectins are natural carbohydrate-binding proteins that can mediate the identification of micro-organisms through the interaction with complex carbohydrates on microbial surfaces (Procópio et al. 2017a, 2017b) promoting host-pathogen communications, immune defence activation and cell-to-cell signalling. The antimicrobial roles of lectins include blockade of invasion and infection, inhibition of growth and germination, regulation of microbial cell adhesion and migration. The carbohydrate-binding properties of lectins are involved in the interactions with eukaryotic cells and pathogens, playing important roles in the immunological defence against pathogens, blocking of viral infections and inhibition of microbial cell adhesion and migration (Chatterjee et al. 2015; Iordache et al. 2015). Lectins have exhibited antibacterial and antifungal activity against Gram-negative, Gram-positive bacteria and fungi, through the interaction with peptidoglycans, polysaccharides, lipopolysaccharides (LPSs), teichoic and teichuronic acids on bacterial and fungal cell walls (Iordache et al. 2015).

Microvirin (MVN)

It is a mannose-specific lectin from the cyanobacterium *Microcystis aeruginosa* PCC7806 able to inhibit the HIV-1 infection, the syncyntium formation between infected and uninfected CD4 T cells, and the HIV-1 transmission (Huskens et al. 2010).

Nostoflan

It is an acidic polysaccharide produced by the edible cyanobacterium *Nostoc flagelliforme*, which has a noticeable inhibitory effect on several enveloped viruses, in particular influenza A virus (Kanekiyo et al. 2007; Ahmadi et al. 2015). The halophilous cyanobacterium *Aphanothece halophytica* releases large amounts of a sulfated exopolysaccharide, composed of arabinose, rhamnose, fucose, mannose, glucose, galactose, glucuronic, and sulfuric acids. Nostoflan shows strong viricidal activity against herpes simplex virus-1. Also have strong antiviral activity against HSV-2, human cytomegalovirus and influenza A virus. Nostoflan has a broad antiviral spectrum against enveloped viruses whose cellular receptors are carbohydrates. Gustafson et al. (1989) reported anti-HIV activity of marine cyanobacterial compounds from *Lyngbya lagerheimii* and *Phormidium tenue*.

Bauerines A-C

The indole alkaloids bauerines A-C (Figure 3) isolated from the terrestrial cyanobacterium *Dichotrix baueriana* GO- 25-5 were screened against Herpes simplex virus 2 (HSV-2), and it was found that bauerine A had anti-HSV-2 activity with an IC_{90} at 2 μg/mL while bauerine B was found to be less active. The most cytotoxic derivative was bauerine C with an IC_{50} at 30 ng/mL (Larsen et al. 1994).

These are indole alkaloids, isolated from *Dichtrix baueriana*, and exhibited activity against the Herpes simplex virus (HSV-2) (Larsen et al. 1994). The extracts of *Lyngbya lagerheimii* and *Phormidium tenue* were found to protect human lymphoblastoid T-cells from the cytopathic effect of HIV infection. Mansour et al. (2011) have found that the polysaccharides isolated from *Gloeocapsa turgidus* and *Synechococcus cedrorum* showed higher antiviral activity against rabies virus in comparison to against herpes-1 virus. The exopolysaccharide from *Aphanothece halophytica* showed an antiviral activity against influenza virus A (H1N1), which showed 30% inhibition of pneumonia in infected mice (Zheng et al. 2006). A new class of HIV-inhibitory compounds called sulfonic acid containing glycolipids has been isolated. The pure compound was found to be active against the HIV virus in the p24 viral protein and syncytium formation assays (Gustafson et al. 1989).

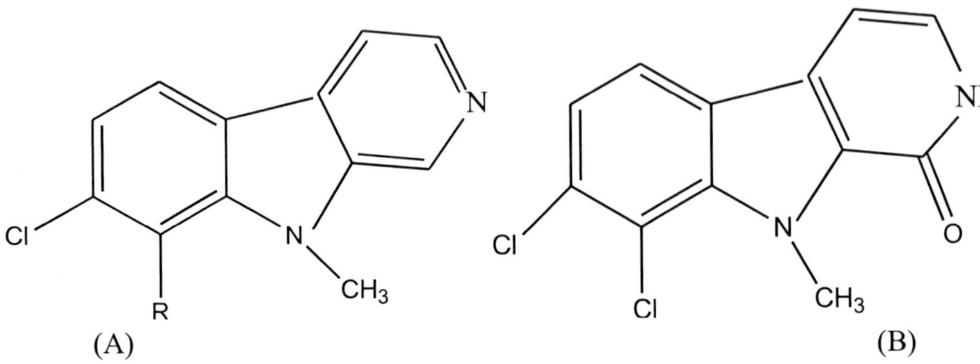

Figure 3. Structure of Bauerine A (R = H) and Bauerine B (R = Cl) (A) Bauerine C (B).

PROTEINS, LIPIDS AND PIGMENTS

Sulfoquinovosyldiacylglycerides (SQDG) from the cyanobacteria *Lyngbya lagerheimii* and *Phormidium tenue* exhibited a high inhibition activity against HIV (Gustafson et al. 1989). Higher inhibition of the tumor promoting stage of the Epstein-Barr virus by MGDG (Mono quinovosyl diayl glycerides) and DGDG in *P. tenue*, compared to SQDG was observed, and Loya et al. (1998) observed that a lipophilic extract containing SQDG in *Scytonema* sp. presented anti-HIV activity. Chirasuwan et al. (2009) demonstrated a pronounced antiviral activity of methanol extract of *Spirulina* on HSV-1, with IC_{50} value of 25.1µg/mL. Allophycocyanin isolated from *S. platensis* exhibited antiviral activity against the enterovirus, which is responsible of severe, sometimes fatal diseases in Taiwan (Shih et al. 2003).

Indolo [2, 3-a]-3-Methoxy-4-Cyanocarbazole and N-Methylindolo [2,3-a]-3-Methoxy-4 Cyanocarbazole

These compounds are responsible for most of the cytotoxicity and antiviral activity related with the blue-green alga *Nostoc sphaericum* EX-5-1. These compounds are active against HSVII and show weak cytotoxicity against KB and LoVo human carcinoma cell lines and moderate antiviral activity against herpes simplex virus type 2. Glycolipids from *Oscillatoria limnetica* (Reshef et al. 1997) *Oscillatoria trichoides* (Loya et al. 1998) and sulfolipids from *Lyngbya lagerheimii* and *Phormidium tenue* (Gustafson et al. 1989) have potential against HIV. Cyclic peptides (Ichthyopeptin A & B) from *Microcystis ichthyoblabe* shows activity against Influenza A Virus (Zainuddin et al. 2007).

ANTIFUNGAL COMPOUNDS

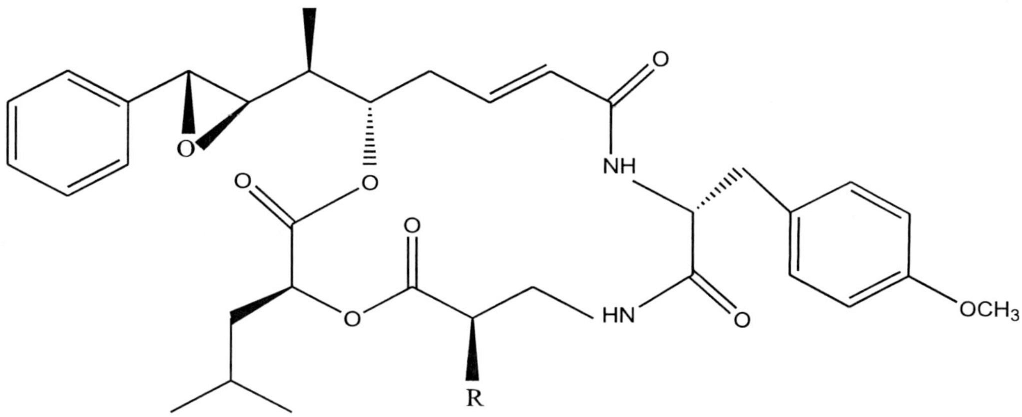

Figure 4. Structure of Cryptophycin B (R = CH_3) and Arenastatin A (R = H).

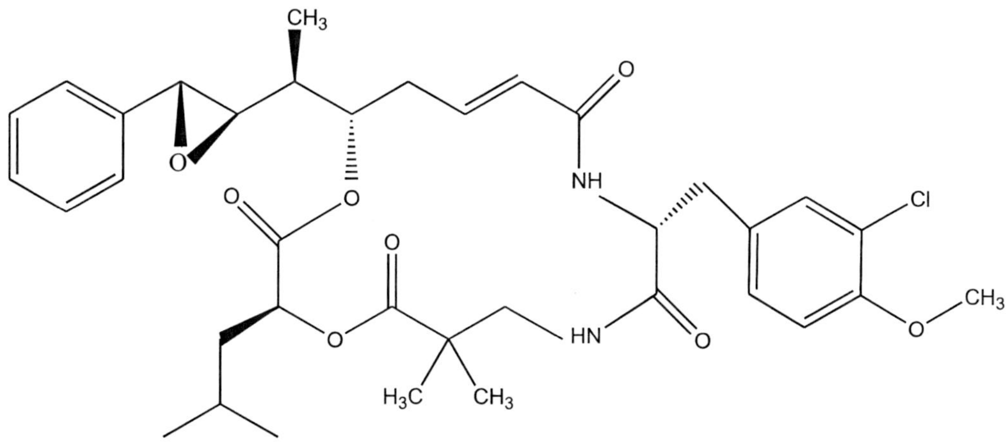

Figure 5. Structure of Cryptophycin 52.

In a study of about 1000 marine cyanobacterial strains, Moore et al. (Moore et al. 1988) had found 9% strains with antifungal activity. Cryptophycin isolated from cyanobacteria were earlier identified as potent fungicide against *Cryptococci* (Schwartz et al. 1990), however later on it was discarded due to its notable toxicity (Figures 4-5). Cyanobacteria produce a broad range of antifungals belonging to structural classes, such as peptides, polyketides and alkaloids.

The indolo [2, 3-a] carbazole scaffold of the tjipanazoles (Figure 6) which is chlorinated to a varying degree and glycosylated 6-deoxy-D-gulose (tjipanazole A1, C1, C2 and G1) and L-rhamnose (tjipanazoles A2, C3, C4 and G2) is characterized by antifungal properties, especially against phytopathogenic fungi causing rice blast and wheat leaf rust infections (Gademann and Portmann 2008). The cyanobacterium *Fischerella ambigua* has been shown to produce a novel antifungal compound, parsinguine. This compound, which was collected after extraction in methanol, inhibited the growth of *Candida krusei*, with a MIC value of 20 μg/mL (Ghasemi et al. 2004). Vestola et al. (2004) studied several cyanobacteria strains from the genera *Anabaena, Nostoc*, *Aphanizomenon,* and *Tolypothrix*, which are known to produce hassallidins, the promising antifungal glycolipopeptides. Eleven strains (3 *Nostoc* sp., *N. calcicula*, 3 *Anabaena*, *A. cf cylindrica*, *Scytonema* sp., *S. hofmanni*, and *Fischerella* sp.) displayed inhibitory activity against *Candida albicans* and/or *Aspergillus flavus.* Shishido et al. (2015) worked on *Anabaena* sp. strains which were able to produce anabaenolysins, lipopeptides and displayed antifungal activity. The cyanobacterial lectin showed antifungal activity against pathogenic *Cryptococcus neoformans* var. *neoformans* and *Cryptococcus gattii*, inhibiting cell growth by binding to the cell wall (Jones et al. 2017).

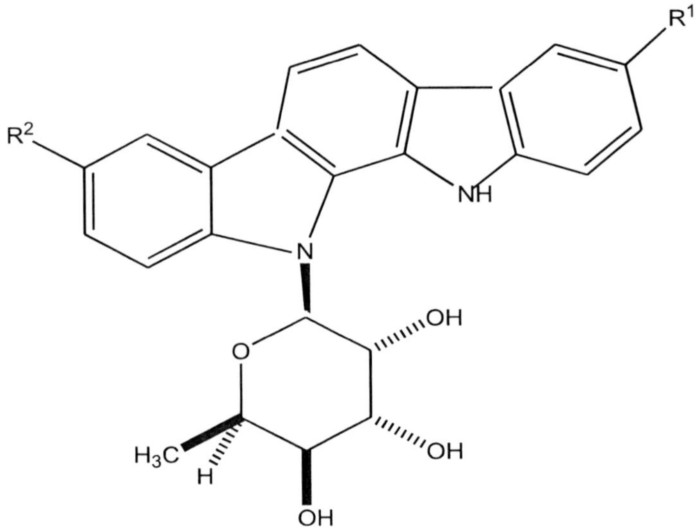

Figure 6. Structure of Tjipanazole A1 (R^1 = Cl, R^2 = Cl), C1 (R^1 = Cl, R^2 = H), G1 (R^1 = H, R^2 = H).

Scytophycin

The potent antifungal macrolide scytophycin was detected in *Anabaena* sp. HAN21/1, *Anabaena cf. cylindrica* PH133, *Nostoc* sp. HAN11/1 and *Scytonema* sp. HAN3/2.

Hassallidin

These were first isolated from *Tolypothrix* (Neuhof et al. 2005, 2006) and are structurally interesting non-ribosomal cyclic depsipeptides, decorated with a sugar and a dihydroxy fatty acid that can also be glycosylated. The MICs of hassallidins A to D against various *Candida* species including *C. albicans* were found to be in the range from 1.5 to 10.5 μM (Neuhof et al. 2006; Vestola et al. 2014). Cytotoxicity of the hassallidins is about 10-folds higher (Neuhof et al. 2006). Shishido et al. (2015) detected antifungal glycol lipopeptide hassallidin production in *Anabaena* spp. BIR JV1 and HAN7/1 and in *Nostoc* spp. 6sf Calc and CENA 219. In addition, three cyanobacterial strains, *Fischerella* sp. CENA 298, *Scytonema hofmanni* PCC 7110 and *Nostoc* sp. N107.3, produced unidentified antifungal compounds that warrant further characterization. Interestingly, all of these strains, which produced antifungal compounds belonged to the cyanobacterial orders Nostocales or Stigonematales. *Synechocystis* and *Synechoccocus* produced substances with inhibitory effects on prokaryotic cells and with apoptotic activity in eukaryote cell lines (Martins et al. 2008). The antibacterial activity of cyanobacteria is mainly directed against Gram-positive bacteria as most Gram-negative bacteria are resistant to toxic agents because of presence of lipopolysaccharides on their outer membrane (Dixon et al. 2004).

Nostofungicidine

It is a lipopeptide derived from the methanolic extract of a field-grown terrestrial *N. commune* (Dembitsky and Rezanka 2005). Nostofungicidine contains a novel b-AA, *viz.* 3-amino-6-hydroxystearic acid. It also exhibited strong antifungal activity against *Aspergillus candidus* and cytotoxicity against NSF-60 cell (Dembitsky and Rezanka 2005; Kim 2006).

Amino-6-Hydroxy Stearic Acid

This novel antifungal comprising cyclic peptide was derived from a terrestrial blue-green alga *Nostoc commune* (Singh et al. 2005).

Microviridins

Microviridins, cyclic peptides with the particular structural characteristic of a tricyclic ring system, are the largest known cyanobacterial oligopeptides (Murakami et al. 1997).

Nostopeptolides

Derived from *Nostoc* sp. showed antifungal activity against *Candida albicans*, antibacterial activity against *Pseudomonas aeruginosa, Mycobacterium tuberculosis*, or interdiction of peptidase activity against trypsin, chymotrypsin, papain, thrombin, plasmin and against leucocyte elastase (Dembitsky and Rezanka 2005).

Cryptophycin-1

An important member of cryptophycin class was first isolated from *Nostoc* ATCC53787 which exhibited antifungal properties (Hirsch et al. 1990; Schwartz et al. 1990).

Ambigols

They are polychlorinated aromatic compound, isolated from terrestrial cyanobacterium *Fischerella ambigua*, having potent activity against several fungi (Falch et al. 1993) and other compound named parsiguine (Ghasemi et al. 2004).

Hapalolindoles

Cyanobacterium *Hapalosiphon fontinalis* produced a group of indole alkaloids, hapalolindoles, which acted as antifungal agent (Moore et al. 1987a).

Ambiguine Isonitrils

These are hapalolindole type alkaloids, with antifungal properties, produced by *Fischerella ambigua, Hapalosiphon hibernicus* and *Westiellopsis prolifica* (Smitka et al. 1992).

Majusculoic Acid

It is fatty acid isolated from *Lyngbya majuscula* which showed antifungal activity against *Candida albicans* ATCC 14500 with the MIC value of 8 µM (MacMillan and Molinski 2005).

Majusculamide C

It is a metabolite isolated from *Lyngbya majuscula/M. producens*, which was found to be active against several plant pathogenic fungi such as *Rhizoctonia solani, Pythium aphanidermatum, Aphanomyces euteiches*, and *Phytophthora infestans* at low µM concentrations Moore and Mynderse 1982; Carter et al. 1984).

Lobocyclamides

The lobocyclamides are cyclic lipopeptides from *Lyngbya confervoides*. A mixture of lobocyclamides A and B had higher activity against a fluconazole-resistant *Candida albicans* than the separate compounds (MIC of thc mixture 10 μM) (MacMillan et al. 2002).

Fischerindole L

Fischerella muscicola, a terrestrial cyanobacterium, has been reported to produce an antifungal compound, fischerindole L, which is chemically related to hapalindoles (Park et al. 1992).

Tjipanazoles

These compounds from *Tolypothrix tjipanasensis* and *Fischerella ambigua* were found to be active against a range of phytopathogenic fungi, which was in contrast to other indolocarbazoles which showed only weak cytotoxicity and no inhibition of protein kinase C (Sonjoukhan et al. 1991).

Antimalarial Compounds

Calothrixins A and B derived from *Calothrix* sp. inhibited the FAF6 strain of *Plasmodium falciparum* with IC_{50} values of roughly 60 and 180 nM (Rickards et al. 1999).

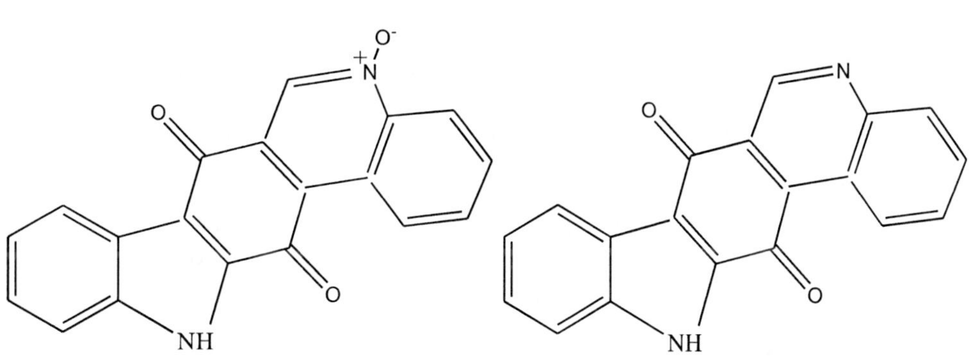

(A) (B)

Figure 7. Structure of Calothrixin A (A) Calothrixin B from *Calothrix* sp. (B).

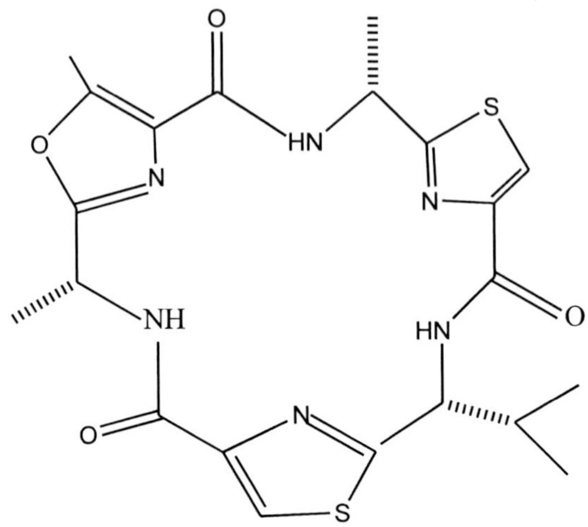

(A)

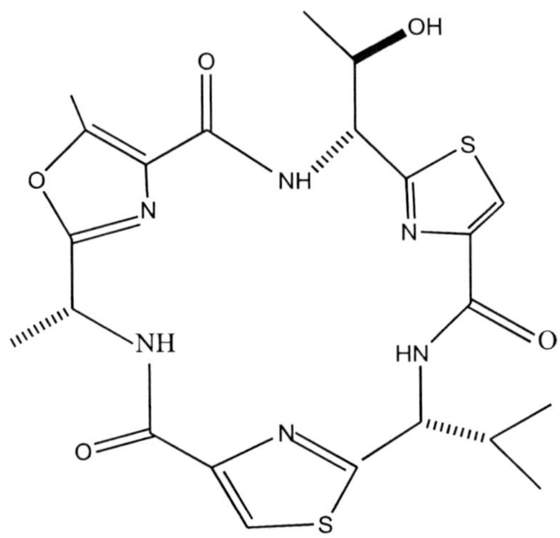

(B)

Figure 8. Structure of Venturamides A (A) Venturamides B (B) from marine *Oscillatoria* sp.

Venturamides A and B (Figure 8) have been reported from marine *Oscillatoria* sp. which showed low micro molar activity in the antimalarial assay against *P. falciparum* W2 (Linington et al. 2007).

Symplocamide A (Figure 9A) obtained from a cyanobacterium *Symploca* sp. displays an active role against parasite of malaria (*P. falciparum* W2; IC_{50} = 0.95 μM), Chagas disease (*Trypanasoma cruzi*; IC_{50}> 9.5 μM) and leishmaniasis (*Leishmania donovani*; IC_{50} > 9.5 μM) (Linington et al. 2007).

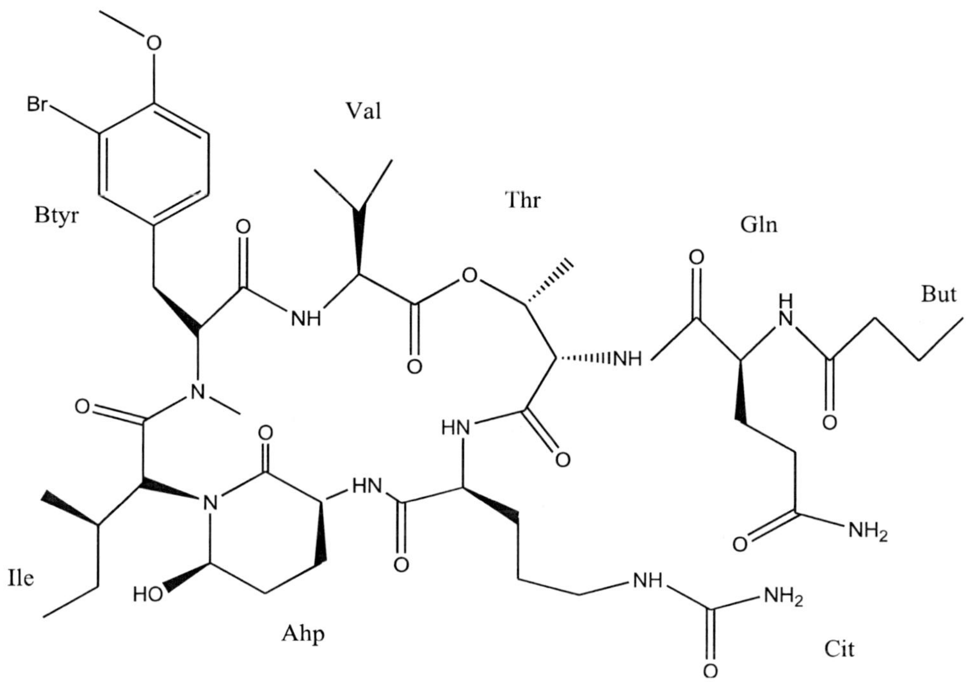

(A)

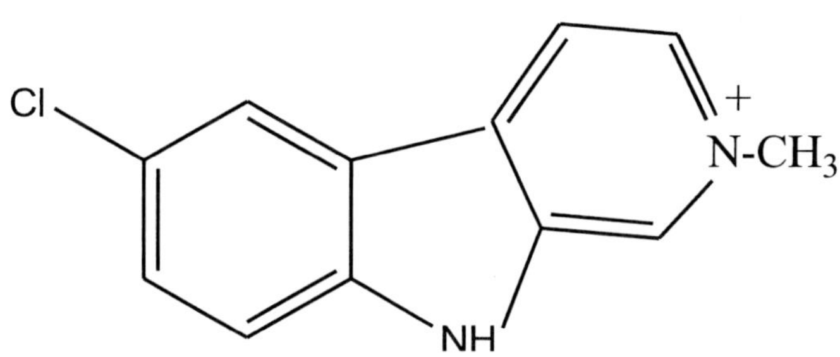

(B)

Figure 9. Structure of Symplocamide A (A) and Nostocarboline (B).

Nostocarboline (Figure 9B) and its symmetrical homodimers are potent and selective inhibitors of *P. falciparum* and also display exciting activity against *L. donovani* (Barbaras et al. 2008). Nostocarboline isolated from the freshwater cyanobacterium *Nostoc* 78-12A, is a potent cholinesterase inhibitor, which is an enzyme targeted at the treatment of Alzheimer's disease (Blom et al. 2006). The organic extracts of the marine cyanobacterium *L. majuscula* displayed an active role against chloroquine-resistant *P. falciparum* (McPhail et al. 2007).

Two new lipopeptides, dragomabin (Figure 10A) and dragonamide (Figure 10 B-E), and the related known metabolites carmabin A were isolated from *L. majuscula* Gomont (Oscillatoriaceae) which showed good antimalarial activity.

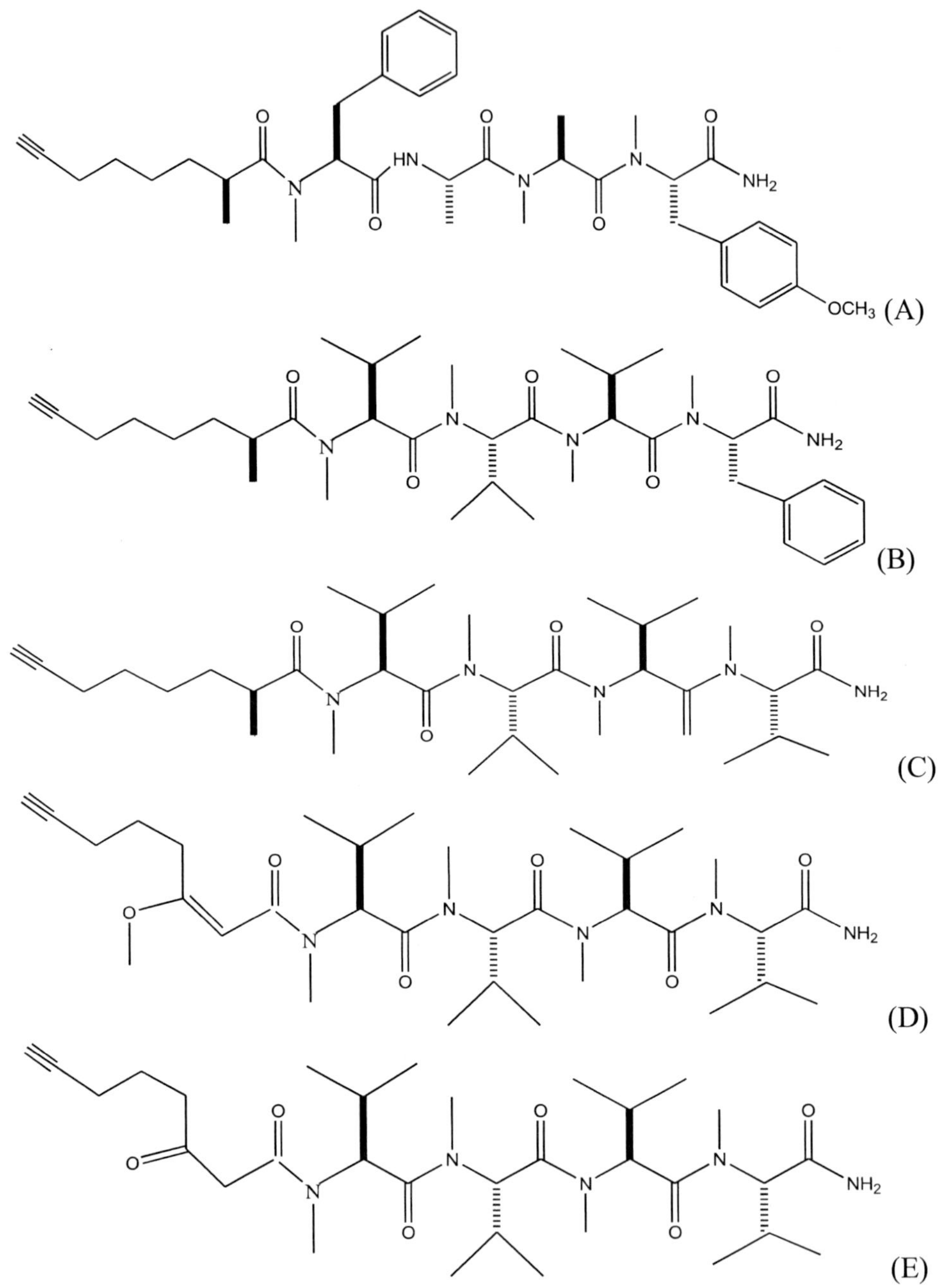

Figure 10. Structure of Dragomabin (A) Dragonamide A (B) Dragonamide B (C) Dragonamide C (D) Dragonamide D (E).

Gupta et al. (2014) reported aplysiatoxin, debromo aplysiatoxin and anhydride bromo aplysiatoxin, as well as two new analogues, 3-methoxyaplysiatoxin and 3-methoxydebromoaplysiatoxin from the marine cyanobacterium *Trichodesmium erythraeum*. The antiviral activities of these marine toxins were assessed using Chikungunya virus (CHIKV)-infected cells. Debromo aplysiatoxin, anhydrodebromoaplysiatoxin and 3-methoxydebromoaplysiatoxin displayed dose-dependent inhibition of CHIKV. Furthermore, debromo aplysiatoxin and 3-methoxydebromoaplysiatoxin exhibited significant anti-CHIKV activities. Similarly, bioassay-guided isolation efforts led to the discovery of several unique linear peptides with antiparasitic activity, such as modified lipopeptide gallinamide A, dragonamide E and almiramides (Figure 11) with anti-leishmanial activity (Linington et al. 2008; Balunas et al. 2009; Sanchez et al. 2010).

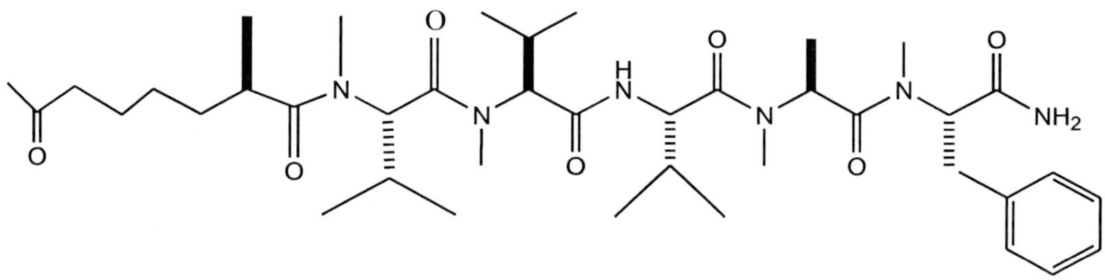

Figure 11. Structure of Almiramide A.

Cyanotoxins

Numorous toxic metabolites have been reported from different species/strains of cyanobacteria in marine as well as freshwater ecosystems that may be harmful to several aquatic and terrestrial organisms including humans (Krienitz et al. 2002; Dittmann and Wiegand 2006; Žegura et al. 2011). On the basis of mechanism of action on terrestrial vertebrates these toxins are classified as hepatotoxins, neurotoxins, dermatotoxins, cytotoxins, toxins with irritating potential (lipopolysaccharides) and toxins acting on the gastrointestinal system. These toxins can also be categorised according to their chemical structures such as cyclic peptides, alkaloids or lipopolyssacharides (Carmichael 1992; Chorus and Bartram 1999; Codd et al. 1999; Wiegand and Pflugmacher 2005; Rastogi and Sinha 2014). Some of the important cyanobacterial toxins have been discussed below:

Microcystins (MCs)

MCs are the most frequently occurring cyanotoxins in surface as well as drinking water and are the most studied cyanobacterial hepatotoxins. These are found within the cyanobacterial cells and are dissolved in the water as well, depending on the nature of the toxin and the growth stage of cyanobacteria (Van Apeldoorn et al. 2007). MCs are found in freshwaters as well as marine habitats. The occurrence of MCs has been reported in several cyanobacterial genera such as *Anabaena*, *Anabaenopsis*, *Aphanocapsa*, *Aphanizomenon*, *Cylindrospermopsis*, *Fischerella*, *Hapalosiphon*, *Lyngbya*, *Microcystis*, *Nostoc*, *Oscillatoria* (*Planktothrix*), *Phormidum*, *Rivularia* and *Synechococcus* but most frequently in the strains of *Anabaena* as well as *Microcystis* sp. (Sivonen et al. 1992;

Krienitz et al. 2002; Hodoki et al. 2012). MCs are cyclic heptapeptides (Figure 12) consisting of seven amino acids, including several D-amino acids and two unusual amino acids such as N-methyldehydroalanine (Mdha) and hydrophobic β-amino acid, 3-amino-9-methoxy-2,6,8-trimethyl-10-phenyldeca-4,6-dienoic acid (Adda) (Botes et al. 1985; Rao et al. 2002).

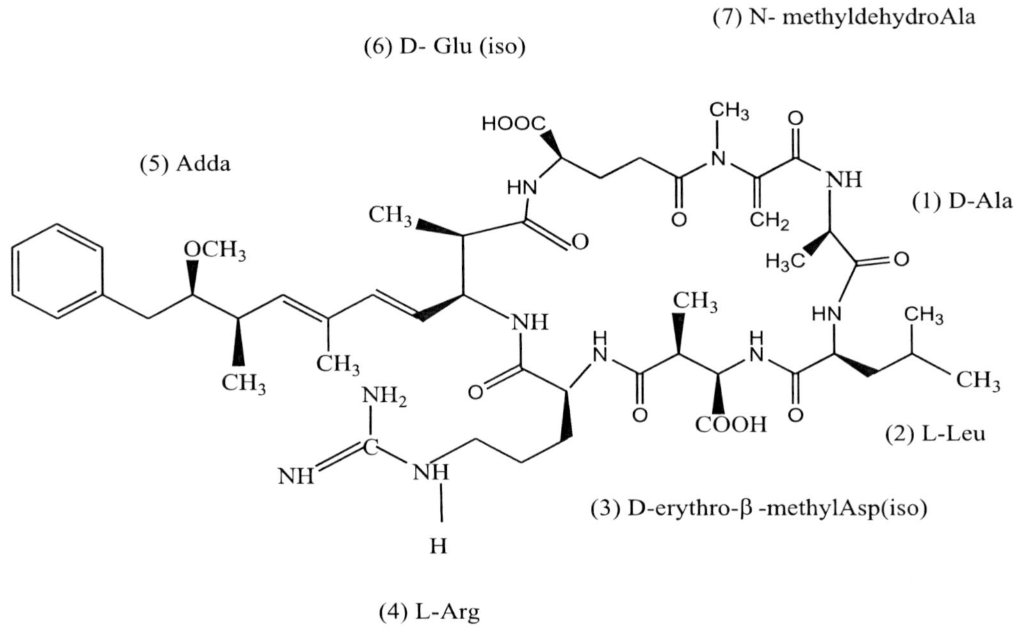

Figure 12. Chemical structure of Microcystin-LR showing leucine (L: Leu) and arginine (R: Arg) at 2 and 4 positions respectively.

Nodularin (NODLN)

NODLN, another class of hepatotoxins, was recognized by Rinehart et al. (1988) from *Nodularia spumigena* living in brackish water (e.g., Baltic Sea). It is a cyclic pentapeptide composed of Adda and D-erythro-β-methylaspartic acid (D-Masp) as well as N-methyl dehydrobutyrine (Mdhb) (Figure 13) (Rinehart et al. 1994). A modified structure of NODLN i.e., NODLN-Har was reported from *Nodularia* PCC7804, that consisted of a homoarginine instead of an arginine (Saito et al. 2001) and two dihydrono dularins have been chemically synthesized (Namikoshi et al. 1993). The mode of action of NODLN in the inhibition of the catalytic subunits of serine/threonine-specific protein phosphatases 1 (PP1) and 2a (PP2a) showed similarity with MCs, except NODLN does not bind covalently to PP1 or PP2a (Bagu et al. 1997). Nodularins exhibit cumulative toxicity and are tumor promoters without any initiation capability (Sueoka et al. 1997).

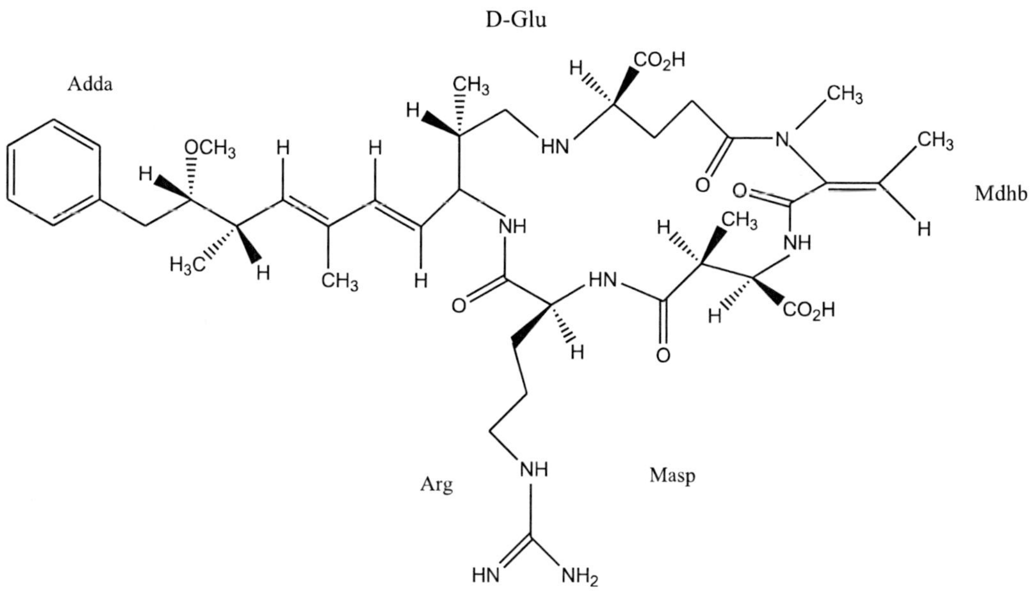

Figure 13. Chemical structure of Nodularin.

Saxitoxins (STXs)

STXs (Figure 14A) are tricyclic alkaloids, consisting of a tetrahydropurine group and two guanidine subunits and this neurotoxic is produced by a number of cyanobacteria such as *Aphanizomenon* sp., *Anabaena* sp., *Lyngbya* sp. and *Cylindrospermopsis raciborskii* (Humpage et al. 2000). These STXs may block neuronal communication by binding to the voltage-gated Na^{+} channels in neurons and discontinue the influx of sodium leading to paralysis of muscles and death by respiratory arrest in mammals hence are also known as paralytic shellfish poisons (PSPs) (Strichartz et al. 1986; Su et al. 2004).

Anatoxins (ANTX)

Anatoxin-a (antx-a) (Figure 14B) (165 Da) and homoanatoxin-a (Figure 14C) (179 Da) are the two common ANTX generally known as fast death factors. The nervous system, the skin (swimmer's itch) or the gastro internal tracts are affected by these neurotoxic alkaloids (Mansell 1996). The toxicological properties of homoanatoxin-a and its structurally related molecules antx-a are similar (Namikoshi et al. 2002). Several species of cyanobacteria such as *Anabaena flos-aquae*, *A. circinalis*, *Aphanizomenon* sp., *Cylindrospermum* sp., *Planktothrix* sp. and *M. aeruginosa* produce anatoxin-a (antx-a) (Edwards et al. 1992; Park et al. 1993). An anatoxin, anatoxin-a(s) (Figure 14D) is the phosphate ester of cyclic N-hydroxyguanidine (Matsunaga et al. 1989). Anatoxin-a(s), isolated from *Anabaena flos-aquae* and *A. lemmermannii* is a potent acetylcholinesterase

(AChE) inhibitor (Matsunaga et al. 1989). With similar action to the nerve agent sarin and is 10 times more lethal than antx-a (Matsunaga et al. 1989; Sivonen et al. 1989).

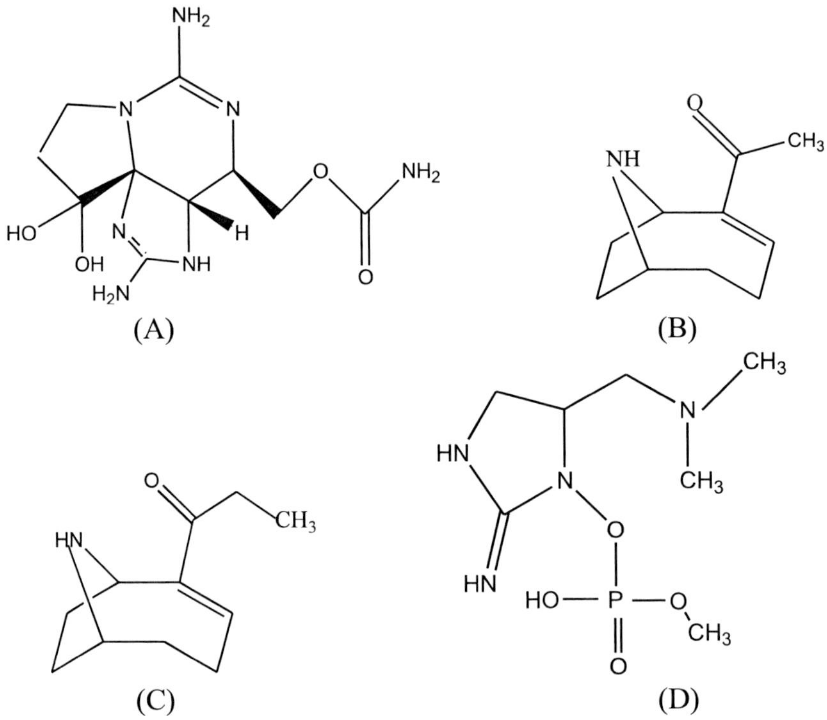

Figure 14. Chemical structures of Saxitoxin (A) Anatoxin-a (B) Homoanatoxin (C) Anatoxin-a(s) (D).

Cylindrospermopsin (CYN)

The alkaloidal polyketide CYN is produced by a number of fresh water cyanobacterial species (Li et al. 2001; Bernard et al. 2003). It contains a hydroxymethyl uracil moiety attached to a sulfated guanidine group (Figure 15A). Deoxy cylindrospermopsin (Figure 15B) that may also be obtained from *C. raciborskii*, is less toxic. Mutagenic as well as carcinogenic activities such as DNA adducts formation, DNA strand breaks, disruption of the kinetochore spindle followed by chromosome loss (aneuploidy) may be induced by CYN (Li et al. 2001; Liu et al. 2009; Humpage et al. 2000). CYN and its analog deoxy cylindrospermopsin (DeoxyCYN) do not stall protein phosphatases but are significant and permanent inhibitors of protein biosynthesis (Froscio et al. 2003; Neumann et al. 2007) and liver and kidney are the major targets for CYN-induced toxicity.

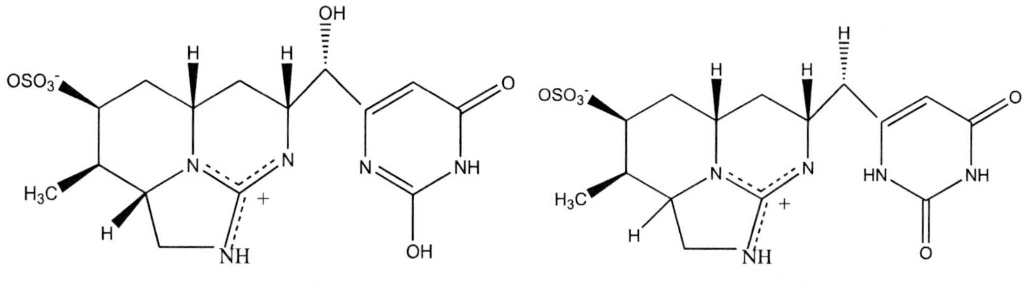

Figure 15. Chemical structure of Cylindrospermopsin (A) Deoxycyclindrospermopsin (B).

Lipopolysaccharides (LPS)

Cyanobacteria possess a unique cell wall that combines the presence of an outer membrane and lipopolysaccharides, as in Gram-negative bacteria, with a thick and highly cross-linked peptidoglycan layer similar to Gram-positive bacteria (Hoiczyk and Hansel 2000; Stewart et al. 2006). Cyanobacterial lipopolysaccharides (LPS) are generally classified as endotoxins, located within or at the surface of the outer cell layer, and are comprised of a carbohydrate polymer, a core oligosaccharide and an acylated glycolipid (≈lipid-A) (Stewart et al. 2006). Cyanobacterial extracellular polymeric substances (EPS) are mainly composed of high-molecular-mass heteropolysaccharides, with variable composition and roles according to the microorganism and the environmental conditions (Pereira et al. 2009). *Microcystis*, *Anabaena*, *Spirulina* and *Oscillatoria* are the major toxic LPS-producing cyanobacterial genera (Smith et al. 2008). The presence of a large variety of long-chained, unsaturated fatty acids and hydroxy fatty acids and the lack of phosphate makes cyanobacterial LPS different from LPS of enteric bacteria. Glucosamine forms the backbone sugar and there is variation in the amounts of 2-keto-3-deoxyoctonate (KDO), galactoses and heptoses (Martin et al. 1989). Inhalation of LPS might lead to "flue-like symptoms," characterized by cough, chills and sore throat. It may also cause fever in mammals and may be involved in the "septic shock syndrome" which may worsen toxicant-induced liver damage (Choi and Kim 1998) by the release of inflammatory mediators such as tumor necrosis factor-a (TNF-a), interferon-g, interleukin (IL) 1 and 6, leukotrienes, prostanoids and nitric oxide (Roth and Hewett 1993). LPS can also inflict the biotransformation enzyme systems, directly or mediated by factors of the immune reaction. In addition to its action on the immune system, the endotoxic LPS from enteric bacterial (e.g., *E. coli, Salmonella typhimurium*, *Vibrio cholerae*, etc.) as well as cyanobacterial sources also affects the detoxication system of diverse organisms (Wiegand and Pflugmacher 2005).

Kalkitoxin

Kalkitoxin (Figure 16) is a lipopeptide neurotoxin produced by some species of cyanobacteria such as *L. majuscule* (Berman et al. 2009) that block the nerve cell sodium channels leading to paralysis and eventually death by respiratory distress. It is a potent brine shrimp toxin with an LD_{50} at 0.17 μM and LD_{25} at 10 ppm, respectively (Wu et al. 2000; Edwards et al. 2004). Kalkitoxin also inhibits cell division in sea urchin embryos (IC_{50}≈0.025 μM).

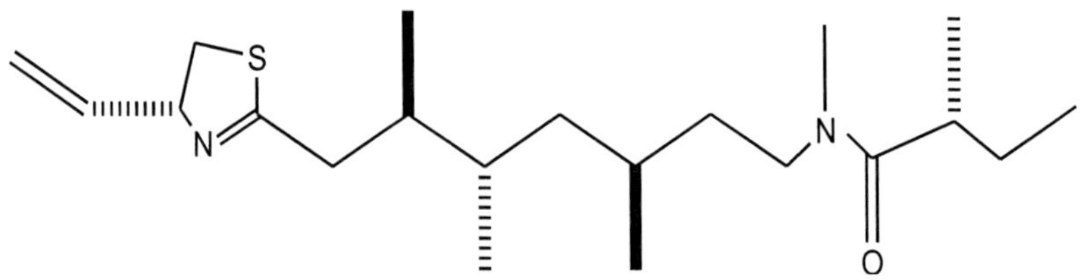

Figure 16. Chemical structure of Kalkitoxin.

Besarhanamides A

Two fatty acid amide-type bioactive secondary metabolites such as besarhanamides A (Figure 17A) and B, (Figure 17B) have been isolated from the marine cyanobacterium, *L. majuscule,* and it was found that besarhanamides A exhibited moderate toxicity with an LD_{50} at 13 μM in the brine shrimp toxicity assay (Tan 2007).

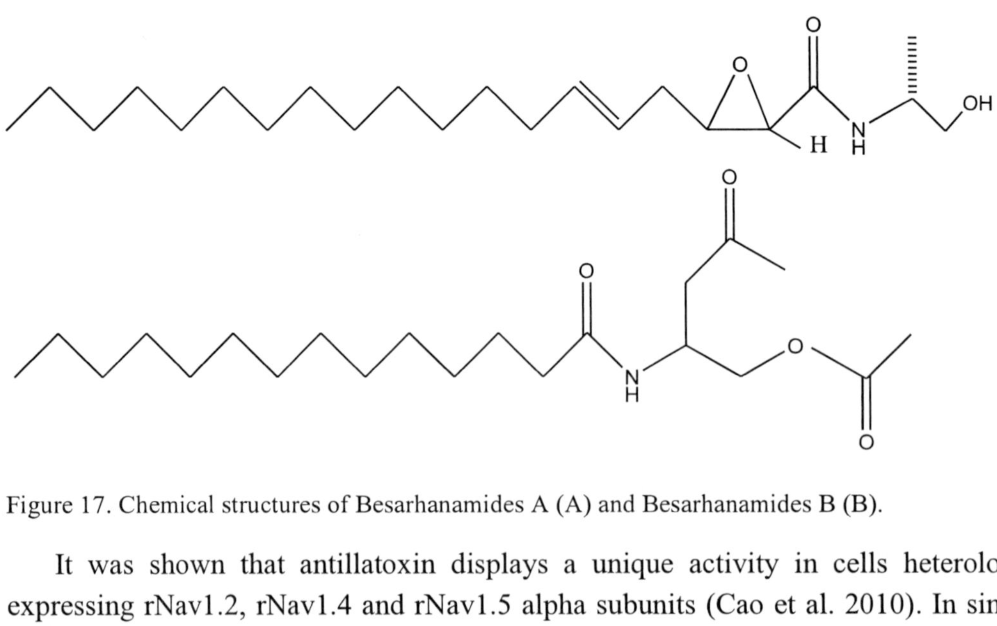

(A)

(B)

Figure 17. Chemical structures of Besarhanamides A (A) and Besarhanamides B (B).

It was shown that antillatoxin displays a unique activity in cells heterologously expressing rNav1.2, rNav1.4 and rNav1.5 alpha subunits (Cao et al. 2010). In similarity with the marine toxin brevetoxin B, antillatoxin was also shown to enhance neurite outgrowth in immature cerebrocortical neurons through activation of VGSCs (Jabba et al. 2010), suggesting a potential therapeutic application of these marine toxins in the treatment of spinal cord injury.

Aplysiatoxin

Aplysiatoxin (Figure 18) derived from *Lyngbya majuscula*, *Oscillatoria* (Mynderse et al. 1977) and certain red algae (e.g., *Gracilaria coronopifolia*) (Ito and Nagai 2000) is a phenolic bislactone that shows several lethal effects such as dilation of the lymphatic vessel and congestion of capillaries, diarrhea and fibrin deposition in the dilated pulmonary artery followed by bleeding, depending upon administration of different doses.

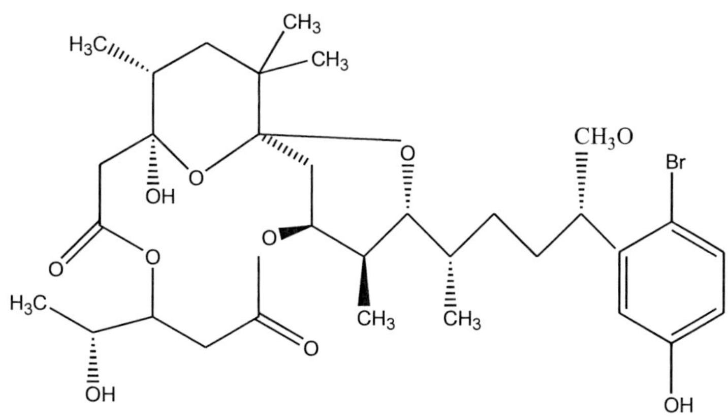

Figure 18. Chemical structure of Aplysiatoxin.

Hoiamides

Hoiamides, the potent neurotoxins were discovered in several mixed collections of *Symploca* and *Oscillatoria* (Pereira et al. 2009; Choi et al. 2010). The hoiamides are characterized by a remarkable integration of PKS and NRPS subunits, characterized by an unusual 15-carbon long linear polyketide chain, a tri-heterocyclic ring system composed of two a-methyl thiazolines and one thiazole, a PKS extended isoleucine residue and a variable number of additional amino- or hydroxyacids, usually forming an overall cyclic structure. Hoiamides A and B stimulate sodium influx with EC_{50} values of 1.7 and 3.9 mM, respectively, in mouse neocortical neurons. Hoiamide A is a sodium channel neurotoxin site 2 partial agonist. Both hoiamides A and B (Figure 19) were also found to potently suppress spontaneous Ca oscillations in neocortical neurons with EC_{50} values of 45.6 and 79.8 nM, respectively.

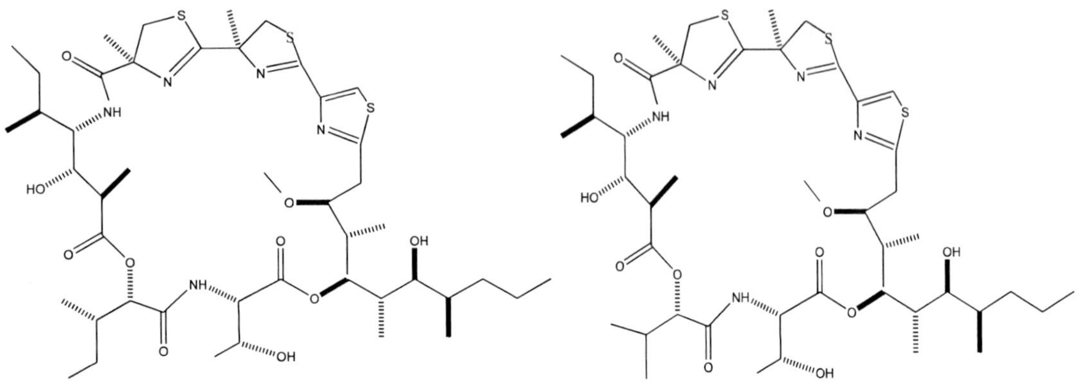

(A) (B)

Figure 19. Chemical structures of Hoiamide B (A) and Hoiamide A (B).

Antillatoxin

Antillatoxin is another cyanobacterial neurotoxin with exceptional levels of methylation featuring seven methyl groups (Figure 20) equivalents on the PKS section (Orjala et al. 1995). Previously, it was shown that this metabolite is a powerful VGSC activator at a site distinct from, yet interactive with, the brevetoxin and batrachotoxin sites (Li et al. 2001).

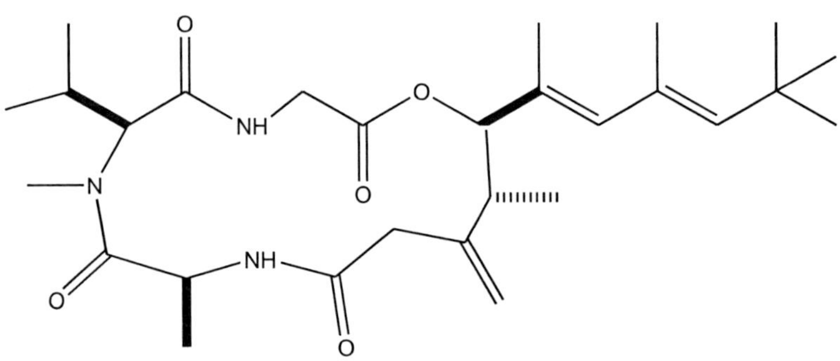

Figure 20. Chemical structure of Antillatoxin.

Palmyrolide A

A Palmyra collection of cyanobacteria composed of *Leptolyngbya* cf. and *Oscillatoria* spp., was found to possess a moderately potent modulator of sodium channel function (Pereira et al. 2010). The active compound, palmyrolide A (Figure 21), was defined as a cyclic lipopeptide with enamide and lactone connections, similar in overall structure to madangolide and laingolide (Klein et al. 1999; Matthew et al. 2010). The C-14 secondary methyl group configoration was determined by chemical degradation studies; however, despite exploration of a number of strenuous conditions, it was not possible to hydrolyze the lactone ring, and thus the configurations at C-5 and C-7 could not be determined. In this regard, the tertiary butyl groups found adjacent to lactone junctures in a number of cyanobacterial metabolites are an intriguing structural adaptation for their chemically stabilizing properties. In pharmacological assays, palmyrolide A inhibited Ca^{2+} oscillations in murine cerebrocortical neurons with an IC_{50} of 3.7 mM, blocked VGSCs in neuro-2a cells with an IC_{50} of 5.2 mM and was not cytotoxic up to 20 mM.

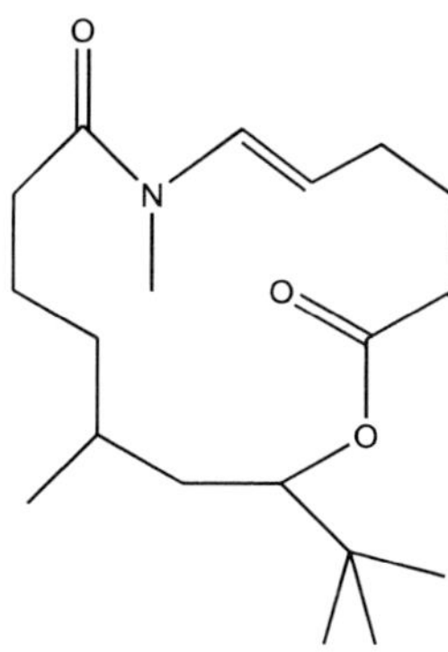

Figure 21. Chemical structure of Palmyrolide.

CYANOTOXINS AS INSECTICIDAL AND LARVICIDAL COMPOUNDS

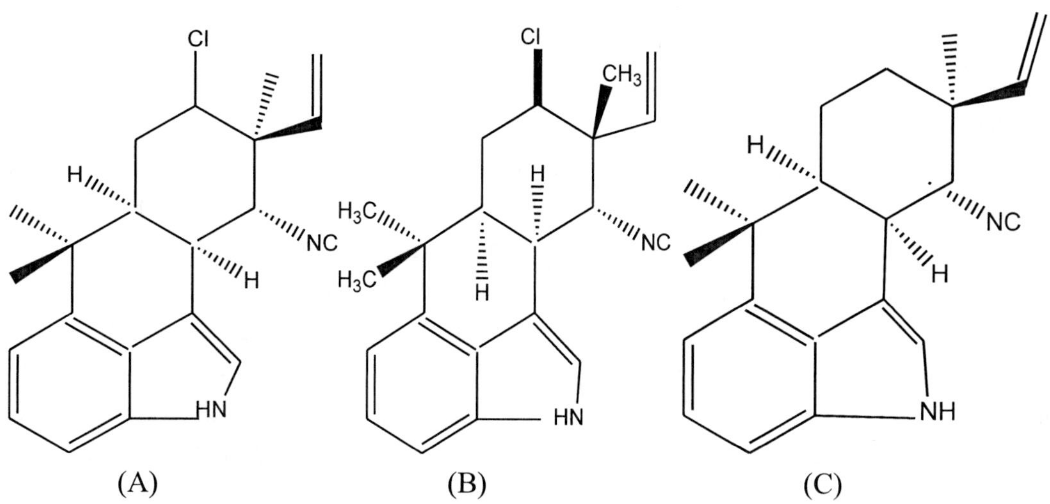

Figure 22. Chemical structures of Hopalindole L (A) Hopalindole A from *Hapalosiphon fontinalis* (B) 12-epi-hapalindole J isonitrile (C).

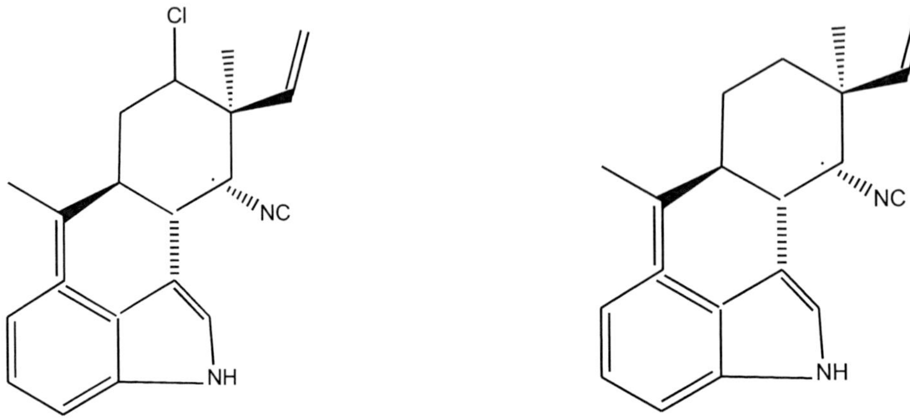

(A) (B)

Figure 23. Chemical structures of 12-epi- Hapalindole E isonitrile from *Fischerella* ATCC 43239 (A) 12-epi- Hapalindole C isonitrile (B).

Fischerella ATCC 43239 produces indole alkaloids called hapalindoles (e.g., 12-epi-hapalindole J isonitrile, 12-epihapalindole C isonitrile, hapalindole L and 12-epi-hapalindole E isonitrile) (Figures 22-23) which act as potent insecticidal alkaloids against dipterans (Becher et al. 2007).

Welwitindolinones, derivatives of the hapalindole series, isolated from *Hapalosiphon welwitschi* (Stratmann et al. 1994) showed multiple-drug resistance reversing activity along with insecticidal properties. The extracts of about 76 cyanobacterial isolates were screened (Kiviranta et al. 1993). Compounds from several isolates were larvicidal to *Aedes aegypti*, however microcystins and anatoxin-a showed maximum inhibition, suggesting commercial potentials for development of larvicides from these strains. Several strains of cyanobacteria such as *Oscillatoria agardhii* and *Westiellopsis* sp. display larvicidal properties to several mosquito species such as *Aedes albopictus*, *A. aegypti*, *Anopheles stephensi*, *Culex tritaeniorhyncus* and *C. quinquefaciatus* (Rao et al. 1999; Harada et al. 2000). Pahayokolides (Figure 24) isolated from *Lyngbya* sp. was found to be effective against microalgae and mosquito larvae (An et al. 2007; Gantar et al. 2008).

It has been found that the cyanotoxins anatoxin-a, microcystins and cylindro spermopsin obtained from *Anabaena*, *Microcystis* and *Cylindrospermopsis,* respectively, show larvicidal activity with 50% mortality (Berry et al. 2008). Antifeedant pesticidal activity of a glycine-rich peptide toxin from cyanobacterium *Scytonema* MKU 106 against two cotton pests, *Heliothis armigera* and *Stylepta derogate* was studied. Toxins from both of the above cyanobacteria exhibit effective antifeedant and larvicidal activity against two cruciferous pests, *Pieris brassicae* and *Plutella xylostella* (Sathiyamoorthy and Shanmugasundaram 1996).

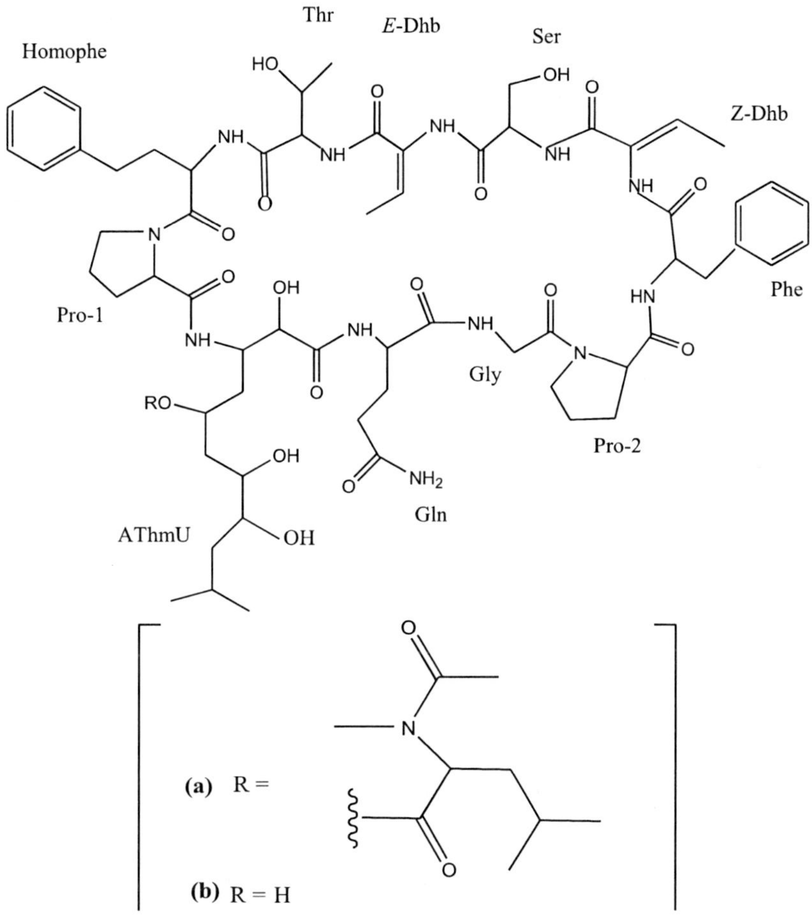

Figure 24. Structures of cyclic peptides Pahayokolides A and B from *Lyngbya* sp.

CYANOBACTERIAL TOXINS AS ALLELOPATHIC COMPOUNDS

Hapalindole A, isolated from *Hapalosiphon fontinalis* (Moore et al. 1984) shows anti algal and anti-mycotic properties. Norharmane (Figure 25) isolated from *Nodularia harveyana* exhibits anti cyanobacterial activity against both filamentous and unicellular cyanobacteria (Volk 2006).

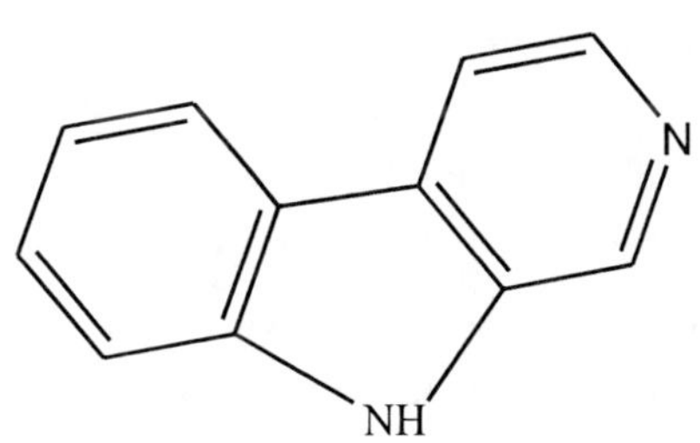

Figure 25. Structures of Norharmane.

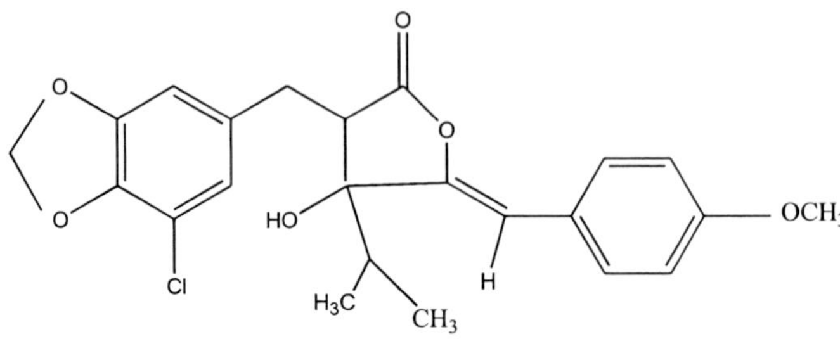

(A)

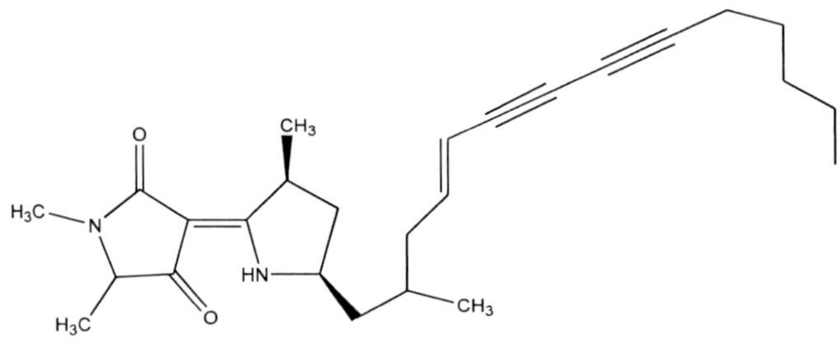

(B)

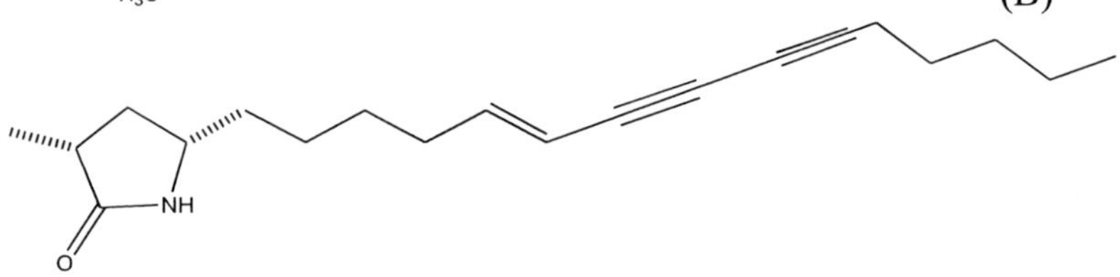

(C)

Figure 26. Structures of Cyanobacterin (A) Fischerellins A (B) and Fischerellins B (C) from *Fischerella* sp.

Cyanobacterin (chlorinated γ-lactone) (Figure 26A), derived from a freshwater *Scytonema* sp. inhibits photosystem II (Gleason and Case 1986). Cyanobacterins LU-1 and LU-2 (Gromov et al. 1991; Vepritskii et al. 1991; Ishibashi et al. 2005) reported from *Nostoc linckia* are structurally different from cyanobacterin but specifically inhibit electron transport in photosystem II. Other CSMs such as nostocyclamide, nostocine A, nostocarboline, etc. have been reported from different *Nostoc* species that are associated with algicidal activity (Todorova et al. 1995; Hirata et al. 2003; Blom et al. 2006). Calothrixins may also inhibit RNA polymerase and DNA synthesis and act as allelopathic compounds (Rickards et al. 1999). Microcystins (microcystin-LR) inhibit the growth of several aquatic plants such as *Ceratophyllum*, *Myriophyllum*, *Cladophora*, *Elodea*, *Lemna* and *Phragmites* etc. (Casanova et al. 1999; Romanowska-Duda and Tarczyńska 2002; Jang et al. 2007). Hapalindoles as well as other metabolites such as fischerellins A and B (Figure 26B-C) from *Fischerella* sp. may inhibit photosystem II and may also exhibit an alternate pathway of allelopathy by inhibiting the RNA synthesis (Gantar et al. 2008; Doan et al. 2001). Nostocarboline, an acetyl- and butyrylcholinesterase and trypsin inhibitor (Becher et al. 2009) from *Nostoc* 78-12A, has been derived that show strong algicidal activity against both eukaryotic and prokaryotic phototrophs (Blom et al. 2006). A brief account of various cyanobacterial toxins and their possible effects has been given in Table 1.

Table 1. Cyanobacterial toxins and their probable effects

CSMs	**Source**	**Effects**	**References**
Microcystins	*Microcystis* sp., *Anabaena* sp., *Oscillatoria* sp., *Anabaenopsis* sp., *Nostoc* sp., *Hapalosiphon* sp.	Inhibits serine/ threonine protein phosphatises 1 and 2a (PP1/2a)	Dittmann et al. (2006)
Nodularin	*Nodularia* sp.	Inhibits the catalytic subunits of serine/threonine-specific protein phosphatises 1 (PP1) and 2a (PP2a)	Gulledge et al. (2002)
Saxitoxins	*Aphanizomenon* sp., *Anabaena* sp., *Lyngbya* sp., *C. raciborskii*	Blocks neuronal communication by binding to the voltage-gated Na^+ channels in nerve cells	Humpage et al. (2000)
Anatoxin-a	*A. flos-aquae*, *A. circinalis*, *Aphanizomenon* sp., *cylindrospermum* sp., *Planktothrix* sp., *M. aeruginosa*	Binding irreversibly to the nicotinic acetylcholine receptors	Namikoshi et al. (2003)
Anatoxins-a(s)	*A. flos-aquae*, *A. lemmermannii*	Blocks acetylcholinesterase activity	Matsunaga et al. (1989)
Homoanatoxin-a	*P. formosa*	Similar to antx-a	Namikoshi et al. (2003)
Cylindrospermopsin	*C. raciborskii*, *U. natans*, *A. ovalisporum*, *R. curvata*, *A. bergii*	Inhibitor of protein biosynthesis, may induce DNA damage	Humpage et al. (2000)
Lipopolysaccharides	Cyanobacterial sp.	Potential irritant, affects any exposed tissue	Stewart et al. (2006)
Kalkitoxin	*L. majuscula*	Block nerve cell sodium channels	Wu wt al. (2000)
Antillatoxin	*L. majuscula*	Potent ichthyotoxin, stimulates voltage-gated sodium channels	Berman et al. (1999)
Aplysiatoxins	*L. majuscula*	Protein kinase C activators and tumor promoters	Fujiki et al. (1986)
Lyngbyatoxin A (LTA)	*L. majuscula*	Causes erythema (dermatitis), blisters, and necrosis in mammals and are potent tumor promoters	Fujiki et al. (1986)

UV-Screening Compounds

Cyanobacteria have evolved mechanisms to counteract the toxicity of UVR or high photosynthetically active radiation by synthesizing UV-absorbing/screening compounds such as mycosporine-like amino acids (MAAs) and scytonemin (Fleming and Castenholz 2007; Sinha et al. 2007; Soule et al. 2007; Singh et al. 2008; Sinha and Häder 2008; Singh et al. 2010; Pathak et al. 2015a, 2015b, 2017a, 2017b, 2018, 2019; Richa and Sinha 2015; Richa et al. 2018).

Scytonemin

Scytonemin is highly stable, lipid soluble, yellow-brown hydrophobic pigment, reported from extracellular polysaccharide sheaths of about 300 cyanobacteria. It is a dimer of indole and phenolic subunits and has a molecular mass of 544 Da. The two subunits in scytonemin are linked through an olefinic carbon atom which is unique among natural products (Figure 27). This linkage provides scytonemin a new ring system and has been termed as 'the scytoneman skeleton' (Proteau et al. 1993). Scytonemin has an *in vivo* absorption maximum at 370 nm whereas purified scytonemin shows maximum absorption at 386 nm, but it also absorb significantly at 252, 278 and 300 nm.

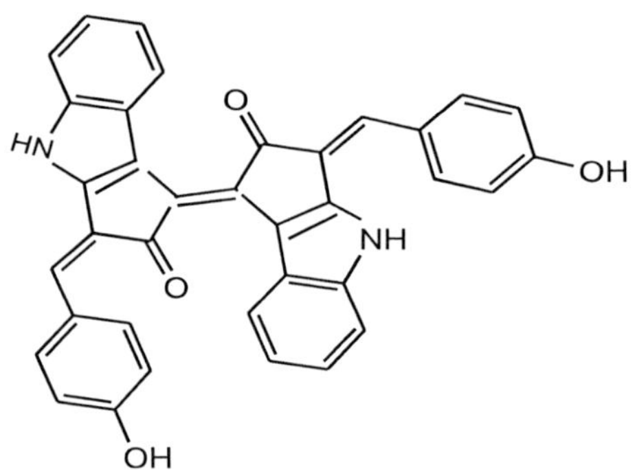

Figure 27. Chemical structure of reduced form of Scytonemin.

Mycosporine-Like Amino Acids (MAAs)

Mycosporine-like amino acids (MAAs) are a diverse group of UV-absorbing, small (<400 Da), colorless, water-soluble, secondary metabolites composed of a cyclohexenone or cyclohexenimine chromophore conjugated with the nitrogen substituent of an amino acid or its imino alcohol (Singh et al. 2008). Generally, the ring system contains a glycine subunit at the third carbon atom (Figure 28). Some MAAs also contain sulfate esters or glycosidic linkages through the imine substituent. MAAs can sometimes be glycosylated or covalently bound to oligosaccharides. Differences between the absorption spectra of MAAs are due to variations in the attached side groups and nitrogen substituents. Several characteristics of MAAs, such as strong UV absorption maxima between 309 and 362 nm and high molar extinction coefficients (ε = 28,100-50,000 M^{-1} cm^{-1}) suggest their role in

UV-screening. In addition, photo stability in both distilled and sea water in the presence of photosensitizers and high resistance against physico-chemical stressors like temperature, strong UVR, various solvents as well as pH make them successful photo protectants in various habitats and organisms. They are usually accumulated as solutes in the cytoplasm, but derivatives covalently bound to oligosaccharides can be excreted into the 'sheath' (glycocalyx), as in *Nostoc commune* (Bohm et al. 1995).

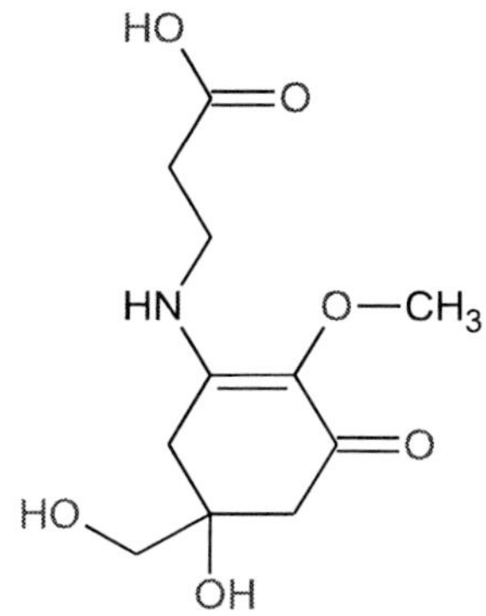

Figure 28. Chemical structure of Mycosporine-glycine.

SIDEROPHORES: THE IRON CHELATORS

Siderophores are low-molecular weight, high-affinity Fe (III)-binding ligands secreted by bacteria under conditions of iron stress to scavenge and transport iron. Cyanobacteria have also been shown to produce low molecular-weight (400 to 1200 Da) ferric-specific chelators (siderophores) under iron-limiting culture conditions (Trick and Kerry 1992; Wilhelm and Trick 1994). Siderophores bind to Fe^{3+} to form a ferrisiderophore complex which facilitates the transport of ferric ions into cells during periods of iron deficiency (Matzanke 1991). Schizokinen, a citrate derivative siderophore was first reported from the fresh water cyanobacterium *Anabaena* PCC 7120 (Goldman et al. 1983). Compounds, synechobactins from the marine cyanobacterium *Synechococcus* PCC 7002, have been isolated that are considered as schizokinen derivatives with one of the hydroxamic acids replaced by a long fatty acid and perform as siderophores (Ito and Butler 2005). It has been suggested that due to the amphiphilic structure of these siderophores they can integrate into the membrane of cyanobacteria and thus prevent its loss by dilution in marine environments (Martinez et al. 2003). Two complex siderophores, anachelin H containing a terminal salicylamide and anachelin 1, terminated by an oxazoline ring were isolated from the biofilm-forming cyanobacterium *Anabaena cylindrica* 1403-2a (Beiderbeck et al. 2000).

It has been shown that the anachelin chromophore recognizes metal oxide surfaces as found in minerals (Beiderbeck et al. 2000; Zürcher et al. 2006). Siderophore photore activity is largely determined by the nature of the Fe (III) binding groups (Barbeau et al.

2003).The thermodynamic stability constants for the complexation of iron(III) with catecholate-type siderophores isolated from the marine bacterium *Alteromonas iuteouiolacea* and the cyanobacterium *Synechococcus* sp. PCC 7002 was estimated and it was suggested that catecholate siderophores may play a role in the solubilization and biological uptake of iron in the marine environment (Lewis et al. 1995). Furthermore, siderophores are of particular interest as possible drugs for use both in the treatment of iron overload and for conditions of both acute and chronic metal toxicity (Neilands 1981) as well as in areas ranging from biomaterials to biosensors (Gademann 2007).

PHYTOHORMONES

Auxin, (indole-3-acetic acid,(IAA) is synthesized by all higher plants (Ljung et al. 2002) and has also been demonstrated in symbiotic nitrogen fixing cyanobacteria (Sergeeva et al. 2002). IAA has recently been identified in the cyanobacterium *Nostoc* sp. inhabiting symbiotic or parasitic environments. The symbiotic forms of cyanobacteria are competent in both accumulating and releasing IAA, and it was also found that accumulation of IAA is stimulated by exogenous tryptophan that may proceed by means of the indole-3-pyruvic acid pathway (Sergeeva et al. 2002). During the evaluation of several cyanobacterial isolates of *Anabaena* sp. for the development of biocontrol agents Prasanna et al. (2008) observed the production of IAA by these cyanobacterial strains. Rodríguez et al. (2006) indicated the occurrence of gibberellin-like plant growth regulators (PGRs) in the cyanobacterium *Scytonema hofmanni*. PGRs such as abscisic acid, ethylene, jasmonic acid, and cytokinin-like substances can be found in cyanobacteria (Ördög and Pulz 1996). Cytokinin-like activity has already been detected in a cyanobacterium *A. africanum* using the cucumber cotyledon expansion bioassay and soybean callus bioassay (Stirk et al. 1999).

ANTIBACTERIAL ACTIVITIES OF CYANOBACTERIA

Cyanobacteria are prokaryotic, oxygenic, phototrophic microorganisms with several species shown to possess biological activity against human pathogenic bacteria. Extracts from *Phormidium autumnale, Myxosarcina* sp., *Gloeotrchia* sp., *Cyanobium* sp. displayed the most important antibacterial activity. Secondary metabolites such as aeruginosin, cyanopeptolin, microcystin, and cylindro spermopsin are found in these cyanobacteria (Silva-Stenico et al. 2013). Exometabolites produced by the cyanobacteria *Nodularia harveyana* and *N. insulare* were tested against four different pathogenic bacteria: *E. coli, Staphylococcus aureus, Bacillus cereus,* and *Pseudomonas aeruginosa*. These exometabolites were characterized as norharmane (9H-pyrido (3,4-b)indole) and 4,40-

dihydroxybiphenyl and demonstrated a good antibacterial activity against *S. aureus* but a lack of effect was observed against *B. cereus* (Volk and Furker 2006). The n-hexane extracts of *Oscillatoria redekeii* showed an inhibitory activity against the pathogenic bacteria *S. aureus, B. subtilis,* and *Micrococcus flavus* with a MIC lower than 100 μg/mL for *S. aureus* (Mundt et al. 2003). Ghasemi et al. (2004) worked on parsiguine, a novel compound from the soil-isolated cyanobacteria, *Fischerella ambigua* had an antibacterial activity against *S. epidermis*. Ethanolic extracts of the cyanobacteria *Anabaena wisconsinense* and *Oscillatoria curviceps* demonstrated activity against *Pseudomonas anguilliseptica* and *Aeromonas hydrophylla.*

Comnostins

Comnostin A, B, C, D and E, have been derived from *Nostoc commune*. All comnostins A-E represented antibacterial activity.

Muscoride A

It is an oligopeptide alkaloid derived from terrestrial freshwater *Nostoc muscorum*. Muscoride A was the first composite possessing N-(2-methylbut-3-en-2-yl) valine and two contiguous methyloxazoles. Muscoride A showed feeble antibacterial activity against *B. subtilis* and *E. coli* (Dembitsky and Rezanka 2005).

Noscomin

Noscomin, has been derived from the culture medium of the terrestrial *N. commune* (Dembitsky and Rezanka 2005). Noscomin represented antibacterial activity against *B. cereus*, *S. epidermidis* and *E. coli.*

Carbamidocyclophanes

Carbamidocyclophanes are paracyclophanes derived from *Nostoc* sp. CAVN 10. It showed middling antibacterial activity against *S. aureus* (Singh et al. 2005).

Dodecahydrophenanthrene, 4-Methylchrysazin and 4-Hydroxy-7-Methylindan-1-One

It was isolated from the cells of *Nostoc.* Dodecahydrophenanthrene showed a selective strong antibacterial activity against *S. epidermidis* equal to chloramphenicol. Middling antibacterial activity against *S. epidermidis* was also found for 4-methylchrysazin and 4-hydroxy-7-methylindan-1-one against *B. cereus* for all 3 composites (Dembitsky and Rezanka 2005).

Hormothamnions

These are cyclic undecapeptides, produced from marine cyanobacterium *Hormothamnion enteromorphoides*, which showed activity against bacteria (Gerwick et al. 1989) and cyclic undecapeptides schizotrin form *Schizothrix* sp. (Pergament and Carmeli 1994).

Malyngolides

These are a group of antibacterial compounds produced by marine cyanobacterium *Lyngbya majuscula* (Gerwick et al. 1987; Burja et al. 2001) (Figure 29) that are active against Gram-positive microorganisms such as *Mycobacterium smegmatis*, *Streptococcus pyogenes*, *Staphylococcus aureus*, and *B. subtilis*, and inactive against the Gram-negative *Salmonella enteritidis*, *E. coli*, and *P. aeruginosa*. It was found that malyngolide interferes with bacterial quorum sensing. The compound acted by blocking the expression of lasR (a homologue of the luxR gene encoding the AHL sensing receptor protein) but does not interfere with the AHL-binding domain of the respective protein.The freshwater cyanobacterium *Oscillatoria redeki* produced an unsaturated hydroxyl fatty acid α-dimorphenolic acid and coriolic acid which reduced the bacterial infections (Mundt et al. 2003).

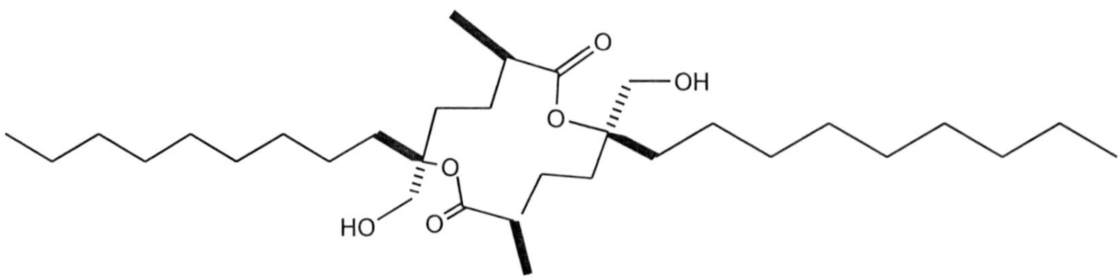

Figure 29. Chemical structures of Malyngolide dimer.

Ambiguine 1 Isonitrile

It was isolated from *Fishcerella* sp. and was found to be active against *E. coli* ESS k-12, *Staphylococcus albus* and *B. subtilis* (Raveh and Carmeli 2007).

Ambiguine

It is hapalindole-type alkaloid from *Fischerella ambigua* and showed antibacterial activity against *Mycobacterium tuberculosis* and against *B. anthracis* (Raveh and Carmeli 2007).

Norbiestane

It is diterpenoid from *Microcoleus lacustris* and was found to be active against *Staphylococcus aureus* (Gutierrez et al. 2008).

Phayokolide A

It is produced by *Lyngbya* sp. and showed activity against *Bacillus* sp. (Berry et al. 2004).

Hapalindoles

These are a large compound family of related indole alkaloids that were first found in *Hapalosiphon* species (Moore et al. 1984; Moore et al. 1987a, 1987b; Klein et al. 1995), and were later also isolated from *Fischerella* (Asthana et al. 2006; Kim et al. 2012) and *Westelliopsis* (Kim et al. 2012). Hapalindoles showed activity against a range of microorganisms including *S. aureus, B. subtilis, Salmonella gallinarum*, and *E. coli* (Bonjouklian et al. 1989). Interestingly, the various hapalindole congeners feature highly differing activity against a panel of microorganisms. The most potent compound, hapalindole A showed MICs at sub-μM concentrations (Kim et al. 2012). While having high antibacterial activity, they later have also been shown to be moderately cytotoxic (Kim et al. 2012) and highly insecticidal (Becher et al. 2007). Their mode of action against eukaryotic cells is likely based on the modulation of sodium channels (Cagide et al. 2014).

Hapalindole T

It is derived from *Fischerella* sp. active against *Staphylococcus aureus, Pseudomonas aeruginosa, Salmonella typhi* and *E. coli* (Asthana et al. 2006).

Scytoscalarol

It is guanidine-sesterterpene isolated from *Scytonema* sp. (Mo et al. 2009) and showed activity against the Gram-positive *Bacillus anthracis*, *S. aureusthan* and *E. coli*. The compound also showed weak activity against *Mycobacterium tuberculosis*. Based on *in silico* docking studies, its mechanism of action against the latter microorganism has been proposed to be the inhibition of the arabinosyltransferase Mtb EmbC (Espinoza-Moraga et al. 2013).

Eucapsitrione

It is an anthraquinone derivative isolated from *Eucapsis* sp. (Sturdy et al. 2010). It was found to selectively inhibit *M. tuberculosis* in assays testing both the activity against fast growing and nonreplicating persistent states. The compound was inactive against *M. smegmatis, S. aureus, E. coli*, and the yeast *Candida albicans*, and also not cytotoxic in the concentrations tested.

Nostocarboline

It is isolated from a *Nostoc* species is a quaternary indole alkaloid (Becher et al. 2005) (Figure 30). It was found to inhibit the enzyme butyryl choline esterase. First thought to be suitable as a lead for the treatment of Alzheimer's disease, it was soon realized that the compound or derivatives of it are also algicidal against photosynthetic organisms (Blom et al. 2006), and active against *M. tuberculosis* and the malaria parasite *Plasmodium falciparum* (Bonazzi et al. 2010). Synthetic studies on Nostocarboline resulted in several nostocarboline dimers (e.g., NCD9, 6) showed their activity against *S. aureus, E. coli*, or *C. albicans* (Locher et al. 2010). It was found to be active opposed *Trypanosoma brucei, Trypanosoma cruzi, Leishmania donovani* and *Plasmodium falciparum* with IC_{50} values ranging from 0.5 to 0.194 mM (Singh et al. 2005)

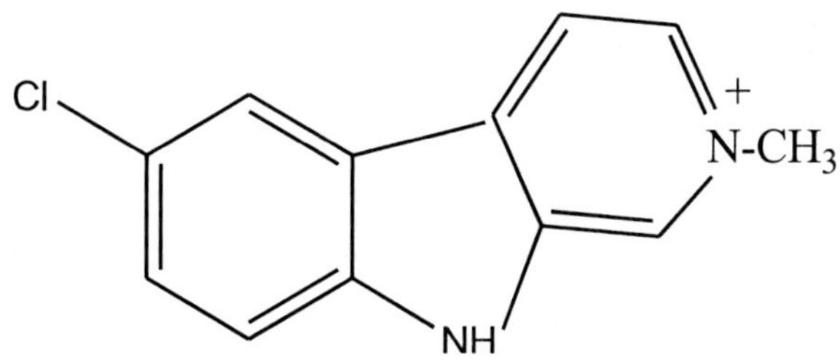

Figure 30. Structure of Nostocarboline isolated from *Nostoc* 78-12A.

ANTIPROTOZOAL COMPOUNDS

Almiramide B and C

Almiramide B and C are liner lipopeptides, isolated from *Lyngbya majuscula*, have potential to inhibit the growth of *Leishmania donovani* (Sanchez et al. 2010) (Figure 31).

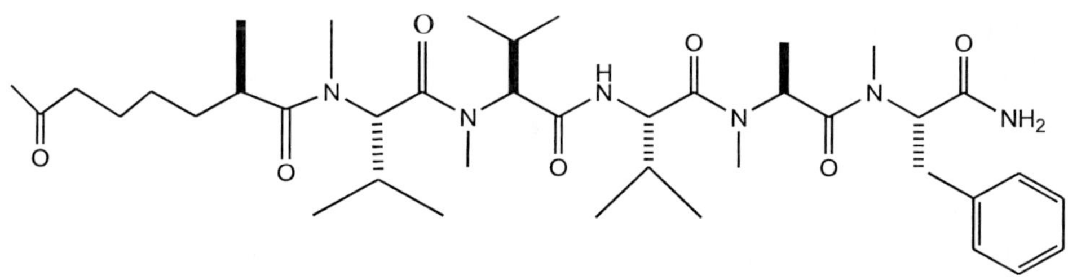

Figure 31. Structure of Almiramide A.

Calothixin A and B

Calothrixin A and B (Indolophenanthridine alkaloids) from *Calothrix* sp. suppress the growth of chloroquine-resistant *Plasmodium falciparum* (Rickards et al. 1999).

Ambigol C

Ambigol C from *Fischerella ambigua* inhibits the growth of *Trypanosoma rhodesiense* and *Plasmodium falciparum* (Wright et al. 2005).

Bastimolide A

It is a natural polyhydroxymacrolide recently isolated from a cyanobacterium whose structure has been firmly established by X-ray analysis.

ANTI-INFLAMMATORY COMPOUNDS

γ-Linolenic Acid

Gamma-linolenic acid (GLA) is a rare polyunsaturated fatty acid which is 170-fold more effective than linoleic acid, the major constituent of most polyunsaturated oils. Cyanobacteria such as *Spirulina platensis* and *Arthrospira* sp. are relatively rich in GLA, which in the human body is easily converted into arachidonic acid and thence into prostaglandin E2 (Euler and Eliassen 1967). Prostaglandins, the master hormones, control many body functions. Prostaglandin E2 is involved in contraction of smooth muscles. GLA is effective for the treatment of atopic eczema (Biagi et al. 1988), premenstrual syndrome (Horrobin 1983), heart diseases (Kernoff et al. 1977) Parkinson's disease, and multiple sclerosis. It lowers low-density lipoprotein in hyper-cholesterolemic patients (Cohen 1997). *Aphanizomenon flos-aquae* contain significant amounts of ω-3-α-linoleic acid, which inhibits the formation of inflammatory prostaglandins and arachidonic acid (Kushak et al. 1999).

Cyanobacterial phycocyanin (CPC), which represents up to 15% of the dry weight of a cyanobacterium also exhibits neuroprotective and hepato-protective effects, besides its anti-inflammatory activity (Reddy et al. 2000). CPC selectively inhibits cytochrome oxidase 2 (COX-2), which is overexpressed in breast cancer cells; it has no effect on COX-1 (Reddy et al. 2000). CPC can be used in the treatment of Alzheimer's and Parkinson's diseases (Rimbau et al. 2001), and have prevented oral and skin cancers. Moreover, it can boost the immune system and possibly lowers the risk of heart disease, prevent onset of cancers, and protect against age-related diseases such as cataracts, macular degradation, and multiple sclerosis (Rissanen et al. 2002).

Table 2. Bioactive compounds produced by cyanobacteria

Compound	Source	Biological activities	References
Venturamide A, B	*Oscillatoria* sp.	Antimalarial	Linington et al. (2007)
Symplocamide A	*Symploca* sp.	Antimalarial, Antileishmaniasis	Linington et al. (2008)
Nostocarboline	*Nostoc* sp.	Antimalarial, Antileishmaniasi, Cholinesterase inhibitor	Barbaras et al. (2008)
Dolastatins	*Lyngbya* sp., *Symploca* sp.	Antimalarials	Fennell et al. (2008)
Calothrixins A,B	*Calothrix* sp.	Antimalarial	Rickards et al. 1999)
Carmabin A,B	*L. majuscule*	Antimalarial, antiproliferative	McPhail et al. 2007)
Dragonamide A, B	*L. majuscule*	Antimalarial	McPhail et al. 2007)
Dragomabin	*L. majuscula*	Antimalarial	McPhail et al. 2007)

Compound	Source	Biological activities	References
Sulfoglycolipid	*Scytonema* sp.	Anti-HIV	Loya et al. 1998)
Scytovirin	*Scytonema varium*	Anti-HIV	Xiong et al. 2006)
Cyanovirin	*Nostoc ellipsosporum*	Anti-HIV	Dey et al. 2000)
Calcium spirulan	*Spirulina platensis*	Anti-HIV	Hayashi et al. 1996a)
Sulfolipids	*Lyngbya lagerhimii, Phormidium tenue*	Anti-HIV	Gustafson et al. 1989)
Ambiguines	*Fischerella* sp.	Antibacterial	Raveh and Carmeli 2007)
Ambiguine	*Fischerella ambigua*	Antibacterial	Raveh and Carmeli 2007)
Muscoride	*Nostoc muscorum*	Antibacterial	Nagatsu et al. 1995)
Tolyporphin	*Tolypothrix nodosa*	Antibacterial	Prinsep et al. 1992)
Didehydromirabazole	*Scytonema mirabile*	Antibacterial	Stewart et al. 1998)
Noscomin	*Nostoc commune*	Antibacterial	Jaki et al. 2000)
Nostocine A	*Nostoc spongiaeforme*	Antibacterial	Hirata et al. 2003)
Abietane diterpenes	*Microcoleous lacustris*	Antibacterial	Gutiérrez et al. 2008)
Hapalindole	*Nostoc CCC537, Fischerella* sp.	Antibacterial	Asthana et al. 2009)
Bis-(v-butyrolactones)	*Anabena variabilis*	Antibacterial	Ma and Led 2000)
Bastadin	*Anabaena basta*	Antibacterial	Miao et al. 1990)
Grassypeptolide	*L. confervoides*	Antiproliferative	
Bauerines A–C	*Dichotrix baueriana*	Anti-HSV-2	Kwan et al. 2008)
Aeruginosins	*Microcystis, Nodularia, Oscillatoria*	Serine proteases inhibitor	Larsen et al. 1994)
Lyngbyastatin	*Lyngbya confervoides*	Serine protease inhibitor	Shin et al. 1997)
Radiosumin	*Plectonema radiosum*	Enzyme inhibitor	Taori et al. 2008)
Circinamide	*Anabaena circinalis*	Enzyme inhibitor	Mooberry et al. 1995)
Agardhipeptin	*Oscillatoria agardhii*	Enzyme inhibitor	Negri and Jones 1995
Microcystin	*Microcystis aeruginosa, Calothrix* sp.	Algicidal/ larvicidal/ herbicidal	Luukkainen et al. 1993)
Fischerllin	*Phormidium ectocarpi*	Antifungal	Berry et al. 2008)
Cyanobacterin	*Scytonema hofmanni*	Algicidal	
Anatoxin-a	*Anabaena* sp.	Larvicides	Srivastava et al. 1999)
Pahayokolides	*Lyngbya* sp.	Antialgal/ larvicidal	Gleason and Case 1986)
Cylindrospermopsin	*Cylindrospermopsis* sp.	Larvicidal	Berry et al. 2008)
Majusculamide A–D	*L. majuscule*	Antifungal	Gantar et al. 2008)
Tanikolide	*L. majuscule*	Antifungal	Berry et al. 2008)
Toyocamycin	*Tolypothrix tenuis*	Antifungal	Moore and Mynderse 1981)
Anhydrohapaloxindole	*Hapalosiphon fontinalis*	Antifungal	Singh et al. 1999)

Oral doses of *Aphanizomenon flos-aquae* to healthy humans produce a slight decrease in the phagocytic activity of polymorph nucleated cells (Jensen et al. 2000). When taken with or before a potential antigen threat, cyanobacteria may enhance IgA levels to protect against food allergies (Hayashi et al. 1998). This property could be useful in the treatment of diseases involving suppressed immunesystems and responses such as allergies, autoimmune diseases, and chronic inflammatory conditions. Table 2 gives an overview of different bioactive compounds produced by cyanobacteria.

CONCLUSION

Cyanobacteria are an important source of novel and valuable bioactive natural compounds that need to be exploited biotechnologically for the benefit of humans. Several cyanobacterial secondary metabolites have been shown to have significant pharmaceutical potentials such as antimicrobial, anticancer, antiviral, anti-inflammatory agents and enzyme inhibiting activities that forms the basis of biomedical research. A diverse range of toxic or otherwise bioactive metabolites from cyanobacteria may be employed ecologically as allelochemicals as well as commercial development and application of these compounds as algicides, herbicides and insecticides. During the last few years diverse compounds with relevant pharmaceutical activities have been discovered from cyanobacteria which indicate that cyanobacteria have a good potential for providing useful substances for biotechnological production of drugs or drug candidates. Cyanobacterial secondary metabolites include a wide range of chemical classes, including diverse nitrogen-rich alkaloids and peptides which have been suggested to pose threats to human and environmental health worldwide, and equally hold considerable potential for the development of pharmaceuticals and other biomedical applications. However, utilization of these compounds for the biotechnological production of pharmacologically important drugs are not economically feasible in many cases, hence, more research and advanced strategies are needed for better results. Iron chelators such as schizokinen and anachelin produced by certain cyanobacteria may be used to develop promising medicines for the treatment of persistent metal toxicity. MAAs well as scytonemin synthesized by cyanobacteria are considered as multipurpose secondary metabolites that act as antioxidants, preventing cellular damage from UV-induced ROS and may be used in the development of artificial human sunscreens. Scytonemin has a great pharmacological potential due to its anti-inflammatory as well as anti-proliferative activities. Thus, cyanobacterial secondary metabolites have great biotechnological and industrial significance which may prove to be a boon for biomedical research.

ACKNOWLEDGMENTS

Jainendra Pathak and Neha Kumari are thankful to CSIR, New Delhi, India, for providing financial assistance in the form of senior research fellowship and junior research fellowship respectively. Haseen Ahmed acknowledges University Grants Comission, New Delhi, India, for providing senior research fellowship.

REFERENCES

Abed, R. M., Dobretsov, S., Sudesh, K. (2009) Applications of cyanobacteria in biotechnology. *J Appl Microbiol* 106: 1-12.

Ahmadi, A., Moghadamtousi, S. Z., Abubakar, S., Zandi, K. (2015) Antiviral potential of algae polysaccharides isolated from marine sources: a review. *Bio Med Res In*: 825203.

An, T., Kumar, T. K. S., Wang, M., Liu, L., Lay, J. O., Liyanage, R., Berry, J., Gantar, M., Marks, V., Gawley, R. E. (2007) Structures of pahayokolides A and B, cyclic peptides from a *Lyngbya* sp. *J Nat Prod* 70: 730-735.

Arment, A. R., Carmichael, W. W. (1996) Evidence that microcystin is a thio-template product. *J Phycol* 32: 591-597.

Asthana, R. K., Srivastava, A., Singh, A. P., Deepali, Singh S. P., Gopal, Nath, Srivastava R., Srivastava B. S. (2006) Identification of an antimicrobial entity from *Fischerella* sp. colonizing neem tree bark. *J App Phycol* 18: 33-39.

Asthana, R. K., Tripathi, M. K., Srivastava, A., Singh, A. P., Singh, S. P., Nath, G., Srivastava, R., Srivastava, B. S. (2009) Isolation and identification of a new antibacterial entity from the Antarctic cyanobacterium *Nostoc* CCC 537. *J Appl Phycol* 21: 81-88.

Bagu, J. R., Sykes, B. D., Craig, M. M., Holmes, C. F. (1997) A molecular basis for different interactions of marine toxins with protein phosphatase-1 Molecular models for bound motuporin, microcystins, okadaic acid, and calyculin A. *J Biol Chem* 272: 5087-5097.

Balunas, M. J., Linington, R. G., Tidgewell, K., Fenner, A. M., Ureña, L-D., Togna, G. D., Kyle, D. E., Gerwick, W. H. (2009) Dragonamide E, a modified linear lipopeptide from *Lyngbya majuscula* with antileishmanial activity. *J Nat Prod* 73: 60-66.

Barbaras, D., Kaiser, M., Brun, R., Gademann, K. (2008) Potent and selective antiplasmodial activity of the cyanobacterial alkaloid nostocarboline and its dimers. *Bioorg Med Chem Lett* 18: 4413-4415.

Barbeau, K., Rue, E. L., Trick, C. G., Bruland, K. W., Butler, A. (2003) Photochemical reactivity of siderophores produced by marine heterotrophic bacteria and cyanobacteria based on characteristic Fe (III) binding groups. *Limnol Oceanogr* 48: 1069-1078.

Barrientos, L. G., Gronenborn, A. M. (2005) The highly specific carbohydrate-binding protein cyanovirin-N: structure, anti-HIV/ Ebola activity and possibilities for therapy. *Mini Rev Med Chem* 5: 21-31.

Becher, P. G., Baumann, H. I., Gademann, K., Jüttner, F. (2009) The cyanobacterial alkaloid nostocarboline: an inhibitor of acetylcholinesterase and trypsin. *J Appl Phycol* 21: 103-110.

Becher, P. G., Beuchat, J., Gademann, K., Jüttner, F. (2005) Nostocarboline: isolation and synthesis of a new cholinesterase inhibitor from *Nostoc* 78-12A. *J Nat Prod* 68: 1793-1795.

Becher, P. G., Keller, S., Jung G, Süssmuth, R. D., Jüttner, F. (2007) Insecticidal activity of 12-epi-hapalindole J isonitrile. *Phytochemistry* 68: 2493-2497.

Beiderbeck, H., Taraz, K., Budzikiewicz, H., Walsby, A. E. (2005) Anachelin, the siderophore of the cyanobacterium *Anabaena cylindrica* CCAP 1403/2A. *Z Naturforsch C* 55: 681-687.

Berman, F., Gerwick, W., Murray, T. (1999) Antillatoxin and kalkitoxin, ichthyotoxins from the tropical cyanobacterium *Lyngbya majuscula,* induce distinct temporal patterns of NMDA receptor-mediated neurotoxicity. *Toxicon* 37: 1645-1648.

Bernard, C., Harvey, M., Briand, J., Biré, R., Krys, S., Fontaine, J. (2003) Toxicological comparison of diverse *Cylindrospermopsis raciborskii* strains: evidence of liver damage caused by a French *C. raciborskii* strain. *Environ Toxicol* 18: 176-186.

Bernillon, J., Parussini, E., Letoublon, R., Favre-Bonvin, J., Arpin, N. (1990) Flavin-mediated photolysis of mycosporines. *Phytochemistry* 29: 81-84.

Berry, J., Gantar, M., Gawley, R. E., Wang, M., Rein, K. S. (2004) Pharmacology and toxicology of phayokolide A, a bioactive metabolite from a fresh water species of *Lyngbya* isolated from the Florida everglades. *Comp Biochem Physiol C* 139: 231-238.

Berry, J. P., Gantar, M., Perez, M. H., Berry, G., Noriega, F.G. (2008) Cyanobacterial toxins as allelochemicals with potential applications as algaecides, herbicides and insecticides. *Mar Drugs* 6: 117-14.

Biagi, P. L., Bordoni, A., Masi, M., Ricci, G., Fanelli, C., Patrizi, A., Ceccolini, E. (1988) A long term study on the use of evening primrose oil (Efamol) in atopic children. *Drugs Exp Clin Res* 14: 285-290.

Blom, J. F., Brütsch, T., Barbaras, D., Bethuel, Y., Locher, H. H., Hubschwerlen, C., Gademann, K. (2006) Potent algicides based on the cyanobacterial alkaloid nostocarboline. *Org Lett* 8: 737-740.

Böhm, G. A., Pfleiderer, W., Böger, P., Scherer, S. (1995) Structure of a novel oligosaccharide-mycosporine-amino acid ultraviolet A/B sunscreen pigment from the terrestrial cyanobacterium *Nostoc commune*. *J Biol Chem* 270: 8536-8539.

Bonazzi, S., Barbaras, D., Patiny, L., Scopelliti, R., Schneider, P., Cole, S. T., Kaiser, M., Brun, R., Gademann, K. (2010) Antimalarial and antitubercular nostocarboline and eudistomin derivatives: synthesis, *in vitro* and *in vivo* biological evaluation. *Bioorg Med Chem* 18: 1464-1476.

Bonjouklian, R., Moore, R. E., Mynderse, J. S., Patterson, G. M. L., Smitka, T. A. (1989) Hapalindoles. US Patent 4870185.

Botes, D. P., Wessels, P. L., Kruger, H., Runnegar, M. T., Santikarn, S., Smith, R. J., Barna, J. C., Williams, D. H. (1985) Structural studies on cyanoginosins-LR,-YR,-YA, and-YM, peptide toxins from *Microcystis aeruginosa*. *J Chem Soc Perkin Trans* 1: 2747-2748.

Burja, A. M., Banaigs, B., Abou-Mansour, E., Burgess, J. G., Wright, P. C. (2001) Marine cyanobacteria- a prolific source of natural products. *Tetrahedron* 57: 9347-9377.

Cagide, E., Becher, P. G., Louzao, M. C., Espiña, B., Vieytes, M. R., Jüttner, F., Botana, L. M. (2014) Hapalindoles from the cyanobacterium *fischerella*: potential sodium channel modulators. *Chem Res Toxicol* 27: 1696-1706.

Cao, Z., Gerwick, W. H., Murray, T. F. (2010) Antillatoxin is a voltage-gated sodium channel activator that displays unique efficacy in heterologously expressed rNav1.2, rNav1.4 and rNav1.5 alpha subunits. *J Pharmacol Exp Ther* 11: 154.

Carmichael, W. (1992) Cyanobacteria secondary metabolites-the cyanotoxins. *J Appl Bacteriol* 72: 445-459.

Carr N. G., Whitton, B. A. (1982) *The Biology of Blue-green Algae*. Blackwell Scientific Publications, Oxford.

Carte, B. K. (196) Biomedical potential of marine natural products. *Biosci* 46: 271-286.

Carter, D. C., Moore, R. E., Mynderse, J. S., Niemczura, W. P., Todd, J. S. (1984) Structure of majusculamide C, a cyclic depsipeptide from *Lyngbya majuscula*. *J Org Chem* 49: 236-241.

Casanova, M. T., Burch, M. D., Brock, M. A., Bond, P. M. (1999) Does toxic *Microcystis aeruginosa* affect aquatic plant establishment? *Environ Toxicol* 14: 97-109.

Chatterjee, A., Ratner, D. M., Ryan, C. M., Johnson, P. J., O'Keefe, B. R., Secor, W. E., Anderson, D. J., Robbins, P. W., Samuelson, J. (2015) Anti-retroviral lectins have modest effects on adherence of *Trichomonas vaginalis* to ephitelial cells *in vitro* and on recovery of *Trichomonas* foetus in mouse vaginal model. *PLoS One* 10: e0135340.

Chirasuwan, N., Chaiklahan, R., Kittakoop, P., Chanasattru, W., Ruengjitchatchawalya, M., Tanticharoen, M., Bunnag, B. (2009) Anti HSV-1 activity of sulphoquinovosyl diacylglycerol isolated from *Spirulina platensis*. *Science Asia* 35: 137-141.

Choi, H, Pereira, A. R., Cao, Z., Shuman, C. F., Engene, N., Byrum, T., Matainaho, T., Murray, T. F., Mangoni, A., Gerwick, W. H. (2010) The hoiamides, structurally intriguing neurotoxic lipopeptides from Papua New Guinea marine cyanobacteria. *J Nat Prod* 73: 1411-1421.

Choi, S. H., Kim, S. G. (1998) Lipopolysaccharide inhibition of rat hepatic microsomal epoxide hydrolase and glutathione S-transferase gene expression irrespective of nuclear factor-κB activation. *Biochem Pharmacol* 56: 1427-1436.

Chorus, I., Bartram, J. (1999) *Toxic cyanobacteria in water: a guide to their public health consequences, monitoring and management*. E & FN Spon: London p 1-400.

Codd, G. A., Bell, S. G., Kaya, K., Ward, C. J., Beattie, K. A., Metcalf J. S. (1994) Cyanobacterial toxins, exposure routes and human health. *Eur J Phycol* 34: 405-415.

Cohen, Z. (1997) The Chemicals of Spirulina In: *Spirulina Platensis (Arthrospira): Physiology, Cell-biology and Biotechnology* (Ed. Vonshak A), Taylor and Francis, London, pp. 175-204.

Dembitsky, V. M., Rezanka, T. (2005) Metabolites produced by nitrogen-fixing *Nostoc* species. *Folia Microbiol* 5: 363-391.

Dey, B., Lerner, D. L., Lusso, P., Boyd, M. R., Elder, J. H., Berger, E. A. (2000) Multiple antiviral activities of cyanovirin-N: blocking of human immunodeficiency virus type 1 gp120 interaction with CD4 and coreceptor and inhibition of diverse enveloped viruses. *J Virol* 74: 4562-4569.

Dittmann, E., Wiegand, C. (2006) Cyanobacterial toxins-occurrence, biosynthesis and impact on human affairs. *Mol Nutr Food Res* 50: 7-17.

Dixon, R., Al-Nazawi, M., Alderson, G. (2004) Permeabilising effects of sub-inhibitory concentrations of microcystin on the growth of *Escherichia coli*. *FEMS Microbiol Lett* 230: 167-170.

Doan, N. T., Stewart, P. R., Smith, G. D. (2001) Inhibition of bacterial RNA polymerase by the cyanobacterial metabolites 12-epi-hapalindole E isonitrile and calothrixin A. *FEMS Microbiol Lett* 196: 135-139.

Edwards, C., Beattie, K. A., Scrimgeour, C. M., Codd, G. A. (1992) Identification of anatoxin-a in benthic cyanobacteria (blue-green algae) and in associated dog poisonings at Loch Insh, Scotland. *Toxicon* 30: 1165-1175.

Edwards, D. J., Marquez, B. L., Nogle, L. M., McPhail, K., Goeger, D. E., Roberts, M. A., Gerwick, W. H. (2004) Structure and biosynthesis of the jamaicamides, new mixed polyketide-peptide neurotoxins from the marine cyanobacterium *Lyngbya majuscula*. *Chem Biol* 11: 817-833.

Espinoza-Moraga, M., Njuguna, N. M., Mugumbate, G., Caballero, J., Chibale, K. (2013) *In silico* comparison of antimycobacterial natural products with known antituberculosis drugs. *J Chem Inf Model* 53: 649-660.

Euler, U. S., Eliassen, R. (1967) *Prostaglandins*, Academic Press, New York.

Falch, B. S., König, G. M., Wright, A. D., Sticher, O., Rüegger, H., Bernardinelli, G. (1993) Ambigol A and B: new biologically active polychlorinated aromatic compounds from the terrestrial blue green alga *Fischeralla ambigua*. *J Org Chem* 58: 6570-6575.

Feldmann, S. C., Reynaldi, S., Stortz, C. A., Cerezo, A. S., Damont, E. B. (1999) Antiviral properties of fucoidan fractions from *Leathesia difformis*. *Phytomedicine.* 6: 335-340.

Fennell, B., Carolan, S., Pettit, G., Bell, A. (2003) Effects of the antimitotic natural product dolastatin 10, and related peptides, on the human malarial parasite *Plasmodium falciparum*. *J Antimicrob Chemother* 51: 833-841.

Fleming, E. D., Castenholz, R. W. (2007) Effects of periodic desiccation on the synthesis of the UV-screening compound, scytonemin, in cyanobacteria. *Environ Microbiol* 9: 1448-1455.

Froscio, S. M., Humpage, A. R., Burcham, P. C., Falconer, I. R. (2003) Cylindrospermopsin-induced protein synthesis inhibition and its dissociation from acute toxicity in mouse hepatocytes. *Environ Toxicol* 18: 243-251.

Fujiki, H., Suganuma, M., Nakayasu, M., Hakii, H., Horiuchi, T., Takayama, S., Sugimura, T. (1986) Palytoxin is a non-12-O-tetradecanoylphorbol-13-acetate type tumor promoter in two-stage mouse skin carcinogenesis. *Carcinogenesis* 7:707-710.

Gademann, K. (2007) Cyanobacterial natural products for the inhibition of biofilm formation and biofouling. *CHIMIA* 61: 373-377.

Gademann, K., Portmann, C. (2008) Secondary metabolites from cyanobacteria: complex structures and powerful bioactivities. *Curr Org Chem* 12: 326-341.

Gantar, M., Berry, J. P., Thomas, S., Wang, M., Perez, R., Rein, K. S. (2008) Allelopathic activity among cyanobacteria and microalgae isolated from Florida freshwater habitats. *FEMS Microbiol Ecol* 64: 55-64.

Garrison, A. R., Giomarelli, B. G., Lear-Rooney, C. M., Saucedo, C. J., Yellayi S., Krumpe L. R. H., Rose, M., Paragas, J., Paragas J., Bray, M., Olinger Jr, G. G., McMahon, J. B. (2014) The cyanobacterial lectin scytovirin displays potent *in vitro* and *in vivo* activity against *Zaire Ebola* virus. *Antiviral Res* 112: 1-7.

Gerwick, W. H., Morzek, C., Moghadam, M. F., Agarwal, S. K. (1989) Novel cytotoxic peptides from the tropical marine cyanobacterium *Hormothamnion enteromorphoides* Discovery, isolation and initial chemical and biological characterization of the hormothamnins from wild and cultured material. *Experientia* 45: 115-121.

Gerwick, W. H., Proteau, P. J., Nagle, D. G., Hamel, E., Blokhin, A., Slate, D. L. (1994) Structure of curacin A, a novel antimitotic, antiproliferative and brine shrimp toxic natural product from the marine cyanobacterium *Lyngbya majuscula*. *J Org Chem* 59: 1243-1245.

Gerwick, W. H., Reyes, S., Alvarad, B. (1987) Two malyngamides form the Caribbean *Lyngbya majuscula. Phytochemistry* 26: 1701-1704.

Ghasemi, Y., Yazdi, M. T., Shafie, A., Amini, M., Shokravi, S., Zarrini, G. (2004) Parsiguine, a novel antimicrobial substance from *Fischerella ambigua. Pharm Biol* 42: 318-322.

Gleason, F. K., Case, D. E. (1986) Activity of the natural algicide, cyanobacterin, on angiosperms. *Plant Physiol* 80: 834-837.

Goldman, S., Lammers, P., Berman, M., Sanders-Loehr, J. (1983) Siderophore-mediated iron uptake in different strains of *Anabaena* sp. *J Bact* 156: 1144-1150.

Gromov, B. V., Vepritskiy, A. A., Titova, N. N., Mamkayeva, K. A., Alexandrova, O. V. (1991) Production of the antibiotic cyanobacterin LU-1 by *Nostoc linckia* CALU 892 (cyanobacterium). *J Appl Phycol* 3: 55-59.

Gulledge, B., Aggen, J., Huang, H., Nairn, A., Chamberlin, A. (2002) The microcystins and nodularins: cyclic polypeptide inhibitors of PP1 and PP2A. *Curr Med Chem* 9: 1991-2003.

Gupta, D. K., Kaur, P., Leong, S. T., Tan, L. T., Prinsep, M. R., Chu, J. J. H. (2014) Anti-Chikungunya viral activities of aplysiatoxin-related compounds from the marine cyanobacterium *Trichodesmium erythraeum*. *Mar Drugs* 12: 115-127.

Gustafson, K. R., Cardellina, J. H., Fuller, R. W., Weislow, O. S., Kiser, R. F., Snader, K. M., Patterson, G. M. L., Boyd, M. R. (1989) AIDS antiviral sulfolipids from cyanobacteria (BGA). *J Natl Canc Inst* 81: 1254-1258.

Gutierrez, R. M. P., Flores, A. M., Solis, R. V., Jimenez, J. C. (2008) Two new antibacterial norbietane diterpenoids from cyanobacterium *Micrococcus lacustris*. *J Nat Med* 62: 328-331.

Harada, Ki, Suomalainen, M., Uchida, H., Masui, H., Ohmura, K., Kiviranta, J., Niku-Paavola, M. L., Ikemoto T. (2000) Insecticidal compounds against mosquito larvae from *Oscillatoria agardhii* strain 27. *Environ Toxicol* 15: 114-119.

Hayashi, K., Hayashi, T., Kojima, I. (1996a) A natural sulphated polysaccharide, calcium spirulan, isolated from: *in vitro* and *ex vivo* evaluation of anti-herpes simplex virus and anti-human immunodeficiency virus. *AIDS Res Hum Retrovirus* 12: 1463-1471.

Hayashi, T., Hayashi, K., Maeda, M., Kojima, I. (1996b) Calcium spirulan, an inhibitor of enveloped virus replication, from a blue-green alga *Spirulina platensis*. *J Nat Prod* 59: 83-87.

Hayashi, O., Hirahashi, T., Katoh, T., Miyajima, H., Hirano, T., Okuwaki, Y. (1998) Class specific influence of dietary *Spirulina platensis* on antibody production in mice. *J Nutr Sci Vitaminol* 44: 841-851.

Hewett, J. A., Roth, R. A. (1993) Hepatic and extrahepatic pathobiology of bacterial lipopolysaccharides. *Pharmacol Rev* 45: 382-411.

Hirata, K., Yoshitomi, S., Dwi, S., Iwabe, O., Mahakhant, A., Polchai, J., Miyamoto, K. (2003) Bioactivities of nostocine A produced by a freshwater cyanobacterium *Nostoc spongiaeforme* TISTR 8169. *J Biosci Bioeng* 95: 512-517.

Hirsch, C. F., Liesch, J. M., Slvatore, M. J., Schwartz, R. E., Sesin, D. F. (1990) *Antifungal fermentation product and method*. US Patent 4946835.

Hodoki, Y., Ohbayashi, K., Kobayashi, Y., Okuda, N., Nakano, S. I. (2012) Detection and identification of potentially toxic cyanobacteria: Ubiquitous distribution of *Microcystis aeruginosa* and *Cuspidothrix issatschenkoi* in Japanese lakes. *Harmful Algae* 16: 49-57.

Hoiczyk, E., Hansel, A. (2000) Cyanobacterial cell walls: news from an unusual prokaryotic envelope. *J Bact* 182: 1191-1199.

Horrobin, D. F., Huang, Y. S. (1983) Schizophrenia: the role of abnormal essential fatty acid and prostaglandin metabolism. *Med Hypotheses* 10: 329-336.

Humpage, A. R., Fenech, M., Thomas, P., Falconer, I. R. (2000) Micronucleus induction and chromosome loss in transformed human white cells indicate clastogenic and aneugenic action of the cyanobacterial toxin, cylindrospermopsin. *Mutat Res Genet Toxicol Environ Mutagen* 472: 155-161.

Huskens, D., Férir, G., Vermeire, K., Kehr, J. C., Balzarini, J., Dittmann, E., Schols, D. (2010) Microvirin, a novel (1,2) Mannose-specific lectin isolated from *Microcystis aeruginosa*, has anti-HIV-1 activity comparable with that of cyanovirin-N but a much higher safety profile. *J Biol Chem* 285: 24845-24854.

Iordache, F., Ionita, M., Mitrea, L. I., Fafaneata, C., Pop, A. (2015) Antimicrobial and antiparasitic activity of lectins. *Curr Pharm Biotechnol* 16: 152-161.

Issa, A. A. (1999) Antibiotic production by the cyanobacteria *Oscillatoria angustissima* and *Calothrix parietina*. *Environ Toxicol Pharmacol* 8: 33-37.

Ito, E., Nagai, H. (2000) Bleeding from the small intestine caused by aplysiatoxin, the causative agent of the red alga *Gracilaria coronopifoli*a poisoning. *Toxicon* 38: 123-132.

Ito, Y., Butler, A. (2005) Structure of synechobactins, new siderophores of the marine cyanobacterium *Synechococcus* sp. PCC 7002. *Limnol Oceanogr* 50: 1918-1923.

Jabba, S. V., Prakash, A., Dravid, S. M., Gerwick, W. H., Murray, T. F. (2010) Antillatoxin, a novel lipopeptide, enhances neurite outgrowth in immature cerebrocortical neurons through activation of voltage-gated sodium channels. *J Pharmacol Exp Ther* 332: 698-709.

Jaki, B., Orjala, J., Heilmann, J., Linden, A., Vogler, B., Sticher, O. (2000) Novel extracellular diterpenoids with biological activity from the cyanobacterium *Nostoc commune*. *J Nat Prod* 63: 339-343.

Jang, M-H., Ha, K., Takamura, N. (2007) Reciprocal allelopathic responses between toxic cyanobacteria (*Microcystis aeruginosa*) and duckweed (*Lemna japonica*). *Toxicon* 49: 727-733.

Jensen, G. S., Ginsberg, D. I., Huerta, P., Citton, M., Drapeau, C. (2000) Consumption of *Aphanizomenon flos-aquae* has rapid effects on the circulation and function of immune cells in humans. A novel approach to nutritional mobilization of the immune system. *JANA* 2: 50-58.

Jones, T. H., McClelland, E. E., McFeeters, H., McFeeters, R. L. (2017) Novel antifungal activity for the lectin scytovirin: inhibition of *Cryptococcus neoformans* and *Cryptococcus gattii*. *Front Microbiol* 8: 755.

Kanekiyo, K., Hayashi, K., Takenaka, H., Lee J-B., Hayashi, T. (2007) Anti-herpes simplex virus target of an acidic polysaccharide, nostoflan, from the edible blue-green algae *Nostoc flagelliforme*. *Biol Pharm Bull* 30: 1573-1575.

Karl G, Cyril P (2008) Secondary metabolites from cyanobacteria: complex structures and powerful bioactivities. *Curr Org Chem* 12: 326-341.

Kernoff, P. B., Willis, A., Stone, K. J., Davies, J. A., McNicol, G. P. (1997) Antithrombotic potential of dihomo-gamma-linolenic acid in man. *Br Med J* 2:1441-1444.

Kim, H., Lantvit, D., Hwang, C. H., Kroll, D. J., Swanson, S. M., Franzblau, S. G., Orjala, J. (2012) Indole alkaloids from two cultured cyanobacteria, *Westiellopsis* sp. and *Fischerella muscicola*. *Bioorg Med Chem* 20: 5290-5295.

Kim, J. D. (2006) Screening of cyanobacteria (Blue-Green algae) from rice paddy soil for antifungal activity against plant pathogenic fungi. *Mycobiology* 34: 138-142.

Kiviranta, J., Abdel-Hameed, A., Sivonen, K., Niemelä, S., Carlberg, G. (1993) Toxicity of cyanobacteria to mosquito larvae-screening of active compounds. *Environ Toxicol Water Qual* 8: 63-71.

Klein, D., Braekman, J. C., Daloze, D., Hoffmann, L., Castillo, G., Demoulin, V. (1999) Madangolide and laingolide A, two novel macrolides from *Lyngbya bouillonii* (Cyanobacteria). *J Nat Prod* 62: 934-936.

Klein, D., Daloze, D., Braekman, J. C., Hoffmann, L., Demoulin, V. (1995) New apalindoles from the cyanophyte *Hapalosiphon laingii*. *J Nat Prod* 58: 1781-1785.

Krienitz, L., Ballot, A., Wiegand, C., Kotut, K., Codd, G. A., Pflugmacher, S. (2002) Cyanotoxin-producing bloom of *Anabaena flos-aquae*, *Anabaena discoidea* and *Microcystis aeruginosa* (Cyanobacteria) in Nyanza Gulf of Lake Victoria, Kenya. In: *Cyanobacteria in Kenyan Rift Valley lakes*, Freie Universität Berlin: Berlin, Germany, Vol. Thesis collection, pp. 40-50.

Kushak, R., VanCott, E., Drapeau, C., Winter, H. (1999) Effect of algae, *Aphanizomenon flos-aquae* on digestive enzyme activity and polyunsaturated fatty acids level in blood plasma. *Gastroenterology* 116: A559.

Kwan, J. C., Rocca, J. R., Abboud, K. A., Paul, V. J., Luesch, H. (2008) Total structure determination of grassypeptolide, a new marine cyanobacterial cytotoxin. *Org lett* 10: 789-792.

Larsen, L. K., Moore, R. E., Patterson, G. M. L. (1994) β-Carbolines from the blue-green alga, *Dichothrix baueriana*. *J Nat Prod* 57: 419-421.

Lewis B. L., Holt P. D., Taylor S. W., Wilhelm S. W., Trick C. G., Butler A., Luther G.W. (1995) Voltammetric estimation of iron (III) thermodynamic stability constants for catecholate siderophores isolated from marine bacteria and cyanobacteria. *Mar Chem* 50: 179-188.

Li, W., Berman, F., Okino, T., Yokokawa, F., Shioiri, T., Gerwick, W., Murray, T. (2001) Antillatoxin is a marine cyanobacterial toxin that potently activates voltage-gated sodium channels. *Proc Nat Acad Sci USA* 98: 7599-7604.

Linington, R. G., Clark, BR, Trimble, E. E., Almanza, A., Ureña, L-D., Kyle, D. E., Gerwick, W. H. (2008) Antimalarial peptides from marine cyanobacteria: isolation and structural elucidation of gallinamide A. *J Nat Prod* 72: 14-17.

Linington, R. G., Edwards, D. J., Shuman, C. F., McPhail, K. L., Matainaho, T., Gerwick, W. H. (2007) Symplocamide A, a potent cytotoxin and chymotrypsin inhibitor from the marine cyanobacterium *Symploca* sp. *J Nat Prod* 71: 22-27.

Linington, R. G., González, J., Ureña, L-D., Romero, L. I., Ortega-Barría, E., Gerwick, W. H. (2007) Venturamides A and B: Antimalarial constituents of the Panamanian marine cyanobacterium *Oscillatoria* sp. *J Nat Prod* 70: 397-401.

Liu, Y., Tan, W., Wu, X., Wu, Z., Yu, G., Li, R. (2011) First report of microcystin production in *Microcystis smithii* Komárek and *Anagnostidis* (Cyanobacteria) from a water bloom in Eastern China. *J Environ Sci* 23: 102-107.

Ljung, K., Hull, A. K., Kowalczyk, M., Marchant, A., Celenza, J., Cohen, J. D., Sandberg, G. (2002) Biosynthesis, conjugation, catabolism and homeostasis of indole-3-acetic acid in *Arabidopsis thaliana*. In: *Auxin Molecular Biology* (Perrot-Rechenmann C, Hagen G), Springer, pp. 249-272.

Locher, H. H., Ritz, D., Pfaff, P., Gaertner, M., Knezevic, A., Sabato, D., Schroeder, S., Barbaras, D., Gademann, K. (2010) Dimers of nostocarboline with potent antibacterial activity. *Chemotherapy* 56: 318-324.

Loya, S., Reshef, V., Mizrachi, E., Silbersten, C., Rachamim, Y., Carmeli, S., Hizi, A. (1998) The inhibition of the reverse transcriptase of HIV-1 by the natural sulfoglycolipids from cyanobacteria: contribution of different moieties to their high potency. *J Nat Prod* 61: 891-895.

Luescher-Mattli M (2003) Algae, a possible source for new drugs in the treatment of HIV and other viral diseases. *Curr Med Chem Anti-Infect Agents* 2: 219-225.

Luukkainen, R., Sivonen, K., Namikoshi, M., Färdig, M., Rinehart, K., Niemelä, S. (1993) Isolation and identification of eight microcystins from thirteen *Oscillatoria agardhii* strains and structure of a new microcystin. *Appl Env Microbiol* 59: 2204-2209.

Ma, L., Led, J. J. (2000) Determination by high field NMR spectroscopy of the longitudinal electron relaxation rate in Cu (II) plastocyanin from *Anabaena variabilis*. *J Am Chem Soc* 122: 7823-7824.

MacMillan, J. B., Ernst-Russell, M. A., de Ropp, J. S., Molinski, T. F. (2002) Lobocyclamides A-C, lipopeptides from a cryptic cyanobacterial mat containing *Lyngbya confervoides*. *J Org Chem* 67: 8210-8215.

MacMillan, J. B., Molinsky, T. F. (2005) Majusculoic acid, a brominated cyclopropyl fatty acid from a marine cyanobacterial mat assemblage. *J Nat Prod* 68: 604-606.

Malloy, K. L., Villa, F. A., Engene, N., Matainaho, T., Gerwick, L., Gerwick, W. H (2010) Malyngamide 2, an oxidized lipopeptide with nitric oxide inhibiting activity from a Papua New Guinea marine cyanobacterium. *J Nat Prod* 74: 95-98.

Mansell, H. L. (1996) Synthetic approaches to anatoxin-a. *Tetrahedron* 52: 6025-6061.

Mansour, H. A., Shoman, S. A., Kdodier, M. H. (2011) Antiviral effect of edaphic cyanophytes on rabies and herpes-1 viruses. *Acta Biol Hung* 62: 194-203.

Martin, C., Codd, G. A., Siegelman, H. W., Weckesser, J. (1989) Lipopolysaccharides and polysaccharides of the cell envelope of toxic *Microcystis aeruginosa* strains. *Arch Microb* 152: 90-94.

Martinez, J. S., Carter-Franklin, J. N., Mann, E. L., Martin, J. D., Haygood, M. G., Butler, A. (2003) Structure and membrane affinity of a suite of amphiphilic siderophores produced by a marine bacterium. *Proc Nat Acad Sci USA* 100: 3754-3759.

Martins R. F., Ramos M.F, Herfindal L, Sousa JA, Skærven K, Vasconcelos VM (2008) Antimicrobial and cytotoxic assessment of marine cyanobacteria- *Synechocystis* and *Synechococcus*. *Mar Drugs* 6: 1-11.

Matsunaga, S., Moore, R. E., Niemczura, W. P., Carmichael, W. W. (1989) Anatoxin-a (s), a potent anticholinesterase from *Anabaena flos-aquae*. *J Am Chem Soc* 111: 8021-8023.

Matthew, S., Salvador, L. A., Schupp, P. J., Paul, V. J., Luesch, H. (2010) Cytotoxic halogenated macrolides and modified peptides from the apratoxin-producing marine cyanobacterium *Lyngbya bouillonii* from Guam. *J Nat Prod* 73: 1544-1552.

Matzanke, B. F. (1991) Structures, coordination chemistry and functions of microbial iron chelates. In: *CRC Handbook of Microbial Iron Chelates (Ed.* Winkelmann G) CRC Press, Boston, pp. 15-64.

Mayer, A. M., Rodríguez, A. D., Berlinck, R. G., Hamann, M. T. (2009) Marine pharmacology in 2005-6: Marine compounds with anthelmintic, antibacterial, anticoagulant, antifungal, anti-inflammatory, antimalarial, antiprotozoal, anti-tuberculosis, and antiviral activities; affecting the cardiovascular, immune and nervous systems, and other miscellaneous mechanisms of action. *Biochim Biophysic Acta (BBA)-General Subjects* 1790: 283-308.

McPhail, K. L., Correa, J., Linington, R. G., González, J., Ortega-Barría, E., Capson, T. L., Gerwick, W. H. (2007) Antimalarial linear lipopeptides from a Panamanian strain of the marine cyanobacterium *Lyngbya majuscula*. *J Nat Prod* 70: 984-988.

Miao, S., Andersen, R. J., Allen, T. M. (1990) Cytotoxic metabolites from the sponge *Lanthella basta* collected in Papua New Guinea. *J Nat Prod* 53: 1441-1446.

Mo, S., Krunic, A., Pegan, S. D., Franzblau, S. G., Orjala, J. (2009)An antimicrobial guanidine-bearing sesterterpene from the cultured cyanobacterium *Scytonema* sp. *J Nat Prod* 72: 2043-2045.

Mooberry, S. L., Stratman, K., Moore, R. E. (1995) Tubercidin stabilizes microtubules against vinblastine-induced depolymerization, a taxol-like effect. *Cancer Lett* 96: 261-266.

Moore, R. E., Cheuk, C., Patterson, G. M. L. (1984) Hapalindoles: new alkaloids from the blue-green alga *Hapalosiphon fontinalis*. *J Am Chem Soc* 106: 6456-6457.

Moore, R. E., Cheuk, C., Yang, X. Q. G., Patterson, G. M. L., Bonjouklian, R., Smitka, T. A., Mynderse, J. S., Foster, R. S., Jones, N. D., Swartzendruber, J. K., Deeter, J. B.

(1987a) Hapalindoles, antibacterial and antimycotic alkaloids from the cyanophyte *Hapalosiphon fontinalis*. *J Org Chem* 52: 1036-1043.

Moore, R. E, Yang, X. Q. G., Patterson, G. M. L. (1987b) Fontonamide and anhydro hapaloxinodole A, two new alkaloids from the blue-green alga *Hapalosiphon fontinalis*. *J Org Chem* 52: 3773-3777.

Moore, R. E, Mynderse, J. S. (1982) *Majusculamide*, C. US Patent 4342751.

Moore, R. E., Patterson, G. M., Carmichael, W. W. (1988) New pharmaceuticals from cultured blue-green algae. *Biomedical Importance of Marine Organisms* 13: 143-150.

Mundt, S., Kreitlow, S., Jansen, R. (2003) Fatty acids with antibacterial activity from the cyanobacterim *Oscillatoria redekei* HUB051. *J Appl Phycol* 15: 263-267.

Murakami, M., Sun, Q., Ishida, K., Matsuda, H., Okino, T., Yamaguchi, K. (1997) Microviridins, elastase inhibitors from the cyanobacterium *Nostoc minutum* (NIES-26). *Phytochemistry* 45: 1197-1202.

Mynderse, J. S., Moore, R. E., Kashiwagi, M., Norton, T. R. (1977) Antileukemia activity in the osillatoriaceae: isolation of debromoaplysiatoxin from *Lyngbya*. *Science* 196: 538-540.

Nagatsu, A., Kajitani, H., Sakakibara, J. (1995) Muscoride A: a new oxazole peptide alkaloid from freshwater cyanobacterium *Nostoc muscorum*. *Tetrahedron lett* 36: 4097-4100.

Namikoshi, M., Carmichael, W. W., Sakai, R., Jares-Erijman, E. A., Kaup, A. M., Rinehart, K. L. (1993) 9-Deazaadenosine and its 5'-. alpha.-D-glucopyranoside, isolated from the cyanobacterium *Anabaena affinis* strain VS-1. *J Am Chem Soc* 115: 2504-2505.

Namikoshi, M., Murakami, T., Watanabe, M. F., Oda, T., Yamada, J., Tsujimura, S., Nagai, H., Oishi, S. (2003) Simultaneous production of homoanatoxin-a, anatoxin-a, and a new non-toxic 4-hydroxyhomoanatoxin-a by the cyanobacterium *Raphidiopsis mediterranea* Skuja. *Toxicon* 42: 533-538.

Negri, A. P., Jones, G. J. (1995) Bioaccumulation of paralytic shellfish poisoning (PSP) toxins from the cyanobacterium *Anabaena circinalis* by the freshwater mussel *Alathyria condola*. *Toxicon* 33: 667-678.

Neilands, J. (1981) *Microbial iron transport compounds (siderophores) as chelating agents*. *Development of Iron Chelators for Clinical Use*, New York: Elsevier/North Holland 13-31.

Neuhof, T., Dieckmann, R., von Döhren, H., Preußel, K., Seibold, M., Schmieder, P. (2006) *Lipopeptides having pharmaceutical activity*. Patent WO 2006/ 092313 A1.

Neuhof, T., Schmieder, P., Preussel, K., Dieckmann, R., Pham, H., Bartl, F., von Döhren, H. (2005) Hassallidin, A., a glycosylated lipopeptide with antifungal activity from the cyanobacterium *Hassallia* sp. *J Nat Prod* 68: 695-700.

Neuhof, T., Schmieder, P., Seibold, M., Preussel, K., von Döhren, H. (2006) Hassallidin B-second antifungal member of the Hassallidin family. *Bioorg Med Chem Lett* 16: 4220-4222.

Neumann C, Bain P, Shaw G (2007) Studies of the comparative in vitro toxicology of the cyanobacterial metabolite deoxycylindrospermopsin. *J Toxicol Environ Health, Part A* 70: 1679-1686.

Olson JM (2006) Photosynthesis in the Archean era. *Photosynth Res* 88: 109-117.

Ördög V, Pulz O (1996) Diurnal changes of cytokinin-like activity in a strain of *Arthronema africanum* (Cyanobacteria), determined by bioassays. *Algological Studies/Archiv für Hydrobiologie, Supplement Volumes* 57-67.

Orjala, J., Nagle, D. G., Hsu, V., Gerwick, W. H. (1995) Antillatoxin: an exceptionally ichthyotoxic cyclic lipopeptide from the tropical cyanobacterium *Lyngbya majuscula*. *J Am Chem Soc* 117: 8281-8282.

Park, H. D., Watanabe, M. F., Harada, K. I., Nagai, H., Suzuki, M., Watanabe, M., Hayashi, H. (1993) Hepatotoxin (microcystin) and neurotoxin (anatoxin-a) contained in natural blooms and strains of cyanobacteria from Japanese freshwaters. *Nat Toxins* 1: 353-360.

Pathak, J., Rajneesh, Richa, Kannaujiya, V. K., Sonker, A. S., Sinha, R. P. (2015a) Diverse functions and applications of novel and unique cyanobacterial sheath pigment, scytonemin. In: *Biological Sciences: Innovations and Dynamics* (Eds. Sinha RP, Richa, Rastogi RP) New India Publishing Agency, New Delhi, pp. 237-262.

Pathak, J., Richa, Rajneesh, Sonker, A. S., Kannaujiya, V. K., Sinha, R. P. (2015b) Isolation and partial purification of scytonemin and mycosporine-like amino acids from biological crusts. *J Chem Pharma Res* 7: 362-371.

Pathak, J., Sonker, A. S., Richa et al. (2017a) Screening and partial purification of photo protective pigment, scytonemin from cyanobacteria dominated biological crusts dwelling on the historical monuments in and around Varanasi, India. *Microbiol Res (Pavia)* 8: 6559.

Pathak, J., Rajneesh, Ahmed, H., Richa, Sinha, R. P. (2017b) Metabolomic profiling of cyanobacterial UV-protective compounds. *Curr Metabolomics* 5: 138-163.

Pathak, J., Rajneesh, Maurya, P., Singh, S. P., Häder, D-P., Sinha, R. P. (2018a) Cyanobacterial farming for environment friendly sustainable agriculture practices: innovations and perspectives. *Front Environ Sci* 6: 7.

Pathak, J., Rajneesh, Kannaujiya, V.K., Richa, Singh, V., Kumar, D., Singh, S. P., Sinha, R. P. (2018b) Enzymatic and non-enzymatic protection strategies of cyanobacteria against ultraviolet radiation. In: *Trends in Life Sci* (Eds. Sinha RP, Srivastava UP) Nova Publishers, New York NY 1-28.

Pathak, J., Ahmed, H., Singh, P. R., Singh, S. P., Häder, D-P., Sinha, R. P., (2019) Mechanisms of Photo protection in Cyanobacteria. In: *Cyanobacteria* (Eds. Mishra AK, Tiwari DN, AN Rai) Elsevier Inc., Academic Press, pp. 145-171.

Pereira, A., Cao, Z., Murray, T. F., Gerwick, W. H. (2009) Hoiamide a, a sodium channel activator of unusual architecture from a consortium of two papua new Guinea cyanobacteria. *Chemistry and Biology* 16: 893-906.

Pereira, A. R., Cao, Z., Engene, N., Soria-Mercado, I. E., Murray, T. F., Gerwick, W. H. (2010) Palmyrolide A, an unusually stabilized neuroactive macrolide from Palmyra Atoll cyanobacteria. *Org Lett* 12: 4490-4493.

Pergament, I., Carmeli, S. (1994) Schizotrin, A: a novel antimicrobial cyclic peptide from a cyanobacterium. *Tetrahedron Lett* 35:8473-8476.

Prasanna R., Nain, L., Tripathi, R., Gupta, V., Chaudhary, V., Middha, S., Joshi, M., Ancha, R., Kaushik, B. D. (2008) Evaluation of fungicidal activity of extracellular filtrates of cyanobacteria-possible role of hydrolytic enzymes. *J Basic Microbiol* 48: 186-194.

Prinsep M. R., Moore R. E., Levine I. A., Patterson G. M. (1992) Westiellamide, a bistratamide-related cyclic peptide from the blue-green alga *Westiellopsis prolifica*. *J Nat Prod* 55: 140-142.

Procópio, T. F., Moura, M. C., Albuquerque, L. P., Gomes, F. S., Santos, N. D., Coelho, L. C., Pontual, E. V., Paiva, P. M., Napoleão, T. H. (2017a) Antibacterial lectins: action mechanisms, defensive roles and biotechnological potential. In *Antibacterials: Synthesis, Properties and Biological Activities* (ed. Collins E) New York, NY: Nova Science Publishers, Inc., pp. 69-89.

Procópio, T. F., de Siqueira Patriota, L. L., de Moura, M. C., da Silva, P. M., de Oliveira A. P., do Nascimento Carvalho, L. V., de Albuquerque Lima, T., Soares, T., da Silva T. D., Coelho, L. C., da Rocha Pitta, M. G. (2017b) CasuL: a new lectin isolated from *Calliandra surinamensis* leaf pinnulae with cytotoxicity to cancer cells, antimicrobial activity and antibiofilm effect. *Int J Biol Macromol* 98: 419-429.

Proteau, P. J., Gerwick, W. H., Garcia-Pichel, F., Castenholz, R. (1993) The structure of scytonemin, an ultraviolet sunscreen pigment from the sheaths of cyanobacteria. *Experientia* 49: 825-829.

Quesada, A., Vincent, W. F. (1997) Strategies of adaptation by Antarctic cyanobacteria to ultraviolet radiation. *Eur J Phycol* 32: 335-342.

Raja, R., Hemaiswarya, S., Kumar, N. A., Sridhar, S., Rengasamy, R. A. (2008) perspective on the biotechnological potential of microalgae. *Crit Rev Microbiol* 34: 77-88.

Rajneesh, Pathak, J., Richa, Hader, D-P., Sinha, R. P. (2019) Impacts of ultraviolet radiation on certain physiological and biochemical processes in cyanobacteria inhabiting diverse habitats. *Environ Exper Bot* 161: 375-387.

Rajneesh, Singh, S. P., Pathak, J., Sinha, R. P. (2017) Cyanobacterial factories for the production of green energy and value-added products: An integrated approach for economic viability. *Renew Sust Energ Rev* 69: 578-595.

Rao, C. S. V. R. (1994) Antimicrobial activity of cyanobacteria. *I J Mar Sci* 23: 55-56.

Rao, D. R., Thangavel, C., Kabilan, L., Suguna, S., Mani, T., Shanmugasundaram, S. (1999) Larvicidal properties of the cyanobacterium *Westiellopsis* sp.(blue-green algae) against mosquito vectors. *Trans R Soc Trop Med Hyg* 93: 232.

Rao, P., Gupta, N., Bhaskar, A., Jayaraj, R. (2002) Toxins and bioactive compounds from cyanobacteria and their implications on human health. *J Environ Biol/Acad Environ Biol India* 23: 215-224.

Rastogi, R. P., Sinha, R. P. (2009) Biotechnological and industrial significance of cyanobacterial secondary metabolites. *Biotechnol Adv* 27: 521-539.

Rastogi, R. P., Sinha, R. P., Incharoensakdi, A. (2014) The cyanotoxin-microcystins: current overview. *Rev Environ Scie Bio Technol* 13: 215-249.

Raveh, A., Carmeli, S. (2007) Antimicrobial ambiguines from the cyanobacterium *Fischerella* sp. collected in Israel. *J Nat Prod* 70: 196-201.

Reddy, C. M, Bhat, V. B, Kiranmai, G., Reddy, M. N., Reddanna, P., Madyastha, K. M. (2000) Selective inhibition of cyclooxyge-nase-2 by C-phycocyanin, a biliprotein from *Spirulina platensis*. *Biochem Biophys Res Commun* 277: 599-603.

Reshef, V., Mizrachi, E., Maretzki, T., Silberstein, C., Loya, S., Hizi, A, Carmeli (1997) New acylated sulfoglycolopids and digalactolipids and related known glycolipids from cyanobacteria with a potential to inhibit the reverse transcriptase of HIV-1. *J Nat Prod* 60: 1251-1260.

Richa, Sinha, R. P (2015) Biochemical characterization of sunscreening mycosporine-like amino acids from two *Nostoc* species inhabiting diverse habitats. *Protoplasma* 252: 199-208.

Richa, Pathak J., Sonker, A. S. et al. (2018) Potential applications of natural bioactive cyanobacterial UV protective compounds. In: *Blue technologies: production and uses of marine molecules* (Eds. La Barre S, Bates S) Wiley VCH, Germany, pp. 693-717.

Rickards, R. W., Rothschild, J. M., Willis A. C., de Chazal, N. M , Kirk, J., Kirk, K., Saliba, K. J., Smith, G. D. (1999) Calothrixins A and B, novel pentacyclic metabolites from *Calothrix* cyanobacteria with potent activity against malaria parasites and human cancer cells. *Tetrahedron* 55: 13513-13520.

Rimbau, V., Camins, A., Pubill, D., Sureda, F. X., Romay, C., González, R., Jiménez, A., Escubedo, E., Camarasa, J., Pallàs, M. (2001) C-phycocyanin protects cerebellar granule cells from low potassium/serum deprivation-induced apoptosis. *Arch Pharmacol* 364: 96-104.

Rinehart, K. L., Harada, K., Namikoshi, M., Chen, C., Harvis, C. A., Munro, M. H., Blunt, J. W., Mulligan, P. E., Beasley, V. R. (1988) Nodularin, microcystin, and the configuration of Adda. *J Am Chem Soc* 110: 8557-8558.

Rinehart, K. L., Namikoshi, M., Choi, B. W. (1994) Structure and biosynthesis of toxins from blue-green algae (cyanobacteria). *J Appl Phycol* 6: 159-176.

Rissanen T., Voutilainen S., Nyyssonen K., Salonen J. T. (2002) Lycopene, atherosclerosis and coronary heart disease. *Exp Biol Med* 227: 900-907.

Rodríguez, A., Stella A., Storni, M., Zulpa, G., Zaccaro, M. (2006) Effects of cyanobacterial extracellular products and gibberellic acid on salinity tolerance in *Oryza sativa* L. *Saline Systems* 2: 1.

Roger, P. A., Watanabe, I. (1986) Techniques for utilizing biological nitrogen fixation in wetland rice: potentialities, current usage and limiting factors. *Fert Res* 9: 39-77.

Romanowska-Duda, Z., Tarczyńska, M. (2002) The influence of microcystin-LR and hepatotoxic cyanobacterial extract on the water plant *Spirodela oligorrhiza*. *Environ Toxicol* 17: 434-440.

Saito, K., Konno, A., Ishii H., Saito, H., Nishida, F., Abe, T., Chen, C. (2001) Nodularin-Har: a new nodularin from *Nodularia*. *J Nat Prod* 64: 139-141.

Sanchez, L. M., Lopez, D., Vesely, B. A., Della, Togna, G., Gerwick, W. H., Kyle, D. E., Linington R. G. (2010) Almiramides A-C: Discovery and development of a new class of leishmaniasis lead compounds. *J Med Chem* 53: 4187-4197.

Sanchez-Baracaldo, P., Hayes, P., Blank, C. (2005) Morphological and habitat evolution in the cyanobacteria using a compartmentalization approach. *Geobiology* 3: 145-165.

Santoyo, S., Herrero, M., Senorans, F. J., Cifuentes, A, Ibanez, E., Nad Jaime, L. (2006) Functional characterization of pressurized liquid extracts of *Spirulina platensis. Eur Food Res Technol* 224: 75-81.

Sathiyamoorthy, P., Shanmugasundaram, S. (1996) Preparation of cyanobacterial peptide toxin as a biopesticide against cotton pests. *Appl Microbiol Biotech* 46: 511-513.

Schaeffer, D. J., Krylov, V. S. (2000) Anti-HIV activity of extracts and compounds from algae and cyanobacteria. *Ecotoxicol Environmen Saf* 45: 208-227.

Schlegel, I., Doan, N. T., de Chazal, N., Smith, G. D. (1998) Antibiotic activity of new cyanobacterial isolates from Australia and Asia against green algae and cyanobacteria. *J Appl Phycol* 10: 471-479.

Schwartz, R. E., Hirsch, C. F., Sesin, D. F., Flor, J. E., Chartrain, M., Fromtling, R. E., Harris, G. H., Salvatore, M. J., Liesch, J. M., Yudin, K. (1990) Pharmaceuticals from cultured algae. *J Ind Microbiol* 5: 113-123.

Sergee, V. N., Gerasimenko, L. M., Zavarzin, G. A. (2002) The proterozoic history and present state of cyanobacteria. *Microbiology English Translation* 71: 623-637.

Sergeeva, E., Liaimer A., Bergman, B. (2002) Evidence for production of the phytohormone indole-3-acetic acid by cyanobacteria. *Planta* 215: 229-238.

Shih, S. R., Tsai K. N., Li Y. S., Chueh C. C., Chan E. C. (2003) Inhibition of enterovirus 71 induced apoptosis by allophycocyanin isolated from a blue green alga *Spirulina platensis*. *J Med Virol* 70: 119-125.

Shin, H. J., Matsuda, H., Murakami, M., Yamaguchi, K. (1997) Aeruginosins 205A and-B, serine protease inhibitory glycopeptides from the cyanobacterium *Oscillatoria agardhii* (NIES-205). *The J Org Chem* 62: 1810-1813.

Shishido, T. K., Humisto, A., Jokela, J., Liu, L., Wahlsten, M., Tamrakar, A., Fewer, D. P., Permi, P., Andreote, A. P., Fiore, M. F. (2015) Antifungal compounds from cyanobacteria. *Mar Drugs* 13: 2124-2140.

Sielaff, H., Christiansen, G., Schwecke, T. (2006) Natural products from cyanobacteria: Exploiting a new source for drug discovery. *Investig Drugs J* 9: 119-127.

Silva-Stenico, M. E., Silva, C. S. P., Lorenzi, A. S., Shishido, T. K., Etchegaray, A., Lira, S. P., Moraes L. A. B., Fiore M. F. (2011) Non-ribosomal peptides produced by Brazilian cyanobacterial isolates with antimicrobial activity. *Microbiol Res* 166: 161-175.

Singh, R. N. (1961) *Role of Blue-Green Algae in Nitrogen Economy of Indian Agriculture.* Indian Council of Agricultural Research, New Delhi, India 1-175.

Singh, S, Kate, B. N, Banerjee, U. C. (2005) Bioactive compounds from cyanobacteria and microalgae: an overview. *Crit Rev Biotechnol* 25: 73-95.

Singh, S. P., Klisch, M, Häder, D-P., Sinha R. P. (2008) Role of various growth media on shinorine (mycosporine-like amino acid) concentration and photosynthetic yield in *Anabaena variabilis* PCC 7937. *World J Microbiol Biotechnol* 24: 3111-3115.

Singh, S. P., Kumari, S., Rastogi, R. P., Singh, K. L., Sinha, R. P. (2008) Mycosporine-like amino acids (MAAs): Chemical structure, biosynthesis and significance as UV-absorbing/screening compounds. *Indian J Exp Biol* 46: 7-17.

Sinha, R. P., Häder D-P. (2008) UV-protectants in cyanobacteria. *Plant Sci* 174: 278-289.

Sinha, R. P., Klisch, M., Gröniger, A., Häder D-P. (1998) Ultraviolet-absorbing/screening substances in cyanobacteria, phytoplankton and macroalgae. *J Photochem Photobiol B* 47: 83-94.

Sinha, R. P., Kumari, S., Rastogi, R. P. (2008) Impacts of ultraviolet-B radiation on cyanobacteria: Photo protection and repair. *J Sci Res* 52: 125-142.

Sinha, R. P., Singh, S. P., Häder, D-P. (2007) Database on mycosporines and mycosporine-like amino acids (MAAs) in fungi, cyanobacteria, macroalgae, phytoplankton and animals. *J Photochem Photobiol B* 89: 29-35.

Sivonen, K., Jones, G. (1999) Cyanobacterial toxins. In: *Toxic Cyanobacteria in Water: a Guide to their Public Health Consequences, Monitoring and Management* (Eds. Chorus I, Bartram J), E and FN Spon: London, pp. 41-111.

Sivonen, K., Kononen, K., Carmichael, W., Dahlem, A., Rinehart, K., Kiviranta, J., Niemela, S. (1989) Occurrence of the hepatotoxic cyanobacterium *Nodularia spumigena* in the Baltic Sea and structure of the toxin. *Appl Env Microbiol* 55: 1990-1995.

Sivonen, K., Namikoshi, M., Evans, W., Carmichael, W., Sun, F., Rouhiainen, L., Luukkainen, R., Rinehart, K. (1992) Isolation and characterization of a variety of microcystins from seven strains of the cyanobacterial genus *Anabaena*. *Appl Env Microbiol* 58: 2495-2500.

Smith, J. L., Boyer, G. L., Zimba, P. V. (2008) A review of cyanobacterial odorous and bioactive metabolites: impacts and management alternatives in aquaculture. *Aquac* 280: 5-20.

Smitka, T. A., Bonjouklian, R., Doolin, L., Jones, N. D., Deeter, J. B., Yoshida, W. Y., Prinse, M. R., Moore, R. E., Patterson, G. M. L. (1992) Ambiguine isonitriles, fungicidal hapalindole type alkaloids from three genera of blue green algae belonging to the Stigonemataceae. *J Org Chem* 57: 857- 861.

Sonjoukhan, R., Smltka, T. A., Doolin, L. E., Molloy, R. M., Debono, M., Shaffer, S. A., Moore, R. E., Stewart, J. B., Patterson, G. M. L. (1991) Tjipanazoles, new antifungal agents from the blue-green alga *Tolypothrix tjipanasensis*. *Tetrahedron* 47: 7739-7750.

Soule, T., Stout, V., Swingley, W., Meeks, J., Garcia-Pichel, F. (2007) Molecular genetics and genomic analysis of scytonemin biosynthesis in *Nostoc punctiforme* ATCC 29133. *J Bact* 189: 4465-4472.

Srivastava, V. C., Manderson, G., Bhamidimarri, R. (1999) Inhibitory metabolites production by the cyanobacterium *Fischerella muscicola*. *Microbiol Res* 153: 309-317.

Stewart, I., Schluter, P. J., Shaw, G. R. (2006) Cyanobacterial lipopolysaccharides and human health-a review. *Environ Health* 5: 1.

Stewart, J. B., Bornemann, V., Chen, J. L., Moore, R. E., Caplan, F. R., Karuso, H., Larsen, L. K., Patterson, G. M. (1988) Cytotoxic, fungicidal nucleosides from blue green algae belonging to the Scytonemataceae. *J Antibiot* 41: 1048-1056.

Stirk, W. A., Ördög, V., Van Staden, J. (1999) Identification of the cytokinin isopentenyladenine in a strain of *Arthronema africanum* (Cyanobacteria). *J Phycol* 35: 89-92.

Stratmann, K., Belli, J., Jensen, C. M., Moore, R. E., Patterson, G. M. (1994) Aulosirazole, a novel solid tumor selective cytotoxin from the blue-green alga *Aulosira fertilissima*. *J Org Chem* 59: 6279-6281.

Strichartz, G., Rando, T., Hall, S., Gitschier, J., Hall, L., Magnani, B., Bay, C. H. (1986) On the mechanism by which saxitoxin binds to and blocks sodium channels. *Ann N Y Acad Sci* 479: 96-112.

Sturdy, M., Krunic, A., Cho, S., Franzblau, S., Orjala, J. (2010) Eucapsitrione, an anti-*Mycobacterium tuberculosis* anthraquinone derivative from the cultured freshwater cyanobacterium *Eucapsis* sp. *J Nat Prod* 73: 1441-1443.

Su, Z., Sheets, M., Ishida, H., Li, F., Barry, W. H. (2004) Saxitoxin blocks L-type ICa. *J Pharmacol Exp Ther* 308: 324-329.

Takebe, Y., Saucedo, C. J., Lund, G., Uenishi, R., Hase, S., Tsuchiura, T., Kneteman, N., Ramessar, K., Tyrrell, D. L., Shirakura, M., Wakita, T. (2013) Antiviral lectins from red and blue-green algae show potent *in vitro* and *in vivo* activity against Hepatitis C virus. *PLoS ONE* 8: e64449.

Tan, L. T. (2007) Bioactive natural products from marine cyanobacteria for drug discovery. *Phytochemistry* 68: 954-979.

Taori, K., Paul, V. J., Luesch, H. (2008) Structure and activity of largazole, a potent antiproliferative agent from the Floridian marine cyanobacterium *Symploca* sp. *J Am Chem Soc* 130: 1806-1807.

Thajuddin, N., Subramanian, G. (2005) Cyanobacterial biodiversity and potential applications in biotechnology. *Curr Sci* 89: 47-57.

Todorova, A. K., Juettner, F., Linden, A., Pluess, T., von Philipsborn, W. (1995) Nostocyclamide: a new macrocyclic, thiazole-containing allelochemical from *Nostoc* sp. 31 (cyanobacteria). *J Org Chem* 60: 7891-7895.

Trick, C. G., Kerry, A. (1992) Isolation and purification of siderophores produced by cyanobacteria, *Synechococcus* sp. PCC 7942 and *Anabaena variabilis* ATCC 29413. *Curr Microbiol* 24: 241-245.

Tyagi, M. B., Singh, D. P., Kumar, A., Jha, P. N., Sinha, R. P., Kumar, A. (2006) Hepatotoxicity of *Microcystis aeruginosa* strains growing as blooms in certain eutrophic ponds. *EXCLI J* 5: 66-78.

Van Apeldoorn, M. E., Van Egmond, H. P., Speijers, G. J., Bakker, G. J. (2007) Toxins of cyanobacteria. *Mol Nutr Food Res* 51: 7-60.

Van Wagoner, R. M., Drummond, A. K., Wright, J. L. (2007) Biogenetic diversity of cyanobacterial metabolites. *Adv Appl Microbiol* 61: 89-217.

Vepritskii, A., Gromov, B., Titova, N., Mamkaeva, K. (1991) Production of the antibiotic-algicide cyanobacterin LU-2 by the filamentous cyanobacterium *Nostoc* sp. *Microbiol* 60: 675-679.

Vestola, J., Shishido, T. K., Jokela, J., Fewer, D. P., Aitio, O., Permi, P., Wahlsten, M.,Wang, H., Rouhiainen, L., Sivonen, K. (2014) Hassallidins, antifungal glycolipopeptides, are widespread among cyanobacteria and are the end-product of a non ribosomal pathway. *Proc Natl Acad Sci USA* 111: E1909-E1917.

Villa, F. A., Lieske, K., Gerwick, L. (2010) Selective MyD88-dependent pathway inhibition by the cyanobacterial natural product malyngamide F acetate. *Eur J Pharmacol* 629: 140-146.

Vincent, W. F., Quesada, A. (1994) Ultraviolet radiation effects on cyanobacteria: implication for Antarctic microbial ecosystems. In: *Antarctic Research Series: Ultraviolet Radiation in Antarctica: Measurements and Biological Effects* (Eds. Weiler CS, Penhale PA) American Geophysical Union, Washington, Vol. 62, pp. 111-124.

Volk, R. B. (2006) Antialgal activity of several cyanobacterial exometabolites. *J Appl Phycol* 18: 145-151.

Volk, R. B., Furkert F. H. (2006) Antialgal, antibacterial and antifungal activity of two metabolites produced and excreted by cyanobacteria during growth. *Microbiol Res* 161:110 -186.

Wase N. V., Wright P. C. (2008) Systems biology of cyanobacterial secondary metabolite production and its role in drug discovery. *Expert Opin Drug Discov* 3: 903-929.

Wiegand, C., Pflugmacher, S. (2005) Ecotoxicological effects of selected cyanobacterial secondary metabolites a short review. *Toxicol appl pharmacol* 203: 201-218.

Wilhelm, S. W., Trick, C. G. (1994) Iron-limited growth of cyanobacteria: Multiple siderophore production is a common response. *Limnol Oceanogr* 39: 1979-1984.

Wright, A. D., Papendrof, O., Koing, G. M. (2005) Ambigol C and 2,4 dichlorobenzoic acid, natural products produced by terrestrial cyanobacterium *Fischerella ambigua. J Nat Prod* 68: 459-461.

Wu, M., Okino, T., Nogle, L. M., Marquez, B. L., Williamson, R. T., Sitachitta, N., Berman, F. W., Murray, T. F., McGough, K., Jacobs, R. (2000) Structure, synthesis, and biological properties of kalkitoxin, a novel neurotoxin from the marine cyanobacterium *Lyngbya majuscula. J Am Chem Soc* 122: 12041-12042.

Xiong, C., O'Keefe, B. R., Byrd, R A., McMohan, J. B. (2006) Potent anti-HIV activity of scytovirin domain 1 peptide. *Peptides* 27: 1668-1675.

Yadav S, Sinha R, Tyagi M, Kumar A (2011) Cyanobacterial secondary metabolites. *Int J Pharm Bio Sci* 2: 144-167.

Yadav S, Sinha RP (2015) A review on toxicology, biosynthesis and detection of cyanobacterial toxins. In: *Biological Sciences: Innovations and Dynamics* (Eds. Sinha RP, Richa, Rastogi RP) New India Publishing Agency, New Delhi, India, pp. 157-181.

Zainuddin, E. N., Mentel, R., Wray, V, Jansen, R., Nimtz, M., Lalk, M., Mundt, S. (2007) Cyclic depsipeptides, ichthyopeptins A and B, from *Microcystis ichthyoblabe. J Nat Prod* 70: 1084-1088.

Žegura, B., Štraser, A., Filipič, M. (2011) Genotoxicity and potential carcinogenicity of cyanobacterial toxins-a review. *Mutat Res Rev Mut Res* 727: 16-41.

Zheng, W., Chen, C., Cheng, Q., Wang, Y., Chu, C. (2006) Oral administration of exopolysaccharide from Aphanothece halophytica (chroococcales) significantly inhibits influenza virus (H1N1)-induced pneumonia in mice. *Int Immuno Pharmacol* 6: 1093-1099.

Zurawell, R. W., Chen, H., Burke, J. M., Prepas, E. E. (2005) Hepatotoxic cyanobacteria: a review of the biological importance of microcystins in freshwater environments. *J Toxicol Environ Health Part B* 8: 1-37.

Zürcher, S., Wäckerlin, D., Bethuel, Y., Malisova, B., Textor, M., Tosatti, S., Gademann, K. (2006) Biomimetic surface modifications based on the cyanobacterial iron chelator anachelin. *J Am Chem Soc* 128: 1064-1065.

In: Innovations in Life Science Research ISBN: 978-1-53615-868-7
Editors: R. P. Sinha, S. Pandey-Rai and N. Ghoshal ©2019 Nova Science Publishers, Inc.

Chapter 3

GENOMICS AND PROTEOMICS OF PHOTO PROTECTIVE COMPOUNDS MYCOSPORINE-LIKE AMINO ACIDS IN CYANOBACTERIA

Haseen Ahmed, Jainendra Pathak, Rajneesh, Abha Pandey, Prashant R. Singh and Rajeshwar P. Sinha*

Laboratory of Photobiology and Molecular Microbiology,
Centre of Advanced Study in Botany, Institute of Science,
Banaras Hindu University, Varanasi, India

ABSTRACT

Mycosporine-like amino acids (MAAs) are the most common group of transparent ultraviolet radiation absorbing intracellular secondary metabolites synthesized by cyanobacteria as part of an overall strategy to diminish the direct and indirect damaging effects of environmental ultraviolet radiation (UVR). MAAs are a family of more than 30 compounds and characterized by a cyclohexenone or cyclohexenimine chromophore core conjugated with amino acids or imino alcohol substituents, having absorption maxima ranging from 268 to 362 nm. These are attached to the core through imine linkages, leading to a combination of resonance tautomers which facilitates absorption of UV light. These UV screening compounds are highly photostable photoprotectant and also serve as potent antioxidants. Considerable interest has been centered on MAAs because experimental evidence indicated that in cyanobacteria the major functions of MAAs are to act as photo-protective UV filters and/or to act as antioxidants. In the present chapter we also describe the structure and physicochemical properties of MAAs, including the recently discovered MAAs and the modern methods used for their isolation and identification. Multiple environmental signals influence MAAs biosynthesis and regulation of induction of these UV screening compounds is a part of complex stress response pathway. This chapter

*Corresponding Author's Email: rpsinhabhu@gmail.com.

provides insight into the genes of cyanobacteria involved in MAA biosynthesis and thus widens the field of research for molecular, bioinformatics and phylogenetic analysis of these evolutionary and industrially important compounds. Study of these sunscreens has also led to the discovery of new classes of compounds, new metabolic pathways, a deeper understanding of microbial photobiology and the potential for dermatological or biomedical applications.

Keywords: cyanobacteria, ultraviolet radiation (UVR), mycosporine-like amino acids (MAAs), physicochemical properties

INTRODUCTION

In cyanobacteria, the main steps in the biosynthesis of MAAs and their genetic basis in cyanobacteria were elucidated recently. A gene cluster with four genes ranging from *ava_3855* to *ava_3858* in cyanobacterium *Anabaena variabilis* ATCC 29413 was found to be responsible for the biosynthesis of an MAA, shinorine (Balskus and Walsh 2010; Singh et al. 2010a). In addition, complete in-frame gene deletion of the O-MT gene from *A. variabilis* ATCC 29413 reveals that, although biochemically distinct; the pentose phosphate and shikimate pathways are inextricably linked to MAA biosynthesis in this cyanobacterium. Furthermore, proteomic data reveal that the shikimate pathway is the predominate route for UV-induced MAA biosynthesis (Pope et al. 2015). Recently, the initial steps in the biosynthesis of mycosporines and MAAs in *A. variabilis* were elegantly elucidated (Balskus and Walsh 2010). A cluster of four genes was found to be responsible for conversion of the common pentose phosphate pathway intermediate sedoheptulose 7-phosphate into shinorine. In the first steps, a dehydroquinate synthase (DHQS) homologue 2-*epi*-5-*epi*-valiolone synthase and an *O*-methyltransferase convert the precursor into 4-deoxygadusol. Cyanobacteria represent a diverse group of photosynthetic bacteria - from tiny single cells to large filamentous forms, and have great potential for biotechnological applications in agriculture, mariculture, biomedicals, food and fuel (Singh et al. 2016). Absorption of solar energy to drive photosynthesis exposes photosynthetic organisms to potentially damaging UVR in their natural habitats due to continuous depletion of stratospheric ozone caused by anthropogenically released pollutants such as chlorofluorocarbons (CFCs), chlorocarbons (CCs) and organobromides (OBs). As a result, the UV-B intensity of can be up to 1.5 $W.m^{-2}$ in the temperate zone or up to 2 $W.m^{-2}$ in the equatorial zone, while the UV-A intensity of and visible light (400-700 nm) is approximately 50-60 $W.m^{-2}$ and 500 $W.m^{-2}$, respectively (Björn 2007). UV-A and UV-B radiation are harmful to living systems, causing damage directly to biological macromolecules such as DNA, proteins including lipids and indirectly by producing reactive oxygen species (ROS). Cyanobacteria tolerate exposure to high doses of UVR in order to access visible solar radiation for photosynthesis (Balskus and Walsh 2010). During evolution, they have developed various strategies to counteract the damaging effects of UV-B radiation for protection from solar UVR (Castenholz and Garcia-Pichel 2000) that vary from behavioral to molecular. One such adaptation is the synthesis of UV-screening secondary metabolites that can be used to screen incoming radiation. Among such metabolites, the MAAs are remarkable for their wide phylogenetic distribution and their unique chemical structures. MAAs are a diverse family of compounds that serve this role in cyanobacteria (Karsten 2002; Liu et al. 2004; Sinha et al. 2007). MAAs production is not restricted to cyanobacteria with production also observed amongst eukaryotic microalgae, yeasts, fungi, corals, and other marine organisms (Piiparinen et al. 2015). To date, more than 30 different compounds in the mycosporine family have been identified from natural sources (Bandaranayake 1998). The physiological role of MAAs as sunscreen compounds to protect against environmental UVR is well established in the producing organisms and the organism that ingests and accumulates MAAs (Shick and Dunlap 2002; Carreto and Carignan 2011). In addition to their photo protective function, MAAs appears to play important roles as antioxidant molecules, compatible solutes, intracellular nitrogen reservoir and also have roles in defense against thermal or desiccation stress (Bandaranayake 1998; Oren and Gunde-Cimerman 2007). Moreover, MAAs have attracted attention in the field of cosmetics and skin care due to their high-UV absorption coefficients (e.g., as the products Helioguard 365 and Helionori), effective UV-filtering capacity and ability to prevent UV-induced skin damage (de la Coba et al. 2009).

The MAAs biosynthesis is probably derived from the shikimate pathway for the synthesis of aromatic amino acids, but one's understanding of the enzymatic and genetic basis of their synthesis is still very limited. Multiple lines of evidence provide that MAAs are biosynthesized the shikimic acid pathway. Complementary evidence was first provided by specific uptake of the shikimate intermediate [U-^{14}C] 3-dehydroquinic acid by the plant fungal pathogen *Trichothecium roseum* to give the structurally related, fungal mycosporines (Favre-Bonvin et al. 1987). Similarly, [^{14}C]pyruvate phosphoenol pyruvate, an obligate precursor of the shikimic acid pathway, yielded radiolabeled MAAs with high specificity in the cyanobacterium *Chlorogloeopsis* sp. PCC 6912, whereas [^{14}C]acetate (polyketide pathway) did not (Portwich and Garcia-Pichel 2003). In addition, MAA biosynthesis was repressed in this cyanobacterium by enhancing exogenous tyrosine, a known feedback inhibitor of the shikimic acid pathway. The same was observed when the cyanobacterium *Nostoc commune* was exposed to the well-known shikimic acid pathway inhibitor, glyphosate (Sinha et al. 2003). Collectively, these observations provide considerable evidence that microbial MAA biosynthesis proceeds from a branch-point intermediate of the shikimic acid pathway.

Thus, developments in the elucidation of the biosynthesis of microbial sunscreens not only provide scientific insight in a major class of secondary metabolites but also open the door for discovery and development of a new generation of sunscreens (Aboulmagd et al. 2001). In this chapter, we cover the group of UV-protective MAAs and summarize (1) their detailed molecular structure and physicochemical properties; (2) their biosynthetic pathways (3) their occurrence and distribution in cyanobacteria and (4) the techniques used for characterization of the most common MAAs.

Molecular Structures and Properties of UV-Absorbing Mycosporine-Like Amino Acids in Cyanobacteria

MAAs and a related group of compounds called mycosporines represent a large family of colorless, low-molecular-weight ranging from 188 to 1050 Da, water-soluble, usually intracellular secondary metabolites and widespread in the biological world (Dunlap and Chalker 1986; Carreto et al., 1990; Rosic and Dove 2011, Wada et al. 2015). Chemically MAAs are alicyclic compounds sharing a central 5-dihydroxy, 5-hydroxymethyl, cyclohex-1, 2-ene ring, with a methoxy group at C-2. This core structure is substituted in C-3 with an amino compound (usually an amino acid or amino alcohol) to form oxo-mycosporines. Ketone replacement with a second amino compound defines the imino-mycosporine group. The N-substitution on C-3 with different amino acids or amino alcohols is what gives the diversity of molecular structures within both groups (Korbee et al. 2006; Sinha et al. 2007; Carreto and Carignan 2011). Within the MAAs group, the most common amino acid on the C-3 position is glycine, whereas they also have a second amino acid, amino alcohol or an enaminone system attached at the C-1 position (D'Agostino et al. 2016). Photo protective roles have especially been attributed to UV-absorbing MAAs since they don't seem to contribute to photosynthesis at all (Moisan and Mitchell 2001). MAAs present typical UV absorption spectra with a single, narrow, and strongly absorbing band that has a maximum around 310 nm for oxo-mycosporines and around 330 nm for simple imino-mycosporines (Table 1) (Gao and Garcia-Pichel 2011).

This unique molecular structuring and bonding provide them unique spectral properties. MAAs are examined to be one of the strongest UV-A/UV-B-absorbing substances in nature (Daniel et al. 2004); with wavelength absorption maxima (λ_{max}) in the 310-362 nm interval and molar attenuation coefficients (ε) between 28100 and 50000 $M^{-1}cm^{-1}$ (Dunlap and Schick 1998; Carreto et al. 2005; Gao and Garcia-Pichel 2011). Their absorption spectra are characterized by a single sharp λ_{max} with a bandwidth of approximately 20 nm and only about 2-3 nm apart from the λ_{max} of structurally similar MAAs which makes it very difficult to distinguish these compounds based merely on their absorption spectra (Karentz 1994; Carroll and Shick 1996). The values of λ_{max} and ε are dependent on the degree of derivatization of the central ring and the nature of the nitrogenous side groups (in particular the presence of additional conjugated double bonds) (Wada et al. 2015). Smaller, mono-substituted mycosporines (typically oxo-mycosporines) have their λ_{max} values shifted to shorter wavelengths in the UV-B and usually have lower ε values; whereas MAAs (imino-mycosporines) are normally bi substituted, with higher ε

values and λ_{max} values shifted to longer wavelengths in the UV-A region (Portwich and Garcia-Pichel 2003).

Table 1. Molecular structure of MAAs with their corresponding absorption maxima (Adapted from Sinha et al. 2007)

MAAs	Molecular structure	λ_{max} (nm)
Mycosporine-glycine	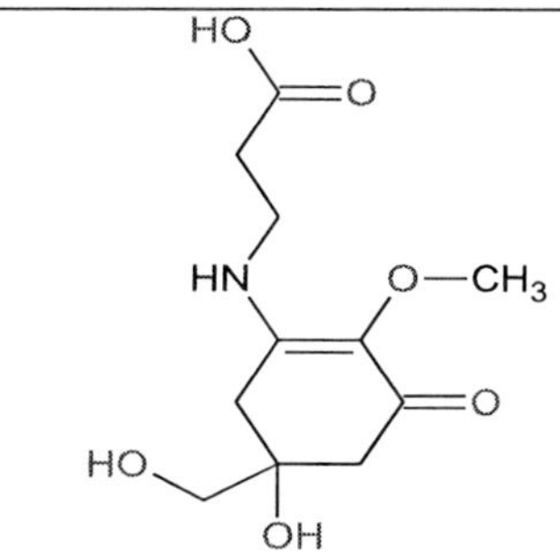	310
Mycosporine-2-glycine	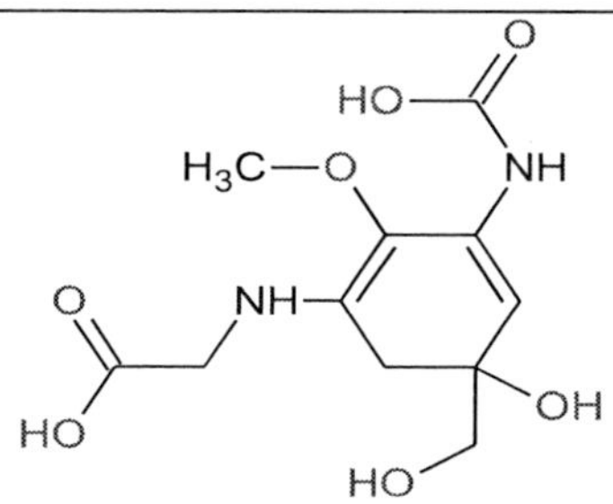	334
Shinorine	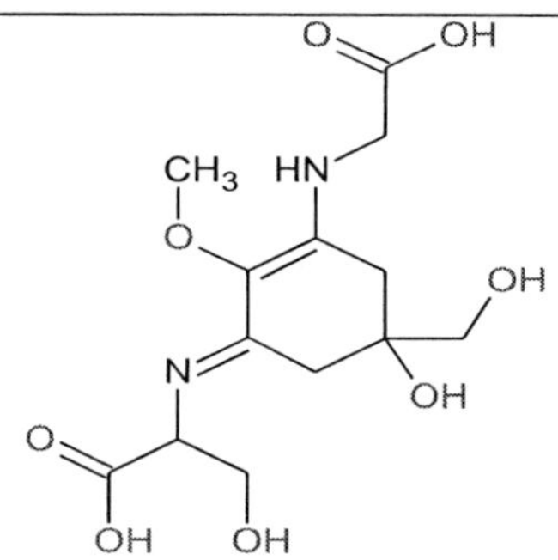	334
Porphyra-334	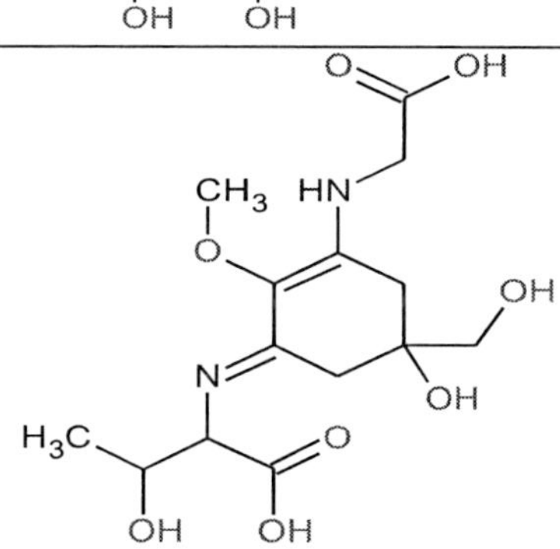	334

Table 1. (Continued)

MAAs	Molecular structure	λ_{max} (nm)
Mycosporine-taurine	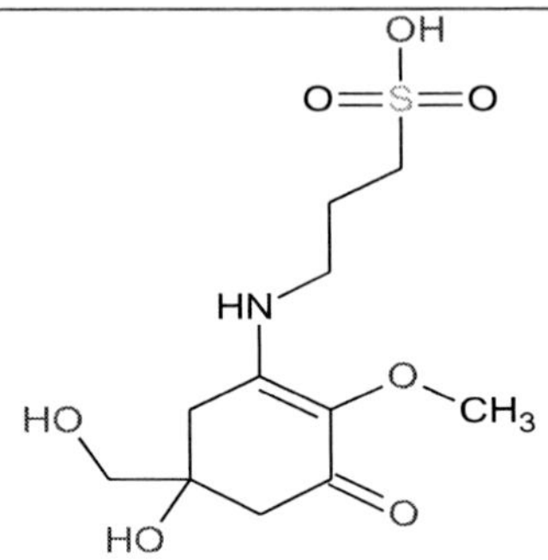	309
Palythine	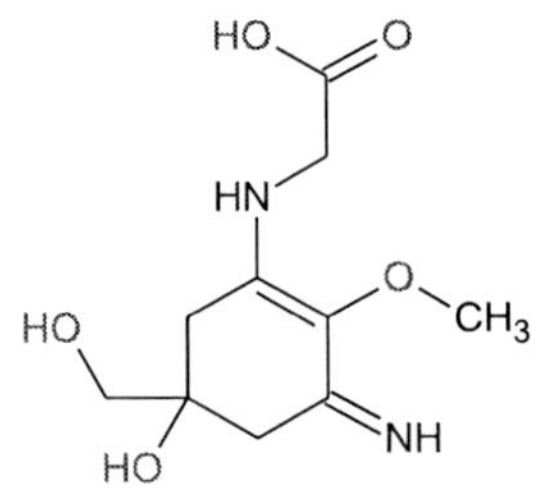	320
Mycosporine-glycine-valine	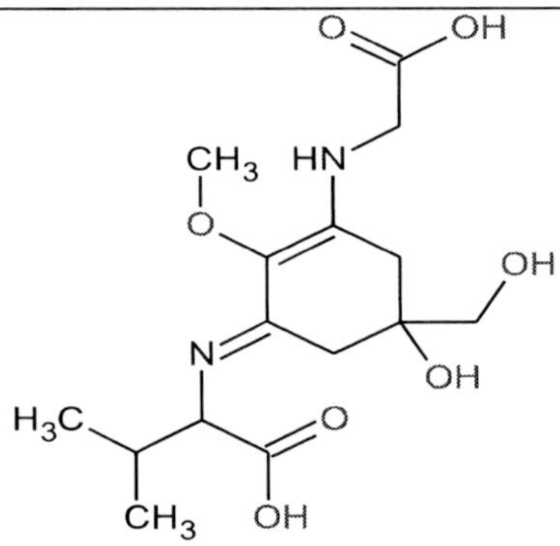	335
Palythene	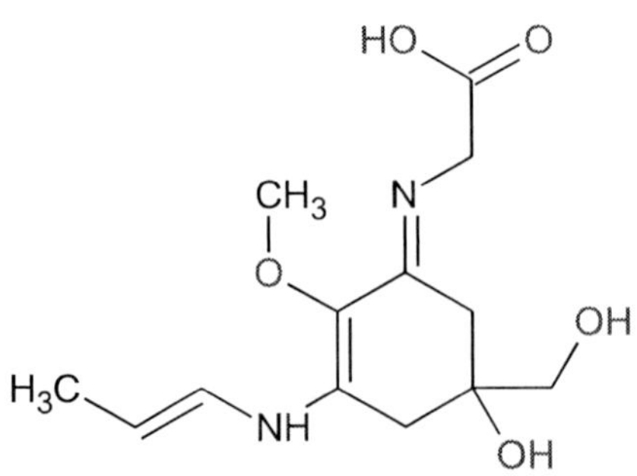	360

MAAs	Molecular structure	λ_{max} (nm)
Euhalothece-362	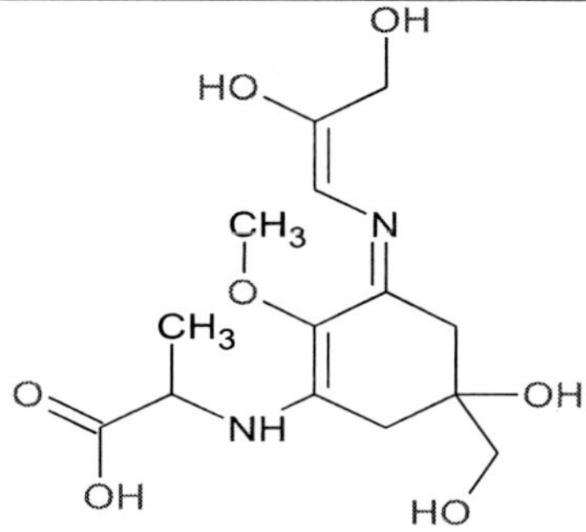	362
Usujirene	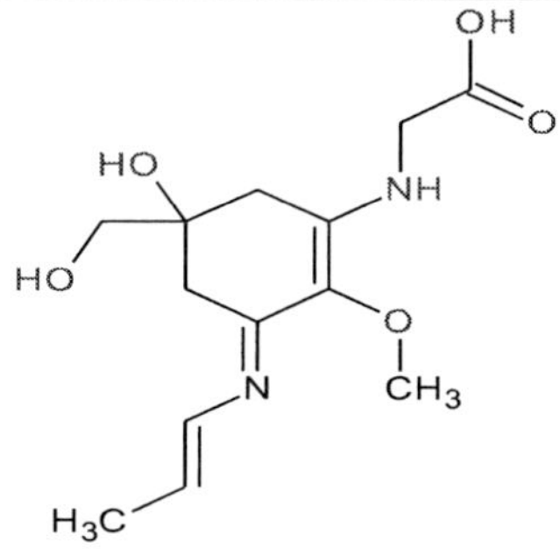	357
Palythine-serine-sulfate	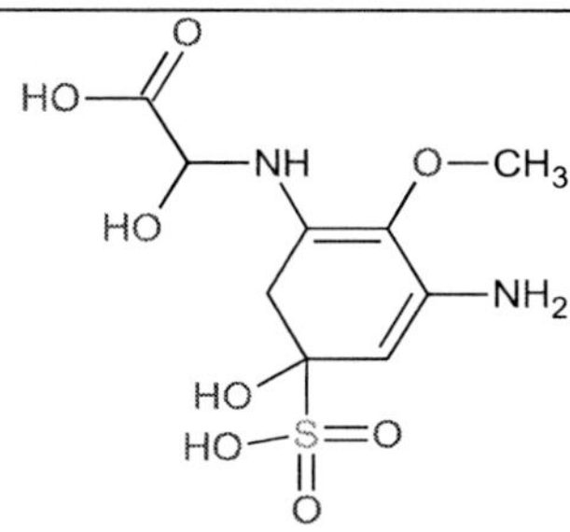	320
Mycosporine-glutamic acid-glycine	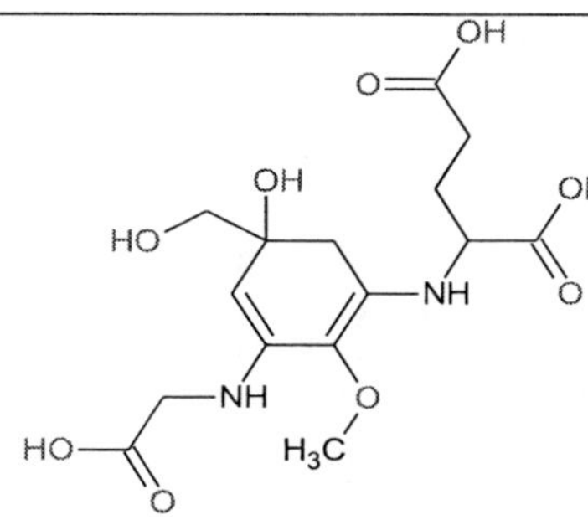	330
Mycosporine-methylamine-threonine	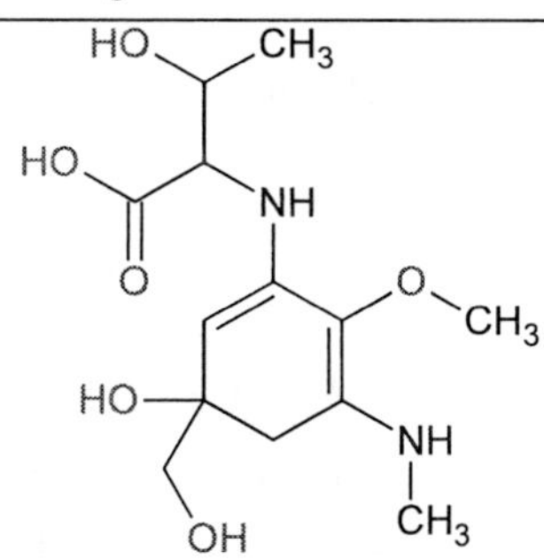	327

Table 1. (Continued)

MAAs	Molecular structure	λ_{max} (nm)
Asterina-330	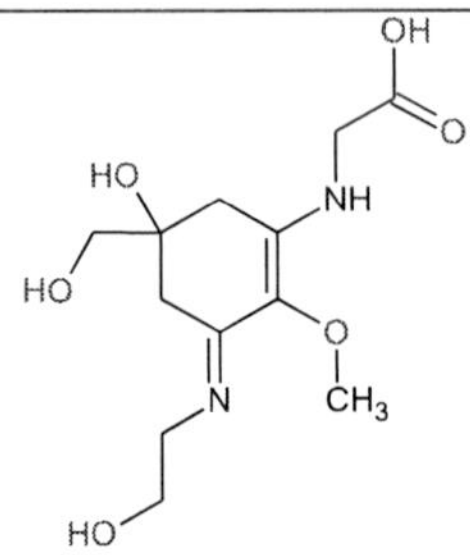	330
Mycosporine-methylamine-serine	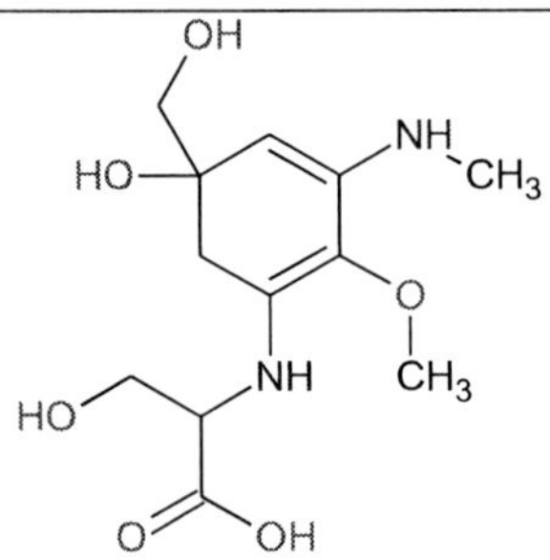	327
Palythine-serine	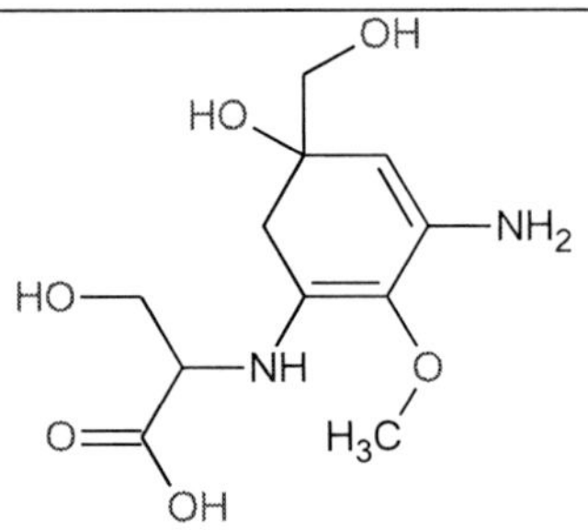	320

For example, the direct metabolic precursor of all mycosporines, 4-deoxygadusol, which has the minimal level of derivatization, has λ_{max} =268 nm at acidic pH, and λ_{max} = 294 nm at basic pH; mycosporine-glycine, the simplest oxo-mycosporine and direct precursor of all other MAAs has a λ_{max} = 310 nm, whereas palythine, a simple mono-substituted MAA (amino-MAA), has λ_{max} = 320 nm and ε = 36200 $M^{-1}cm^{-1}$ (Carreto et al. 2005; Gao and Garcia-Pichel 2011). Palythene, a bi-substituted MAA with an additional conjugated double bond, has one of the most red-shifted bands of all known MAA species, with λ_{max} = 360 nm and ε = 50000 $M^{-1}cm^{-1}$ (Uemura et al. 1980). The observed red-shift in λ_{max} is a consequence of the degree of resonance delocalization inside the molecules; the more efficient is the electron delocalization (i.e., the stronger the π-conjugation character) the lower is the energy requirement for an electronic transition and consequently the higher are the λ_{max} and ε values (Carreto and Carignan 2011; Wada et al. 2015). From a thermodynamic perspective, the fate of the electronic excitation energy is also a very relevant aspect of the absorption event since it is directly linked to the amount of entropy produced by the dissipative microscopic structure (i.e., the polyatomic molecule).

Nonradiative, vibrational relaxation pathways of the excited state lead to more efficient energy dissipation and higher entropy production when compared to the fluorescent or phosphorescent radiative decay channels (Michaelian 2011, 2012, 2016). MAAs have the ability to protect the cell by absorbing UVR and dissipating the energy as heat without generating ROS (Conde et al. 2004). In addition to their importance in determining the ecology and adaptations of cyanobacteria and other microbes to UV exposure (Finking and Marahiel 2004), MAAs have become promising candidates for their use in pharmaceutical and cosmetic applications and have been commercialized due to their high-UV absorption coefficients (Cardozo et al. 2007) and their ability to protect skin from UV mediated damage (De la Coba et al. 2009; Oyamada et al. 2008). In this respect MAAs prove to be very efficient dissipative structures, although all studies hitherto have discussed these thermodynamically relevant characteristics only from the standpoint of photostability and UV-Photo protection.

Genetics and Proteomics of MAAs Biosynthesis in Cyanobacteria

The occurrence, distribution, and biology of MAAs have been subjects of intense investigation, but elucidation of the precise biochemical mechanisms underlying their functions has been hampered by a lack of knowledge regarding the molecular and genetic basis for their biosynthesis. Several lines of evidence now suggest that the progenitor of the enzymatic machinery for MAA biosynthesis was probably a cyanobacterium or the cyanobacterial ancestor, while endosymbiotic events and prokaryote-to-eukaryote lateral gene transfer events during evolution account for their prevalence among all other biological taxa (Rozema et al. 2002; Waller et al. 2006; Starcevic et al. 2008; Singh et al. 2010a, 2012). Previous studies with cyanobacteria and fungi revealed that the amino acid substituents were added in the later stages of biosynthesis and suggested that the MAAs core might originate from shikimate pathway intermediate 3-dehydroquinate (Favre-Bonvin et al. 1987; Portwich and Garcia-Pichel 2003). Identification of a MAA biosynthetic gene cluster in a cyanobacterium and the discovery of analogous pathways in other sequenced organisms have revealed two unprecedented enzymatic strategies for imine formation. The cyclohexenone core of MAAs is derived from intermediates of two fundamental anabolic pathways; the shikimate pathway (Favre-Bonvin et al. 1987; Shick et al. 1999; Portwich and Garcia-Pichel 2003) and the pentose phosphate pathway (Balskus and Walsh 2010), with the shikimate pathway being the predominant route for UV-induced MAA biosynthesis (Pope et al. 2015).

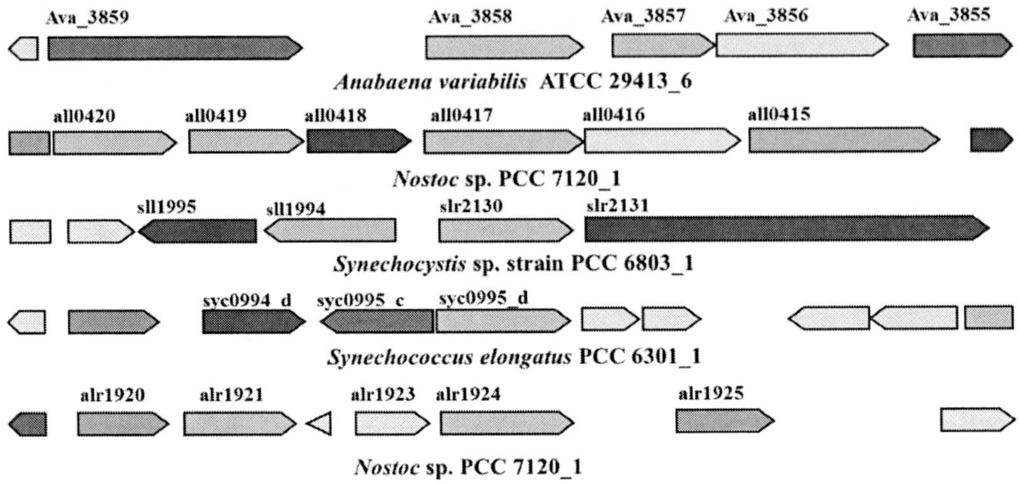

Figure 1. Genomic region comparison of *A. variabilis* PCC 7937 (ATCC 29413), *Anabaena* (*Nostoc*) sp. PCC 7120, *Synechocystis* sp. PCC 6803 and *Synechococcus* sp. PCC 6301 for the YP_324358 (Ava_3858) gene (Light green arrow) of *A. variabilis* PCC 7937 (Modified from Singh et al. 2010a).

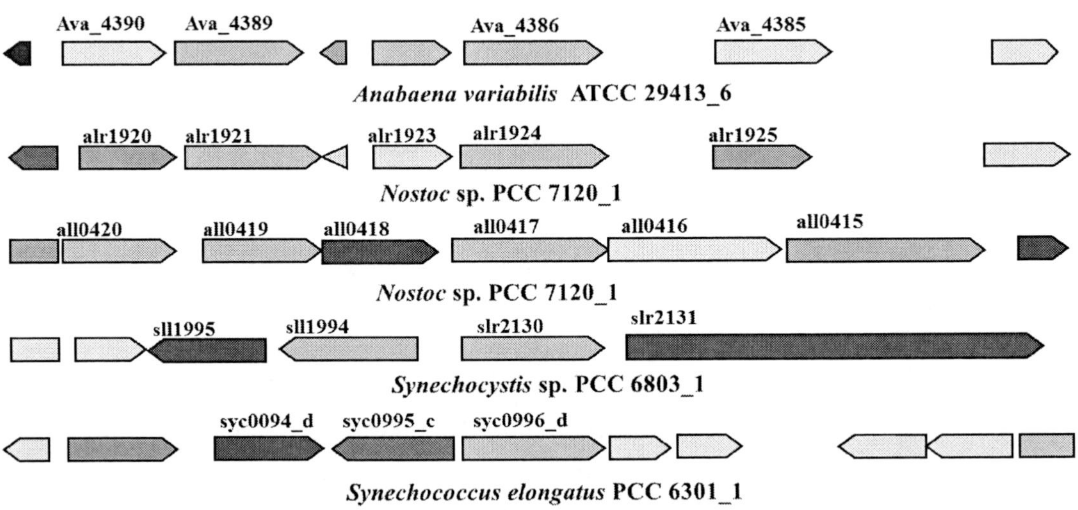

Figure 2. Genomic region comparison of *A. variabilis* PCC 7937 (ATCC 29413), *Anabaena* (*Nostoc*) sp. PCC 7120, *Synechocystis* sp. PCC 6803 and *Synechococcus* sp. PCC 6301 for the YP_324879 (Ava_4386) gene (light green arrow) of *A. variabilis* PCC 7937 (Modified from Singh et al. 2010a).

Radioisotope tracer studies of the cyanobacterium *Chlorogloeopsis* sp. strain PCC 6912 have suggested that the central ring of the MAAs has its origin in the shikimate pathway and that direct amino acid condensation upon the core accounts for the origin of the variable substituents of MAAs and that mycosporine-glycine is the direct precursor of bi substituted imino-mycosporines such as shinorine, suggesting that the biosynthesis of the MAAs would be very similar or possibly identical in eukaryotic and prokaryotic organisms (Portwich and Garcia-Pichel 2003). Similar results were observed recently in the cyanobacterium *Anabaena doliolum*. In this cyanobacterium, mycosporine-glycine acts as precursor for the biosynthesis, induced by UV-B radiation, of shinorine and porphyra-

334 (Singh et al. 2008). MAA biosynthesis in this organism is dependent on photosynthesis only for the carbon source, since the inhibitory effect of DCMU on MAA biosynthesis (Carreto et al. 1990) was overcome by externally added fructose. The main steps of the MAA biosynthetic pathway and their genetic basis have been elucidated by Balskus and Walsh (2010) and Singh et al. (2010a). A bioinformatics study was conducted by Singh et al. (2010a) to identify the possible genes involved in MAA biosynthesis by analyzing the genome of MAA-synthesizing and non-synthesizing cyanobacteria. They tested four cyanobacteria, namely *A. variabilis* PCC 7937, *Anabaena* sp. PCC 7120, *Synechocystis* sp. PCC 6803 and *Synechococcus* sp. PCC 6301, for their ability to synthesize MAA (Figure 1). Out of the four cyanobacteria studied only *A. variabilis* PCC 7937 was able to synthesize MAA. Genome mining of these four cyanobacteria revealed that *A. variabilis* PCC 7937 possess two sets of the 3-dehydroquinate synthase (DHQS) genes YP-324358 and YP-324879 (Figure 2).

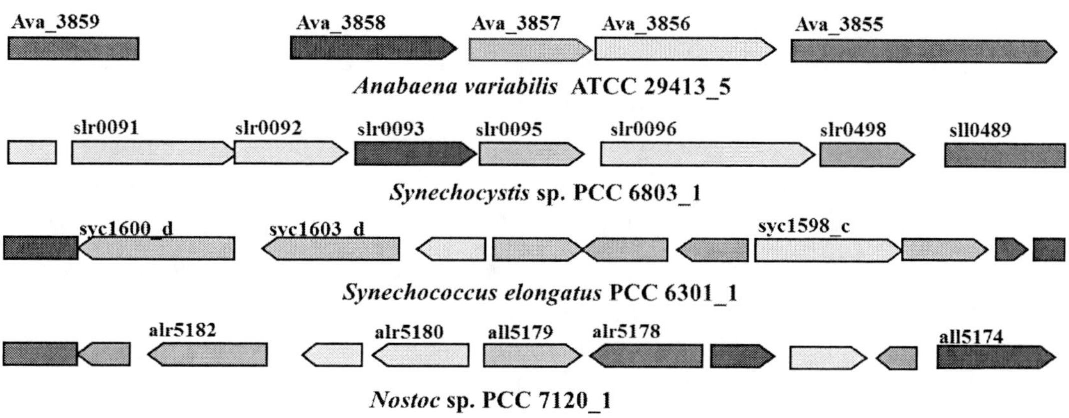

Figure 3. Genomic region comparison of *A. variabilis* PCC 7937 (ATCC 29413), *Synechocystis* sp. PCC 6803, *Synechococcus* sp. PCC 6301 and *Anabaena* (*Nostoc*) sp. PCC 7120 for the YP_324357 (Ava_3857) gene (light green arrow) of *A. variabilis* PCC 7937 (Modified from Singh et al. 2010a).

Genomic region analysis for these genes revealed that gene YP-324358 has O-methyltransferase gene YP-324357 downstream to it while in other cyanobacteria this combination was absent (Figure 3). This study provided the first insight into the genes of cyanobacteria involved in MAA biosynthesis and thus widens the field of research for molecular, bioinformatics, and phylogenetic analysis of the MAAs. Based on the results, they proposed that YP_324358 and YP_324357 gene products are involved in the biosynthesis of the common core (deoxygadusol) of all MAAs (Singh et al. 2010a).

However, the results of Balskus and Walsh (Balskus and Walsh 2010) contradict the long standing assumption that MAA biosynthesis involved a Shikimate pathway intermediate. These authors failed to observe any 6-deoxygadusol production when the putative substrate 3-dehydroquinate was incubated with the dehydroquinate synthase (DHQS) and O-methyltransferase (O-MT) homologues (NpR 5600 and NpR 5599 respectively) from *Nostoc punctiforme* (ATCC 29133) and the typical cofactors S-adenosylmethionine (SAM), nicotinamide adenine dinucleotide (NAD^+) and Co^{2+}. In addition, treatment of the intermediate of the pentose phosphate pathway: Sedoheptulose 7-phosphate (SH7-P) with NpR 5600 and NpR 5599 in the presence of SAM, NAD^+ and Co^{2+}, produce a single product, 6-deoxygadusol. The results of Balskus and Walsh (Balskus and Walsh 2010), confirms that the expression of the gene Ava_3856 from *A. variabilis* can convert 6-deoxygadusol and glycine into mycosporine-glycine in the presence of Adenosine Triphosphate (ATP) and Mg^{2+} cofactors. These authors also confirmed the biosynthetic role of a non-ribosomal peptide synthetase (NRPS)-like enzyme Ava_3855, in the ATP-dependent conversion of mycosporine-glycine and serine to shinorine. In contrast, nothing is known about the chemically plausible condensation of amino acids with mycosporine-taurine, the other known oxo-carbonyl MAAs.

The main steps in the biosynthesis of MAAs and their genetic basis were also elucidated by Spence et al. 2012 in the cyanobacterium *A. variabilis* ATCC 29413. Heterologous expression of a putative gene cluster consisting of four open reading frames (ORFs) ranging from *ava_3855* to *ava_3858* in *Escherichia coli* produces shinorine as the sole MAA (Balskus and Walsh 2010). Ava_3858 encodes a sedoheptulose 7-phosphate cyclase called demethyldeoxygadusol synthase (DDG synthase) catalyzes the conversion of sedoheputulose-7-phosphte, an intermediate of the pentose phosphate pathway, to demethyl 4-deoxygadusol (DDG) and Ava_3857 encodes the enzyme O-methyltransferase (OMT) catalyse the biosynthesis of the parent precursor of mycosporines, 4-deoxygadusol (4-DG) (Pope et al. 2015). The product of Ava_3856, an ATP-grasp amino acid Ligase-catalyses the addition of glycine to 4-DG to yield mycosporine-glycine. The product of gene Ava_3855 - a nonribosomal peptide synthetase (NRPS) homolog catalyses the subsequent condensation of serine with mycosporine-glycine via imine linkages to produce shinorine (Balskus and Walsh 2010).

The genetic basis for the pathway leading to the formation of the imino-mycosporine shinorine does not apparently share such consistency among cyanobacteria, in that the gene demonstrably responsible for the condensation of serine onto mycosporine-glycine in *A. variabilis*, Ava_3855 (Balskus and Walsh 2010), has no corresponding homologues in many other cyanobacteria, including *N. punctiforme*. In the cyanobacterium, *N. punctiforme* ATCC 29133, MAA biosynthesis is similar to that of *A. variabilis* and involves homologous genes (NpR5598-5600) (Katoch et al. 2016). Ava_3856 also has a homologue in *N. punctiforme*, NpR5598, but its activity has not been determined directly. In contrast, homologues of Ava_3885 are missing from the genome of *N. punctiforme* and also from the MAA cluster present in many other cyanobacteria. Heterologous expression in *E. coli* of three contiguous genes (NpR5600, NpR5599 and NpR5598) named *mysA*, *mysB* and *mys C* respectively led to the production of mycosporine glycine (an oxo-mycosporine) (Figure 4).

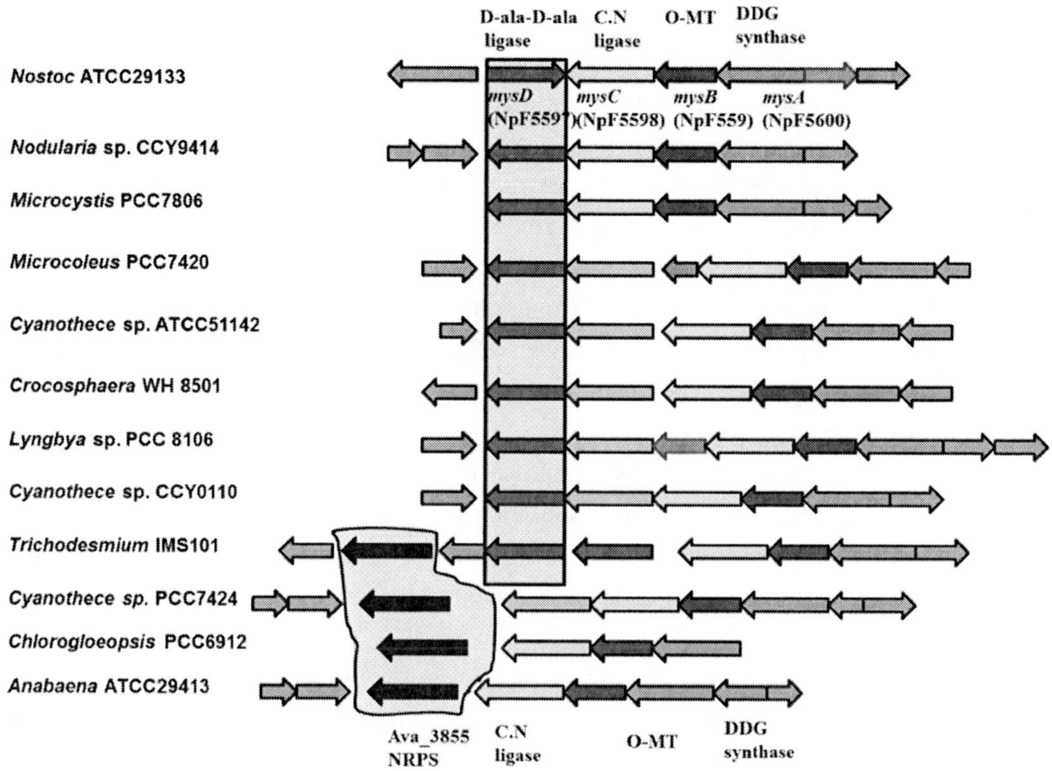

Figure 4. Comparative genomic analyses of the mycosporine cluster region among cyanobacteria whose genomes are completely sequenced, with relevant assignments made to *N. punctiforme.* The ability of cyanobacteria to synthesize mycosporines is common, but not universal, among cyanobacteria. Strains *N. punctiforme* ATCC 29133 and *A. variabilis* ATCC 29413, *Microcoleus* strain PCC 7420, *Microcystis* strain PCC 7806, *Chlorogloeopsis* sp. PCC 6912 and *Trichodesmium* strain IMS101 are known producers. The core *mysA-mysC* locus is conserved throughout, but strains cluster into two separate groups on the basis of having a homologue of either Ava_3855 or *mysD*. *Trichodesmium* has both types. The split does not apparently coincide with taxonomic or phylogenetic boundaries (Modified from Gao and Garcia-Pichel 2011).

However, one exception is the last step (condensation of serine to mycosporine glycine), which is catalyzed by a D-ala-D-ala ligase like protein (NpF5597) rather than a non-ribosomal peptide synthetase those in *A. variabilis* (Balskus and Walsh 2010; Singh et al. 2010a) has been pointed out as one of the main paradoxes in the biosynthesis of imino-mycosporines (Gao and Garcia-Pichel 2011). The logical implication is that, in *N. punctiforme*, shinorine must be synthesized by an alternative enzyme or pathway. It has been suggested that the product of gene NpF5597 of *N. punctiforme* might be responsible for this activity (Gao and Garcia-Pichel 2011), given that it is spatially associated with the MAA cluster, has conserved homologues spatially associated with the MAA cluster in various cyanobacteria, and presents homologies to known amino acid ligating enzymes. Interestingly, it was found that the biosynthesis of shinorine has been found to occur two distinct yet functionally identical enzymes, a D-ala D-ala ligase (*mysD*; *Nostoc* type *mys* cluster; (Gao and Garcia-Pichel 2011) or by a non-ribosomal peptide synthetase (NRPS) enzyme (*mysE*; *Anabaena*-type *mys* cluster).

After 2010, it is well known that 4-deoxygadusol is the parent core structure of MAAs, which, in cyanobacteria, is derived from the conversion of the pentose phosphate pathway intermediate sedoheptulose-7-phosphate by the enzymes 2-epi-5-epivaliolone synthase (EVS), and O-MT. Genome mining has revealed homologues of the EVS gene to be typically present in cyanobacteria known to produce MAAs, and accordingly this gene is normally absent in non-producers (Singh et al. 2010a; Rosic 2012). However, *Synechocystis* sp. PCC6803, which lacks an EVS gene, is controversial because it is reported that exposing this cyanobacterium to UVR for just a few hours produced three novel MAAs (Zhang et al. 2007).

However, it was found out that deletion of the EVS gene from *A. variabilis* ATCC 29413 have little effect on MAA production, thus suggesting that MAA biosynthesis is not exclusive to the pentose phosphate pathway (Spence et al. 2012). By using pathway-specific inhibitors, Pope et al. in 2015 demonstrated that MAA biosynthesis occurs also via the shikimate pathway.

From the results of this knockout experiment conducted by Spence et al. 2012, they propose that functionally duplicate yet distinct convergent pathways exist for the biosynthesis of MAAs in *A. variabilis* ATCC 29413; they involve both the pentose phosphate pathway and, probably, the shikimic acid pathway to produce shinorine the convergent intermediate 4-deoxygadusol. Complete in-frame gene deletion of the O-MT gene from *A. variabilis* ATCC 29413 revealed that the pentose phosphate and shikimate pathways are inextricably linked to MAA biosynthesis in *A. variabilis* ATCC 29413. Furthermore, their proteomic data showed that the shikimate pathway is the main route for UV induced MAA biosynthesis (Pope et al. 2015) (Figure 5).

Occurrence and Distribution of MAAs in Cyanobacteria

In order to survive, almost all organisms have developed a variety of natural defense strategies to overcome the detrimental effects of UVR exposure, including biosynthesis of UV-screening compounds such as MAAs (Bandaranayake 1998; Callone et al. 2006; Sommaruga et al. 2006). Details of the pathway and the enzymes involved in the biosynthesis and transformations of various MAAs in cyanobacteria remain a topic of discussion. MAAs are distributed widely in the environment (McClintock and Karentz 1997). For the first time, their presence has been detected in cyanobacterium and several species of corals from the Great Barrier Reef (Shibata 1969). Rare novel MAA, containing the amino acid alanine (2-(e)-2,3-dihydroxipro-1-enylimino-mycosporine-alanine) was isolated from the unicellular cyanobacterium *Euhalothece* sp. (Volkmann et al. 2006).

Another novel MAA tentatively identified as dehydroxyl-usujirene was isolated from the cyanobacteria *Synechocystis* sp. (Zhang et al. 2007) (Figure 6).

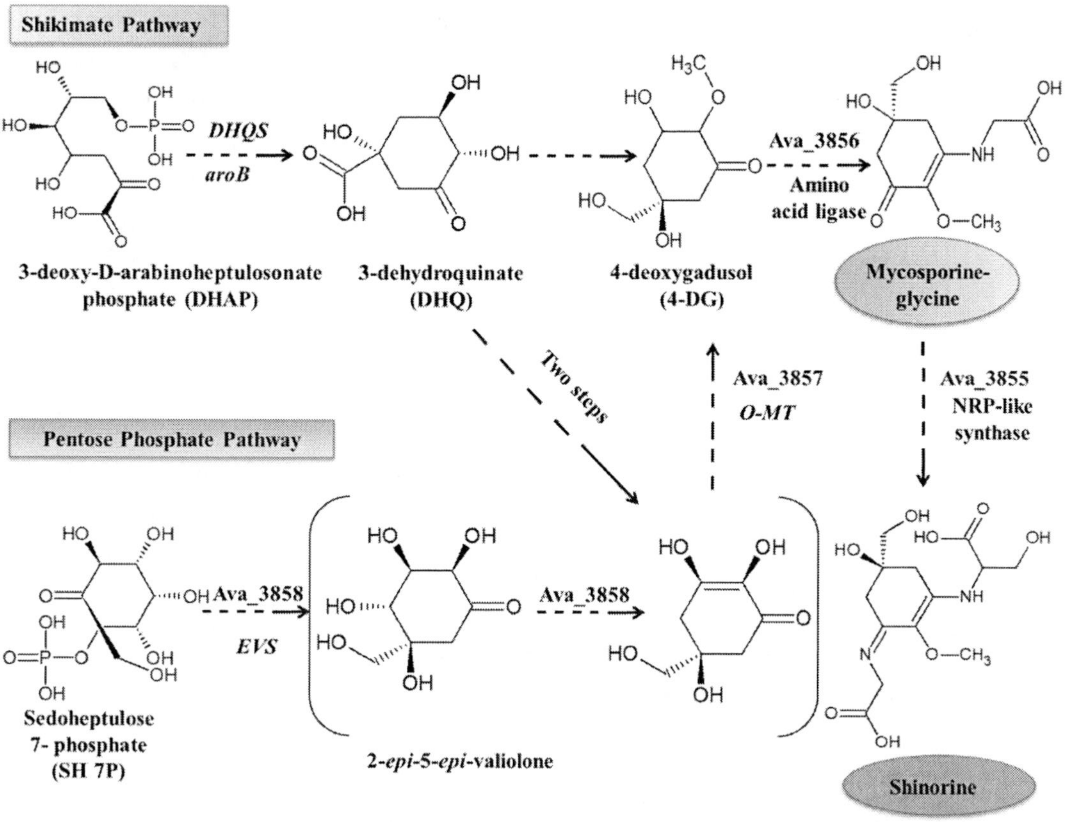

Figure 5. Two possible pathways for the biosynthesis of mycosporine-glycine in the cyanobacterium *A. variabilis* ATCC 29413. 4-deoxygadusol can be produced through either the shikimate pathway or the pentose pathway, and is then combined with a glycine molecule to produce mycosporine-glycine (Modified from Pope et al. 2015).

There are reports that some *Anabaena* spp. only synthesize the MAA shinorine (Sinha et al. 1998; Sinha et al. 2007; Sinha and Häder 2008) while other *Anabaena* spp. isolated from the same habitat lack the ability to synthesize shinorine and other MAAs (He et al. 2002; Gao et al. 2007; Singh et al. 2008). *Anabaena* spp., the MAA profile of a rice-field cyanobacterium, *A. doliolum*, is unique as it reveals the biosynthesis of three different types of MAAs, shinorine, porphyra-334 and mycosporine-glycine (Singh et al. 2008). Recently four cyanobacteria, e.g., *Anabaena* sp. PCC 7120, *A. variabilis* PCC 7937, *Synechococcus* sp. PCC 6301 and *Synechocystis* sp. PCC 6803, were tested for their ability to synthesize MAAs, and genomic and phylogenetic analysis was conducted to identify the possible set of genes that might be involved in the biosynthesis of these compounds. Out of the four investigated species, only *A. variabilis* PCC 7937 was able to synthesize MAA (Singh et al. 2010a). These authors were not able to detect any trace of MAAs from *Synechocystis* sp. PCC 6803, however, Zhang et al. (2007) have reported the presence of some unusual and new MAAs from this cyanobacterium (dehydroxylusujirene, mycosporine-taurine), different from the normally reported MAAs (mycosporine-glycine, palythene, palythinol, shinorine, asterina-330, porphyra-334 and euhalothece-362) in cyanobacteria (Sinha et al. 2007; Sinha and Häder 2008). However, as has been previously noted by Singh et al. (2010a) the identity of these compounds is questionable. Rastogi et al. (2014a) first reported on the occurrence of MAA mycosporine-glycine (M-Gly) in the cyanobacterium *Arthrospira* sp. CU2556 strains. Three species of *Nodularia* (*N. harveyana, N. baltica*, and *N. spumigera* synthesized porphyra-334 and shinorine (Sinha et al. 2003; Wulff et al. 2007). A symbiotic cyanobacterium *Prochloron* sp., exclusively found in association with tropical colonial ascidians, also has been reported to contain high amounts of MAAs.

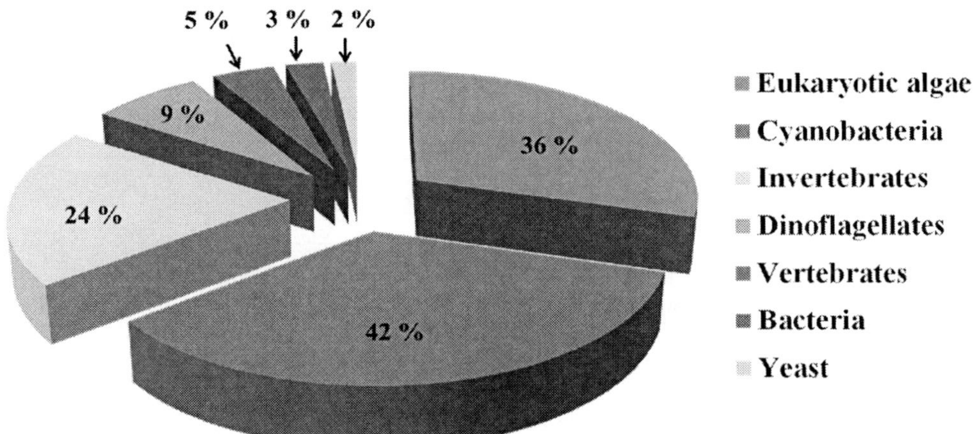

Figure 6. Various organisms studied for the synthesis/production of MAAs. The data are derived from published articles between 2001 and 2016. The figure represents that cyanobacteria are the most studied candidates for MAAs production (Modified from Jain et al. 2017).

Isolated *Prochloron* cells from *Lissoclinum patella* contained shinorine, which was also dominant in the host tunic together with minor amounts of mycosporine-glycine and palythine (Dionisio-Sese et al. 1997). The same species was reported to contain mycosporine-glycine and shinorine (Dunlap and Yamamoto 1995) or only mycosporine-glycine (Lesser et al. 1990). However, some cyanobacteria groups contain multiple MAAs, but with very few exceptions their structures have not been elucidated. Among all the *Microcoleus* strain studied by Karsten and García-Pichel (1996) four different MAAs were detected, one of them shinorine and three unidentified compounds showed absorption maxima at 332, 344 and 346 nm. Rastogi et al. (2016) characterized the occurrence of four different MAAs, Palythine, Asterina Porphyra and palythene in the cyanobacterium *Nostoc* sp. R76DM. They also reported three different MAAs with UV-absorption maxima at 326, 331 and 345 nm (Rastogi et al. 2014b). In addition, mycosporine-2-glycine and a rare novel MAAs containing the amino acid alanine (euhalothece-362) was described in the unicellular cyanobacterium *Euhalothece* sp., inhabiting in a hypersaline saltern pond (Kedar et al. 2002; Volkmann et al. 2006).

A high concentration of MAAs including shinorine and several unknown MAAs was reported in the freshwater bloom forming cyanobacteria *Microcystis aeruginosa* (Liu et al. 2004). Ocean bloom forming cyanobacteria *Trichodesmium* spp. were also reported to contain high amounts of MAAs and asterina-330 and shinorine were the most abundant. They also contain porphyra-334, mycosporine-glycine and palythene (Subramanian et al. 1999). *A. variabilis* PCC 7937 synthesizes the primary MAA shinorine under normal conditions, however under sulfur deficiency, a secondary MAA palythine-serine appears. Addition of methionine to sulfur-deficient cultures resulted in the disappearance of palythine-serine, suggesting the role of primary MAA under sulfur deficiency (Singh et al. 2010b). This is the first time that palythine-serine was found to be synthesized by cyanobacteria. Results also support that palythine-serine is synthesized from shinorine after decarboxylation and demethylation of the glycine subunit as proposed earlier (Carreto et al. 2005; Sinha et al. 2008; Carignan et al. 2009). More unusual MAAs include those where MAAs are covalently linked to oligosaccharides through the imine substituents as reported in the terrestrial cyanobacterium *N. commune* (Böhm et al. 1995) and Nazifi et al. (2013) reported the novel glycosylated Porphyra-334 and Palythine-threonine in the same cyanobacterium *N. commune*. Waditee-Sirisattha et al. (2014) identified mycosporine-2-glycine in halotolerant cyanobacterium *Aphanothece halophytica*. Rastogi et al. (2014c) identified two MAAs production in cyanobacteria *Gloeocapsa* sp. CU-2556 were, shinorine (λmax 333 nm) and an unknown MAA designated as mycosporine-307 (λmax 307 nm) in UVR. Biological crusts collected from various sites of Varanasi, India such as roof top, bark of *Mangifera indica*, window ledge and agricultural field were studied for the presence of MAAs by Pathak et al. (2017a).

Techniques Used for Characterization of MAAs

Emerging developments in analytical technologies for characterization of secondary metabolites can provide more information from the experimental data generated, leading to identification of metabolites. These techniques include fast and high-resolution separation system such as Ultra Performance Liquid Chromatography (UPLC), high-mass accuracy and large dynamic-range MS instruments and higher sensitivity NMR systems (Moco et al. 2007). Techniques such as NMR, FTIR, DIMS involve high- throughput, short analysis time strategies hence are more suitable for larger replicate numbers than approaches such as chromatographic techniques which involves longer analysis times (Dunn et al. 2005). Spectroscopic measurements are sensitive and non-destructive which require only small amounts of samples for analysis. UV-Visible spectrophotometry is commonly used for analysis of MAAs. It measures the amount of visible radiation or UVR absorbed by a substance in solution. Quantitative spectrophotometric analysis is used to ascertain the quantity of molecular species absorbing the radiation (Behera et al. 2012). It is a totally non-invasive, label-free technique which excites vibrations of molecular bonds and has also allowed for rapid, non-destructive, reagent-less, and high throughput analysis for a diverse range of sample types (Downes 2010).

MAAs are usually identified based on their retention time during HPLC and their characteristic UV absorption spectra obtained diode array detection (DAD). Although DAD allows the fast acquisition of UV-VIS absorption spectra, the lack of fine spectral absorption, the influence of pH and, for some specific MAAs, wavelength absorption maxima identical or only a few nm apart, makes it very difficult to distinguish MAA compounds based on absorption spectra only. In addition, a common difficulty has been the lack of commercially available standards, the preparation of which is costly and time consuming (Carreto et al. 2011). As a result, detection, identification and quantification are most usually based on co-elution with secondary standards provided by other researchers and/or comparison with published UV spectral data and HPLC retention times. Such characterization of these compounds should be treated with caution (Bandaranayake 1998; Whitehead and Hedges 2002). However, certain closely related unknown compounds have similar absorption maxima and RT, which still cause difficulties in the identification process. Hence NMR and LC coupled with ESI-MS (LC-ESI-MS) are commonly utilized to examine structural diversity of MAAs (Pathak et al. 2017b).

Metabolites of interest are extracted by liquid extraction with either aqueous solvent (for water soluble compounds such as MAAs) or organic solvent (for lipid soluble compounds such as scytonemin), or with a combination of solvents. The type of metabolites extracted depends on the chemical properties of the solvent used (Moco et al. 2007). Hence, for characterization of MAAs, UV-VIS spectroscopy followed by HPLC, ESI-MS/MS, FTIR and NMR constitute major analytical techniques. Mass spectrometry detection (HPLC/MS) using pure reference compounds, can make an invaluable contribution in the identification of MAAs because of its high sensitivity, and the availability of powerful tandem mass spectrometric techniques (Yoshiki et al. 2009; Volkmann and Gorbushina 2006; Carignan et al. 2009).

However, only the crystal and molecular structure of palythine (Furusaki et al. 1980) and palythene (Uemura et al. 1980) were unambiguously determined by X-ray analysis, including its absolute configuration (Table 2). The structure of both compounds was defined as an inner salt and the features of delocalization of the positive and negative charge were elucidated. For example, in palythene, the positive and negative charges in the inner salt are favourable to exist in the near position each other, indicating that in the representation of this compound as a resonance hybrid, the canonical form 1 contributes more effectively than the form 2. In the crystalline state (as trihydrate) palythine also exists as a zwitterion which is stabilized by resonance between two nearly-equivalent structures.

Table 2. Techniques used for characterization of the most common MAAs (Adapted from Pathak et al., 2017b)

MAAs	Chemical assay	Optical Activity	UV-Vis Spectro-scopy	IR Spectro-scopy	X Ray Diffraction	MS Spectroscopy (ESI-MS/MS^n)	NMR Spectroscopy (^1H-NMR, ^{13}C-NMR)	References
Mycosporine-glycine	Amino acid composition after neutral or alkaline hydrolysis Methylation	+	+	-	-	ESI-MS (high resolution)	^1H-NMR ^{13}C-NMR	Ito and Hirata (1977), Bandaranayake et al. (1996), Balskus and Walsh (2010)
Palythine	Amino acid composition after alkaline hydrolysis Interchange H/D	+	+	+	+	ESI-MS (high resolution) ESI-MS^n	^1H-NMR ^{13}C-NMR	Furusaki et al. (1980), Cardozo et al. (2006), Whitehead et al. (2001), Takano et al. (1978), Tsujino et al. (1978)
Mycosporine-taurine	Amino acid composition after hydrolysis (not detailed)	-	+	-	-	ESI-MS (high resolution)	-	Stochaj et al. (1994)
Palythine-serine	No	-	+	-	-	ESI-MS^n	^1H-NMR	Teai et al. (1997)
Palythine-serine sulfate	No	-	-	-	-	ESI-MS (high resolution)	^1H-NMR ^{13}C-NMR	Wu et al. (1997)
Mycosporine-methylamine-serine	No	-	+	-	-	ESI-MS^n	^1H-NMR	Teai et al. (1997)
Mycosporine-methylamine-threonine	Amino acid composition after hydrolysis (not detailed)	-	+	-	-	ESI-MS (high resolution)	^1H-NMR ^{13}C-NMR	Wu et al. (1995)

Table 2. (Continued)

MAAs	Chemical assay	Optical Activity	UV-Vis Spectro-scopy	IR Spectro-scopy	X Ray Diffraction	MS Spectroscopy (ESI-MS/MS^n)	NMR Spectroscopy (^{1}H-NMR, ^{13}C-NMR)	References
Asterina-330	Methylation H/D Interchange	-	+	-	-	EI-MS (high resolution) ESI-MS (high resolution) ESI-MS^n	^{1}H-NMR	Bandaranayake (1998), Nakamura et al. (1981), Cardozo et al. (2006), Whitehead et al. (2001)
Palythinol	Amino acid composition after acid hydrolysis Methylation Interchange H/D	+	+	+	-	ESI-FTICR-MS^n ESI-MS^n	^{1}H-NMR ^{13}C-NMR	Cardozo et al. (2006), Whitehead et al. (2001), Takano et al. (1978), Cardozo et al. (2008)
Mycosporine-glutamic acid-glycine	Amino acid composition after alkaline hydrolysis	-	+	-	-	SI-MS	^{1}H-NMR ^{13}C-NMR	Bandaranayake et al. (1996)
Shinorine	Amino acid composition after acid hydrolysis Interchange H/D	+	+	+	-	ESI-MS ESI-MS^n FD-MS	^{1}H-NMR ^{13}C-NMR	Cardozo et al. (2006), Whitehead et al. (2001), Tsujino et al. (1980)
Mycosporine-2-glycine	Amino acid composition after acid hydrolysis	-	+	-	-	ESI-MS^n (high resolution)	^{1}H-NMR ^{13}C-NMR	Volkmann et al. (2006), Stochaj et al. (1994), Liu et al. (2004)
Mycosporine-glycine-valine	Amino acid composition after alkaline hydrolysis	-	-	-	-	-	-	Karentz et al. (1991)

MAAs	Chemical assay	Optical Activity	UV-Vis Spectro-scopy	IR Spectro-scopy	X Ray Diffraction	MS Spectroscopy (ESI-MS/MS^n)	NMR Spectroscopy (^{1}H-NMR, ^{13}C-NMR)	References
Porphyra-334	Amino acid composition after acid hydrolysis Methylation	+	+	+	-	ESI-MS^n	^{1}H-NMR ^{13}C-NMR	Klisch et al. (2007), Takano et al. (1979), Whitehead et al. (2001)
Z-palythenic acid	Amino acid composition after hydrolysis (not detailed)	+	+	-	-	FD-MS ESI-MS^n	+	Kobayashi et al. (1981), Whitehead et al. (2001)
Usujirene	Amino acid composition after alkaline hydrolysis MAA residue after acid hydrolysis Catalytic hydrogenation	+	+	+	-	FD-MS (high resolution) FAB-MS (high resolution)	^{1}H-NMR ^{13}C-NMR	Nakayama et al. (1999), Sekikawa et al. (1986)
Palythene	MAA residue after acid hydrolysis Catalytic hydrogenation	+	+	+	+	ESI-MS^n	^{1}H-NMR ^{13}C-NMR	Vemura et al. (1980), Whitehead et al. (2001), Takano et al. (1978)
Euhalothece 362	Amino acid composition after acid hydrolysis	-	+	-	-	ESI-MS^n (high resolution)	+	Volkmann et al. (2006)
Dehydroxylusujirene	No	-	+	-	-	ESI-MS/MS	-	Zhang et al. (2007)
Palythine-threonine	Amino acid composition after alkaline hydrolysis Interchange H/D	-	+	-	-	ESI-MS (high resolution) ESI-MS^n	-	Carignan et al. (2009)
Palythine-threonine sulfate	No	-	-	-	-	ESI-MS (high resolution)	^{1}H-NMR ^{13}C-NMR	Wu et al. (1997)

In addition, the X-ray analysis showed that palythine and water molecules are connected by twelve types of hydrogen bonds, forming a three-dimensional hydrogen-bonded structure (Furusaki et al. 1980). Elucidation of MAAs has been achieved by chemical degradation, spectroscopic techniques (UV-spectra, IR-spectra, ESI-MSn; ^{1}H and ^{13}C-NMR, etc.) and X-ray analysis. However, in many cases the MAAs was isolated in small amounts and the structure characterization was only based from the molecular mass of the isolated compound and the amino acids yielded by their alkaline hydrolysis. Laser-flash photolysis experiments have lead to the characterization of the excited triplet state of the MAAs shinorine, porphyra-334 and palythine (Conde et al. 2000, 2004, 2007).

Data for the triplet-triplet energy transfer from 1-naphthalene-methanol to porphyra-334 was used to assess the quantum yield of the MAA triplet production in N_2-purged solutions amounting *ca.* 3% (Conde et al. 2004). This result accounts for the low probability of triplets to generate reactive intermediates which might start further chemical reactions, in agreement with the lack of radical detection reported by Shick et al. (Shick et al. 2000). The energy levels of the triplet states (ET) were estimated from the evaluation of the rate constants for the energy transfer processes with donors of variable triplet energies. Thus, ET for porphyra-334 in water solution is delimited by a value of <250 kJ·mol^{-1} whereas ET for palythine yields approximately 330 kJ·mol^{-1} (Conde et al. 2000, 2007). On this basis, porphyra-334 but not palythine may be able to effectively quench the excited state of thymine with a triplet energy in DNA estimated in 270 kJ·mol^{-1} (Bosca et al. 2006). This quenching reaction by MAAs has been proposed by Misonou et al. (2003) as a protection mechanism against UV-induced damage in the red alga *Porphyra yezoensis*, in addition to the simple filtering effect (Misonou et al. 2003).

Conclusion and Future Perspectives

Taxonomically diverse aquatic and terrestrial cyanobacteria have evolved the capacity to synthesize and accumulate a variety of MAAs as part of an overall strategy to diminish the direct and indirect damaging effects of environmental UVR. The biotechnological exploitation of cyanobacteria as a source of bioactive compounds for the cosmetic industry represents a promising area of research. This group of microorganisms produce UV absorbing/screening compounds MAAs that serve as natural photo protectants and can be suitable alternatives to currently used synthetic UV filters. Details of the pathway and the enzymes involved in the biosynthesis, transference, accumulation and transformations of more than 30 fully characterized MAAs and of the recently discovered compounds require further studies. Nevertheless, there is biochemical evidences to assume that the high diversity of MAAs present in cyanobacteria is principally derived from the synthesis and transformation of the so-called -primary MAAs. This genomics and proteomics study will definitely provide foundation for further studies (e.g., mutant development, promoter analysis) of these evolutionary important compounds which was so far completely clueless for workers interested in this field. Genome mining approach reveals the presence of a unique combination of two genes DHQS and O-MT in MAAs-synthesizing cyanobacteria which are missing in other non-MAA-synthesizing cyanobacteria. DHQS gene product together with thc O-MT gene product catalyzes the formation of deoxygadusol which is the core of all MAAs. Characterization of each MAA in cyanobacteria is of crucial importance to scientific progress in the field. However, the major difficulty has been the lack of commercially available standards and therefore, identification and quantification are most usually based on co-elution with secondary standards and/or comparison with published UV spectral data and HPLC retention times. Mass spectrometry coupled to chromatography, calibrated with certified references, can provides high sensitivity and selectivity, making these assays superior to the ones depending on conventional detection technologies. This study provides insight the genetics and proteomics of the genes involved in the MAAs biosynthesis and opens the field for molecular, bioinformatics and phylogenetic analysis of these evolutionary important photo protective compounds in cyanobacteria and other organisms.

ACKNOWLEDGMENTS

Haseen Ahmed acknowledges University Grants Comission, New Delhi, India, for providing senior research fellowship. Jainendra Pathak, Abha Pandey and PR Singh are thankful to CSIR, New Delhi, India, for providing financial assistance in the form of senior research fellowships.

REFERENCES

Aboulmagd, E., Sanio, F. B. O., Steinbuchel, A. (2001) Purification of *Synechocystis* sp. strain PCC6308 cyanophycin synthetase and its characterization with respect to substrate and primer specificity. *Appl Environ Microbiol* 67: 2176-2182.

Balskus, E. P., Walsh, C. T. (2010) The genetic and molecular basis for sunscreen biosynthesis in cyanobacteria. *Science* 329: 1653-1656.

Bandaranayake, W. M. (1998) Mycosporines: are they nature' s sunscreems? *Nat Prod Rep* 15: 159-172.

Bandaranayake, W. M., Bemis J. E., Bourne, D. J. (1996) Ultraviolet absorbing pigments from the marine sponge *Dysidea herbacea:* Isolation and structure of a new mycosporine. *Comp Biochem Physiol* 115: 281-286.

Behera, S., Ghanty, S., Ahmad, F., Santra, S., Banerjee, S. (2012) UV-Visible spectrophotometric method development and validation of assay of Paracetamol tablet formulation. *J Anal Bioanal Techniques* 3: 1-6.

Björn, L. O. (2007) Stratospheric ozone, ultraviolet radiation, and cryptogams. *Biol Conserv* 135: 326-333.

Böhm, G. A., Pfleiderer, W., Böger, P., Scherer, S. (1995) Structure of a novel oligosaccharide-mycosporine-amino acid ultraviolet A/B sunscreen pigment from the terrestrial cyanobacterium *Nostoc commune*. *J Biol Chem* 270: 9-17.

Bosca, F, Lhiaubet-Vallet, V., Cuquerella, M. C., Castell, J. V., Miranda, M. A. (2006) The triplet energy of thymine in DNA. *J Am Chem Soc* 128: 6318-6319.

Callone, A. I., Carignan, M., Montoya, N. G., Carreto, J. I. (2006) Biotransformation of mycosporine like amino acids (MAAs) in the toxic dinoflagellate *Alexandrium tamarense*. *J Photochem Photobiol Biol* 84: 204-212.

Cardozo, H. M., Guaratini, T., Barros, M. P., Falcao, V. R., Tonon, A. P., Lopes, N. P., Campos, S., Torres, M. A., Souza, A. O., Colepicolo, P., Pinto, E. (2007) Metabolites from algae with economic impact. *Comp Biochem Physiol* 146: 60-78.

Cardozo, K. H., Vessecchi, V. M., Carvalho, V. M., Pinto, E., Gates, P. J., Colepicolo, P., Galembeck, S. E., Lopes, N. P. (2008) A theoretical and MAAs spectrometry study of the fragmentation of mycosporine-like amino acids. *Int J Mass Spectrom* 273: 11-19.

Carignan, M. O., Cardozo, K. H. M., Oliveira-Silva, D., Colepicolo, P., Carreto, J. I. (2009) Palythine-treonine, a major novel mycosporine-like amino acid (MAA) isolated from the hermatypic coral *Pocillopora capitata*. *J Photochem Photobiol B* 94: 191-200.

Carreto, J. I., Carignan, M. O. (2011) Mycosporine-like amino acids: relevant secondary metabolites. Chemical and ecological aspects. *Mar Drugs* 9: 387-446.

Carreto, J. I., Carignan, M. O., Daleo, G., De Marco, S. G. (1990) Occurrence of mycosporine-like amino acids in the red-tide dinoflagellate *Alexandrium excavatum*: UV photo protective compounds? *J Plankton Res* 12: 909-921.

Carreto, J. I., Carignan, M. O., Montoya, N. G. (2005) A high-resolution reverse-phase liquid chromatography method for the analysis of mycosporine-like amino acids (MAAs) in marine organisms. *Mar Biol* 146: 237-252.

Carreto, J. I., Roy, S., Whitehead, K., Llewellyn, C., Carignan, M. O. (2011) UV-absorbing -pigments: Mycosporine-like Amino Acids. In: *Phytoplankton Pigments in Oceanography* (Eds. Roy S, Egeland ES, Llewellyn C, Johnsen G, Wright S), Cambridge University Press: Cambridge, UK, pp. 412-444.

Carroll, A. K., Shick J. M. (1996) Dietary accumulation of UV-absorbing mycosporine-like amino acids (MAAs) by the green sea urchin (*Strongylocentrotus droebachiensis*). *Mar Biol* 124: 561-569.

Castenholz, R. W., Garcia-Pichel, F. (2000) Cyanobacterial responses to UV-radiation. In: *The ecology of cyanobacteria* (Ed. Whitton BA), Kluwer Academic Publishers, Dordrecht, Netherlands. pp. 591-611.

Conde, F. R, Churio, M. S., Previtali, C. M. (2000) The photoprotector mechanism of mycosporine-like amino acids. Excited state properties and photostability of porphyra-334 in aqueous solution. *J Photochem Photobiol B* 56: 139-144.

Conde, F. R., Churio, M. S., Previtali, C. M. (2004) The deactivation pathways of the excited-states of the mycosporine-like amino acids shinorine and porphyra-334 in aqueous solution. *Photochem Photobiol Sci* 3: 960-967.

Conde, F. R., Churio, M. S., Previtali, C. M. (2007) Experimental study of excited-state properties and photostability of the mycosporine-like amino acid palythine in water solution. *Photochem Photobiol Sci* 6: 669-674.

D'Agostino, P. M., Javalkote, V. S., Mazmouz, R., Pickford, R., Puranik, P. R., Neilan, B. A. (2016) Comparative profiling and discovery of novel glycosylated mycosporine-like amino acids in two strains of the cyanobacterium *Scytonema* cf. *crispum. Appl Environ Microbiol* 82: 5951-5959.

Daniel, S., Cornelia, S., Fred, Z. (2004) UV-A sunscreen from red algae for protection against premature skin aging. *Cosmet Toilet Manufact World* 129: 139-143.

De la Coba, F., Aguilera, J., Figueroa, F. L., de Gálvez, M. V., Herrera, E. (2009) Antioxidant activity of mycosporine-like amino acids isolated from three red macroalgae and one marine lichen. *J Appl Phycol* 21: 161-169.

Dionisio-Sese, M. L., Ishikura, M., Maruyama, T., Miyachi, S. (1997) UV-absorbing substances in the tunic of a colonial ascidian protect its symbiont, *Prochloron* sp., from damage by UV-B radiation. *Mar Biol* 128: 455-461.

Downes, A., Elfick, A. (2010) Raman spectroscopy and related techniques in biomedicine. *Sensors* 10: 1871-1889.

Dunlap, W. C., Chalker, B. E. (1986) Identification and quantitation of near-UV absorbing compounds (S-320) in a *hermatypic scleractinian. Coral Reefs* 5: 155-159.

Dunlap, W. C., Shick, J. M. (1998) Ultraviolet radiation-absorbing mycosporine-like amino acids in coral reef organisms: a biochemical and environmental perspective. *J Phycol* 34: 418-430.

Dunlap, W. C., Yamamoto, Y. (1995) Small-molecule antioxidants in marine organisms: antioxidant activity of mycosporine-glycine. *Comp Biochem Physiol* 112: 105-114.

Dunn, W. B., Bailey, N. J. C., Johnson, H. E. (2005) Measuring the metabolome: current analytical technologies. *Analyst* 130: 606-625.

Favre-Bonvin, J., Favre-Bonvin, J., Bernillon, J., Salin, N., Arpin, N. (1987) Biosynthesis of mycosporines: Mycosporine glutaminol in *Trichothecium roseum. Phytochemistry* 26: 2509-2514.

Finking, R., Marahiel, M. A. (2004) Biosynthesis of nonribosomal peptides. *Annu Rev Microbiol* 58: 453-488.

Furusaki, A., Matsumoto, T., Tsujino, I., Sekikawa, I. (1980) The crystal and molecular structure of palythine, trihydrate. *Bull Chem Soc Jpn* 53: 319-323.

Gao, K., Yu, H., Brown, M. T. (2007) Solar PAR and UV radiation affects the physiology and morphology of the cyanobacterium *Anabaena* sp. PCC 7120. *J Photochem Photobiol B* 89: 117-124.

Gao, Q., Garcia-Pichel, F. (2011) An ATP-grasp ligase involved in the last biosynthetic step of the imino-mycosporine shinorine in *Nostoc punctiforme* ATCC 29133. *J bacteriol* 193: 5923-5928.

He, Y. Y., Klisch, M., Häder, D-P. (2002) Adaptation of cyanobacteria to UV-B stress correlated with oxidative stress and oxidative damage. *Photochem Photobiol* 76: 188-196.

Ito, S., Hirata, Y. (1977) Isolation and structure of a mycosporine from the zoanthidian *Palythoa tuberculosa*. *Tetrahedron Lett* 28: 2429-2430.

Jain, S., Prajapat, G., Abrar, M., Ledwani, L., Singh, A., Agrawal, A. (2017) Cyanobacteria as efficient producers of mycosporine-like amino acids. *J basic microbiol* 57: 715-727.

Karentz D (1994) Ultraviolet tolerance mechanisms in Antarctic marine organisms. *Ultraviolet radiation in Antarctica: measurements and biological effects*. 62: 93-110.

Karentz, D., McEuen, F. S., Land, M. C., Dunlap, W. C. (1991) Survey of mycosporine-like amino acid compounds in Antarctic marine organisms: potential protection from ultraviolet exposure. *Mar Biol* 108: 157-166.

Karsten, U. (2002) Effects of salinity and ultraviolet radiation on the concentration of mycosporine-like amino acids in various isolates of the benthic cyanobacterium *Microcoleus chthonoplastes*. *Phycological Res* 50: 129-134.

Karsten, U., García Pichel, F. (1996) Carotenoids and mycosporine-like amino acids compounds in members of the genus *Microcoleus* (Cyanobacteria): A chemosystematic study. *System Appl Microbiol* 19: 285-294.

Katoch, M., Mazmouz, R., Chau, R., Pearson, L. A. (2016) Heterologous production of cyanobacterial mycosporine-like amino acids mycosporine-ornithine and mycosporine-lysine in *Escherichia coli*. *Appl Environ Microbiol* 82: 6167-6173.

Kedar, L., Kashman, Y., Oren, A. (2002) Mycosporine-2-glycine is the major mycosporine-like amino acid in a unicellular cyanobacterium (*Euhalothece* sp.) isolated from a gypsum crust in a hypersaline saltern pond. *FEMS Microbiol Lett*. 208: 233-237.

Klisch, M., Richter, P., Puchta, R., Häder D-P, Bauer, W. (2007) The Stereostructure of Porphyra-334: An Experimental and Calculational NMR Investigation. Evidence for an Efficient 'Proton Sponge'. *Helvetica Chim Acta* 90: 488-511.

Kobayashi, J., Nakamura, H., Hirata, Y. (1981) Isolation and structure of a UV-absorbing substance 337 from the ascidian *Halocynthia roretzi*. *Tetrahedron Lett* 22: 3001-3002.

Korbee, N., Figueroa, F. L., Aguilera, J. (2006) Accumulation of mycosporine-like amino acids (MAAs): biosynthesis, photocontrol and ecophysiological functions. *Revista Chilena de Historia Natural* 79: 119-132.

Lesser, M. P., Stochaj, W. R., Tapley, D. W., Shick, J. M. (1990) Bleaching in coral reef anthozoans: Effects of irradiance, ultraviolet radiation and temperature on the activities of protective enzymes against active oxygen. *Coral Reefs* 8: 225-232.

Liu, Z., Häder D-P, Sommaruga, R. (2004) Occurrence of mycosporine-like amino acids (MAAs) in the bloom-forming cyanobacterium *Microcystis aeruginosa*. *J Plankton Res* 26: 963-966.

McClintock, J. B., Karentz, D. (1997) Mycosporine-like amino acids in 38 species of subtidal marine organisms from McMurdo Sound, Antarctica. *Antarct Sci* 9: 392-398.

Michaelian, K. (2011) Thermodynamic dissipation theory for the origin of life. *Earth Syst Dynam* 2: 37-51.

Michaelian, K. (2012) HESS Opinions "Biological catalysis of the hydrological cycle: life's thermodynamic function". *Hydrol Earth Syst Sci* 16: 2629-2645.

Michaelian, K. (2016) *Thermodynamic Dissipation Theory of the Origin and Evolution of Life: Salient characteristics of RNA and DNA and other fundamental molecules suggest an origin of life driven by UV-C light*, Self-published. Printed by Create Space, Mexico City.

Misonou, T., Saitoh, J., Oshiba, S., Tokimoto, Y., Maegawa, M., Inoue, Y., Hori, H., Sakurai, T. (2003) UV-absorbing substance in the red alga *Porphyra yezoensis* (Bangiales, Rhodophyta) block thymine photodimer production. *Mar Biotechnol* 5: 194-200.

Moco, S., Bino, R. J., De Vos, R. C. H., Vervoort, J. (2007) Metabolomics technologies and metabolite identification. *Trends Anal Chem* 26: 855-866.

Moisan, T. A., Mitchell, B. G. (2001) UV absorption by mycosporine-like amino acids in *Phaeocystis antarctica* Karsten induced by photosynthetically available radiation. *Mar Biol* 138: 217-227.

Nakamura, H., Kobayashi, J., Hirata, Y. (1981) Isolation and structure of a 330 nm UV-absorbing substance, asterina-330 from the starfish *Asterina pectinifera*. *Chem Lett* 28: 1413-1414.

Nakayama, R., Tamura, Y., Kikuzaki, H., Nakatani, N. (1999) Antioxidant effect of the constituents of susabinori (*Porphyra yezoensis*). *JAOCS* 76: 649-653.

Nazifi, E., Wada, N., Yamaba., M., Asano, T., Nishiuchi, T., Matsugo, S., Sakamoto, T. (2013) Glycosylated porphyra-334 and palythine-threonine from the terrestrial cyanobacterium *Nostoc commune*. *Mar drugs* 11: 3124-3154.

Oren, A., Gunde-Cimerman, N. (2007) Mycosporines and mycosporine-like amino acids: UV protectants or multipurpose secondary metabolites? *FEMS Microbiol Lett* 269: 1-10.

Oyamada, C., Kaneniwa, M., Ebitani, K., Murata, M., Ishihara, K (2008) Mycosporine-like amino acids extracted from Scallop (*Patinopecten yessoensis*) ovaries: UV protection and growth stimulation activities on human cells. *Mar Biotech* 10: 141-150.

Pathak, J., Ahmed, H., P Sinha, R. (2017b) Metabolomic profiling of cyanobacterial UV-protective compounds. *Curr Metabolomics* 5: 138-163.

Pathak, J., Sonker, A. S., Richa, Rajneesh, Kannaujiya, V. K., Singh, V., Ahmed, H., Sinha, R. P, (2017a) Screening and partial purification of photo protective pigment scytonemin from cyanobacterial crusts dwelling on the historical monuments in and around Varanasi, India. *Microbiol Res (Pavia)* 8: 6559.

Piiparinen, J., Enberg, S., Rintala, J-M., Sommaruga, R., Majaneva, M., Autio, R., Vahatalo, A. V. (2015) The contribution of mycosporine-like amino acids, chromophoric dissolved organic matter and particles to the UV protection of sea-ice organisms in the Baltic Sea. *Photochem Photobiol Sci* 14: 1025-1038.

Pope, M. A., Spence, E., Seralvo, V., Gacesa, R., Heidelberger, S., Weston, A. J., Dunlap, W. C., Shick, J. M., Long, P. F. (2015) O-Methyltransferase Is Shared between the Pentose Phosphate and Shikimate Pathways and Is Essential for Mycosporine-Like Amino Acid Biosynthesis in *Anabaena variabilis* ATCC 29413. *Chem Bio Chem* 16: 320-327.

Portwich, A., Garcia-Pichel, F. (2003) Biosynthetic pathway of mycosporines (mycosporine-like amino acids) in the cyanobacterium *Chlorogloeopsis* sp. strain PCC 6912. *Phycologia* 42: 384-392.

Rastogi, R. P., Incharoensakdi, A. (2014a) Analysis of UV-absorbing photoprotectant mycosporine-like amino acid (MAA) in the cyanobacterium *Arthrospira* sp. CU2556. *Photochem Photobiol Sci* 13: 1016-1024.

Rastogi, R. P., Incharoensakdi, A. (2014b) UV radiation-induced biosynthesis, stability and antioxidant activity of mycosporine-like amino acids (MAAs) in a unicellular cyanobacterium *Gloeocapsa* sp. CU2556. *J Photochem Photobiol B Biol* 130: 287-292.

Rastogi, R. P, Incharoensakdi, A., Madamwar D. (2014) Responses of a rice-field cyanobacterium *Anabaena siamensis* TISTR-8012 upon exposure to PAR and UV radiation. *J Plant hysiol* 171: 1545-1553.

Rastogi, R. P., Sonani, R. R., Madamwar, D., Incharoensakdi, A. (2016) Characterization and antioxidant functions of mycosporine-like amino acids in the cyanobacterium *Nostoc* sp. R76DM. *Algal Res* 16: 110-118.

Rosic, N. N. (2012) Phylogenetic analysis of genes involved in mycosporine-like amino acid biosynthesis in symbiotic dinoflagellates. *Appl Microbiol Biotechnol* 94: 29-37.

Rosic N. N., Dove, S. (2011) Mycosporine-like amino acids from coral dinoflagellates. *Appl Environ Microbiol* 77: 8478-8486.

Rozema, J., Björn L. O., Bornman, J. F., Gaberščik, A., Häder D-P, Trošt, T., Germ, M., Klisch, M., Gröniger, A., Sinha, R. P., Lebert, M., He, Y. Y., Buffoni-Hall, R., de Bakke, N. V. J., van de Staaij, J., Meijkamp, B. B. (2002) The role of UV-B radiation in aquatic and terrestrial ecosystems- an experimental and functional analysis of the evolution of UV-absorbing compounds. *J Photochem Photobiol B* 66: 2-12.

Sekikawa, I., Kubota, C., Hiraoki, T., Tsujino, I. (1986) Isolation and structure of a 357 nm UV absorbing substance, usujirene, from the red alga *Palmaria palmata* (L.) O. Kuntze *Jpn J Phycol* 34: 185-188.

Shibata, K. (1969) Pigments and a UV-absorbing substance in corals and a blue-green alga living in the Great Barrier Reef. *Plant Cell Physiol* 10: 325-335.

Shick, J. M., Dunlap, W. C. (2002) Mycosporine-like amino acids and related gadusols: Biosynthesis, accumulation, and UV-protective functions in aquatic organisms. *Annu Rev Physiol* 64: 223-262.

Shick J. M., Dunlap W. C., Buettner, R. (2000) Ultraviolet (UV) protection in marine organisms II. Biosynthesis, accumulation, and sunscreening function of mycosporine-like amino acids. In: *Free Radicals in Chemistry, Biology and Medicine* (Eds. Yoshikawa T, Toyokuni S, Yamamoto Y, Naito Y), OICA International: London, UK, pp. 215-228.

Shick, J. M., Romaine-Lioud, S., Ferrier-Pages, C., Gattuso, J. P. (1999) Ultraviolet-B radiation stimulated shikimate pathway dependent accumulation of mycosporine-like amino acids in the coral *Stylophora pistillata* despite decrease in its population of symbiotic dinoflagellates. *Limnol Oceanogr* 44: 1667-1682.

Singh, J. S., Kumar, A., Rai, A. N., Singh, D. P. (2016) Cyanobacteria: a precious bio-resource in agriculture, ecosystem, and environmental sustainability. *Front Microbiol* 7: 529.

Singh, S. P., Häder, D-P., Sinha, R. P. (2012) Bioinformatics evidence for the transfer of mycosporine-like amino acid core (4-deoxygadusol) synthesizing gene from cyanobacteria to dinoagellates and an attempt to mutate the same gene (YP 324358) in *Anabaena variabilis* PCC 7937. *Gene* 500: 155-163.

Singh, S. P., Klisch M., Sinha, R. P., Häder, D-P. (2010) Genome mining of mycosporine-like amino acid (MAA) synthesizing and non-synthesizing cyanobacteria: a bioinformatics study. *Genomics* 95: 120-128.

Singh, SP, Klisch, M., Sinha, R. P., Häder, D-P. (2010b) Sulfur deficiency changes mycosporine-like amino acid (MAA) composition of *Anabaena variabilis* PCC 7937: a possible role of sulfur in MAA bioconversion. *Photochem Photobiol* 86: 862-870.

Singh, S. P., Sinha, R. P., Klisch M., Häder, D. P. (2008) Mycosporine-like amino acids (MAAs) profile of a rice-field cyanobacterium *Anabaena doliolum* as influenced by PAR and UVR. *Planta* 229: 225-33.

Sinha, R. P., Ambasht, N. K., Sinha, J. P., Klisch, M., Häder, D-P. (2003) UV-B-induced synthesis of mycosporine-like amino acids in three strains of *Nodularia* (cyanobacteria). *J Photochem Photobiol B* 71: 51-58.

Sinha, R. P., Ambasht N. K., Sinha J. P., Klisch M., Häder, D-P (2003) UV-B-induced synthesis of mycosporine-like amino acids in three strains of *Nodularia* (cyanobacteria). *J Photochem Photobiol B* 71: 51-58.

Sinha R. P., Häder, D-P. (2008) UV-protectants in cyanobacteria. *Plant Sci* 174: 278-289.

Sinha, R. P., Klisch, M., Gröniger, A., Häder, D-P (1998) Ultraviolet absorbing/screening substances in cyanobacteria, phytoplankton and macroalgae. *Photochem Photobiol* 47: 83-94.

Sinha, R. P., Kumari, S., Rastogi, R. P. (2008) Impacts of ultraviolet-B radiation on cyanobacteria Photo protection and repair. *J Sci Res* 52: 125-142.

Sinha, R. P., Singh, S. P., Häder, D-P (2007) Database on mycosporines and mycosporine-like amino acids (MAAs) in fungi, cyanobacteria, macroalgae, phytoplankton and animals. *J Photochem Photobiol B* 89: 29-35.

Sommaruga, R., Whitehead, K., Shick, J. M., Lobban, C. S. (2006) Mycosporine-like amino acids in the zooxanthella-ciliate symbiosis *Maristentor dinoferus*. *Protist* 157: 185-191.

Spence, E., Dunlap, W. C., Shick, J. M., Long, P. F. (2012) Redundant pathways of sunscreen biosynthesis in a cyanobacterium. *Chem biochem* 13: 531-533.

Starcevic, A., Akthar, S., Dunlap, W. C., Shick, J. M., Hranueli, D., Cullum, J., Long, P. F. (2008) Enzymes of the shikimic acid pathway encoded in the genome of a basal metazoan, *Nematostella vectensis*, have microbial origins. *Proc Natl Acad Sci USA* 105: 2533-2537.

Stochaj, W. R., Dunlap, W. C., Shick, J. M. (1994) Two new UV-absorbing mycosporine-like amino acids from the sea anemone *Anthopleura elegantissima* and the effects of zooxanthellae and spectral irradiance on chemical composition and content. *Mar Biol* 118: 149-156.

Subramanian, A., Carpenter, E. J., Karentz, D., Falkowski, P. G. (1999) Bio-optical properties of the marine diazotrophic cyanobacteria *Trichodesmium* spp. I. Absorption and photosynthetic action spectra. *Limnol Oceanogr* 44: 608-617.

Takano, S., Uemura, D., Hirata, Y. (1978) Isolation and structure of two new amino acids, palythinol and palythene, from the zoanthid *Palythoa tuberculosa*. *Tetrahedron Lett* 49: 4909-4912.

Takano, S., Uemura, D., Hirata, Y. (1979) Isolation and structure of a 334 nm UV absorbing substance, porphyra-334 from the red algae *Porphyra tenera* Kjellman. *Chem Lett* 26: 419-420.

Teai, T., Raharivelomanana, P., Bianchini, J. P., Faura, R., Martín, P. M. V., Cambon, A. (1997) Structure de deux nouvelles imino-mycosporines isolées de *Pocillopora eydouxy*. *Tetrahedron Lett* 38: 5799-5800.

Tsujino, I., Yabe, K., Sekikawa, I. (1980) Isolation and structure of an new amino acid, shinorine from the red alga *Chondrus yendoi* Yamada et Mikami. *Bot Mar* 23: 65-68.

Tsujino, I., Yabe, K., Sekikawa, I., Hamanaka, N. (1978) Isolation and structure of a mycosporine from the red alga *Chondrus yendoi*. *Tetrahedron Lett* 16: 1401-1402.

Uemura, D., Katayama, C., Wada, A., Hirata, Y. (1980) Crystal and molecular structure of palythene possessing a novel 360 nm chromophore. *Chem Lett* 9: 755-756.

Uemura, D., Katayama, C., Wada, A., Hirata, Y. (1980) Crystal and molecular structure of palythene possessing a novel 360 nm chromophore. *Chem Lett* 9: 755-756.

Volkmann, M., Gorbushina, A. A. (2006) A broadly applicable method for extraction and characterization of mycosporines and mycosporine-like amino acids of terrestrial, marine and freshwater origin. *Microbiol Lett* 255: 286-295.

Volkmann, M., Gorbushina, A. A., Kedar, L., Oren, A. (2006) Structure of euhalothece-362, a novel red-shifted mycosporine-like amino acid, from a halophilic cyanobacterium (*Euhalothece* sp). *Microbiol Lett* 258: 50-54.

Wada, N., Sakamoto, T., Matsugo, S. (2015) Mycosporine-like amino acids and their derivatives as natural antioxidants. *Antioxidants* 4: 603-646.

Waditee-Sirisattha, R., Kageyama, H., Sopun, W., Tanaka, Y. (2014) Identification and upregulation of biosynthetic genes required for accumulation of mycosporine-2-glycine under salt stress conditions in the halotolerant cyanobacterium *Aphanothece halophytica*. *Appl Environ Microbiol* 80: 1763-1769.

Waller, R. F., Slamovits, C. H., Keeling, P. J. (2006) Lateral gene transfer of a multigene region from cyanobacteria to dinoagellates resulting in a novel plastid-targeted fusion protein. *Mol Biol Evol* 23: 1437-1443.

Whitehead, K., Hedges, J. I. (2002) Analysis of mycosporine-like amino acids in plankton by liquid chromatography electrospray ionization mass spectrometry. *Mar Chem* 80: 27-39.

Whitehead, K., Karentz, D., Hedges, J. I. (2001) Mycosporine-like amino acids (MAAs) in phytoplankton, a herbivorous pteropod (*Limacina helicina*), and its pteropod predator (*Clione antarctica*) in Mc Murdo Bay, Antarctica. *Mar Biol* 139: 1013-1019.

Cardozo, K. H. M., Carvalho, V. M., Pinto, E. (2006) Fragmentation of mycosporine-like amino acids by hydrogen/deuterium exchange and electrospray ionisation tandem mass spectrometry. *Rapid Comm Mass Spectrom* 20: 253-258.

Wu, J. J., Chalker, B. E., Rideout, J. A. (1997) Two new UV-absorbing compounds from *Stylophora pistillata*: Sulfate esters of mycosporine-like amino acids. *Tetrahedron Lett* 38: 2525-2526.

Wu, J. J., Rideout, J. A., Chalker, B. E. (1995) Isolation and structure of a novel mycosporine-like amino acid from the reef-building corals *Pocillopora damicornis* and *Stylophora pistillata*. *Tetrahedron Lett* 36: 5255-5256.

Wulff, A., Mohlin, M., Sundback, K. (2007) Intraspecific variation in the response of the cyanobacterium *Nodularia spumigena* to moderate UV-B radiation. *Harmful Algae* 6: 388-399.

Yoshiki, M., Tsuge, K., Tsuruta, Y., Yoshimura, T., Koganemaru, K., Sumi, T., Matsui, T., Matsumoto, K. (2009) Production of new antioxidant compound from mycosporine-like amino acid, porphyra-334 by heat treatment. *Food Chem* 113: 1127-1132.

Zhang, L., Li, L., Wu, Q. (2007) Protective effects of mycosporine-like amino acids of *Synechocystis* sp. PCC 6803 and their partial characterization. *J Photochem Photobiol B* 86: 240-245.

In: Innovations in Life Science Research ISBN: 978-1-53615-868-7
Editors: R. P. Sinha, S. Pandey-Rai and N. Ghoshal

Chapter 4

TARGETED GENOME EDITING TOOLS IN PLANTS

Sushma Mishra*

Department of Botany, Dayalbagh Educational Institute, Dayalbagh, Agra, Uttar Pradesh, India

ABSTRACT

In the past few decades, there have been several advances in plant transformation methods enabling stable integration of foreign DNA into plant genome. However, the random integration of transgenes, leading to position effects and unintended gene expression changes in genome, has been cited as the major drawback. A significant breakthrough in plant research has been the advent of genome editing technology, which aims at making site-specific, precise alterations in genome. This technology involves creation of double strand breaks at predestined sites of genome by sequence-specific nucleases, followed by its repair either by homologous or non-homologous recombination, resulting in targeted editing of genome. Non-homologous end joining, the prevalent mechanism of DNA repair in higher eukaryotes, simply involves joining of the broken ends of chromosomes, concomitant with random insertion or deletion of nucleotides. This error-prone mechanism of DNA repair has been exploited to generate gene knockouts in plants. In contrast, the homologous recombination mechanism of DNA repair is fairly accurate, as it uses a homologous sequence as template for regeneration of missing DNA sequence at the break point. The drawback of inherent low frequency of homologous recombination in plant cells could be alleviated by the introduction of double strand breaks by sequence-specific nucleases. This book chapter presents an overview of different types of engineered nucleases available for genome editing, followed by some of the successful studies done in the field of gene targeting in plants, and finally the regulatory status of genome edited plants.

Keywords: CRISPR/Cas9, engineered nucleases, genome editing, TALENs, zinc finger nucleases

---

* Corresponding Author's Email: sushmamishra87@gmail.com.

INTRODUCTION

Genetic modification in plants has primarily focused on two objectives: first, dissecting out gene function; and second, introduction of desired traits into plants. The availability of enormous amount of sequence data through high throughput sequencing technologies, has presented the herculean task of functional characterization of novel genes. Investigation of gene function generally involves creation of loss-of-function or gain-of-function mutants. Some of the commonly used approaches for functional genomics include T-DNA and transposon mutagenesis, activation tagging, RNA interference etc. In addition to this, plant biologists are also interested in introduction of desired/novel traits into plants like conferring stress tolerance, biofortification, production of edible vaccines, phytoremediation and many more (Salt et al. 1995; Bollineni et al. 2014; Mishra et al. 2019). Among the various methods used for crop improvement like classical plant breeding, mutation breeding, protoplast fusion and polyploidy; the genetic transformation of plants by *Agrobacterium*-mediated transformation has been the most commonly used approach. However, it suffers from a major limitation of random integration of foreign DNA into plant genome, resulting into unintended gene inactivation or alteration in expression of host genes. To circumvent this problem, researchers are gradually switching over to one of the major breakthrough technologies of recent times, called genome editing or genome engineering, which aims at making specific, targeted changes in genome.

Genome Editing

Genome editing is a process of making specific, precise, targeted modifications in genome using engineered nucleases, also called "molecular scissors". These modifications include *gene disruption* - the targeted induction of minor insertions and deletions; *gene correction* - the introduction of discrete base substitutions specified by a homologous donor DNA construct; and targeted *gene addition* - the transfer of entire transgenes into a native genomic locus (Carroll 2011; Curtin et al. 2012; Zhang et al. 2013). This technology is based on a conserved pathway in cells, in which a double strand break in DNA triggers cellular repair enzymes to initiate the repair either by error-prone Non-Homologous End Joining (NHEJ), or by high fidelity homology-directed repair (HDR). In the earliest example of genome editing in higher plants, oligonucle-otide-mediated mutagenesis was used to cause site-specific alterations in the genome. This involved the use of chemically synthesized oligonucleotides with the specific base replacement or addition caused by endogenous DNA-repair enzymes (Beetham et al.1999). Oligonucleotide-mediated mutagenesis differs from genome editing approaches in that the former does not deliver a nuclease to the site of action.

Genome Editing by NHEJ

NHEJ is the principal means by which broken chromosomes are repaired in somatic cells of multicellular eukaryotes. This DNA double strand break (DSB) repair mechanism does not require a template, and simply involves joining the broken ends inaccurately, often resulting in deletions, insertions and substitutions at the break site (Puchta 1998). If the random insertion or deletion occurs in the coding sequence, it might result in frameshift mutations, thus creating a *gene knock out*. Also, if insertion or deletion mutations occur in promoter regions, it can disrupt regulatory sequences and hence, can alter gene expression. Therefore, although error-prone, this method has been widely used in mammalian cells and few plants (Baltes and Voytas 2015).

Genome Editing by HR

In the other method of DNA repair mechanism which involves homologous recombination (HR), the information is copied from a DNA template, which could either be a homologous chromosome, sister chromatid or user-supplied DNA with homology to break site. This method of DNA integration commonly occurs during meiosis and mating-type switching in yeast (Aylon and Kupiec 2004). The user supplied DNA template can harbor SNPs or one or more transgenes, which could be incorporated at the break site after homologous recombination (Mohanta et al. 2017). Although genome editing is applicable to multiple organisms like *Drosophila melanogaster*, *Arabidopsis thaliana*, *Ceanorhabditis elegans*, zebrafish, rats, humans and many more (Urnov et al. 2010; Miller et al. 2011; Wood et al. 2011; Puchta and Fauser 2013), the targeted genome editing in crop plants would be the focus of this chapter.

GENE TARGETING IN PLANTS

Gene targeting (GT) is defined as the process that uses HR pathway to introduce changes in DNA sequence within genomes (Baltes and Voytas 2015; Rinaldo and Ayliffe 2015). Although the first report of gene targeting in plants came just 5 years after the first transgenic plant was generated; this method of gene transfer was not preferred due to very low frequency of HR (typically around 10^{-5}) in plants (Paszkowski et al. 1988). In order to increase the transformation efficiency of GT in plants, the components of HR pathway from bacteria and yeast have been overexpressed in plants. However, approaches involving over-expression of *RecA* gene of *Escherichia coli* in transgenic tobacco, and *RAD54* gene of *Saccharomyces cerevisiae* in transgenic *Arabidopsis*, have met with little success (Reiss et al. 2000). In another approach, Terada et al. (2002) reported an efficient GT method in rice involving HR using positive/negative selection marker. The negative selection marker was used to select against random NHEJ events, so that these are eliminated from the transgenic pool. The researchers obtained specific substitution mutations of the targeted gene, *Waxy*, by precise HR in about 1% of the survivors. However, the major breakthrough in the field of genome editing occurred when it was reported that the introduction of DSB at the target site increases the frequency of GT by a magnitude of 2-3 fold (Jasin 1996; Baltes and Voytas 2015).

Engineered Nucleases for Gene Targeting

Targeted genome editing requires a set of custom designed nucleases/engineered nucleases for the introduction of DSBs with high precision at pre-determined target locus (Puchta and Fauser 2013). Some of the commonly used site-specific nucleases for genome engineering include Homing Endonucleases (HE), Zinc Finger Nucleases (ZFNs), Transcription Activator-like Effector Nucleases (TALENs) and Clustered Regularly Interspaced Short Palindromic Repeats/ CRISPR-associated protein (CRISPR/Cas 9).

The prerequisites for performing genome editing are availability of genome sequence for designing nucleases specific for the target DNA, and efficient transformation system for delivery of engineered nucleases into plant genome. Some of the commonly used delivery methods of engineered nucleases include *Agrobacterium*-mediated transformation, Transfection, PEG-mediated uptake in protoplast, and viral vector-mediated transfer. In addition, sequence-specific nucleases (from small mega nucleases to large TALENs) have been shown to be delivered to plant cells as purified protein (Luo et al. 2015). However, compared to transformation of DNA by other methods, delivery of purified nucleases results in lower mutagenesis frequencies.

Homing Endonucleases

Homing Endonucleases (HE) are enzymes that have long DNA recognition sequences, typically 18–30 base pairs, and generate double-strand DNA breaks (Figure 1). HE are also called mega nucleases due to their lengthy recognition sites. Unlike restriction endonucleases that are very specific for their short, symmetrical sequences, mega nucleases tolerate some sequence degeneracy in their recognition sequence (Belfort and Bonocora 2014). Despite their name as mega nucleases, these are small proteins (less than 300 amino acids) encoded within the introns (non coding sequences of mRNA) and inteins (non coding portion of polypeptide that is spliced out after translation) of bacteria and bacteriophages (Stoddard 2005) These enzymes are involved in mobilizing introns/inteins to an intron less/intein less allele through the introduction of DSB at the site of intron/intein insertion (Jasin 1996). These were the first class of sequence-specific nucleases to be used in genome editing in plants (Stephens et al. 1997; Seligman et al. 2002). Some of the commonly used mega nucleases in plants are *I-SceI* and *I-CreI*.

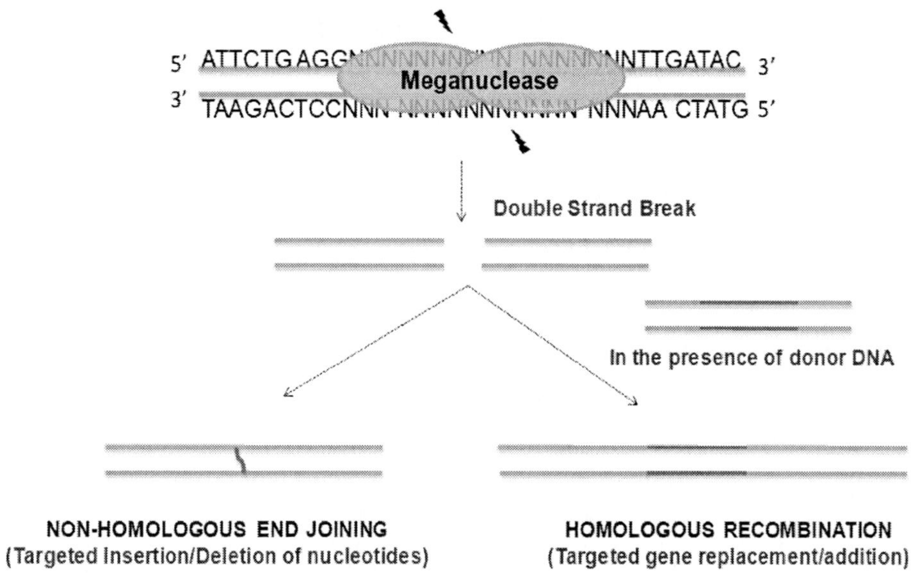

Figure 1. Schematic representation of Genome Editing in plants by Hybrid Mega nucleases/Homing Endonucleases. Meganucleaes have large, asymmetric recognition sites (12-40 bp); their catalytic domain is not very clearly demarcated from its DNA binding domain. The donor template DNA, containing stretches of sequences homologous to that flanking the target DNA site, may contain transgene(s) or modified gene sequence to bring about gene addition or replacement, respectively.

I-SceI is a HE encoded by mitochondrial genome of *Saccharomyces cerevisiae* that recognizes an 18 bp site. Siebert and Puchta (2002) reported the elimination of a negative selectable marker gene, *codA*, flanked by *I-SceI* cleavage sites, from tobacco plant genome by *Agrobacterium*-mediated transient expression of *I-SceI*. The resulting DSB and consecutive repair resulted in loss of marker gene function; the cells become selectable by their resistance to 5-fluorocytosine. Hence, this study highlighted the role of DSB-mediated genome editing in obtaining marker-free transgenics. In another set of experiments, *I-CreI*, the HE encoded by chloroplast genome of *Chlamydomonas reinhardtii*, was engineered to recognise a 22 bp sequence in the maize *liguless1* gene (Gao et al. 2010). The endonuclease was delivered to immature embryos of maize by *Agrobacterium*-mediated transformation. This system produced targeted mutations at the *liguless1* locus in about 3% of the transformants.

An advantage of mega nucleases is their small size (about 165 amino acids), which makes them amenable to plant transformation by a number of delivery methods, including direct protein delivery (Luo et al. 2015). However, HE have not been widely used in genome editing mainly due to three reasons: (i) the lack of modular DNA-binding domain architecture (as found in other engineered nucleases like ZFNs and TALE proteins) and unclear demarcation of DNA binding and catalytic domains that make their protein engineering difficult (Stoddard 2011); (ii) they can tolerate some sequence degeneracy in their target site, which increases the frequency of off-target binding (Argast et al. 1998);

(iii) Their long recognition sites had to be inserted in the genome before they could be used to effect recombination, thereby limiting their utility as nucleases in genome editing.

Zinc Finger Nucleases (ZFNs)

This class of engineered nucleases are derived from zinc finger proteins, the class of transcription factors commonly found in eukaryotes (Takatsuji 1998). ZFNs are composed of two functional domains: a DNA binding domain, typically composed of 3-6 individual zinc finger repeats, each recognizing a set of 3 nucleotides; and a DNA cleaving domain, composed of endonuclease domain of *Fok*I, a type IIS restriction endonuclease (Carroll 2011; Mohanta et al. 2017). When the DNA binding and DNA cleaving domains are fused together, a highly specific pair of "genomic scissors" are created (Figure 2). A tandem array of three to six ZFs corresponds to target sites of 9–18 bp which is sufficient to target unique sites in complex genomes. One of the most commonly used DNA binding domain in ZFN is Cys_2His_2 zinc finger motif, which was first discovered in transcription factor IIIA from *Xenopus laevis*. ZFNs bind as dimers to their target sites, with each half on either side of target sequence at the correct orientation and spacing. ZFNs cause DSB at specific target site, which stimulates the cell's natural DNA repair processes (HR and NHEJ) at targeted genome loci (Gupta et al. 2012; Petolino 2015).

In a study lead by Fyodor D. Urnov's group, ZFNs have been successfully used to reduce phytate levels in maize seeds by targeting *IPK1* (encoding for inositol-1,3,4,5,6-pentakisphosphate-2-kinase) involved in phytate biosynthesis pathway. Phytate accounts for roughly 75% of seed phosphorous content, which is an anti-nutritional component in animal feed and also contributes to water pollution. The insertional disruption of *IPK1* was concomitant with targeted addition of herbicide tolerance gene (introduced as a heterologous donor DNA molecule), thereby delivering a *stacked trait* through targeted manipulation of a single locus (Shukla et al. 2009). This study also demonstrated that the ZFN-edited plants were fertile and the transgene was transmitted to the next generation in normal Mendelian pattern. In another report, the endogenous acetolactate synthase genes (involved in branched-chain amino acid biosynthesis) in tobacco, namely *SuRA* and *SuRB*, were targeted by ZFNs to confer resistance to imidazolinone and sulphonylurea herbicides (Townsend et al. 2009). ZFN-encoding constructs were electroporated into tobacco protoplasts, and the target sites in *SuRA* and *SuRB* were amplified by PCR and subjected to high throughput pyrosequencing. Herbicide-resistance mutations were introduced into *SuR* loci by ZFN-mediated gene targeting at frequencies exceeding 2% of transformed cells.

However, the major constraints in using ZFNs in genome engineering include complicated designing of ZFNs, low targeting efficiency and off-target effects. Since each zinc finger repeat binds to a codon, the designing of these nucleotides is complicated. Often, multiple pair of ZFNs needs to be designed for a single target and test them extensively as substantial proportion of ZFNs pairs fail. For instance, a panel of 66 ZFNs was designed against distinct portions of a single target gene, *IPK1* (Shukla et al. 2009). Further, the ZFN technology has been hampered by cytotoxicity, possibly due to off-target cleavage at multiple genomic sites (Cornu et al. 2008).

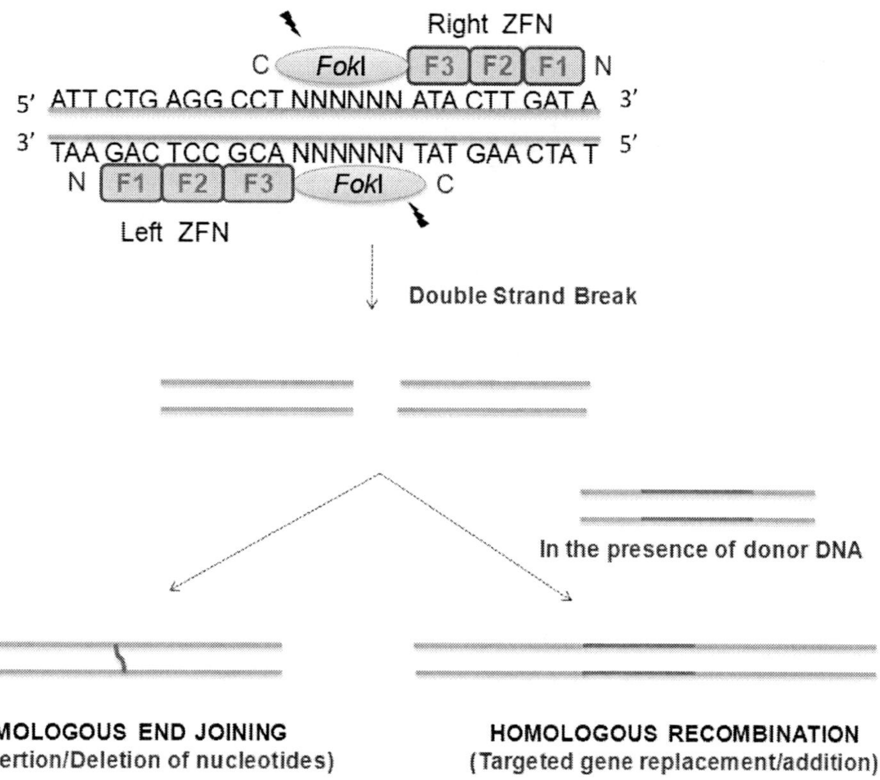

Figure 2. Schematic representation of Genome Editing using Zinc Finger Nucleases (ZFNs). Here, the DNA binding domain comprises of an array of three zinc fingers, each specific for a set of 3 nucleotides in target DNA. Dimerizaton of these engineered nucleases is mediated by the *Fok*I cleavage domain, which cleaves within 5-7 bp "spacer" sequences that separates the two inverted half sites. The donor template DNA, containing stretches of sequences homologous to that flanking the target DNA site, may contain transgene(s) or modified gene sequence to bring about gene addition or replacement, respectively.

Transcription Activator-Like Effector Nucleases (TALENs)

TALENs are another class of engineered nucleases that have been used for genome engineering. These are derived from virulence determinants/effector proteins called Transcription Activator-like Effector (TALEs), which are secreted by pathogenic bacteria of genus *Xanthomonas* into plant cells (Scholze and Boch 2011). These effector proteins bind DNA and alter transcription of genes in host cells, thereby increasing its susceptibility to pathogen. The DNA-binding specificity of TALs is determined by a set of tandem repeats of amino acids, each repeat conferring recognition to one base pair in DNA. Rearrangement of repeat modules allows design of proteins with desired DNA-binding specificities (Bogdanove et al. 2010; Bogdanove and Voytas 2011). Like ZFNs, TALENs are also made by fusing an engineered DNA-binding domain with the catalytic domain of *Fok*I endonuclease which produces a targeted DSB in the specific site of genome (Christian et al. 2010; Li et al. 2011). In addition, just like ZFNs, TALENs also bind as dimers and cleave on target DNA site, with each monomer in proper orientation and proper spacing (Figure 3).

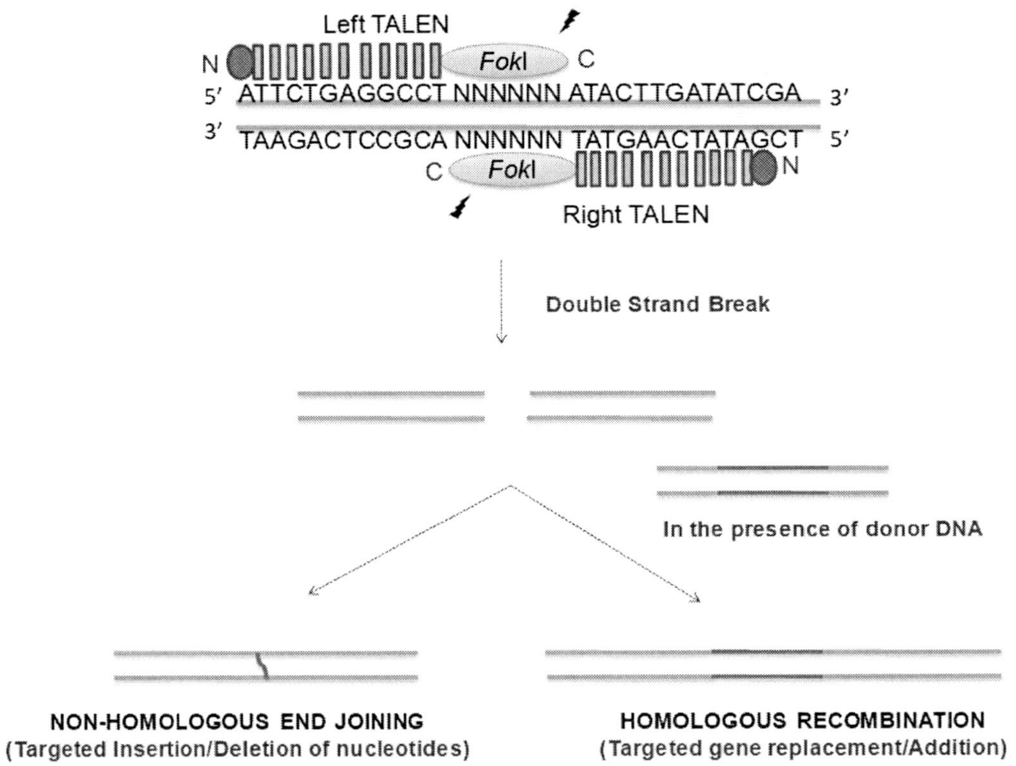

Figure 3. Schematic representation of Genome Editing using Transcription Activator-like Effector Nucleases (TALENs). Like ZFNs, TALENs too bind as dimers to the target site, with each half bound on either side of target sequence. The DNA binding domain of TALENs is composed of a series of 33-35 amino acid repeats, each recognizing a specific nucleotide. The DNA cleavage domain comprises of *Fok*I endonuclease. The donor template DNA, containing stretches of sequences homologous to that flanking the target DNA site, may contain transgene(s) or modified gene sequence to bring about gene addition or replacement, respectively.

In one of the reports, TALENs were designed specifically against the acetolactate synthase with the aim of creating specific amino acid substitutions in this gene, in order to confer resistance to sulfonylurea and imidazolinone herbicides (Zhang et al. 2013). Frequencies of NHEJ-induced mutagenesis in tobacco protoplasts by TALENs were revealed by high-throughput sequencing. TALENs introduced targeted mutations in acetolactate synthase gene in 30% of transformed cells, and the frequency of targeted gene insertion was around 14% (Zhang et al. 2013). In another study led by Bing Yang's group, TALENs have been used to create rice resistant to bacterial blight disease by targeting rice bacterial blight susceptibility gene *OsSWEET14*. Transcription activator–like (TAL) effectors of *Xanthomonas oryzae* pv. *oryzae* contribute to pathogen virulence by transcriptionally activating specific rice disease-susceptibility genes like *OsSWEET14*, which encodes for a sucrose transporter protein. The virulence factors, AvrXa7 and PthXo3, divert sugars from plant cells to the pathogen. In this report, two pairs of TALENs were designed against the regulatory elements present in promoter of *OsSWEET14*, in order to interfere with the virulence function of AvrXa7 and PthXo3, but not its developmental function. The TALEN-modified T2 plants showed strong resistance to bacterial blight infection (Li et al. 2012).

TALENs are considered to be improved nucleases for targeted genome editing in comparison to ZFNs. The binding specificity of a zinc finger array in ZFNs is not completely predictable, as the specificities of neighboring ZFs interdepend, which results in highly laborious screening of libraries to identify suitable candidates. On the other hand, TALENs recognize DNA in an apparently modular fashion; the simple one-to-one recognition rules between protein repeats and nucleotide sequences, also makes them simpler to design (Cermak et al. 2011). Therefore, unlike the context-dependent DNA binding of ZFNs, TALENs can be easily and rapidly constructed to target virtually any DNA sequence due to simple protein-DNA code. Additionally, TALENs are more specific than ZFNs with respect to the target sequence, and hence, the chances of off-site cleavage are highly reduced. This probably occurs as TALs are highly sensitive to changes in their target DNA boxes and three to four mutations typically block recognition. In spite of these advantages, TALENs have their own share of limitations: Since a TAL is needed for each nucleotide, TALENs are quite large in size; approximately six times larger than mega nucleases (~950 amino acids vs. ~165 amino acids) and hence, more difficult to deliver than other nucleases. Also, a new TALEN protein must be generated for each DNA target site, which increases time and costs for development.

Clustered Regularly Interspaced Short Palindromic Repeats/ CRISPR-Associated Protein 9 (CRISPR/Cas9)

CRISPR/Cas9 is the most popular and preferred genome editing tool designed so far. It makes use of an endonuclease protein called Cas9, which is guided by a synthetic RNA called 'guide RNA' (gRNA), to introduce a DSB at a specific location within genome (Bhaya et al. 2011; Noman et al. 2016). The most widely used CRISPR/Cas9 system is derived from defense mechanism of *Streptococcus pyogenes*, which involves Cas9 nuclease, guided by two noncoding RNAs: trans-activating crRNA (tracrRNA) and CRISPR RNAs (crRNA). During viral attack, the defense mechanism of this bacteria initiate a series of steps resulting into cleavage of viral DNA into fragments and insertion into specific loci called CRISPR loci which is interspersed with "spacer DNA". The CRISPR loci is composed of an array of palindromic repeats (short segments of DNA, 20-40 bp), the transcription of CRISPR loci and spacer generates small RNA called CRISPR RNA (crRNA) which is complementary in sequence to that of foreign DNA. The presence of PAM recognition motif (NGG) downstream to the target sequence distinguishes the bacterial encoded RNA from the target sequence. Cas9 binds the PAM sequence present in target mRNA, opens up the DNA due to helicase activity, allowing RNA-DNA hybridization, finally cleaving the target DNA sequence (Figure 4). During subsequent encounter with the invading virus, the CRISPR loci is transcribed and processed to generate small RNA called CRISPR RNA. This defense mechanism of bacteria has been exploited to generate a highly specific pair of Cas9 endonuclease which would introduce double strand in the target site as specified by the gRNA, the chimeric molecule derived from components of crRNA and tracRNA (Figure 4). Meganuclease, ZFNs, and TALE proteins all recognize specific DNA sequences through protein-DNA interactions, whereas, CRISPR/Cas9 does so through RNA-DNA interactions.

The application of CRISPR/Cas9 in imparting resistance against DNA and RNA viruses has been discussed. Ali et al. 2015 have successfully applied this technology to impart molecular immunity against Tomato yellow leaf curl virus (TYLCV), a DNA virus. Engineering resistance into DNA viruses generally involves mutations in viral genome. Using this strategy, gRNA molecules specific for different coding and non-coding sequences of the viral genome were designed and delivered tobacco rattle virus (TRV) into *Nicotiania benthamiana* plants overexpressing Cas9 endonuclease. Next, when these over-expressors were challenged with an infectious TYLCV clone. The plants expressing CRISPR/Cas9 exhibited delayed or reduced accumulation of viral DNA and significantly attenuated symptoms of infection (Ali et al. 2015). The molecular analyses revealed that the CRISPR/Cas9 system targeted the viral genome for degradation and introduced mutations at the target sequences. The sgRNA targeting the stem-loop sequence within the TYLCV origin of replication in the intergenic region, were found to be most effective. Similarly, in another study, this technology has also been used to generate *Nicotiana benthamiana* and *Arabidopsis* plants resistant to RNA virus (Zhang et al. 2018). These plants expressing Cas9 (from *Francisella novicida*) and sgRNA specific for the cucumber mosaic virus (CMV) or tobacco mosaic virus (TMV) exhibited significantly attenuated virus infection symptoms and reduced viral RNA accumulation by 40-80% in comparison to Control (Zhang et al. 2018). The group also reported that the inhibition occurred most likely due to blocking of viral genome replication and protein translation by FnCas9 binding to the viral genome, rather than viral genome cleavage (as was the case with DNA virus resistance).

CRISPR/Cas9 technology has been used to generate parthenocarpic fruits of tomato (Klap et al. 2017). The group targeted Slagamous-like 6 (SlAGL6) gene to produce facultative parthenocarpic fruits under heat stress conditions. Loss-of-function mutations generated by CRISPR/Cas9 technology have also been used to engineer male sterility trait into rice (Zou et al. 2017). *OsLAP6/OsPKS1* is expressed in the tapetum and is involved in sporopollenin biosynthesis, a crucial component of pollen exine formation. The male sterile lines generated by targeting this gene did not show any detectable morphological defects in vegetative growth or floral organ development, except for the abnormality in pollen development. Another agriculturally important trait is increasing seed yield. The generation of complete knock outs is very difficult in polyploids like *Brassica napus*, an allotetraploid. In an effort to increase the number of locules in siliques of *Brassica napus*, CRISPR/Cas9 has been used to effectively knock out three key genes involved in CLV signalling pathway with stable transformation in rapeseed. The double mutation of *BnCLV3* produced more leaves and multilocular siliques with a significantly higher number of seeds per silique and a higher seed weight than the wild-type and single mutant plants, potentially contributing to increased seed production (Yang et al. 2018).

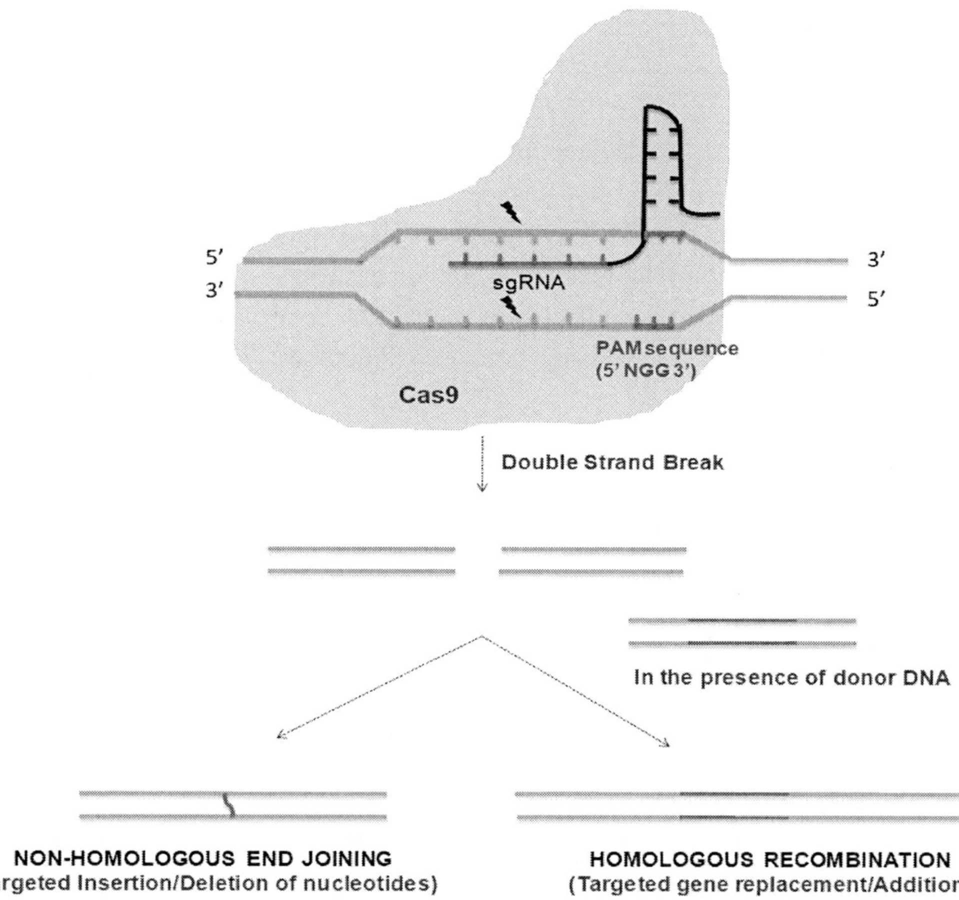

Figure 4. Schematic representation of Genome Editing in Plants by Clustered Regularly Interspaced Short Palindromic Repeats/ CRISPR-associated protein (CRISPR/Cas9). Unlike ZFNs and TALENs, which bind to their target DNA as protein dimers, Cas9 endonuclease is a monomeric protein that brings about the cleavage of target site. The Cas9 cleavage domain is guided to the target site by a short stretch of guide RNA that is complementary in sequence to the target site. The presence of PAM sequence in target site is imperative for inducing double strand break by the Cas9 endonuclease. The donor template DNA, containing stretches of sequences homologous to that flanking the target DNA site, may contain transgene(s) or modified gene sequence to bring about gene addition or replacement, respectively.

Among all the genome editing tools available today, CRISPR/Cas9 is the most powerful and widely used nucleases. The most striking advantage of this system is that the endonuclease could be targeted to any site in the genome, simply by designing a guide RNA complementary for that site. This property eliminates the need to construct a completely customized endonuclease for each target, something that is still required by TALENs and ZFNs. This is because making a sequence of RNA is much easier than designing a specific 3D protein shape, as in ZFNs and TALENs. Genome engineering using CRISPR/Cas9 simply involves the addition of 20 nucleotides to the sgRNA molecule, allowing targeted genome editing and regulation. Moreover, CRIPR/Cas9 could be used to simultaneously target multiple genomic loci, a process also known as multiplexing (Xing et al. 2014). In higher eukaryotes, CpG methylation (involved in regulation of gene expression) often represents a major epigenetic drawback for genome engineering applications (Valton et al. 2012). CRISPR/Cas9 has an added virtue that it has the ability to cleave even methylated loci, something which is not possible with other nucleases (Sander and Joung 2014; Malzahn et al. 2017). However, in spite of all these advantages, CRISPR/Cas9 system does suffer from some potential off-target mismatches, which lead to unintended changes within the genome. Some of the ways to alleviate this drawback include optimization of gRNA length, chemical modification of gRNA, standardizing the concentration of Cas9 nuclease and using paired Cas9 nucleases (Ran et al. 2013; Wolt et al. 2016a).

Regulatory Aspects of Genome Edited Plants

Since genome editing may range from simple random insertions or deletions of nucleotides to additions of single or multiple transgenes at a specific locus, the regulatory mechanisms cannot be generalized. A case-specific study regarding the nature of the edit, specificity, source of editing machinery, and post-editing trait segregation is necessary. If the crop does not contain foreign DNA, either because of the type of editing machinery used (e.g., Cas9 and introduced synthetic gRNA) or because post-editing segregation removes foreign DNA elements, the crop cannot be differentiated from crops derived using traditional plant breeding methods and chemical or radiation mutagenesis. Hence, this class of genome-edited crop should be subject to the same types of regulatory consideration as are traditionally bred crops. On the other hand, if there is evidence for foreign DNA insertion without subsequent removal, the genome-edited crop is indistinguishable from a genetically modified (GM) crop, and therefore, should be given the regulatory consideration as appropriate for a GMO.

The regulatory procedures involving genome edited crops vary depending upon the particular nation, which follows either *product-based* or *process-based* evaluation system. For instance, in countries like Canada, where product-based approach is followed, the plant traits developed from conventional breeding, mutagenesis, transgenesis or genome editing will all be subject to a similar regulatory approval process (Wolt et al. 2016b). This would mean that genome edited products will not be subject to the restrictions imposed on GM products, except in some cases where the mutation involves a large sequence insertion into the genome. In contrast, some countries like USA follow process-based evaluation; this implies, plants transformed by *Agrobacterium tumefaciens* are typically subject to regulatory review by USDA, whereas those transformed by biolistics are not (when the gene donor, recipient and vector are otherwise not known plant pests) (Camacho et al. 2014). According to European Union, genome edited crops which do not add any new sequence from other species, should be non-regulated, as the same mutation could occur naturally, although at low frequency. (Hartung and Schiemann 2014). However, the EU regulations also are process-based, and anti-GMO activists are trying to imply the same regulations on genome edited crops as well. It is also imperative that the distinction between genome edited crops and genetically modified plants be clearly made. Both genetic engineering and genome editing aim towards crop improvement; but, the latter method involves plant genetic modification with much higher precision and increased efficiency, and for the most part, does not involve integration of foreign genes into the plant genome.

Conclusion

Genome Editing has tremendous potential to enhance our understanding of basic and applied plant biology. Some of the major applications of gene targeting in plants include functional genomics, creation of cell lines that produce higher amounts of proteins and antibodies, generation of plants resistant to biotic and abiotic stresses, gene therapy and much more. With genome editing it is possible to determine the function of gene not only through creation of complete knock-outs (not amenable/possible with RNAi), but also through knock-in cell lines with promoters, fusion tags or reporters integrated into endogenous genes. However, there are certain issues that need to be addressed, if the full potential of this technology is to be realized. First, there is a need to limit off-target effects, a problem that exists even with improved engineered nucleases. Second, the regulatory aspects of genome edited crops need to be reconsidered; it would be logical if genome edited crop varieties that do not incorporate foreign DNA, are treated as conventional mutation events. This is crucial so that the humongous challenge of feeding 11.2 bn people by the year 2100 (United Nations 2016) could be accomplished, a task that seems impossible through adoption of conventional breeding methods only.

REFERENCES

Ali, Z., Abul-Faraj, A., Li L., Ghosh, N., Piatek, M., Mahjoub, A., Aouida, M., Piatek, A., Baltes, N. J., Voytas, D. F., Dinesh-Kumar, S. (2015) Efficient virus-mediated genome editing in plants using the CRISPR/Cas9 system. *Mol Plant* 8: 1288-1291.

Argast, G. M., Stephens, K. M., Emond, M. J., Monnat Jr, R. J. (1998) I-*Ppo*I and I-*Cre*I homing site sequence degeneracy determined by random mutagenesis and sequential in vitro enrichment. *J Mol Biol* 280: 345-53.

Aylon, Y., Kupiec, M. (2004) New insights into the mechanism of homologous recombination in yeast. *Mutat Res* 566: 231-248.

Baltes, N. J., Voytas, D. F. (2015) Enabling plant synthetic biology through genome engineering. *Trends Biotechnol* 33: 120-131.

Beetham, P. R., Kipp, P. B., Sawycky, X. L., Arntzen, C. J., May, G. D. (1999) A tool for functional plant genomics: Chimeric RNA/DNA oligonucleotides cause in vivo gene-specific mutations. *Proc Nat Acad Sci USA* 96: 8774-8778.

Belfort, M., Bonocora, R. P. (2014) Homing endonucleases: from genetic anomalies to programmable genomic clippers. In: *Homing endonucleases-methods and protocols* (Ed. Edgell DR), Humana Press, Totowa, NJ, pp. 1-26.

Bhaya, D., Davison, M., Barrangou, R. (2011) CRISPR-Cas Systems in Bacteria and Archaea: Versatile Small RNAs for Adaptive Defense and Regulation. *Annu Rev Genet* 45: 273-297.

Bogdanove, A. J., Schornack, S., Lahaye, T. (2010) TAL effectors: finding plant genes for disease and defense. *Curr Opin Plant Biol* 13: 394-401.

Bogdanove, A. J., Voytas, D. F. (2011) TAL effectors: customizable proteins for DNA targeting. *Science* 333: 1843-1846.

Bollineni, H., Gopala, K. S., Sundaram, R. M., Sudhakar, D., Prabhu, K. V., Singh, N. K., Pal, M., Mishra, S., Khurana, J. P., Singh, A. K. (2014) Marker assisted biofortification of rice with pro-vitamin A using transgenic Golden Rice lines: progress and prospects. *Indian J Genet Plant Breed* 74(4): 624-30.

Camacho, A., Van Deynze, A., Chi-Ham, C., Bennett, A. B. (2014) Genetically engineered crops that fly under the US regulatory radar. *Nat Biotechnol* 32: 1087-1091.

Carroll, D. (2011) Genome engineering with zinc-finger nucleases. *Genetics* 188: 773-782.

Cermak, T., Doyle, E. L., Christian, M., Wang, L., Zhang, Y., Schmidt, C., Baller, J. A., Somia, N. V., Bogdanove, A. J., Voytas, D. F. (2011) Efficient design and assembly of custom TALEN and other TAL effector-based constructs for DNA targeting. *Nucleic Acids Res* 39: e82-e82.

Christian, M., Cermak, T., Doyle, E. L., Schmidt, C., Zhang, F., Hummel, A., Bogdanove, A. J., Voytas, D. F. (2010) Targeting DNA double-strand breaks with TAL effector nucleases. *Genetics* 186: 757-761.

Cornu, T. I., Thibodeau-Beganny, S., Guhl, E., Alwin, S., Eichtinger, M., Joung, J. K., Cathomen, T. (2008) DNA-binding specificity is a major determinant of the activity and toxicity of zinc-finger nucleases. *Mol Therapy* 16: 352-8.

Curtin, S. J., Voytas, D. F., Stupar, R. M. (2012) Genome engineering of crops with designer nucleases. *Plant Genome* 5: 42-50.

Gao H, Smith J, Yang M, Jones S, Djukanovic V, Nicholson MG, West A, Bidney D, Falco SC, Jantz D, Lyznik LA (2010) Heritable targeted mutagenesis in maize using a designed endonuclease. *Plant J* 61: 176-187.

Gupta, A., Christensen, R. G., Rayla, A. L., Lakshmanan, A., Stormo, G. D., Wolfe, S. A. (2012) An optimized two-finger archive for ZFN-mediated gene targeting. *Nat Methods* 9: 588-590.

Hartung, F., Schiemann, J. (2014) Precise plant breeding using new genome editing techniques: opportunities, safety and regulation in the EU. *Plant J* 78: 742-752.

Jasin, M. (1996) Genetic manipulation of genomes with rare-cutting endonucleases. *Trends Genet* 12: 224-228.

Klap, C., Yeshayahou, E., Bolger, A. M., Arazi, T., Gupta, S. K., Shabtai, S., Usadel, B., Salts, Y., Barg, R. (2017) Tomato facultative parthenocarpy results from SlAGAMOUS-LIKE 6 loss of function. *Plant Biotech J* 15: 634-47.

Li, T., Huang, S., Zhao, X., Wright, D. A., Carpenter, S., Spalding, M. H., Weeks, D. P., Yang, B. (2011) Modularly assembled designer TAL effector nucleases for targeted gene knockout and gene replacement in eukaryotes. *Nucleic Acids Res* 39: 6315-6325.

Li T, Liu B, Spalding MH, Weeks DP, Yang B (2012) High-efficiency TALEN-based gene editing produces disease-resistant rice. *Nature Biotech* 30: 390-392.

Luo, S., Li, J., Stoddard, T. J., Baltes, N. J., Demorest, Z. L., Clasen, B. M., Coffman, A., Retterath, A., Mathis, L., Voytas, D. F., Zhang, F. (2015) Non-transgenic plant genome editing using purified sequence-specific nucleases. *Mol Plant* 8: 1425-1427.

Malzahn, A., Lowder, L., Qi, Y. (2017) Plant genome editing with TALEN and CRISPR. *Cell Biosci* 7: 21.

Miller, J. C., Tan, S., Qiao, G., Barlow, K. A., Wang, J., Xia, D. F., Meng, X., Paschon, D. E., Leung, E., Hinkley, S. J., Dulay, G. P. (2011) A TALE nuclease architecture for efficient genome editing. *Nat Biotechnol* 29: 143-148.

Mishra, S., Gupta, D., Ranjan, R. (2019) Molecular Approaches for Enhancing Abiotic Stress Tolerance in Plants. In: *Approaches for Enhancing Abiotic Stress Tolerance in Plants* (Ed. Hasanuzzaman M, Nahar K, Fujita M, Oku H, Islam T), CRC Press, Boca Raton pp. 405-421.

Mohanta, T. K., Bashir, T., Hashem, A., Abd Allah, E. F., Bae, H. (2017) Genome Editing Tools in Plants. *Genes* 8: 399.

Noman, A., Aqeel, M., He, S. (2016) CRISPR-Cas9: tool for qualitative and quantitative plant genome editing. *Front Plant Sci* 7: 1740.

Paszkowski, J., Baur, M., Bogucki, A., Potrykus, I. (1988) Gene targeting in plants. *EMBO J* 7: 4021-4026.

Petolino, J. F. (2015) Genome editing in plants via designed zinc finger nucleases. *Vitr Cell Dev Biol* 51: 1-8.

Puchta, H. (1998) Repair of genomic double-strand breaks in somatic plant cells by one-sided invasion of homologous sequences. *Plant J* 13: 331-339.

Puchta, H., Fauser, F. (2013) Gene targeting in plants: 25 years later. *Int J Dev Biol* 57: 629-637.

Ran, F. A., Hsu, P. D., Lin, C. Y., Gootenberg, J. S., Konermann, S., Trevino, A. E., Scott, D. A., Inoue, A., Matoba, S., Zhang, Y., Zhang, F. (2013) Double nicking by RNA-guided CRISPR Cas9 for enhanced genome editing specificity. *Cell* 154: 1380-1389.

Reiss, B., Schubert, I., Köpchen, K., Wendeler, E., Schell, J., Puchta, H. (2000) RecA stimulates sister chromatid exchange and the fidelity of double-strand break repair, but not gene targeting, in plants transformed by *Agrobacterium*. *Proc Natl Acad Sci USA* 97: 3358-3363.

Rinaldo, A. R., Ayliffe, M. (2015) Gene targeting and editing in crop plants: a new era of precision opportunities. *Mol Breed* 35: 40. https://doi.org/10.1007/s11032-015-0210-z

Salt, D. E., Blaylock, M., Kumar, N. P., Dushenkov, V., Ensley, B. D., Chet, I., Raskin, I. (1995) Phytoremediation: a novel strategy for the removal of toxic metals from the environment using plants. *Nature Biotechnol* 13(5): 468-474.

Sander, J. D., Joung, J. K. (2014) CRISPR-Cas systems for editing, regulating and targeting genomes. *Nat Biotechnol* 32: 347-355.

Scholze, H., Boch, J. (2011) TAL effectors are remote controls for gene activation. *Curr Opin Microbiol* 14: 47-53.

Seligman, L. M., Chisholm, K. M., Chevalier, B. S., Chadsey, M. S., Edwards, S. T., Savage, J. H., Veillet, A. L. (2002) Mutations altering the cleavage specificity of a homing endonuclease. *Nucleic Acids Res* 30: 3870-3879.

Shukla, V. K., Doyon, Y., Miller, J. C., DeKelver, R. C., Moehle, E. A., Worden, S. E., Mitchell, J. C., Arnold, N. L., Gopalan, S., Meng, X., Choi, V. M. (2009) Precise genome modification in the crop species *Zea mays* using zinc-finger nucleases. *Nature* 459: 437-441.

Siebert, R., Puchta, H. (2002) Efficient repair of genomic double-strand breaks by homologous recombination between directly repeated sequences in the plant genome. *Plant Cell* 14: 1121-31.

Stephens, K. M., Monnat Jr, R. J., Heath, P. J., Stoddard, B. L. (1997) Crystallization and preliminary X ray studies of *I CreI*: a group I intron-encoded endonuclease from *C. reinhardtii*. *Proteins* 28: 137-139.

Stoddard, B. L. (2005) Homing endonuclease structure and function. *Q Rev Biophys* 38: 49-95.

Stoddard B. L. (2011) Homing endonucleases: from microbial genetic invaders to reagents for targeted DNA modification. *Structure* 19: 7-15.

Takatsuji, H. (1998) Zinc-finger transcription factors in plants. *Cell Mol Life Sci* 54: 582-596.

Terada, R., Urawa, H., Inagaki, Y., Tsugane, K., Iida, S. (2002) Efficient gene targeting by homologous recombination in rice. *Nat Biotechnol* 20: 1030.

Townsend, J. A., Wright, D. A., Winfrey, R. J., Fu, F., Maeder, M. L., Joung, J. K., Voytas, D. F. (2009) High-frequency modification of plant genes using engineered zinc-finger nucleases. *Nature* 459: 442-445.

Urnov, F. D., Rebar, E. J., Holmes, M. C., Zhang, H. S., Gregory, P. D. (2010) Genome editing with engineered zinc finger nucleases. *Nature Rev Genet* 11: 636-646.

Valton, J., Daboussi, F., Leduc, S., Molina, R., Redondo, P., Macmaster, R., Montoya, G., Duchateau, P. (2012) 5′-Cytosine-phosphoguanine (CpG) methylation impacts the activity of natural and engineered mega nucleases. *J Biol Chem* 287: 30139-30150.

Wolt, J. D., Wang, K., Sashital, D., Lawrence-Dill, C. J. (2016a) Achieving plant CRISPR targeting that limits off-target effects. *Plant Genome* 9. doi: 10.3835/plantgenome 2016.05.0047.

Wolt, J. D., Wang, K., Yang, B. (2016b) The Regulatory Status of Genome-edited Crops. *Plant Biotechnol J* 14: 510-518.

Wood, A. J., Lo, T. W., Zeitler, B., Pickle, C. S., Ralston, E. J., Lee, A. H., Amora, R., Miller, J. C., Leung, E., Meng, X., Zhang, L. (2011) Targeted genome editing across species using ZFNs and TALENs. *Science* 333: 307-307.

"*World Population Prospects-Population Division-United Nations*". esa.un.org. Retrieved 2016-09-15.

Xing, H. L., Dong, L., Wang, Z. P., Zhang, H. Y., Han, C. Y., Liu, B., Wang, X. C., Chen, Q. J. (2014) A CRISPR/Cas9 toolkit for multiplex genome editing in plants. *BMC Plant Biol* 14: 327.

Yang, Y., Zhu K., Li H., Han S., Meng, Q., Khan S. U., Fan C., Xie, K., Zhou, Y. (2018) Precise editing of CLAVATA genes in *Brassica napus* L. regulates multilocular silique development. *Plant Biotechnol J.* https://doi.org/10.1111/pbi.12872.

Zhang, T., Zheng, Q., Yi, X., An, H., Zhao, Y., Ma, S., Zhou, G. (2018) Establishing RNA virus resistance in plants by harnessing CRISPR immune system. *Plant Biotechnol J.* doi: 10.1111/pbi.12881.

Zhang, Y., Zhang, F., Li, X., Baller, J. A., Qi, Y., Starker, C. G., Bogdanove, A. J., Voytas, D. F. (2013) Transcription activator-like effector nucleases enable efficient plant genome engineering. *Plant Physiol* 161: 20-27.

Zou, T., Xiao, Q., Li, W., Luo, T., Yuan, G., He, Z., Liu, M., Li, Q., Xu, P., Zhu, J., Liang, Y (2017) OsLAP6/OsPKS1, an orthologue of PKSA/LAP6, is critical for proper pollen exine formation. *Rice* 10: 53.

In: Innovations in Life Science Research ISBN: 978-1-53615-868-7
Editors: R. P. Sinha, S. Pandey-Rai and N. Ghoshal ©2019 Nova Science Publishers, Inc.

Chapter 5

Molecular Markers: A Tool in Plant Breeding for Pyramiding Resistance Genes

Ruchi Rai[*] and Shilpi Singh
Molecular Biology Section, Laboratory of Algal Biology,
Centre of Advanced Study in Botany, Institute of Science,
Banaras Hindu University, Varanasi, Uttar Pradesh, India

Abstract

Employment of plant biotechnology has significantly accomplished the increasing food demand world wide and show the way towards sustainable agriculture. Molecular markers emerge as boon in plant breeding as it augument the efficiency, speed and accuracy of conventional breeding methods. They are widely applied in estimating genetic identity, genetic diversity, QTL analysis, map-based gene cloning, development of linkage maps, quality control and marker-assisted selection (MAS). They are also effective in introgression of several agronomically important traits such as disease resistance, abiotic stress tolerance and aid in tracking the accumulation of severalgenes in a cultivar which is otherwise difficult to sort out on the basis phenotypic evaluation. Marker-assisted selection leads to successful release of new and better crop varieties in lesser time as it reduced the number of backcross generations required to release a variety. The present chapter is an attempt to review various molecular markers used in breeding starting from hybridization based RFLP to the most advanced Genotype-based-sequencing (GBS). Advanced form of MAS such as microarray, functional diagnostic markers, Genomic Selection (GS) and genome editing are also described which can be integrated to revolutionize the plant breeding programmes and facilitate enhanced adoption of modern genetics and breeding approaches in coming years.

Keywords: molecular markers, RFLP, NGS, MAS, gene pyramiding

---

* Corresponding Author's Email: ruchivrai@yahoo.co.in.

INTRODUCTION

Plant breeder geneticists entail to augment the crop production at least by 1.5 to 2.0 million tonnes (MT) annually in the next 10-15 years due to increase in population size and high demand for food grains. As per world food security assumption, the newer aids in plant breeding like biotechnology and genetic engineering will help to meet the requirement of food grain in accordance with increase in demand (FAO 1996). Although conventional plant breeding techniques are quite useful for augmenting food production but implementation of newer technologies such as molecular markers will certainly maximize the rate of breeding success stories (Ortiz 1998; Ruttan 1999; Huang et al. 2002). Presently, the main focus of plant breeders is introgression of various traits of agronomical interest along with biotic and abiotic stress tolerance, increase in durability towards specific diseases and limited use of water and nutrients. (Mackill et al. 1999; Slafer et al. 2005; Trethowan et al. 2005). These changes which are governed by genetic traits facilitate enhancement of durability, yield and sustainability of economically and agriculturally important crops. Conventional plant breeding involves use of morphological and phenotypic markers for the identification of agronomically important yield related traits. Phenotypic screening is quite labour intensive, time consuming, controlled by developmental stages and environmental conditions, insufficiently polymorphic and dominant in nature. Hence, DNA (molecular) markers emerge as a new versatile tool in plant breeding which make selection more efficient (Bueren et al. 2010; Joshi et al. 2011) precise, rapid (Thottappilly et al. 2000) at an early stage of development. These molecular markers are helpful in synchronized selection of quality traits along with disease resistance. The use of DNA markers in plant breeding is called marker-assisted selection (MAS) and is efficient tool for estimating genetic diversity, hybrid and varietal identification, phylogenetic relationship, gene mapping, tracking quantitative trait loci, pyramiding genes for crop improvement. Molecular markers significantly helps in shortening the time required for development and release of a new elite variety by reducing the number of backcross generation (Hasan et al. 2015). However it is important to validate these markers in different genetic backgrounds before utilizing them in gene pyramiding.

This chapter is a brief overview of the various molecular marker technologies used in plant breeding programme starting from hybridization based RFLP to most advance next generation sequencing (GBS) which revolutionized the plant breeding through development of SNPs. Application of these molecular markers in improving the efficiency and accuracy of conventional plant breeding through marker assisted selection (MAS) and gene pyramiding is also discussed.

TYPES OF MARKERS

Early genetic mapping studies were focused on discrete traits with simple Mendelian inheritance such as shape, colour, size or height. These traits often have one to one correspondence with the genes controlling thc traits. In such cases, morphological characters can be used as a reliable indicator for specific genes and are useful as genetic markers on chromosomes. Polymorphic, highly heritable morphological traits were some of the earliest markers employed in scientific investigations. Markers can be grouped into different categories on the basis of the level of analysis such as phenotype, gene products and level of gene expression (Figure 1).

Morphological Markers

Morphological markers are generally recognized visually (eg, traits such as flower color, seed shape, growth habits or pigmentation). They are known to be linked with other important traits like disease resistance or quality trait and thus indirectly can be used for checking the segregation of genes. These types of marker are usually of limited use as they do not meet the criteria of an ideal marker. They are insufficiently polymorphic and are generally dominant. They are often affected by environmental conditions or developmental growth stages, and need large plant population, time and effort (Stuberet al. 1999).

Biochemical Markers

They are allelic variants of enzymes that are identified by electrophoresis and specific staining. The biochemical markers/isozymes are better over morphological marker as they are not affected by developmental and environmental conditions. The only problem with isozymes in marker assisted selection (MAS) is that they do not produce a great amount of polymorphism between genetically similar cultivars. Isozyme numbers is limited and their expression is often restricted to specific developmental stages and tissue. In addition, the analysis of secondary metabolites (biochemical constituents) for varietal identification is constrained to those plants that produce a suitable range of metabolites (Joshi et al. 2011).

DNA Markers (or Molecular Markers)

Molecular marker may be defined as DNA sequences used for chromosome mapping or tagging useful traits as it can be located at a specific site in a chromosome. They are typically a small region of DNA sequences showing polymorphism in different individuals within a species. Molecular markers can provide genomic information at early generation of plant evaluation due to ease of detection by electrophoresis (Bagge and L″ubberstedt 2008) which is critical for success in plant breeding and also help to recognize polymorphisms without checking its phenotype. If tightly linked to genes of interest, molecular markers can be used to select indirectly for the desirable allele and this represents the simplest form of marker-assisted selection (MAS) (Hoisington et al. 2002).

According to Joshi et al. (2011), an ideal DNA makers should however restrain following properties (i) Highly polymorphism (i) Co dominant inheritance (ii) Frequent occurrence in genome (iii) Selective neutral behavior (the DNA sequences of any organism are neutral to environmental conditions or management practices) (iv) Easy access (availability) (v) Easy and fast assay (vi) Reproducible - highly reproducibility (vii) Easy exchange of data between laboratories.

Classification of Molecular Markers

Molecular markers has evolved enormously in last three decades can be categorized on the basis of degree of multiplexing capacity/throughput as (Figure 1). Some of the important marker systems are described below:

Low-Throughput Hybridization Based Markers/First Generation Markers

Restriction Fragment Length Polymorphism (RFLP)

Restriction fragment length polymorphism (RFLP) is the most primitive hybridization based marker which was first used for human genome mapping (Botstein et al. 1980) and later for plant genomes (Chao et al. 1989; Anderson et al. 1992; Devos et al. 1992; Devos and Gale 1993; Marino et al. 1996). This technique is based upon hybridization of a radio labelled probe to fragments of genomic DNA following digestion with restriction enzymes. Restriction endonucleases cut genomic DNA at specific palindrome recognition sequences, generating thousands of fragments, length of which is defined for a particular individual. Differences in the sequence with which probe hybridizes may result in differences (polymorphism) in the length of the fragments detected by the probe. The main types of probes are cDNA or genomic clones or multiple copy genes such as ribosomal RNA genes and these probes along with restriction enzymes are major determinants of the results. RFLP analysis, however, has some limitations, since it is time-consuming and labour intensive. The RFLP markers have been successfully employed in crops like *Quercusphellos phellos* (Coutinho et al. 2016) *Saccharum* sp. (Nerkar et al. 2015) and *Vignaradiata* (Islam et al. 2015) for assessing genetic diversity, etc. however this approach has been relatively less feasible in crops like wheat due to the polyploid nature, high proportion of repetitive DNA, large genome size and recent origin of wheat (Gupta et al. 1999). Despite of their usefulness the generation of RFLP marker their application is time consuming and requirement of radioactive isotopes further makes the analysis relatively expensive and hazardous. Finally their inability to detect single base changes restricts their use in detecting point mutations occurring within the region.

Medium-Throughput PCR Based Markers/Second Generation Markers

Randomly Amplified Polymorphic DNA (RAPD)

This assay was developed independently by two laboratories (Welsh and McClelland 1990; Williams et al. 1990; Devos and Gale 1992). The RAPD assay is based on the amplification of genomic DNA with single primers of arbitrary nucleotide sequence. In this reaction, a single species of primer binds to genomic DNA at two different sites on opposite strands of the DNA template. If these priming sites are within an amplifiable distance of each other, a discrete DNA product is produced through thermocyclic amplification. On average, each primer will direct the amplification of several discrete loci in the genome, making the assay an efficient way to screen for nucleotide sequence polymorphism between individuals. RAPD polymorphisms are the result of either a nucleotide base change that alters the primer binding site, or an insertion or deletion within amplified region. Therefore, the polymorphisms are usually noted by the presence or absence of an amplification product from a single locus. This also means that the RAPD technique tends to provide only dominant markers. RAPD differs from RFLPs in its simplicity, non-involvement of radioactivity, requirement of small amounts of DNA (10-25ng), and ease of use (Williams et al. 1993). However RAPD technique is very sensitive to methodological variations (MacPherson et al. 1993; He et al. 1994; Jones et al. 1997). Therefore, PCR reactions conditions must be identical to obtain consistent results. Difficulties with assay reproducibility may be overcome by eliminating variation in DNA concentration, consistent reaction components and carefully controlled thermal cycling during amplification. However, once a marker is found linked with trait of interest, it is required to convert the RAPD into more reproducible PCR assay using techniques such as allele-specific PCR or sequence characterized amplified regions (SCAR) assay.

Amplified Fragment Length Polymorphism (AFLP)

AFLP analysis (Vos et al. 1995) combines specificity of restriction analysis with PCR amplification. The genomic DNA is first digested with one or more restriction endonucleases (generally *Eco* RI and *Mse* I). Next an adapter of known sequence is ligated to the digested genomic DNA. DNA amplification is carried out using primers with sequence specificity for the adapter. The primer also contains one or more bases at their 3'ends which provide amplification selectivity by limiting the number of perfect sequence matches between the primer and available DNA templates. The resulting amplification products are separated on polyacrylamide sequencing gels and either silver stained or detected using radioactivity.

In complex genetic organisms, like plants, two amplification steps are performed. The first is a pre-amplification using primer which may or may not contain selective nucleotides. Diluted products from the pre-amplification are used as template for the second amplification along with primers with selective nucleotides. AFLP analysis can be applied regardless of the origin and complexity of the genome.

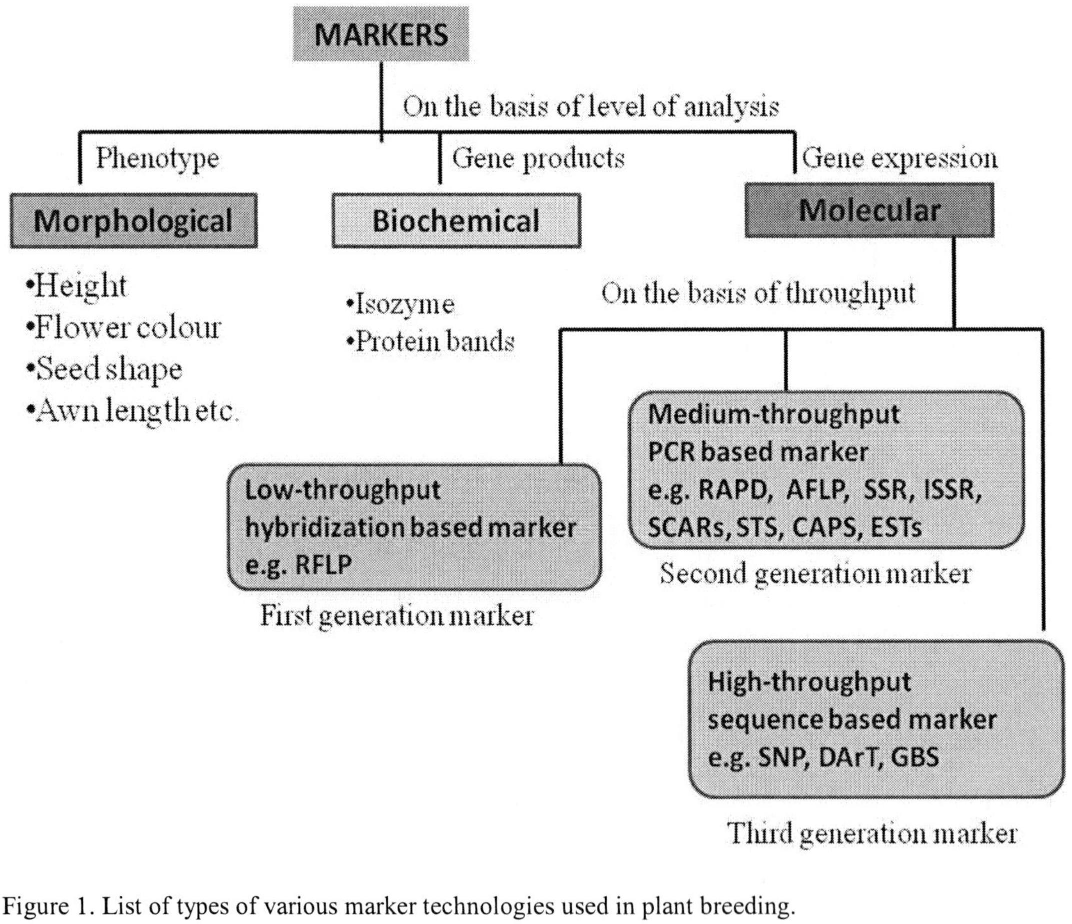

Figure 1. List of types of various marker technologies used in plant breeding.

The AFLP analysis has been widely used in number of crops (Barrett and Kidwell 1998; Barrett et al. 1998; Goodwin et al. 1998; Bohn et al. 1999; Shan et al. 1999) for developing polymorphic markers. It involves the use of fluorescent primers for automated fragment analysis system and software packages which increases its sensitivity and efficiency. The polymorphic bands from AFLP patterns may also be converted into RFLP probes or locus-specific PCR markers, although conversion of AFLP marker into SCAR/STS is not easy (Shan et al. 1999) due to certain difficulties which came across during data generation and analysis.

Microsatellites or Simple Sequence Repeats (SSRs)

Litt and Luty (1989) coined the term microsatellites, while the term minisatellite was introduced by Jeffrey et al. (1985). They are multilocus probes creating complex banding pattern and are usually non-species specific occurring ubiquitously. Simple sequence repeats are composed of tandemly repeated two to five nucleotide DNA core sequences such as (AT)n, (GT)n, (ATT)n or (GACA)n spread throughout eukaryotic genome. The DNA sequences flanking microsatellites are generally conserved within individuals of the same species, allowing the selection of PCR primers that will amplify the intervening SSR in all genotypes. These tandem repeats are highly polymorphic and their differences in length results into PCR products of different base pairs even among closely related cultivars, due to mutations causing variation in number of repeating units. SSR-specific primer pairs (forward and reverse) are designed for PCR amplification based on the sequences flanking the microsatellites.

The SSR assay is increasingly being used for plant breeding application due to its relative advantages and is still considered inexpensive for simply inherited traits. Locus-specific micro-satellite-based markers have been reported from many plant species. They are highly polymorphic, co-dominant markers which have been continuously used for almost all aspects of breeding and genetics research in a wide range of plant species such as lettuce, barley, wheat, rice and maize (Van de Wiel et al. 1999; Gupta and Varshney 2000; Varshney et al. 2005). According to Gao et al. (2016) SSR marker are most widely used in MAS due to their ease in availability, simplicity of technique, low cost and generation of high rate of polymorphism. However, SSR markers are time consuming and expensive to develop because the genomic regions carrying SSRs must be identified and sequenced.

Inter Simple Sequence Repeats (ISSR)

It is a variant of the polymerase chain reaction that uses simple sequence repeat primers (eg. [AC]n) to amplify regions between their target sequences. The technique exploits the abundant and random distribution of SSRs in plant genomes by amplifying DNA sequences between closely related SSRs. These dominant markers sometimes utilize microsatellites attached at the 3' ends are used for amplifying genomic DNA (Zietkiewicz et al. 1994). Gold et al. (1999) identified an ISSR marker linked to rust resistance genes *Sr39* and *Lr35* in wheat. Although ISSRs are highly informative, reproducible, co dominant and locus-specific, they have several drawbacks including high cost and time required for their development.

Sequence Tagged Sites (STS)

In this assay, primers are designed based on a known sequence which defines a specific region of a genome. Therefore, a STS is a short, unique sequence amplified by PCR that identifies a known location on a chromosome. STS primers mostly designed as 20-mers, are longer than RAPD primers, and therefore more stable and less affected by reaction factors than RAPD primers. Since STS primers are designed from low-copy number sequences, they are more likely to identify single or low copy DNA sequences of a genome while RAPD often identifies repetitive DNA sequences. RFLP and RAPD markers can be converted into STS markers. The α-amylase and γ-gliadin genes specific sequences were used as STS primers to detect polymorphism in wheat by Talbert et al. (1994).A disadvantage of STS primers as compared to RAPD is the need for sequence information before primer designing.

Sequence Characterized Amplified Regions (SCARs)

Sequence Characterized Amplified Regions (SCARs) (Paran and Michelmore 1993) are an example of class of sequence-dependent markers derived from sequence-arbitrary marker loci. By cloning and sequencing RAPD amplification products Paran and Michelmore (1993) were able to design sequence-dependent primers and amplify PCR products unique for each marker locus. SCARs are advantageous over RAPD markers as they detect only a single locus, their amplification is less sensitive to reaction conditions and they can be potentially converted into codominant markers.

Cleaved Amplified Polymorphic Sequences (CAPS)

Cleaved Amplified Polymorphic Sequences are PCR amplification products which are subsequently cleaved using restriction endonucleases and size separated in order to reveal polymorphism (Konieczny and Ausubel 1993). It assays for variation in the restriction sites in the PCR amplified products hence also known as PCR-RFLP. It uses battery of restriction enzymes to detect variation in the amplified product obtained from STS and SCAR. Chen et al. (2003) developed CAPS marker for wheat stripe rust resistance gene *Yr5* from an STS marker. Moury et al. (2000) developed CAPS to assist selection of tomato spotted wilt virus (TSWV) resistance in pepper.

Expressed Sequence Tags (ESTs)

cDNA sequences are also potentially available source of molecular marker. An EST is a DNA fragment representing the sequence from a cDNA clone that corresponds to mRNA molecule or part of it. ESTs serve as markers for genetic and physical mapping and as clones for expression analysis. ESTs can be matched with sequences available in nucleotide sequence data bases or Genebank and protein sequence data bases and then assigned to specific genes with the help of softwares. While STSs represent unique sequences which may or may not be transcribed, the ESTs represent only expressed genes. Development of EST-SSR marker has now become common in a wide variety of plant species.

High-Throughput Sequence-Based Markers/Third Generation Markers

Single Nucleotide Polymorphism (SNPs)

Single Nucleotide Polymorphism are the ultimate DNA markers which involves exploitation of polymorphism exists in DNA sequence as single base pair difference (Marshall 1997; Nikiforov 1994). They are small insertions/deletions and are the basis of difference between alleles. Although SNPs suffers from the disadvantage of being biallelic, unlike STMSs which is polyallelic, they are of interest because of their abundance (more than 1 per 1000 bp) and presence in single-copy DNA sequence. They may be found in both transcribed and non-transcribed regions and in some instances are the direct causes for observed phenotypic variation. The prerequisite for locus-specific primers is one the major shortcoming for its large scale utility. The advantages of SNPs is that they are most abundant in numbers even within the gene which would provide ideal marker and do not involve gel electrophoresis. SNP detection is more rapid because it is based on oligonucleotide hybridization analysis. Waxy gene (amylose content) and dwarfing gene in Rice and male sterility in Onion (Alcala et al. 1997) are few of the examples where SNP markers are known to be linked with valuable plant genes. The over whelming attraction towards SNPs is because of two main reasons (i) they offer the potential for highest density of markers and (ii) SNP output is one of the binary type useful for automatic data interpretation.

Diversity Arrays Technology (DArT)

This is a microarray-based technique which involves assaying a specific segment of DNA from representation made of whole genomic DNA. This technique does not involve prior DNA sequence information and involves preparation of genomic representations. Representations were generated by restriction digestion of genomic DNA followed by ligation of restriction fragments to enzyme-specific adaptors. Genome complexity was reduced by PCR amplification of ligated mixture using primers with selective overhangs. The amplified DNA fragments from the representations are cloned, purified and finally arrayed onto a solid support (microarray) forming a diversity panel (Semagn et al. 2006). These Diversity Panels helps in genetic fingerprinting of any organism or group of organisms by hybridizing the unknown representation with arrayed nucleic acid fragments which are labeled with fluorescent probes. The ratio of signal intensity measured at each array is the characteristic feature of genotype and the difference in spot are indicator of polymorphic segments in the individuals. This marker has overcome the various limitations of SNPs which is a sequencing based, high-cost marker system. Diversity Array markers are highly reproducible, quick, low in cost as compared to SNPs and does not require any prior sequence knowledge. However, it requires skilled personnel being microarray-based hybridization technique, more physical energy, high cost and advanced lab facilities which limit the use and application of DArT markers in plant research.

Genotyping-By-Sequencing (GBS)

The GBS approach includes the digestion of genomic DNA with restriction enzymes followed by the ligation of barcode adapter, PCR amplification and sequencing of the amplified DNA pool on a single lane of flow cells. It is a multiplexed system for the Illumina NGS platform developed in the Buckler lab (Elshire et al. 2011) for constructing libraries. It generates large numbers of SNPs for use in genetic analyses and genotyping (Beissinger et al. 2013). Key components of this system include low cost, reduced sample handling, fewer PCR and purification steps, no size fractionation, no reference sequence limits, efficient barcoding and easiness to scale up (Davey et al. 2011). GBS is becoming increasingly important as a cost-effective and unique tool for marker-assisted breeding in a range of plant species. Two types of GBS techniques were developed on the basis of ion PGM system usage (i) Digestion of restriction enzyme for marker-assisted selection (MAS) programmes which weremainly used for the identification of new markers and (ii) Multiplex enrichment PCR in which specific PCR primers are selected for the amplification of points of interest. The high-resolution association in maize was studied using GBS technique along with many other species having a complex genome (Poland et al. 2012). Genotyping-by-sequencing has recently been applied to large complex genomes of barley (*Hordeum vulgare* L.) and wheat (*Triticumaestivum* L.), and shown to be an effective tool for rapid generation of molecular markers (Poland et al. 2012). Genotyping cost of GBS per individual is lowest in comparison to array-based and other NGS-based markers in wheat and other non-model crop species (Bassi et al. 2016). The fraction of the genome covered by GBS can potentially be much greater than the fraction captured by even the dense SNP arrays currently available in crop plants (Gorjanc et al. 2015). Furthermore, unlike SNP arrays that are typically developed from a limited sample of individuals, GBS can capture genetic variation that is specific to a population or family of interest. GBS has the advantage that markers are discovered using the population to be genotyped, thus minimizing ascertainment bias. Hence, the flexibility, low cost and GEBV prediction accuracy of GBS make this an ideal approach for GS (Bhat et. al. 2016).

DNA (molecular) markers described above have been successfully implemented in genetics and breeding activities in various agricultural crops. All of them have certain advantages and disadvantages and we should choose them accordingly on the basis of work to be done and crop plants used in the study.

Marker-Assisted Selection

Marker assisted selection (MAS) involves the use of tightly linked (preferably less than 5 cM) DNA markers to locate genes controlling the agriculturally important trait. It helps in phenotype screening and successful transfer of target genes into crop plants. Apart from the type of marker used in MAS, there are few other concerns regarding use of DNA markers in MAS such as reliability, quantity and quality of DNA required, technical procedure for marker assay, level of polymorphism and cost (Mackill and Ni 2000; Mohler and Singrun 2004). The reliability of the markers in predicting phenotype will increase with the increase in number of flanking markers, as the chance of recombination occurring between multiple markers is much lower than for single marker. Marker assisted selection is applied for both monogenic (qualitative) and polygenic (QTL) traits and the goal of MAS is pyramiding or cumulating desirable combinations of genes in new varieties.

MAS ascertain basic advantages as compared to conventional phenotypic selection: It is simpler and selection of the single plants can be carried out at early stages of development with greater reliability (Collard and Mackill 2006). An additional benefit of marker assisted selection (MAS) is that it can be performed on DNA extracted from leaf tissue and consequently provides a non-destructive, seed quantity independent alternative to phenotypic based selection and hence can be used at any point in a breeding programme. Cultivars developed by MAS are non-transgenic and therefore are away from the controversies associated with transgenic crops. Hence, plant breeders utilize the MAS method in the rapid recovery of the recurrent parent genome and selective transfer of genomic regions of interest resulting into faster development and release of varieties (Wijerathna 2015; Khan et al. 2015).

Prerequisites for an Efficient Marker-Assisted Selection Programme

Populations for Gene Tagging/QTL Mapping

The size of population and type of marker utilized are two essential constituents for efficient MAS. The population for genetic/QTL mapping is developed from crosses between diverse parents which exhibit an adequate level of polymorphism. The mapping populations used in marker assisted selection programme includes near-isogenic lines (NILs), DHs, backcrosses (BCs), recombinant inbred lines (RILs) and F_2 populations (Paterson 1996). Although 50-250 individuals are practically good for mapping population but more than 500-thousands of individuals is preferentially recommended (Mohan et al. 1997;Dhingani et al. 2015) for high-resolution and fine linkage mapping.

Marker Development

Marker development for agricultural important trait such as disease resistance (qualitative/quantitative) requires an accurate phenotyping. The inheritance of a trait was screened on F_1, obtained from cross between two different lines which is then self-pollinated or backcrossed with one of the parents to develop the F_2 or BC_1. Phenotypic evaluation can be done either in the field condition (under heavy disease inoculums condition) or in greenhouse/growth chamber with inoculation of plants with specific pathotypes. Analysis of phenotypic screening is used to estimate the number of genes (monogenic/polygenic) controlling the trait. However, the mode of inheritance is difficult to identify when more than three genes are involved in a particular trait expression. In these cases, quantitative trait locus (QTL) mapping can help in identifying the number of involved genes and their roles. Genotyping for desired trait was done by leaf sampling before disease inoculation and then its DNA extraction followed by marker analysis. DNA samples from the parental lines are screened for polymorphisms, using markers that span the chromosome(s) of interest. The polymorphic markers were then used for genotyping the mapping population (Figure 2).

The accuracy of the linkage map will increase with probable number of marker used. The phenotypic data was then correlated with segregation pattern of marker data to find out the number of recombinant individuals. Inaccurate phenotypic data or genotypic data influence the distance between markers (Hackett 2002). Genetic linkage map is constructed on the basis of recombination frequency i.e., the number of recombinant individuals and taking into consideration that 1% crossing over value is similar 1cM genetic distance between marker and gene of interest. A linkage map consists of linkage groups, the number of the group being equal to the number of haploid chromosome of the species under study.

Genetic/QTL Mapping

The development of molecular marker technology renewed interest in genetic mapping/QTL mapping. QTL mapping is based on the assumption that tightly linked marker will be co-inherited along with desired gene during recombination. However, distance between marker and gene of interest is used to calculate recombination frequency. Simple statistical tests such as a χ^2 analysis are used to determine the number of genes involved in governing each trait. With the help of segregation analysis, the actual distance and relative order of markers can be calculated (Collard et al. 2005). MAPMAKER is used to determine linkage between locus and the marker on the basis of segregation of the marker and the disease phenotype in the mapping population. Study done by this software Mapmaker (Lander et al. 1987) performs multipoint linkage analysis. A LOD score is calculated as the log (to base 10) of the ratio between the odds of the hypothesis that a linkage exists and the odds that a linkage does not exists. The Kosambi mapping function (Kosambi 1944) was used to covert recombination frequency into genetic map distance in centimorgans. The genetic map is constructed and some statistical programs are applied to identify molecular markers linked with the trait of interest (Collard et al. 2005). The analysis can also be done manually by product-ratio method on the basis of F_2 data.

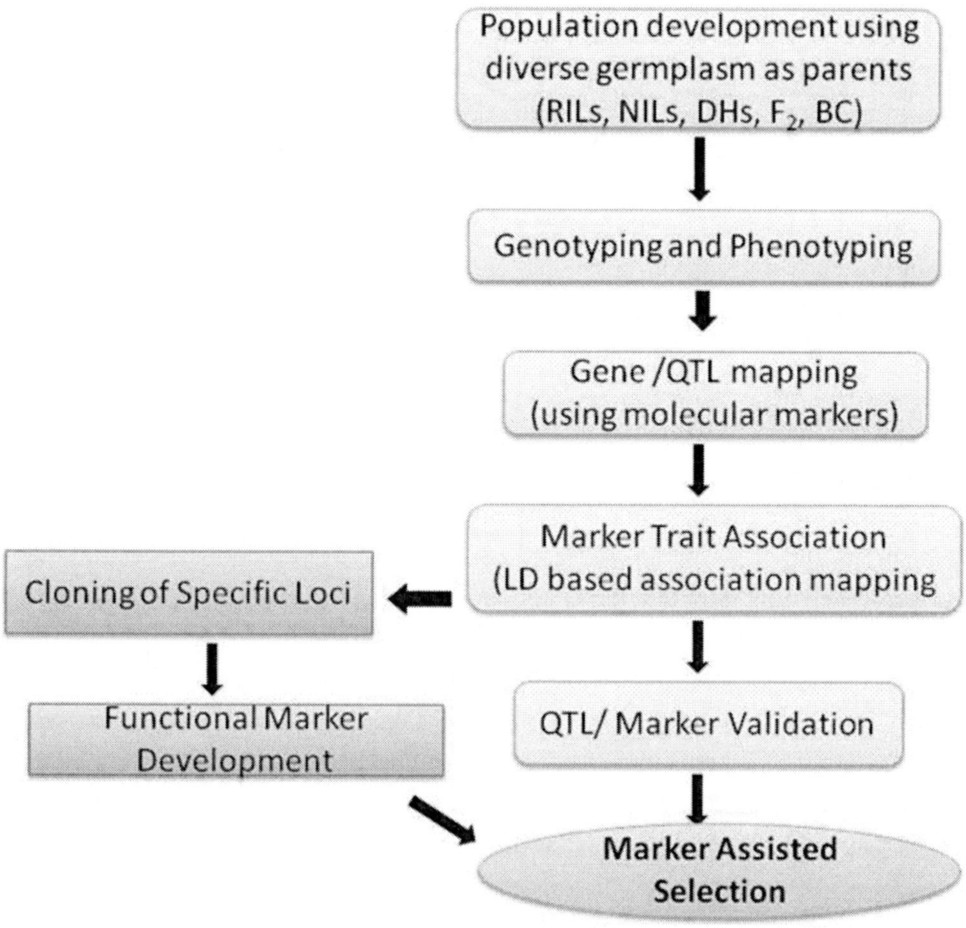

Figure 2. Steps involved in Marker Assisted Selection (MAS) for tagging disease resistance gene.

QTL/Marker Validation

Validation of particular QTL is equally important as QTL detection which can be performed in a segregating populations (Zhou et al. 2003), in NILs (Prasad et al. 2003; Gupta et al. 2006) or among diverse varieties (Mago et al. 2005; Snape et al. 2006) to evaluate the applicability of marker in marker-assisted selection. The validating populations will be developed by crossing different parents in order to check the presence of a particular QTL in other populations with different genetic background. Marker is also validated in varieties with particular desired trait or in some other population made from cross between parents having desired traits. It helps in testing the reliability of markers in different genetic background. Once tightly linked markers that reliably predict a trait phenotype have been identified, they may be used for MAS. An excellent example of successful MAS is the development of an improved version of the pearl millet hybrid HHB 67 with resistance to downy mildew.

Functional/Diagnostic Markers (FMs)

The various PCR-based markers used for MAS in plant breeding have limitations because of false positives produced by genetic recombination during meiosis (Frisch et al. 1999). This leads to development of functional markers (FMs) from cloning of specific loci of the genome which is responsible for variation in phenotypic traits (Risch 1992, Anderson and Lubberstedt 2003, Xu et al. 2005, Nawaz 2017). FMs are directly linked with the genes of interest and recombination between gene and marker is absent. They have a high probability of being co-inherited with the trait of interest hence will avoid false selection. Since these markers don't show any recombination (co-inherited) and located within or tightly linked to the desired gene, they can be used reliably to predict the genes available in segregating populations. Diagnostic (i.e., perfect marker) is directly associated with the desired trait and do not require independent validation and have absolute linkage with the trait being selected. For effective marker-assisted selection in breeding, the marker should be within 10cM of the gene of the interest (Timmerman et al. 1994). A co-dominant marker is preferred as it helps in differentiating the heterozygote in backcross generations (49). Several candidate genes have been now converted into functional markers in various crop species and are quite useful in plant breeding as they help in early screening of large collection of germplasm precisely in very short time.

Marker Assisted Gene Pyramiding

Pyramiding is the process of introgression of multiple genes/QTLs at the same time to combine agronomically important genes (such as resistance to specific pathogens, insects and nematodes, the genes of high-molecular weight proteins, etc.) into one genotype. Stacking several major genes into a single cultivar provides durable resistance for example mutational events at several *Avr* loci would be required to produce a new virulent pathotype. However, introgression of more than one effective R gene using traditional breeding methods is time consuming and often not practical due to epistatic effects. The best strategy for providing durable resistance in a particular genotype is introgression of minor resistance genes (broad-spectrum) in combination with major resistance genes via linked DNA markers. The development of marker-assisted selection (MAS) (Witcombe and Hash 2000) linked to R genes, enables pyramiding of more than one effective gene. It is an attempt to enhance the longevity and durability of cultivars (Kloppers and Pretorius 1997; Shanti et al. 2001; Singh et al. 2001) as it is quite difficult to overcome the resistance controlled by multiple genes through mutation as compared to resistance controlled by a single gene. MAS based gene pyramiding also helps in minimizing the number of backcross generations required by breeders to evaluate the desired gene combination. The incorporation of quantitative resistance controlled by QTLs offers another promising strategy to develop durable disease resistance in several crop plants.

MAS Gene Pyramiding Strategies

The main idea behind gene pyramiding is to create a genotype that is homozygous for the desired alleles at all loci. This is possible through introgression of the transgene into elite cultivars marker assisted backcrossing (MAB). Successful gene pyramiding involves three major steps:

1. Selection of parents containing the desirable genes.
2. The pedigree step aims at cumulating of all target genes in a single genotype called the root genotype through successive crossings.
3. The fixation step aims at fixing the target genes into a homozygous state to avoid their segregation in successive generations. Although the pedigree step may be common, several different procedures can be used to undergo fixation in gene pyramiding.

Double haploid (DH) production and recombinant inbred line (RIL) techniques are commonly used for homozygous line production. However, marker assisted backcross gene pyramiding further enhances the probability of getting the ideal genotype with a combination of useful traits.

Marker Assisted Backcrossing (MAB)

Gene pyramiding through marker assisted backcrossing depends upon three different schemes:

(a) Stepwise transfer- In this method, the recurrent parent (RP1) is crossed with donor parent (DP1) to produce the F1 hybrid which was then backcrossed up to third backcross generation (BC3) to produce the improved recurrent parent (IRP1) (Figure 3A). This improved recurrent parent is then again crossed with other second donor parent (DP2) to pyramid another gene. Though this pyramiding strategy is very accurate even if the method is less accepted as it is time taking.

(b) Simultaneous transfer- In this scheme, the recurrent parent (RP1) is crossed with series of donor parents (DP_1, DP_2, etc.) to get the F_1 hybrids which are then intercrossed to produce improved F_1 (IF1) (Figure 3B).This improved F_1 is then backcrossed with the recurrent parent to get the improved recurrent parent (IRP) lines.

(c) Simultaneous and stepwise transfer-The third strategy involves simultaneous crossing of recurrent parent (RP1) with many donor parents and then backcrossing them up to the BC3 generation. The backcross populations with the individual gene are then intercrossed with each other to get the pyramided lines (Figure 3C). This is the most adequate way as in this method not only time is reduced but fixation of genes is fully assured. Marker assisted backcrossing have been successfully applied to pyramid many desired genes in various crops and these examples are summarized in Table 1.

However, the efficiency of successful marker assisted backcrossing leading to durable crop improvement rely upon several factors such as linkage of markers and the target gene, number of genes to be pyramided, the genetic base of the trait, the number of individuals that can be analyzed and the correct selection of genetic background in which the target gene has to be transferred, the type of molecular marker(s) used, and accessible technical facilities available for crop improvement.

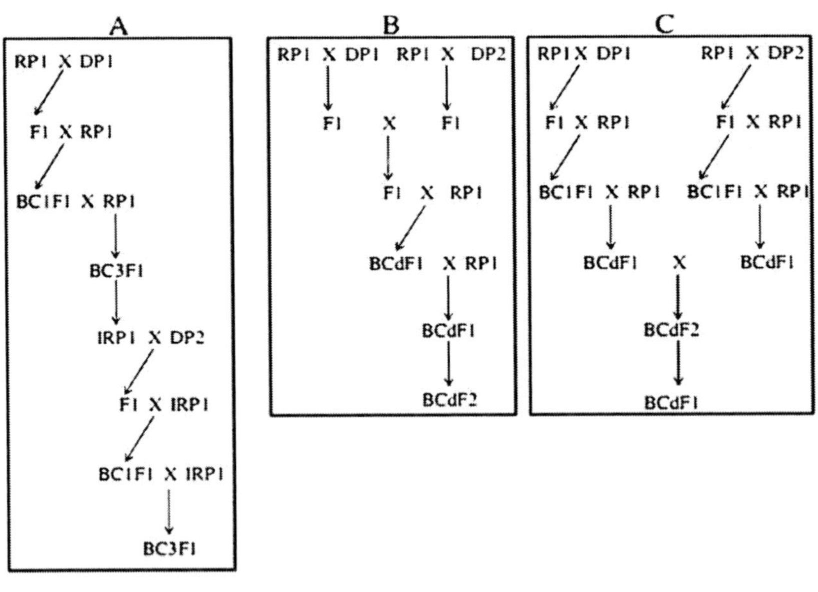

Figure 3. Three different schemes involved in gene pyramiding through marker assisted backcrossing (A) Stepwise transfer (B) Simultaneous transfer (C) Simultaneous and stepwise transfer.

Table 1. Selected examples of MAS based gene pyramiding for important traits in major crops

Crop	Characters/Trait	Pyramided genes	Reference
Rice	Blight resistance	*Xa4,xa5,xa13,Xa21*	Singh et al. (2001) Narayanan et al. (2002) Joseph et al. (2004)
	Blast resistance	*Pi(2)t,Piz5,Pi(t)a*	Hittalmani et al. (2000)
	Gallmidge resistance	*Gm1,Gm4*	Kumaravadivel et al. (2006)
	Multiple resistance	Bacterial blight (Xa21) Sheath b light (RC7) Yellow stem borer Bt fusion gene (cry1AB /cry1Ac)	Datta et al. (2002)
Wheat	Leaf rust resistance	*Lr41, Lr42, Lr43*	Cox et al. (1994)
	Powdery mildew resistance	*Pm-1, Pm-2*	Liu et al. (2000)
	Cereal cyst nematode	CreX and CreY	Dominiques et al. (2007)
Cotton	Insect pest resistance	*Cry 1Ac, Cry 2AcJ*	Gahan et al. (2005)
	Bacterial blight & Sheath blight	chi11, tlp and Xa21	Maruthasalam et al. (2007)
Pea	Nodulation ability	Sym9, Sym10	Schneider et al. (2002)
Barley	Yellow mosaic virus	rym4, rym5, rym9,rym11	Werner et al. (2005)
	Stripe rust resistance	3 QTL	Castro et al. (2003)
Broccoli	Diamond back moths resistance	cry1 A c + cry1c	Cao et al. (2002)
Soyabean	Soybean mosaic virus resistance	Rsv1, Rsv3, Rsv4	Zhu et al. (2006)
	epidopteran resistance	cry1Ac +corn ear worm QTL	Walker et al. (2002)
Chickpea	Lepidopteran resistance	cry1Ac + cry1Ab	Meenakshi et al. (2011)

Source:Malav et al. 2016.

CONCLUSION

Gene pyramiding is one of the most promising approaches for durable crop improvement. With rapid evolution in molecular marker techniques we can significantly accelerate plant breeding research by pyramiding genes. The recent development in NGS based Genotype-based-sequencing (GBS) technology will provide ultimate MAS tool for hastening the marker assisted gene pyramiding. Efficient gene pyramiding requires QTL mapping, detection and application of linked marker after careful validation in different genetic background. In spite of various low cost high throughput marker technologies, inaccurate phenotyping is still a major obstacle in achieving our goals. Development of functional diagnostic markers, Genomic Selection (GS) which is advanced form of MAS integrated with genome editing could help in revolutionizing the plant breeding programmes in coming years.

Acknowledgments

Ruchi Rai wishes to thank DST for WOSA fellowship. Shilpi Singh would like to thank UGC for DS Kothari fellowship.

References

Alcala, J., Giovannoni, J. J., Pike, L. M. & Reddy, A. S. (1997). Application of Genetic Bit Analysis (GBA) for Allelic Selection in Plant Breeding. *Mol Breed*,*3*,495-502.

Andersen, J. R. & Lubberstedt, T. (2003). Functional markers in plants. *Trends Plant Sci*,*8*,554-560.

Anderson, J. A., Ogihara, Y., Sorrells, M. E. & Tanksley, S. D. (1992). Development of a chromosomal-map for wheat based on RFLP markers. *Theor Appl Genet*, *83*,1035-1043.

Bagge, M. & Lubberstedt, T. (2008). Functional markers in wheat: technical and economic aspects. *Mol Breed*, *22*,319-328.

Barrett, B. A., Kidwell, K. K. & Fox, P. N. (1998). Comparison of AFLP and pedigree-based genetic diversity assessment among wheat cultivars from the Pacific Northwest.*Crop Sci*, *38*,1271-1278.

Bassi, F. M., Bentley, A. R., Charmet, G., Ortiz, R. &Crossa, J. (2016). Breeding schemes for the implementation of genomic selection in wheat (*Triticum* sp.). *Plant Sci*, *242*,23-36.

Beissinger, T. M., Hirsch, C. N., Sekhon, R. S., Foerster, J. M., Johnson, J. M., Muttoni, G.,Vaillancourt, B., Buell, C. R., Kaeppler, S. M. & de Leon, N. (2013). Marker density and read depth for genotyping populations using genotyping-by-sequencing. *Genetics*, *193*, 1073-1081.

Bhat, J. A., Salgotra, R. K. &Dar, M. Y. (2015). Phenomics: a challenge for crop improvement in genomic era. *Mol Plant Breed*, *6*,1-11.

Bohn, M., Utz, H. F. & Melchinger, A. E. (1999). Genetic similarities between winter wheat cultivars determined on the basis of RFLPs, AFLPs and SSRs and their use for predicting progeny variance. *Crop Sci*, *39*,228-237.

Botstein, D., White, R. L., Skolnick, M. & Davis, R. W. (1980). Construction of genetic linkage map in man using restriction fragment length polymorphisms. *Am J Hum Genet*, *32*, 314-331.

Bueren, E., Backel, G., Vriend, H. & Ostergard, H. (2010). The role of Molecular markers and marker assisted selection in breeding for organic agriculture. *Euphytica*, *175*, 51-64.

Cao, L. Y., Zhuang, J. Y., Zhan, X. D., Zheng, K. L. & Cheng, S. H. (2003). Hybrid rice resistant to bacterial blight developed by marker assisted selection. *Zhongguo Shuidao Kexue (Chin J Rice Sci)*, *17*, 184-186.

Castro, A. J., Capettini, F., Corey, A. E., Filichkina, T., Hayes, P. M., Kleinhofs, A., Kudrna, D., Richardson, K., Sandoval-Islas, S., Rossi, C. &Vivar, H. (2003). Mapping and pyramiding of qualitative and quantitative resistance to stripe rust in barley. *TheorAppl Genet*, *107*, 922-930.

Chao, S., Sharp, P. J., Worland, A. J., Warham, E. J., Koebner, R. M. D. & Gale, M. D. (1989). RFLP-based maps of wheat homologous group 7 chromosomes. *Theor Appl Genet*, *78*, 495-504.

Chen, X., Soria, M. A., Yan, G., Sun, J. & Dubcovsky, J. (2003). Development of sequence tagged site and cleaved amplified polymorphic sequence markers for wheat stripe rust resistance gene *Yr 5*. *Crop Sci*, *43*, 2058-2064.

Collard, B. & Mackill, D. (2006). *Marker assisted breeding for rice improvement. Invited review paper, Philosophical Transactions of the Royal Society of London Series* London, Springer Publisher.

Collard, B. C., Jahufer, M. Z., Brouwer, J. B. &Pang, E. C. (2005). An introduction to markers., quantitative trait loci (QTL) mapping and marker-assisted selection for crop improvement: the basic concepts. *Euphytica*,*142*,169-196.

Coutinho, J. P., Carvalho, A., Martin, A., Ribeiro, T., Morais-Cecílio, L. & Lima-Brito, J. (2016). Oak Ribosomal DNA: Characterization by FISH and Polymorphism Assessed by IGS PCR-RFLP. *Plant Syst Evol*, *302*,5,27-544.

Cox,T. S., Raupp, W. J. & Gill, B. S. (1993). Leaf rust-resistance genes, *Lr41*, *Lr42* and *Lr43* transferred from *Triticumtauschii*to common wheat. *Crop Sci*, *34*, 339-343.

Datta, K., Baisakh, N., Thet, K. M., Tu, J.& Datta, S. K. (2002). Pyramiding transgenes for multiple resistances in rice against bacterial blight, yellow stem borer and sheath blight. *Theor Appl Genet*, *106*, 1-8.

Davey, J. W., Hohenlohe, P. A., Etter, P. D., Boone, J. Q., Catchen, J. M. & Blaxter, M. L. (2011). Genome-wide genetic marker discovery and genotyping using next-generation sequencing. *Nat Rev Genets*, *12*,499-510.

Devos, K. M. & Gale, M. D. (1992). The use of random amplified polymorphic DNA markers in wheat. *Theor Appl Genet*, *84*,567-572.

Devos, K. M. & Gale, M. D. (1993). Extended genetic maps of the homologous group 3 chromosomes of wheat, rye and barley. *Theor Appl Genet*, *85*,649-652.

Dhingani, R. M., Umrania, V. V., Tomar, R. S.,Parakhia, M. V. & Golakiya, B. A. (2015). Introduction to QTL mapping in plants. *Ann Plant Sci*, *4*,1072-1079.

Dominique, B., Jocelyne, L., Paulette, A., Anne, T., Roger, R. & Joseph, J. (2007). Marker-assisted pyramiding of two cereal cyst nematode resistance genes from *Aegilopsvariabilis*in wheat. *Mol Breed*, *20*, 31-40.

Elshire, R. J., Glaubitz, J. C., Sun, Q., Poland, J. A., Kawamoto, K. & Buckler, E. S. (2011). A robust, simple genotyping-by-sequencing (GBS) approach for high diversity species. *PLoS ONE*,*6*,e19379.

FAO. (1996). *Food, Agriculture and Food Security: Developments since the World Food Conference and Prospects. Technical Background Document No 1 for the World Food Summit, Rome*.

Frisch, M., Bohn, M. & Melchinger, A. A. (1999). Minimum sample size and optimal positioning of flanking markers in marker-assisted backcrossing for transfer of a target gene. *Crop Sci*,*39*, 967-975.

Gahan, L. J., Ma, Y. T., Cobble, M. L. M., Gould, F., Moar, W. J. & Heckel, D. G. (2005). Genetic basis of resistance to Cry 1Ac and Cry 2Aa in *Heliothis virescens* (Lepidoptera: Noctuidae). *J Econ Entomol*, *98*, 1357-1368.

Gao, L., Jia, J. & Kong, X. (2016). A SNP-based molecular barcode for characterization of common wheat. *PLoS ONE*, *11*,e0150947.

Gold, J., Harder, D., Townley-Smith, F., Aung, T. &Procunier, J. (1999). Development of a molecular marker for rust resistance gene *Sr39* and *Lr35* in wheat breeding lines. *Elec J Biotech*, *2*,35-40.

Goodwin, S. B., Hu, X. & Shaner, G. (1998). An AFLP marker linked to a gene for resistance to *Septoriatritici* blotch in wheat. In: *Proc 9th Int Wheat Symp* (Ed. Slinkard AE), Univ Extension Press, Univ of Sakatchewan Saskatoon, pp. 108-110.

Gorjanc, G., Cleveland, M. A., Houston, R. D. & Hickey, J. M. (2015). Potential of genotyping-by-sequencing for genomic selection in livestock populations.*Genet Select Evol*, *47*,12.

Gupta, P. K., Varshney, R. K., Sharma, P. C. & Ramesh, B. (1999). Molecular markers and their application in wheat breeding. *Plant Breed*, *118*, 369-390.

Gupta, S. K., Charpe, A., Koul, S., Haque, Q. M. R. &Prabhu, K. V. (2006). Development and validation of SCAR markers co-segregating with an *Agropyronelongatum* derived leaf rust resistance gene *Lr24* in wheat. *Euphytica*, *150*,233-240.

Gupta, P. K. & Varshney, R. K. (2000). The development and use of microsatellite markers for genetic analysis and plant breeding with emphasis on bread wheat. *Euphytica*, *113*,163-185.

Hackett, C. A. (2002). Statistical methods for QTL mapping in cereals. *Plant Mol Biol*, *48*,585-599.

Hasan, M. M., Rafii, M. Y., Ismail, M. R., Mahmood, M., Rahim, H. A., Alam, M. A., Ashkani, S., Malek, M. A.& Latif, M. A. (2015). Marker-assisted backcrossing: a useful method for rice improvement. *Bio technol Bio technol Equip*, *29*,237-254.

He, Q., Viljanen, M. K. & Mertsola, J. (1994). Effects of thermocyclers and primers on the reproducibility of banding patterns in randomly amplified polymorphic DNA analysis. *Mol Cell Probes*, *8*, 155-160.

Hittalmani, S., Parco, A., Mew, T. V., Zeigler, R. S. & Huang, N. (2000). Fine mapping and DNA marker-assisted pyramiding of the three major genes for blast resistance in rice. *Theor Appl Genet*, *100*, 1121- 1128.

Hoisington, D., Bohorova, N., Fennell, S., Khairallah, M., Pellegrineschi, A. & Ribaut, J. M. (2002). The application of biotechnology in wheat improvement: new tools to improve wheat productivity. In: *Bread wheat: improvement and production* (Eds. Curtics B. C., Rajaram S., Gomez H.), FAO Plant Production Series no. 30, Rome, pp. 175-198.

Huang, J. K., Pray, C. & Rozelle, S. (2002). Enhancing the crops to feed the poor. *Nature*, *418*,678-684.

Islam, S., Mir, J. I. & Kudesia, R. (2015). Evaluation of genetic diversity in *Vignaradiata* (L.) using protein profiling and molecular marker (RFLP). *Int J Plant Breed Genet*, *9*,238-246.

Jeffreys, A. J., Wilson, V. & Thein, S. L. (1985). Hypervariable “minisatellite” regionsin human DNA. *Nature*, *314*, 67-73.

Jones, C. J., Edwards, K. J., Castaglione, S., Winfield, M. O., Sala, F., van de Wiel, C., Bredemeijer, G., Vosman, B., Matthes, M., Daly, A., Brettschneiderm, R., Bettinim, P., Buiatti, M., Maestri, E., Malcevschi, A., Marmiroli, N., Aert, R., Volckaertm, G., Ruedam, J., Linacero, R., Vazquez, A. & Karp, K. (1997). A Reproducibility testing of RAPD, AFLP and SSR markers in plants by a network of European laboratories. *Mol Breed*, *3*,381-390.

Joseph, M., Gopalakrishnan, S., Sharma, R. K., Singh, A. K., Singh, V. P., Singh, N. K. & Mohapatra, T. (2004). Combining bacterial blight resistance and Basmati quality characteristics by phenotypic and molecular marker assisted selection in rice. *Mol Breed*, *13*, 377-387.

Joshi, S. P., Prabhakar, K., Ranjekar, P. K. & Gupta, V. S. (2011). *Molecular markers in plant genome analysis*, 1-19.

Khan, M. H., Dar, Z. A. & Dar, S. A. (2015). Breeding strategies for improving rice yield- a review *Agric Sci*, *6*, 467-478.

Kloppers, F. J. & Pretorius, Z. A. (1997). Effects of combinations amongst genes Lr13, Lr34 and Lr37 on components of resistance in wheat to leaf rust. *Plant Pathol*, *46*,737-750.

Konieczny, A. & Ausubel, F. M. (1993). A procedure for mapping *Arabidopsis* mutations using co-dominant ecotype-specific PCR-based markers. *The Plant J*, *4*,403-410.

Kosambi, D. D. (1944). The estimation of map distances from recombination values. *Ann Eugen*, *12*, 172-175

Kumaravadivel, N., Uma, M. D., Saravana, P. A. & Suresh, H. (2006). Molecular marker-assisted selection and pyramiding genes for gall midge resistance in rice suitable for Tamil Nadu region. *International Rice Congress*, 257p.

Lander, E. S., Green, P., Abrahmson, J., Barlow, A., Daly, M. J., Lincoln, S. E. & Newberg, L. (1987). MAPMAKER: an interactive computer package for constructing primary genetic linkage maps of experimental and natural populations. *Genomics*, *1*, 174-181.

Litt, M. & Luty, J. A. (1989). A hypervariable microsatellite revealed by *in vitro* amplification of a dinucleotide repeat within the cardiac muscle actin gene. *Am J Hum Genet*, *44*,397-401.

Liu, J., Liu, D., Tao, W., Li, W., Wang, S., Chen, P., Cheng, S. &Gao, D. (2000). Molecular marker-facilitated pyramiding of different genes for powdery mildew resistance in wheat. *Plant Breed*, *119*, 21-24.

Mackill, D. J., Nguyen, H. T. & Zhang, J. (1999). Use of molecular markers in plant improvement programs for rain fed low land rice. *Field Crops Res*, *64*,177-185.

Mackill, D. J. & Ni, J. (2000). Molecular mapping and marker assisted selection for major-gene traits in rice. In: *Proc 4th Int Rice Genetics* (Eds. Khush GS, Brar DS, Hardy B), Los Baños, The Philippines: International Rice Research Institute, pp. 137-151.

MacPherson, J. M., Eckstein, P. E., Scoles, G. J. & Gajadhar, A. A. (1993). Variability of the random amplified polymorphic DNA assay among thermal cyclers and effects of primer and DNA concentration. *Mol Cell Probes*, *7*,293-299.

Mago, R., Bariana, H. S., Dundas, I. S., Spielmeyer, W., Lawrence, G. J., Pryor, A. J. & Ellis, J. G. (2005). Development of PCR markers for the selection of wheat stem rust resistance genes *Sr24* and *Sr26* in diverse wheat germplasm. *Theor Appl Genet*, *111*, 496-504.

Malav, A. K. & Indu, Chandrawat K. S. (2016). Gene pyramiding: An overview. *Int J Curr Ress Biosci Plant Biol*, *3*, 22-28.

Marino, C. L., Nelson, J. C., Lu, Y. H., Sorrells, M. E., Leroy, P., Tuleen, N. A., Lopes, C. R. & Hart, G. E. (1996). Molecular genetic maps of the group 6 chromosomes of hexaploid wheat (*Triticum aestivum* L. Em. Thell.). *Genome*, 39, 359-366.

Marshall, E. (1997). The hunting of the SNP. *Science*,*278*,2047.

Maruthasalam, S., Kalpana, K., Kumar, K., Loganathan, M., Poovannan, K., Raja, J. A., Kokiladevi, E., Samiyappan, R., Sudhakar, D. & Balasubramanian, P. (2007). Pyramiding transgenic resistance in elite Indica rice cultivars against the sheath blight and bacterial blight. *Plant Cell Rep*,*26*, 791-804.

Meenakshi, M., Aditya, K. S., Indraneel, S., Illimar, A. &Amla, D. V. (2011). Pyramiding of modified cry1Ab and cry1Ac genes of *Bacillus thuringiensis*in transgenic chickpea (*Cicerarietinum* L.) for improved resistance to pod borer insect *Helicover paarmigera*. *Euphytica*,*182*, 87-102.

Mohan, M., Nair, S., Bhagwat, A., Krishna, T. G., Yano, M., Bhatia, C. R. & Sasaki, T. (1997). Genome mapping, molecular markers and marker-assisted selection in crop plants. *Mol Breed,3*,87-103.

Mohler, V. & Singrun, C. (2004). General considerations: marker-assisted selection In: *Molecular Marker Systems in Plant Breeding and Crop Improvement* (Eds. Lorz H,Wenzel G), Springer, Berlin, Heidelberg,pp. 305-317.

Moury, B., Pflieger, S., Blattes, A., Lefebve, V. & Palloix, A. (2000). A CAPS marker to assist selection of tomato spotted wilt virus (TSWV) resistance in pepper. *Genome,43*,137-142.

Narayanan, N. N., Baisakh, N., Vera, C. M., Gnanamanickam, S. S., Datta, K. & Datta, S. K. (2002). Molecular breeding for the development of blast and bacterial blight resistance in rice cv.IR50. *Crop Sci,42*, 2072- 2079.

Nawaz, M. A., Rehman, H. M., Baloch, F. S., Ijaz, B., Ali, M. A., Khan, I. A., Lee, J. D., Chung, G. & Yang, S. H. (2017). Genome and transcriptome-wide analysis of cellulose synthase gene superfamily in soybean. *J Plant Physiol,215*, 163-175.

Nerkar, G., Farsangi, F. & Devarumath, R. (2015). Organellar Genome Diversity in *Saccharum* and *Erianthus* sp. Revealed by PCR RFLP. *Mol Plant Breed, 6*, 1-11.

Nikiforov, T. T., Rendle, R. B., Goelet, P., Rogers, Y. H., Kotewicz, M. L., Anderson, S., Trainor, G. L. & Knapp, M. R. (1994). Genetic Bit Analysis: a solid phase method for typing single nucleotide polymorphisms. *Nucleic Acids Res,22*, 4167-4175.

Ortiz, R. (1998). Critical role of plant biotechnology for the genetic improvement of food crops: perspectives for thenext millennium. *Electron J Biotechnol, 1*, 16-17.

Paran, L. & Michelmore, R. W. (1993). Development of reliable PCR-based markers linked to downy mildew resistance genes in lettuce. *Theor Appl Genet, 85*, 985-993.

Paterson, A. H. (1996). Making genetic maps. In: *Genome mapping in plants* (Ed. Paterson AH), Austin, (TX): RGLandes Company, pp. 23-39

Poland, J. A. & Rife, T. W. (2012). Genotyping-by-sequencing for plant breeding and genetics. *Plant Genome,5*, 92-102.

Prasad, M., Kumar, N., Kulwal, P., Röder, M. S., Balyan, H., Dhaliwal, H. & Gupta, P. (2003). QTL analysis for grain protein content using SSR markers and validation studies using NILs in bread wheat. *Theor Appl Genet,106*, 659-667.

Risch, N. (1992). Genetic linkage, interpreting LOD scores. *Science,255*, 803-805.

Ruttan, V. W. (1999). The transition to agricultural sustainability. *Proc Natl Acad Sci USA, 96*, 5960-5967.

Schneider, A. (2002). Mapping of a nodulation loci sym9 and sym10 of pea. *Theor Appl Genet, 104*, 1312-1316.

Semagn, K., Bjornstan, A. & Ndjiondjop, M. N. (2006). An Overview of Molecular Marker Methods for Plants. *Afr J Biotechnol,5*, 2540-2568.

Shan, X., Blake, T. K. & Talbert, L. E. (1999). Conversion of AFLP markers to sequence-specific PCR markers in barley and wheat. *Theor Appl Genet, 98*, 1072-1078.

Shanti, M. L., George, M. L. C., Cruz, C. M. V., Bernardo, M. A., Nelson, R. J., Leung, H., Reddy, J. N. & Sridhar, R. (2001). Identification of resistance genes effective against rice bacterial blight pathogen in eastern India. *Plant Dis*,*85*, 506-512.

Singh, S., Sidhu, J. S., Huang, N., Vikal, Y., Li, Z., Brar, D. S., Dhaliwal, H. S. & Khush, G. S. (2001). Pyramiding three bacterial blight resistance genes (xa5, xa13 and Xa21) using marker-assisted selection into Indica rice cultivar PR106. *Theor Appl Genet*,*102*, 1011-1015.

Slafer, G. A., Araus, J. L., Royo, C. & Del Moral, L. F. G. (2005). Promising eco-physiological traits for genetic improvement of cereal yields in Mediterranean environments. *Ann Appl Biol*, *146*, 61-70.

Snape, J. W., Worland, A. J. & Sayers, E. J. (2006). Genetic analysis of orange blossom midge resistance in wheat using chromosome substitution lines and isogfenic lines. In, *Proc 13th Int EWAC Conference* (Eds.Börner A, Pankova K, Snape J W), Prague, Czech Republic, pp.71-74.

Stuber, C., Polacco, M. & Senior, M. (1999). Synergy of Empirical Breeding, Marker Assisted selection and Genomics to Increase Crop Yield Potential. *Crop Sci*, *39*, 1571-1578.

Talbert, L. E., Blake, N. K., Chee, P. W., Blake, T. K. & Magyar, G. M. (1994). Evaluation of sequence-tagged-site-facilitated PCR products as molecular markers in wheat. *Theor Appl Genet*,*87*, 789-794.

Thottappilly, G., Magonouna, H. D. & Omitogun, O. G. (2000). The use of DNA markers for rapid improvement of crops in Africa. *Afric Crop Sci J*, *8*, 99-108.

Timmerman, G. M., Frew, T. J., Weeden, N. F., Miller, A. L. & Goulden, D. S. (1994). Linkage analysis of er-1, a recessive *Pisumsativum* gene for resistance to powdery mildew fungus (*Erysiphepisi* D. C.). *Theor Appl Genet*,*88*, 1050-1055.

Trethowan, R. M., Reynolds, M., Sayre, K. & Ortiz-MonasterioI. (2005). Adapting wheat cultivars to resource conserving farming practices and human nutritional needs. *Ann Appl Biol*, *146*, 405-413.

Van de Wiel, C., Arens, P. & Vosman, B. (1999). Microsatellite retrieval in lettuce (*Lactuca sativa* L.). *Genome*,*42*, 139-149.

Varshney, R. K., Graner, A. & Sorrells, M. E. (2005). Genic microsatellite markers in plants: features and applications. *Trends Biotech*,*23*, 48-55.

Vos, P., Hogers, R., Bleeke,R. M., Reijans,M., van de Lee, T., Fornes, M.,Frijters,A., Pot, J., Peleman, J., Kuiper, M. & Zabeau, M. (1995). AFLP: a new technique for DNA fingerprinting. *Nucleic Acids Res*,*23*, 4407-4414.

Walker, D., Roger Boerma, H. R., All, J. & Parrott, W. (2002). Combining cry1Ac with QTL alleles from PI 229358 to improve soybean resistance to *Lepidopteran* pests. *Mol Breed*,*9*, 43-51.

Welsh, J. & McClelland, M. (1990). Fingerprinting genomes using PCR with arbitrary primers. *Nucleic Acids Res*,*18*, 7213-7218.

Werner, K., Friedt, W. & Ordon, F. (2005). Strategies for pyramiding resistance genes against the barley yellow mosaic virus complex (BaMMV, BaYMV, BaYMV-2). *Mol Breed,16*, 45-55.

Wijerathna, Y. M. A. M. (2015). Marker assisted selection: biotechnology tool for rice molecular breeding. *Adv Crop Sci Technol*, *28*, 1-4.

Williams, J. G., Kubelik, A. R., Livak, K. J., Rafalski, J. A. & Tingey, S. V. (1990). DNA polymorphisms amplified by arbitrary primers are useful as genetic markers. *Nucleic Acids Res,18*, 6531-6535.

Witcombe, J. R. & Hash, C. T. (2000). Resistance gene deployment strategies in cereal hybrids using marker assisted selection: gene pyramiding, three-way hybrids, and synthetic parent populations. *Euphytica,112*, 175-186.

Xu, Y., McCouch, S. R. & Zhang, Q. (2005). How can we use genomics to improve cereals with rice as a reference genome? *Plant Mol Biol,59*, 7-26.

Zhou, W. C., Kolb, F. L., Bai, G. H., Domier, L. L., Boze, L. K. & Smith, N. J. (2003). Validation of a major QTL for scab resistance with SSR markers and use of marker-assisted selection in wheat. *Plant Breed,122*, 40-46.

Zhu, Y., Yu, DY. ., Chen, S. Y. & Gai, J. Y. (2006). Inheritance and gene mapping of resistance to SMV strain SC-7 in soybean. *Acta Agron Sinica,32*, 936-938.

Zietkiwicz, E., Rafalski, A. & Labuda, D. (1994). Genome fingerprinting by simple sequence repeat (SSR)-anchored polymerase chain reaction amplification. *Genome*,20, 176-183.

In: Innovations in Life Science Research ISBN: 978-1-53615-868-7
Editors: R. P. Sinha, S. Pandey-Rai and N. Ghoshal ©2019 Nova Science Publishers, Inc.

Chapter 6

SIGNIFICANCE OF PLOIDY IN GENOME EVOLUTION, SPECIATION AND CROP IMPROVEMENT

Lakee Sharma[1], Sanjay Kumar Rai[2], Apoorva[1] and Shashi Pandey-Rai[1,*]

[1]Laboratory of Morphogenesis, Centre of Advanced Study in Botany, Institute of Science, Banaras Hindu University, Varanasi, Uttar Pradesh, India

[2]Department of Horticulture, Dr Rajendra Prasad Central Agricultural University, Pusa, Samastipur, Bihar

ABSTRACT

Whole genome duplications, i.e., polyploidy, shaped the genome evolution throughout the plant kingdom. This is especially true for all angiospermic plants that have experienced at least minimum two whole genome duplications. Even plants that are nowadays genetically diploid did undergo past paleo-polyploidization events. The omnipresence of polyploidy in plants led to a debate about the influence of whole genome duplication in diversification. In plants, it is apparent that much of this biological diversity is due to both doubling of the chromosomes (polyploidy) as well as hybridization between species. However, it remains unclear how these two processes interact in the evolution of new species. In spite of much improvement and advances in evolutionary biology, major questions remain unanswered regarding the main drivers of speciation and the implications of polyploidy in genome evolution. In particular, knowledge about the interplay of diverse types of reproductive barriers, polyploidy and the role of natural selection in the speciation process remains incomplete. In addition, the role of polyploidy in diversification, speciation and biological novelty is a topic of much current debate. This chapter has been devoted to providing knowledge about the mechanisms underlying diversification inside polyploids and how polyploidy could promote diversification and crop improvement and could enhance metabolites in medicinally important plants.

[*]Corresponding Author's Email: shashi.bhubotany@gmail.com.

Keywords: polyploidy, genome duplication, genome diversification, speciation, crop improvement, metabolites

INTRODUCTION

One generalization that has recently emerged out from the massive effort in the area of genome sequencing and mapping in a diversity of organisms is genome doubling through polyploidy. It is a prominent process and hasrole in plant evolution through interspecific hybridization and speciation. The elucidation of the major causes and consequences of polyploidy has been the main focus of various researches in the last hundred years (Ramsey and Ramsey 2014). The studies related to the polyploidization and its role in speciation have provided a wide range of information about different aspects and origin of polyploidy, including classification, frequency, mechanisms of origin and various events, as well as its ecological, genetic and evolutionary consequences.

Polyploids are responsible for variations through mechanisms of gene flow with diploids and multiple origins of polyploids and through mechanisms that generate 'de novo variation' such as chromosomal rearrangements and epigenetic phenomenon. Polyploidy has long been considered as a well demonstrated example of instant or sympatric speciation and species generated through polyploidy are mostly reproductively isolated from their diploid progenitors.

Polyploidy in Plants: An Overview

Genomes of plant species greatly vary in size and content of the DNA. These variations are driven by various polyploidization processes that have happened over the course of evolution. Plant genomes vary in size enormously, from the smallest known plant genome, found in *Genlisea tuberosa* (1C ≈ 61 Mbp) and largest 150 billion base pair in *Paris japonica*. The characteristic feature of the living material is their ability to reproduce themselves. However, the ever dynamic nature of the surrounding environment also greatly affects the plants, much like other organisms. Several evolutionary and selective bottlenecks are necessary to change/adoption of organisms to keep their "race going on" (Elliot 1958; Grant 1981). New species arise from existing ones in the time course by the process of evolution and genome diversification forms the major basis for speciation. New species arise by their reproductive isolation and genetic diversification involves various genetic events such as fusion, deletion, duplication, elongation, inversion and/or translocation of geneic and/or nongeneic DNA segments. Mutations in the existing genes and horizontal addition of genes (horizontal gene transfer) are the major causes in the creation of new variations and the general recombination process further amplifies the variation created by above processes (Grant 1981).Polyploidization such as autopolyploidy (multiplication of basic set of chromosomes) and allopolyploidy (combination of genetically distinct chromosome sets) are also the other important mechanisms for the generation of genetic variations and speciations, which provide rapid evolution of species. Ancient polyploidy hasdirectly influence in increase in number of plant species and its complexity. However, functional diversification and expansion of specific gene families having duplication events are responsible for fast diversification of angiosperm lineages (Levin 1983; Grant 1981). Mutations occurring in parental and duplicated genes and selection imposed by the environment are the main forces that drive genome diversification of the angiosperms (Lewis 1980; Elliot 1958).

Polyploidization in Plants: An Evolutionary Implication

During the course of plant evolution genome doubling (polyploidy) has been continued to be a pervasive force (Van de Peer Y et al. 2009). While having these processes plant genomes harbor multiple rounds of polyploidization events, often followed by massive silencing and elimination of duplicated genes (Mable 2003). Recent studies have refined our knowledge of the number and timing of polyploidy events and their effects on genome, i.e., alterations in genome structure. Studies reveal that many polyploids perform rapid genomic alterations, some of them arising with the onset of polyploidy. After duplication of genes, different gene classes show different grades of survivorship. That is some polyploids produces resistant varieties and some susceptible, some of them causes speciation and others loss their existence during the course of evolution. Recent theories are now supported by evidence suggesting that duplicated genes that are preserved stably in genomes typically get diversified functionally or undergo sub-functionalization by prolonged neo-functionalization. Polyploidy extensively alters the gene expression and can induce genetic and epigenetic events which include gene silencing and elimination of DNA sequences.

Polyploidy was discovered in 1907 and since then till now it is known as an important phenomenon in vascular plants especially in angiosperms. Several plant species shows to have a polyploidy ancestry. Phylogenetic and cytogenetic data, suggests that 15% of angiosperm and 31% of fern speciation events are accompanied due to increased ploidy level. However, previous knowledge of the frequency of polyploid speciation informed that the formation of neopolyploid species is rare. Thus the polyploidy seems to have substantial contribution to speciation, cladogenesis and diversification.

Mechanisms of Induction of Polyploidy

Polyploidization is a common and well known event in angiospermic plants. However, the success of polyploidization in the speciation is infrequent, which have made the evolution of species a slow process. Survival of polyploids gets affected mainly by various degrees of sterility occurred after the immediate polyploids which get reproductively isolated from their progenitor populations (Matzke and Matzke 1998; Matzke et al.1999; Hieter and Griffith 1999; Hilu 1993; Kihara and Ono 1926). The genomes of polyploids undergo alterations in structure and number using all the possible mechanisms of variation. These alterations tend to obviate the collapse of meiosis in nascent polyploids, due to non-disjunction arising from multivalent chromosome associations. Differentiation in chromosomal structure and number facilitates the formation of bivalent chromosome associations, required for the normal progress of meiosis. Thus successful speciation takes place via polyploidy.

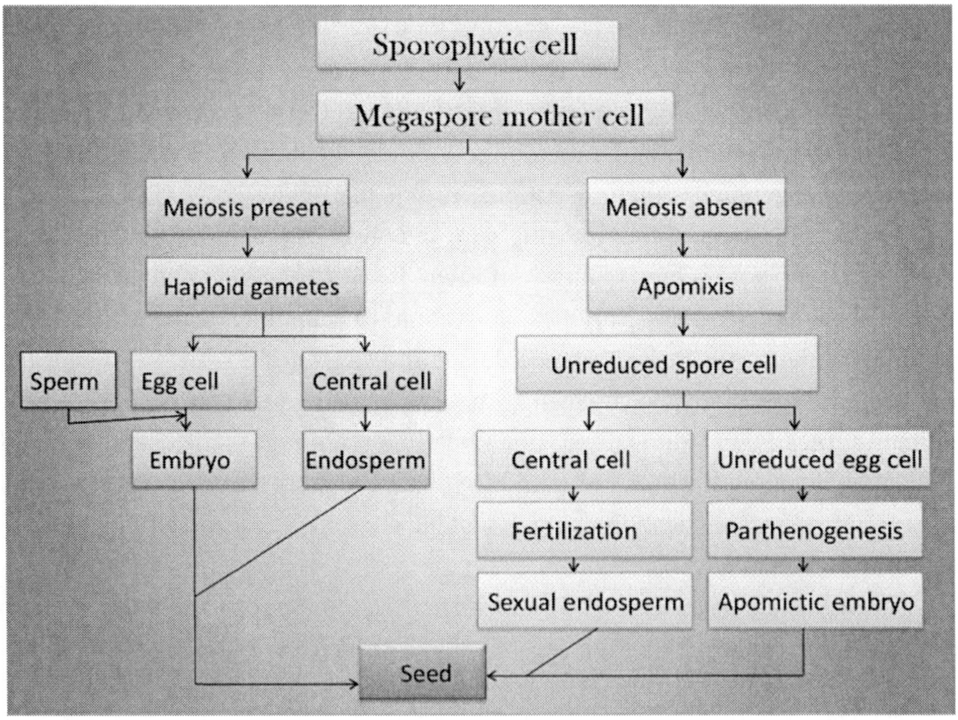

Figure 1. Different events during reproduction in plants.

Besides being reproductively isolated, different species, from other such species, possess one or more unique distinguishing function(s). However, the polyploidization provides raw geneic DNA for the evolution of new genes and imposed redundancy of all genes. The selective pressures for improved fertility also serve as the great cause for the evolution of new genes (improved genes).Cladogenesis reveals that genomes of species clades are known to be syntenous to various degrees, i.e., organisms of recent splitting show similar cluster of genes in the same relative position in the genome. The successful establishment of allopolyploids may be inversely related to synteny between their respective parents. Protoplasts fusion of selective species followed by hybrid regeneration appears to be a rapid means of speciation for developing new crops at commercial level (Larkins et al. 2001; Lawton-Rauh 2000; Leitch and Bennet 1997; Osborn et al. 2003; Otto and Whitton 2000; Stebbins 1971; Stebbins 1985).

In nature plant species can survive easily by being reproductively isolated because pollinations within a species alone result in normal seed formation (Figure 1). Polyploids arise due to a rare mitotic or meiotic catastrophe. The non-disjunction is one of the examples that cause the formation of gametes which have a complete set of duplicate chromosomes where diploid gametes are frequently formed.

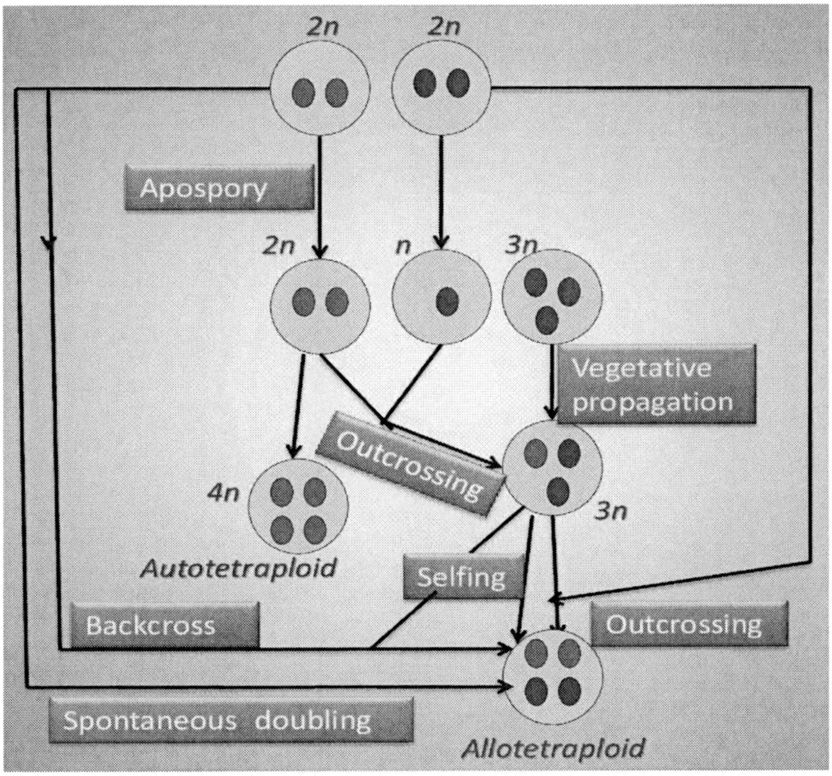

Figure 2. Different natural as well as anthropogenic events to achieve polyploidy in plants.

Ploidy in plants is achieved by many pathways including natural as well as anthropogenic mechanisms. Natural processes are apospory, spontaneous doubling of chromosomes (due to some naturally occurring environmental factors) and by some other processes. These polyploids may be further treated as reservoirs for the production of new progeny by anthropogenic means like, by their backcrossing, vegetative propagation, outcrossing, selfing etc. However, anthropogenically new polyploids can also be produced by diploid plants (Figure 2).

The new species raised by any means of speciation are reproductively isolated and genetically different from their parental species that is they no longer share same gene pool and also cannot interbreed. The reproductive isolation of new species by any event (sympatric, parapatric and/or allopatric) is an account of developmental features which are in turn related to geneic divergence.

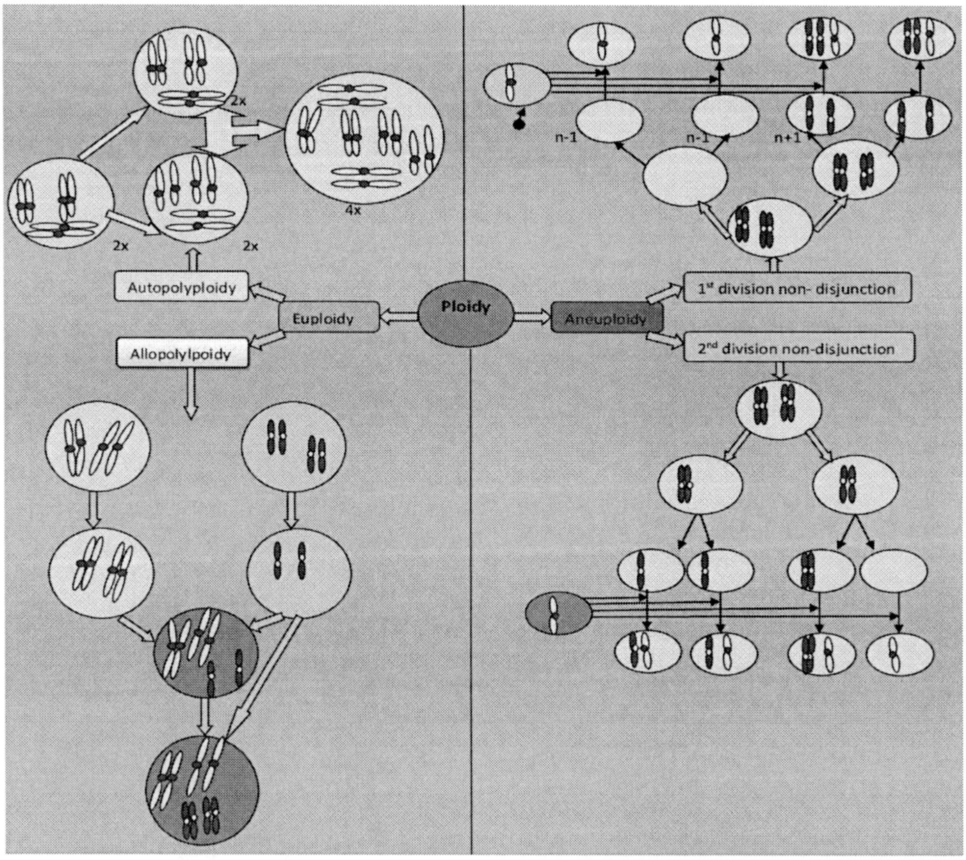

Figure 3. An outline of the mechanism of occurrence of both types of ploidy events, i.e., Euploidy and Aneuploidy.

There are several ways for a cell/organism to double the chromosome number and become polyploid. Most phenomenons are related to incomplete cell division. Cytokinesis normally separates the cell into two after chromosome segregation and failure of cytokinesis produces a single tetraploidbinucleate cell (Figure 3). The failure of the spindle checkpoint to maintain the process ofmitotic arrest in cellsduring cell division with abnormal disrupted mitotic spindles results in cells transiting from M phase to G1 without undergoing normal mitotic cell division and cytokinesis. Nondisjunction can occur during either meiosis I or II, with different results. If homologous chromosomes fail to separate during meiosis I, the result is two gametes that lack that chromosome and two gametes with two copies of the chromosome. If sister chromatids fail to separate during meiosis II, the result is one gamete that lacks that chromosome, two normal gametes with one copy of the chromosome, and one gamete with two copies of the chromosome.

Apospory: A Natural Event for the Induction of Polyploidy

In sexual reproduction, during gametogenesis, lack of meiosis results in the production of diploid gametes. This process of production of diploid gametes is known as apospory. Once the diploid (unreduced) egg is formed it fuses with haploid (reduced) sperm and thus polyploidy (triploid) progeny is formed (Figure 4).

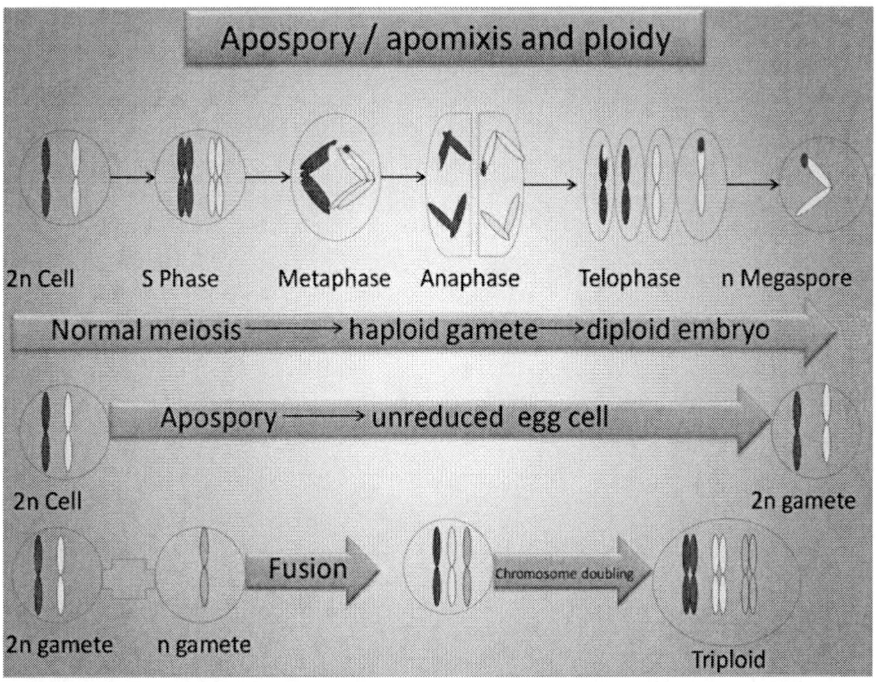

Figure 4. Mechanism of natural production of triploids *via* apospory.

In populations, genetic construction of genes (single copy, duplicated and multicopy) contributes to the evolutionary process. The new species are genetically different from parental species which is the result of interplay and crosstalks of many types of geneic and non-geneic events which produce localized or widespread genomic redundancies. Polyploidization events result in duplication of whole genomes.

Alteration in Genetic Makeup and Ploidy

Change in the genetic makeup of an organism is the basis of induction and production of ploidy. This change in geneic and non-geneic DNA of plant may occur either by mutation or by recombination (Figure 5). Mutation is the certain sudden change in the DNA. It may take either on genomic level or chromosome level or gene level. While during recombination homologous as well as non-homologous (site specific) chromosomes get combined. Retro transpositions are the way by which multiplication of nongeneic DNA within and between generations takes place (Stebbins 1947; Swanson et al. 1981; Wolfe 2001).Duplication of localized DNA causes duplication or elongation of only specific genes. The occurrence of mutation and recombination events (mutation events like chromosomal, genomic and point mutation and recombination events like homologous and site specific recombination) are the cause of divergence of geneic DNA as a result of which new alleles of genes arise.

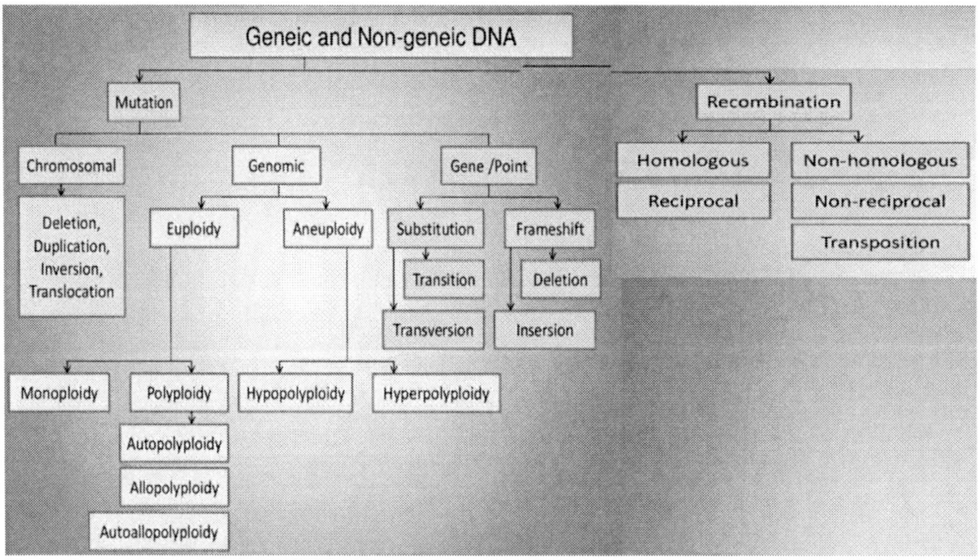

Figure 5. Types of events involved in alteration in the genetic makeup of a particular species that altogether determine population polymorphism at the species level.

Recent genomic sequence data provide substantial evidence that supports the evolution of new genes by duplication because duplication provides genetic redundancy in which the latter copy of the genome is less under the selection pressure and mutation in this copy is not so lethal for the organism. Duplication of localized or regional DNA sequences playscritical role in the evolutionary process because these provide the raw material for the origin of new genes which further leads to speciation.

During meiosis abnormalities like non-disjunction, may produce trisomics in which one chromosome is duplicated (n+1). Errors in crossing over at meiosis can result in duplication of only a sizable part of a chromosome, not the whole chromosome (deletion or duplication of chromosomal segment), a gene or a domain in a gene. Besides such events of point or insertional mutation, an error in crossing over may elongate specific genes (Ramsey and Schemske 1998; Ramsey and Schemske 2002; Rieseberg 2001).

New alleles of genes arise from mutational and recombination processes. Homologous recombination and gene conversion events increase the variability and diversification in the existing genome at any given time scale.

Mutation is the basis of all type of gene/genome variability. Segmental mutations involve alterations in sizable parts of chromosomes and occur in different ways like deletions, insertions or inversions. Point mutations involve changes in individual nucleotide pairs. Frameshift mutation changes the whole reading frame because either addition or deletion of nucleotides takes place in this process. Frameshift and nonsense mutations cause dysfunction of the gene. Synonymous and asynonymous base substitution mutations in nucleotide sequences within a gene are responsible for the formation of proteins normal or abnormal in function, respectively. However, asynonymous mutations that improve the function may get selected while deleterious ones are eliminated. These play a very significant role in the fixation of allelic variation among genes. Thus new alleles are generated from the preexisting genes (Schultz 1980; Soltis and Soltis 1995; Soltis and Soltis 1999; Kihara and Ono 1926; Larkins et al. 2001; Lawton-Rauh 2000).

Polyploidy in organisms occurred by addition of one or more sets of chromosomes to the original set and thus the entire genome gets duplicated. Diploid has two copies of each chromosome, these pair at the first division during meiosis. Diploid parents produce normal gametes and therefore these are fertile. Haploids have only one set of chromosomes thus they cannot undergo normal meiosis and are therefore sterile. Triploids and tetraploids have three and four sets of chromosomes respectively.During meiosis, multivalentsare formed. Errors in disjunction result in the production of gametes that possess either supernumerary or less than normal number of chromosomes. Therefore they may be partially or fully sterile. However, the polyploid genomes are the base for origin of larger diploid genomes in the process of origin of new species.

High levelpolyploids, such as octaploids, are generally stunted and malformed having extreme genetic redundancy and somatic instability but these plants may be valuable in several breeding programmes to enhance the degree of heterozygosity and further for induction and selection of desirable traits. In most cases, it is apparent that inducing autopolyploidy can improve substantial hybrid vigour in crop plants. Polyploidy can also result in a wide range of effects on plants that relate to gene interactions, gene dose effects, gene silencing and regulation of specific characters and processes.

TYPES OF POLYPLOIDY IN PLANTS

Polyploids are of two kinds, autopolyploids and allopolyploids. Autotetraploids possess each chromosome in four copies. While the allopolyploids possess chromosomes from two different species thus each chromosome is present in two copies. The autotetraploids are partially or fully sterile and allopolyploids are partial to fully fertile (Kihara and Ono 1926; Larkins et al. 2001; Lawton-Rauh 2000).

Autotetraploid *versus* Allotetraploids

Autopolyploids contain multiple sets (>2) of the same or similar genomes, i.e., from same species, in their nucleus. Their origin is mostly caused by intraspecific crosses followed by chromosome doubling because of unreduced gametes and their chromosomes may form multivalents during meiosis and exhibit a multisomic inheritance. The significance of autopolyploidy is the capacity to maintain high levels of heterozygosity, i.e., reduction in inbreeding depression and increase in heterosis, with multiple alleles per locus. Rarely, but autopolyploids can reach homozygosity with multiple dosages of a given allele. On the other hand, the formation of multivalents during meiosis is often associated with sterility of organisms harboring duplicated genes. Therefore, it is not surprising that many autopolyploids are perennial species that also perform vegetative propagation. Spontaneous or natural autopolyploids also occur in nature but in a low frequency and can be induced artificially using various ways, such as heat and chemical treatments. The effective method to obtain autopolyploids by artificial methods is by use of colchicines, a spindle fibre poison or suppressant and inhibits the polymerization step of spindle fibres at cell division. This results in multiplication of normal chromosome number. Autotetraploids/autopolyploids are partially sterile because they form gametes with more than normal number of chromosomes (Figure 6). Thus they get reproductively isolated from their diploid progenitors. However, they are raw material for the origin of new diploid species with comparatively larger size genome. For this, they undergo genome modification in terms of loss of some genes, formation of new alleles from duplicated genes, changes in gene order and gene distribution arising from chromosomal aberrations.

Studies reveal that many crop plants are allopolyploids. A good example is bread wheat *T. aestivum*; possess 42 chromosomes, 14 each from three distinct species of family Poaceae. In the past, the genome of *T. urartu* may be combined with the genome of *A. speltoids* or genome of species related to it to give rise to *T. turgidum* with 14A and 14B chromosomes, i.e., 14 chromosomes from each 2 species. Later the genome *T. turgidum* combines with the genome of another past species *T. tauschi* to produce the hexaploid species now named *T. aestivum* which is fully fertile because divalent are formed at meiosis which is easily segregated during anaphase. This is because of extensive chromosome tailoring in its evolution and presence of gene Ph1 whose product suppresses association between homeologous chromosomes (Song et al. 1995; Stebbins 1950; Stebbins 1958).

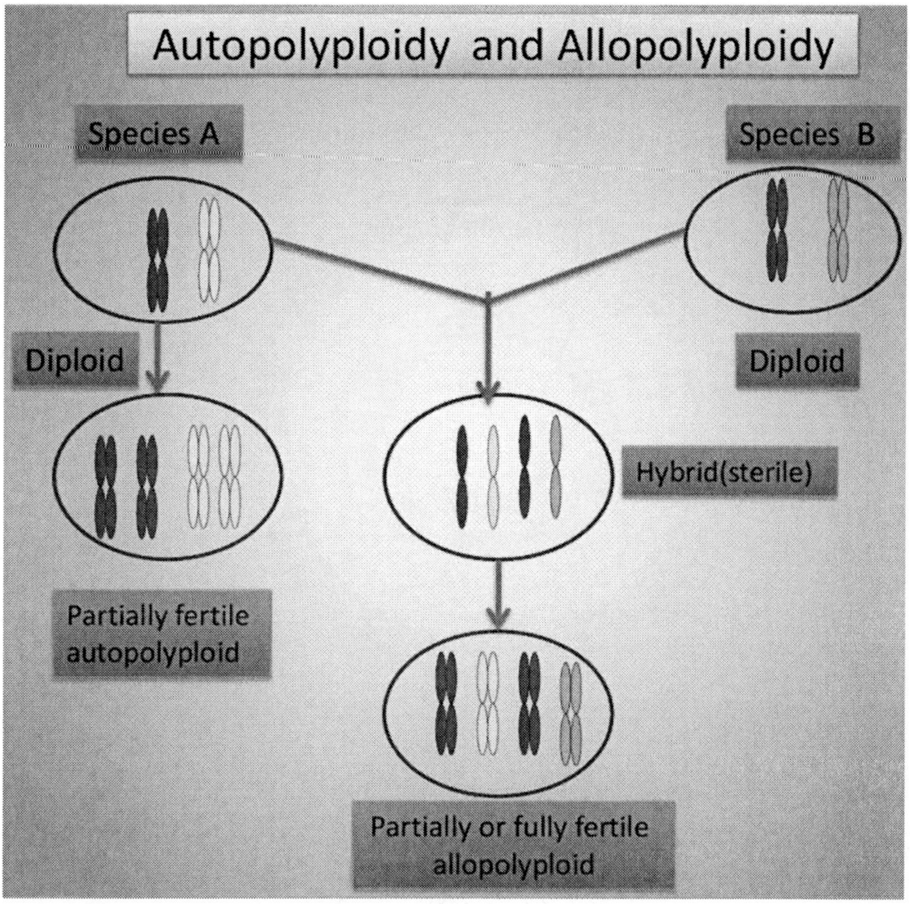

Figure 6. Autoploidy showing duplication of one basic set of chromosomes and allopolyploidy by a combination of two genetically distinct chromosomes set.

PLOIDY AND SPECIATION

Early changes in the chromosome number of organisms can play an important role in diversification and represent an ongoing natural phenomenon also. There are 4 main types of chromosomal aberrations which are responsible for the speciation. These events are duplication, deletion, inversion and translocation (Figure 7). The large prevalence of polyploid ancestry indicates that it is a successful evolutionary transition that has a significant effect on patterns and rates of diversification of different species. The genome sizes of many plant species have been measured till now and complete genome sequencing of many plant species is already done. Comparative whole genome sequencing is highly advanced in the plants of families of Brassicaceae and Poaceae.

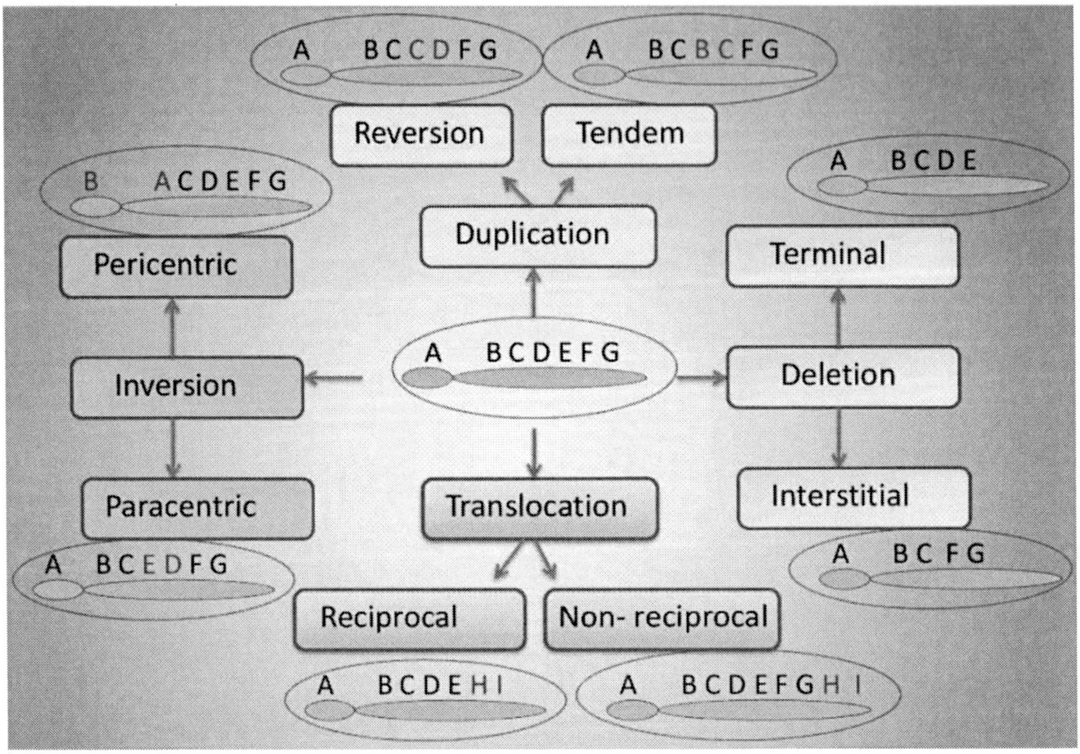

Figure 7. A variety of chromosomal aberrations which are ultimately involved in generation of new species.

The distributions of genome size in Brassicaceae and Poaceae are polymodal, i.e., there are more than two modes of the frequency of genome distribution. They are peaks of genome size that overlap with little less than genome size duplications in Poaceae, same is also observed in Brassicaceae. Genome duplications are supported by evidence like comparative genome sequencing which has shown genome duplications in related species and also of large scale chromosome rearrangements. Even *Arabidopsis thaliana,* the angiosperm possessing small genome has also experienced multiple rounds of genome duplication. Recently Ozkau and Feldman groups have shown that induced allopolyploids of family Poaceae undergo rapid chromosome rearrangements and mutations generating new alleles. Clearly, polyploidy has been shown a major and very important event in genome evolution and all species have experienced polyploidization in the evolutionary history.

One of the major drawbacks in the rapid generation of synthetic polyploids seems to the formation of multivalents at meiosis and allopolyploids are less prone to this problem than autopolyploids. Multivalent formation will be least among allopolyploid hybrids because they harbor genomes of distant species. Protoplast fusion technology is a convenient method and offers great potential for direct regeneration of interspecific allopolyploids. Regeneration of allopolyploids by protoplast fusion technology is highly successful in certain crop species, such as *Brassica*, tomato, potato etc. We suggest that the fusion of protoplasts of these species with protoplast of selected distant species of economic importance offer means of producing viable new crops at commercial level (Stebbins 1971; Stebbins 1985; Stebbins 1981).

KEY ROLE OF POLYPLOIDY IN PLANT BREEDING

In contrast to the natural evolutionary process for generation of new species from isolated populations, new species of plants can also arise abruptly/suddenly through polyploidy. The common mechanism for sudden speciation is through the generation of natural polyploids. Polyploidy is an important phenomenon inforcing the evolution of new species (Figure 8). Autopolyploidy and allopolyploidy should be exploited more and more in crop improvement (Ram 2014). Protoplast fusion technology using protoplasts of highly regenerative species as recipients and protoplast of genetically dissimilar species offer immense scope for generation of new types of crops which are economically and commercially more useful. It is an interesting phenomenon with intriguingly, a lot of practical applications. As mentioned previously it has been an area of interest from the eighteenth century and continues to be a major area of interest as seen from the increasing number of publications being generated continually on various aspects of polyploidy over the last decade (Osborn et al. 2003).

The main goals of polyploidy in crop improvements are:

- Gene buffering: More adaptive potential with slower response to selection.
- Increasing dosage effect: cumulative and additive effect of the alleles increases the number of phenotypes.
- Increasing allele diversity/heterozygosity: different allele combinations and new opportunities for breeding unique traits (Ram 2014).
- Novel phenotypic variation: genome interactions and change in gene expression

Advancement in technologies and various molecular tools like GISH, RFLP, flowcytometry seem to be helping us to gain better understandings into various aspects of polyploidy in plants and animals and also help in remodulating many of the traditionally held tenets Simon (Renny-Byfield et al. 2010). Meanwhile, it is implicated by some the certain factors such as the occurrence of high rates of polyploid in plants, the absence of ethical issues etc. may provide plants as a very good model to elucidate the molecular mechanisms of polyploid production (Larkins et al. 2001). However, others have suggested a 'cross-disciplinary' approach involving the contribution of both plants and animals in looking into the origin and establishment of polyploids lineages as a more efficient strategy (Osborn et al. 2003; Otto and Whitton 2000; Stebbins 1971).

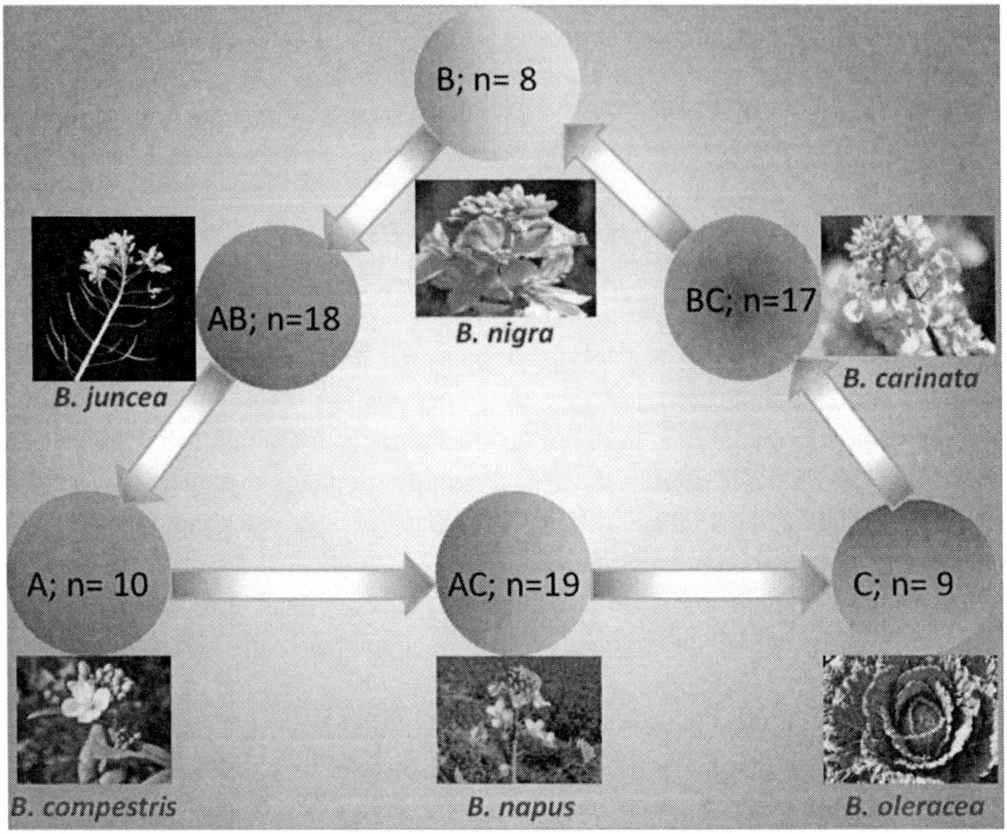

Figure 8. Triangle ploidy of *Brassica* sp.

Improvement of Cultivated Crop Plants *via* Polyploidization

As mentioned earlier in presented article, polyploidy in living beings can be obtained by both natural as well as induced process. Although more than 40% plants are listed since now which possess the process called natural or self polyploidization, polyploidization of cultivated crop plants by artificial or synthetic (induced polyploidy) ways is trending now in the field of cytogenetics, plant breeding as well as in metabolomics. Induced polyploidy at a commercial level by plant breeders is flourishing very rapidly. Table 1 below is listed with some important crop plants harboring either autopolyploidy or allopolyploidy or both.

Ploidy Status and Metabolome Profile (Role in Metabolite Diversity)

Metabolome profile of polyploids reflects the ploidy status of the plant species. In recent years it has been demonstrated that polyploidy enhances both primary and secondary metabolism. Induced polyploidy as well as natural polyploidy enhances the secondary metabolite concentration and play a very significant role in inducing metabolite diversity among plant species. The difference in secondary metabolites concentration of mainly terpenes, phenols, flavonoids, anthocyanins induced by polyploidy in plants (Dhawan and Lavania 1996). Trisomics of *D. stramonium*is causes 136-227% increase in alkaloid content, tetraploidy in *Cinchona succirubra*, for 100% increase in quinine concentration (Figure 9).

Table 1. Examples of some important polyploid crop plants

S. No.	Name of crop plant		Ploidy type	Ploidy level
	Common name	Botanical name		
1-	Apple	*Malus* spp.	Autopolyploidy or Allopolyploidy	3x
2-	Banana	*Musa* spp.	Autopolyploidy or allopolyploidy	3x
3-	Cassava	*Manihotsculenta*	Autopolyploidy	3x
4-	Grapes	*Vitis vinifera*	Autopolyploidy or Allopolyploidy	3x
5-	Lime	*Citrus* spp.	Allopolyploidy	3x
6-	Marigold	*Tageteserecta*	Allopolyploidy	3x
7-	Rose	*Rosa indica*	Autopolyploidy or Allopolyploidy	3x
8-	Tulip	*Tulip spp.*	Autopolyploidy	3x
9-	Watermelon	*Citrullus vulgaris*	Autopolyploidy	3x
10-	Coffee	*Coffea Arabica*	Allopolyploidy	4x
11-	Cotton	*Gossypiumhirsutum*	Autotetraploid	4x
12-	Durum wheat	*Triticumturgidum*	Allopolyploidy	4x
13-	Kiwi	*Actinidiachinensis*	Autopolyploidy	4x
14-	Mustard	*Brassica juncea*	Allopolyploidy	4x
15-	Potato	*Solanum tuberosum*	Autopolyploidy	4x
16-	Rapeseed	*Brassica napus*	Allopolyploidy	4x
17-	Rye	*Secale cereal*	Autopolyploidy	4x
18-	Ipomoea	*Ipomeabatatus*	Autopolyploidy	6x
19-	Kiwi	*Actinidiachinensis*	Autopolyploidy	6x
20-	Oat	*Avena Sativa*	Allopolyploidy	6x
21-	Prunus	*Prunusdomestica*	Allopolyploidy	6x
22-	Rose	*Rosa indica*	Allopolyploidy	6x
23-	Wheat (bread)	*Triticumaestivum*	Allopolyploidy	6x
24-	Chrysanthemum	*Chrysanthemum* spp.	Autopolyploidy or Allopolyploidy	8x
25-	Strawberry	*Fragariaananassa*	Allopolyploidy	8x
26-	Sugarcane	*Saccharumofficinarum*	Allopolyploidy	8x
27-	Cocks comb	*Celosia argentea*	Allopolyploidy	12x
28-	Cordgrass	*Spartinaanglica*	Allopolyploidy	12x

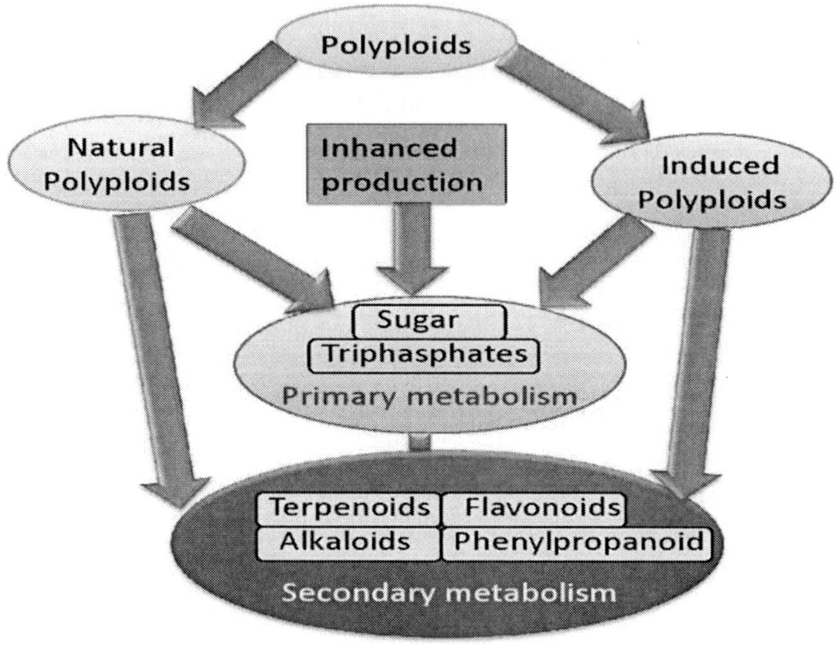

Figure 9. Relationship of metabolome and ploidy level.

FUTURE PROSPECTS AND CONCLUSION

During the course of evolution, near about, all plant species have doubled their genome which is followed by diploidization and then repeated polyploidization (Adams and Wendel 2005; Xiaohan Yang et al.2011). Many of our crop plants are polyploids and a number of them are being synthetically produced. However, it has been impossible to confirm with certainty, which diploid genomes when merged will coexist stably. Till then transforming the present 'hit and miss' process into a more exact science will provide an area of focus in future. Also, a more detailed analysis of gene expression changes in many of the newly formed polyploids could help uncover the impact of new variation on polyploid evolution. Polyploidy lineages mask the deleterious alleles commonly in small and isolated populations (Comai 2005). An understanding the timing and frequency of changes in many of the newly formed polyploids along with the mechanisms involved in diploidization may also facilitate gain a more thorough insight into the evolutionary impact of rapid genome changes in neoploployploids along with strategies possible for manipulation.

ACKNOWLEDGMENTS

Authors are thankful to CAS-Botany (BHU) and DST-FIST for providing funds and internet facilities to the Department of Botany, Institute of Sciences, Banaras Hindu

University. Lakee Sharma is thankful to CSIR-New Delhi, India for providing financial support in the form of fellowship.

REFERENCES

Adams, K. L.,Wendel, J. F. (2005) Polyploidy and genome evolution in plants. *Curr Opin Plant Biol* 8: 135-141.

Comai, L. (2005) The advantages and disadvantages of being polyploid. *Nat Rev Genet* 6:836.

Dhawan, O. P., Lavania, U. C. (1996) Enhancing the productivity of secondary metabolites via induced polyploidy: A review. *Euphytica* 87: 81-89.

Elliot, F. C. (1958) *Plant Breeding and Cytogenetics.* McGraw Hill Book Company Inc., pp.136-178.

Grant, V. P. (1981) *Polyploidy.* Columbia University Press, New York, pp.283-352.

Hieter, P., Griffith, T. (1999) Polyploidy- more is more or less. *Science* 285:210-211.

Hilu, K. W. (1993) Polyploidy and the evolution of domesticated plants. *Am J Bot* 80:1494-1499.

Kihara, H., Ono, T. (1926) Chromosome numbers and systematic grouping of the Rumex species. *Zeitschr. Zellf. Micr. Anat* 4 (3): 475-81.

Larkins, B. A., Dilkes, B. P., Dante, R. A., Coelho, C. M., Woo, Y-M., Liu, Y. (2001) Investigating the hows and whys of DNA endoreduplication. *J Exp Bot* 52:183-192.

Lawton-Rauh, A. (2000) Evolutionary dynamics of duplicated genes in plants. *Mol Phylogenet Evol* 29:396-409.

Leitch, I. J., Bennet, M. D. (1997) Polyploidy in angiosperms. *Trends Plant Sci* 2: 470-476.

Levin, D. A. (1983) Polyploidy and novelty in flowering plants. *Amer Nat* 122:1-24.

Lewis, W. H. (1980) *Polyploidy: biological relevance.* Plenum Press, New York.

Mable, B. K. (2003) Breaking down taxonomic barriers in polyploidy research. *Trends Plant Sci*8: 582-590.

Matzke, M. A., Matzke, A. J. M. (1998) Polyploidy and transposons. *Tree* 13: 241.

Matzke, M. A., Scheid, O. M., Matzke, A. J. M. (1999) Rapid structural and epigenetic changes in polyploid and aneuploid genomes. *Bioessays* 21: 761-767.

Osborn, T. C., Pires, J. C., Birchler, J. A., Auger, D. L., Chen, Z. L., Lee, H. S., Comai, L., Madlung, A., Doerge, R. W., Colot, V., Martienssen, R. A. (2003) Understanding mechanisms of novel gene expression in polyploids. *Trends Genet* 19: 141-147.

Otto, S. P., Whitton, J. (2000) Polyploid incidence and evolution. *Ann Rev Genet* 34: 401-437.

Ram, M. (2014) Polyploidy and distant hybridization in plant breeding. In: *Plant breeding methods* (Ed.Ram M), PHI Learning Pvt Ltd, New Delhi, pp. 423-445.

Ramsey, J., Ramsey, T. S. (2014) Ecological studies of polyploidy in the 100 years following its discovery. *Phil Trans R Soc B* 369: 1-20.

Ramsey, J., Schemske, D. W. (1998) Pathways, Mechanisms and rates of polyploidy formation in flowering plants. *Ann Rev Ecol Syst* 29: 467-501.

Ramsey, J., Schemske, D. W. (2002) Neopolyploidy in Flowering Plants. *Annu Rev Ecol Syst* 33: 589-639.

Renny-Byfield, S., Ainouche, Leitch, I. J., Lim, K. Y., Le, Comber, S. C., Leitch, A. R. (2010) Flow cytometry and GISH reveal mixed ploidy populations and Spartinanonaploids with genomes of *S. alterniflora* and *S. maritima* origin. *Annals of Botany*105: 527-533.

Schultz, R. J. (1980) *Role of polyploidy in the evolution of fishes.*Ref 13, 313-40.

Soltis, D. E., Soltis, P. S. (1995) The dynamic nature of polyploid genomes. *Proc Natl Acad Sci* 92: 8089-8091.

Soltis, D. E., Soltis, P. S. (1999) Polyploidy: recurrent formation and genome evolution. *Tree* 14: 348-352.

Song, Lu. P., Tang, K., Osborn, T. C. (1995) Rapid genome change in synthetic polyploids of *Brassica* and its implications for polyploid evolution. *Proc Natl Acad Sci* 92: 7719-7723.

Stebbins Jr, G. L. (1947) Types of Polyploids: their classification and significance. *Adv Genet* 1: 403-432.

Stebbins Jr, G. L. (1950) *Variation and Evolution in plants*. Columbia University Press, New York.

Stebbins Jr, G. L. (1958) The inviability, sterility and weakness of inter-specific hybrids. *Adv Genet* 9: 147-215.

Stebbins Jr, G. L. (1971) *Processes of Organic Evolution.*Englewood Cliffs, N: Prentice-Hall.

Stebbins Jr. G. L. (1985) Polyploidy Hybridization and the invasion of new habitats. *Ann Missouri Bot Gard* 72: 824-832.

Swanson, C. P., Merz, T., Young, W. J. (1981) *Cytogenetics: The Chromosome in Division, Inheritance and Evolution.* Englewood Cliffs, Prentice-Hall.

Van de Peer, Y., Maere, S., Meyer, A. (2009) The evolutionary significance of ancient genome duplications. *Nat Rev Genet* 10: 725-732.

Wolfe, K. H. (2001) Yesterday'spolyploids and the mystery of diploidization. *Nat Rev Genet* 2: 333-41.

Yang, X., Ye, C. Y., Cheng, Z. M., Tschaplinski, T. J., Wullschleger, S. D., Yin, W., Xia, X., Tuskan, G. A. (2011) Genomic aspects of research involving polyploid plants. *Plant Cell Tiss Org* 104: 387-397.

In: Innovations in Life Science Research ISBN: 978-1-53615-868-7
Editors: R. P. Sinha, S. Pandey-Rai and N. Ghoshal ©2019 Nova Science Publishers, Inc.

Chapter 7

RECENT ADVANCES IN EPIGENETIC RESEARCH IN PLANTS

***Neha Pandey*[1,*], *Anupam Tiwari*[2,*] *and Shashi Pandey-Rai*[3,†]**
[1]Department of Botany, CMP College (A Constituent PG College of University of Allahabad), Prayagraj, Uttar Pradesh, India
[2]Department of Botany, School of Bioengineering and Biosciences, Lovely Professional University, Phagwara, Punjab, India
[3]Laboratory of Morphogenesis, Centre of Advanced Study in Botany, Institute of Science, Banaras Hindu University, Varanasi, Uttar Pradesh, India

ABSTRACT

Epigenetic regulation of transcription is based on three principal aspects that involve DNA methylation, histone modifications and role of small RNAs. These epigenetic mechanisms and their enormous combinations can induce huge number of genotypic and phenotypic changes in the plants. Recent discoveries in this regard have given a new platform for more efficient production of better crops which are high yielding and better adapted to the environmental threats. It is now well established that epigenetic mechanisms are regulators of transcription however our understanding of how exactly these epigenetic modifications bring changes in gene expression is in infancy stage. DNA methylation, histone modifications and small RNAs are now seem to be involved in different aspects of plant's life such as vegetative to reproductive phase transformation and flowering, response towards environmental stimuli, plant pathogen-interaction and many more. The aim of this chapter is to explore recent advancements in the area of plant epigenetics with especial concern for increased crop productivity.

Keywords: epigenetics, DNA methylation, histone modifications, epi-fingerprinting

---

* Equal contribution.
† Corresponding Author's Email: shashi.bhubotany@gmail.com.

INTRODUCTION

A renowned developmental biologist Conrad H. Waddington had coined the term 'epigenetics' in 1942 to address the mechanism of bringing phenotypic changes by genes (Waddington 1942). Further, the term was made more understandable in 1987 by Robin Holliday who proposed epigenetics as the study of heritable changes in gene expression without any changes in DNA sequences (Holliday 1987). Now the term is no longer restricted to DNA and therefore, epigenetics explains a biological phenomenon where biomolecules are structurally or chemically modified without any alteration in their primary sequence (Reik 2007; Berger et al. 2009). This ultimately causes changes in gene expression.

In recent years, chromatin remodeling has become one of the most important factors for explaining the regulation of transcription. Initial level of chromatin organization includes wrapping of nuclear DNA around histone octamer that forms nucleosome and consist of a core histone fold and a histone tail (Luger et al. 1997). Next level of chromatin organization involves H1 binding to the linker regions resulting in the formation of a zig-zag structure. It is the chemical modifications of histone tail, which brings structural and functional rearrangements of chromatin (Kurdistani and Grunstein 2003). Major regulatory epigenetic mechanism known so far are DNA methylation, histone modifications, RNA methylation and regulation by small non coding RNAs (Berger et al. 2009). Intensive researches conducted during last decade have now confirmed that the regulation of gene expression is being controlled by epigenetic modifications which control the accessibility of regulatory complexes to the genome. These modifications are interestingly regulated by various internal and external signals such as many developmental signals or environmental stimuli. However, how these signals ultimately bring epigenetic modifications and how they may contribute to plant developmental and environmental defense is still at its infancy stage. The major hurdle behind this was probably the insufficient tools and techniques to deeply characterize genome-wide epigenetic dynamics of an organism. However, many new tools and technologies is being invented to help researches in identifying new molecular players involved in epigenetic regulation. These information can further be utilized for various crop improvement plans. During the past century, utilization of Mendelian genetic approaches and variation in DNA sequences have been intensively used for production of improved crop varieties. However, climate change and increasing population has led to the discovery of new methods for crop improvement that includes epigenetic mechanisms based methods. Due to its central role in environmental responses and developmental gene regulation, epigenetics may become a potential tool for crop improvement plans.

DIFFERENT TYPES OF EPIGENETIC MODIFICATIONS

Genome Accessibility of transcriptional machinery to genes is a highly regulated phenomenon and is governed by chemical modifications to DNA and N-terminal histone tail. These epigenetic modifications have majorly been categorized in to three types: DNA methylation, histone modifications and RNA modifications (Figure 1). DNA methylation involves transfer of a methyl group to the cytosine (mostly) or adenine (less commonly) base. However N-terminal tail of histones is subjected to many chemical modifications such as methylation, acetylation, phosphorylation, ubiquitylation and sumoylation. RNA modifications are restricted to non-coding small RNAs.

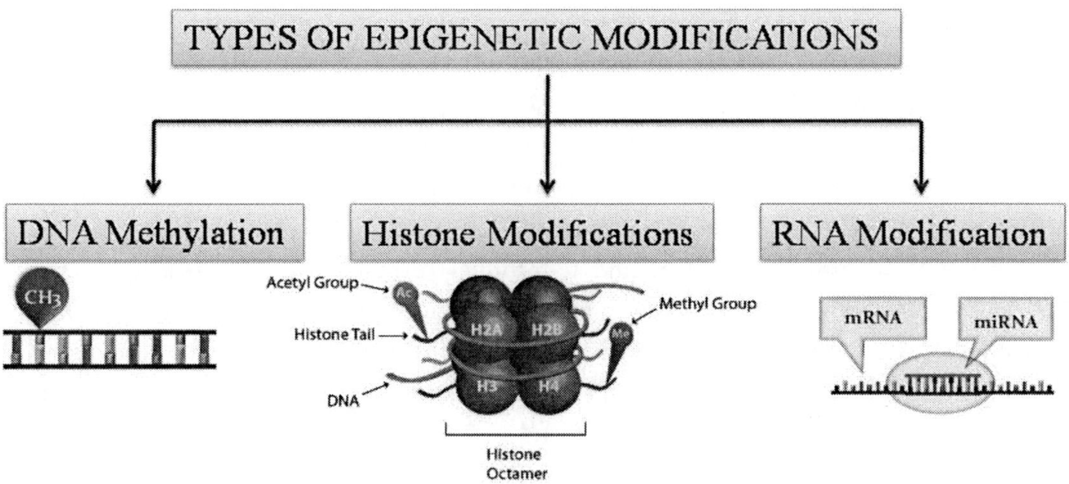

Figure 1. Major types of epigenetic modifications.

DNA METHYLATION

DNA methylation is one of the most common epigenetic mechanisms known to have significant role in various plant biological processes. It involves a process of addition of methyl group (-CH_3) to DNA bases. Methyl group is generally added to only two of the four DNA bases: cytosine and adenine. On cytosine methyl group is added to two sites: C^4 and N^5 while adenine has only one methylation site i.e., N^6. Therefore, DNA methylation process involves three kinds of methylated bases i.e., C^5-methylcytosine, N^4-methyl-cytosine and N^6-methyladenine (Figure 2).

Among the three types of methylated bases, the most common type of methylated base is the 5-methylcytosine which is formed by an enzyme catalyzed reaction involving DNA methyltransferases commonly known as DNA MTase family of enzymes. These enzymes transfer methyl group from S-adenosine-L-methionine (AdoMet; a universal methyl group

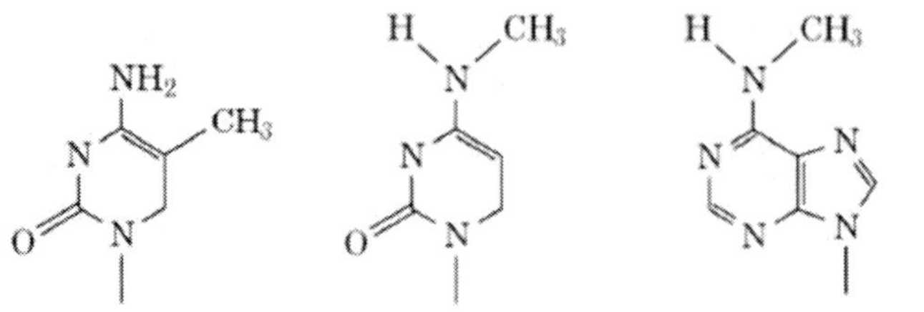

Figure 2. Types of methylated bases.

doner) to DNA base (Dryden 1999;Vilkaitis et al. 2001). While in animals DNA methylation usually take place at only CG dinuceotide sequence, plants posses two extra trinucleotide (CHG and CHH, where H can be any nucleotide) other than CH dinucleotide where methylation usually occurs. This causes high degree of methylation level in plants as compared to animals (Gruenbaum et al. 1981; Chen and Li 2004). Cytosine methylation takes place in two distinct steps, first a cytosine DNA methyltransferases (C5-MTase) recognizes a DNA sequence and then the enzyme transfers a methyl group from S-AdoMet to 5th carbon of cytosine residue (Posfai et al. 1989; Kumar et al. 1994). The enzyme has two different domains (N-terminal and C-terminal respectively) to perform these two functions. DNA methylation in higher plants takes place in two different ways: "*de novo*" methylation where new methylation pattern is establishes by methylation of previously unmethylatedcytosines and "*maintenance*" methylation where the methylation pattern already formed is being maintained (Chen and Li 2004). Plant system possesses different set of enzymes for maintaining methylation pattern of different dinuceotide/trinucleotide sites. For CG methylation plants possess DNA Methyltransferase 1(MET1) enzymes. Similarly for methylation at CHG and CHH trinucleotides plants possess Chromomethylase 3 (CMT 3) and Domains Rearranged Methyltransferase 2 (DRM2) enzymes respectively (Chan et al. 2005; Saze et al. 2003; Cao and Jacobsen 2002; Bartee et al. 2001). DRM2 is also responsible for *de novo* cytosine methylation in all three sequence contexts and is guided through RNA-directed DNA methylation (RdDM) pathway involving small interfering RNAs (siRNA) (Cao and Jacobsen 2002; Chan et al. 2005; Wassenegger et al. 1994; Matzke et al. 2009).

DNA methylation process at any region has been found to be coupled with an equally important phenomenon: DNA demethylation which is the removal of a methyl group from nucleotides. DNA demethylation is an inseparable part of epigenetic reprogramming of genes and it can be passive or active. Passive DNA demethylation is an enzyme catalyzed process involving actions of the enzyme DNMT1. In contrast, active DNA demethylation occur by the direct removal of 5-methylcytosine and involves sequential modification of cytosine bases that have been converted by TET enzyme (ten-eleven translocation (TET) family of 5-mC hydroxylases) that convert 5-mC to 5-hmC, then 5-hmC to 5-fC(5-formylcytosine), and finally 5-fC to5-caC(5-carboxylcytosine). DNA methylation plays significant role in regulation of the whole plant life cycle starting from plant growth and development to apoptosis. The link between DNA methylation and gene expression has been well established. More the extent of DNA methylation, less is the expression of concerned genomic region.

Pattern of cytosine methylation in genomic DNA and more specifically in promoter region of genes have significant impact on overall genome function (Hasbun et al. 2008). Inactivation of genes is generally associated with cytosine methylation while their activation requires cytosine demethylation (Zemach et al. 2010). A recent study has suggested that selective methylation of certain TF binding sites (TFBS) may prevent TF binding and conversely demethylation may facilitate TF binding. Although this surely is not a general mechanism of transcription regulation but probably affects gene expression to a great extent.

HISTONE MODIFICATION

Genomic DNA in eukaryotes is packaged with histone proteins to form chromatin, a more complex structure. DNA (146 base pairs) is wrapped with octomeric unit of histones involving two copies of H2A, H2B, H3 and H4 and a connector H1 histone protein (Figure 3). This forms a nucleosome which is the fundamental unit of chromatin. Based on the spacing between nucleosomes, chromatin structure can be divided into two broad catagories: heterochromatin where nucleosomes are densely packaged and euchromatin where nucleosomes are less densely packaged.

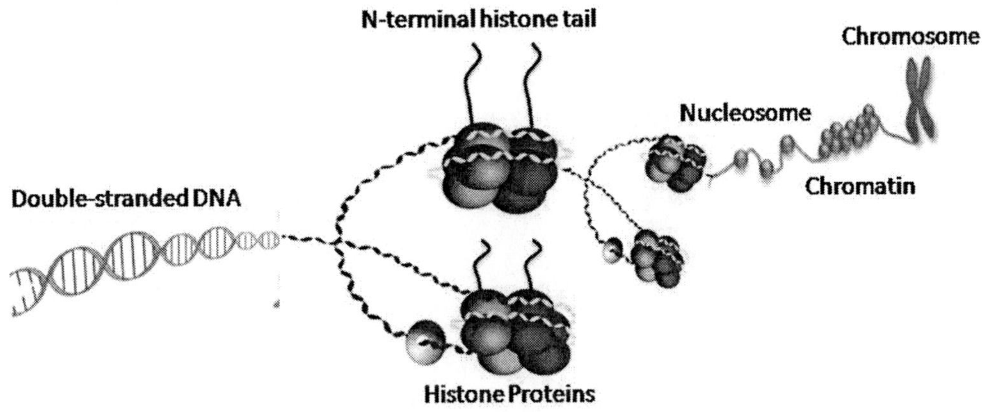

Figure 3. Schematic representation of organization and assembly of genetic material. Double-stranded DNA is wrapped around histone octamer (involving H2A, H2B, H3 and H4) to form nucleosome. Protruding N-terminal histone proteins are subjected to various epigenetic modifications.

Table 1. Types of histone modifications and their role

S. No.	Modification pattern	Histone type	Proposed role
1	Acetylation	H2A	Transcriptional activation
		H2B	Transcriptional activation
		H3	Transcriptional activation, DNA repair, Histone deposition
		H4	Histone deposition, transcriptional activation, DNA repair
2	Phosphorylation	H1	transcriptional activation, chromatin decondensation
		H2A	mitosis, chromatin assembly, transcriptional repression, DNA repair
		H2B	Apoptosis, DNA repair, transcriptional activation
		H3	mitosis, meiosis, immediate-early gene activation, transcriptional activation
		H4	mitosis, chromatin assembly, DNA repair
3	Methylation	H1	transcriptional silencing
		H3	transcriptional activation (Lys4, Arg17), transcriptional repression (Arg8, Lys27)
		H4	transcriptional activation (Arg3), transcriptional repression (Lys59)
4	Ubiquitylation	H2A	Spermatogenesis (mammals)
		H2B	Meiosis, transcriptional activation, euchromatin
		H3	
		H4	transcriptional repression
5	Sumoylation	H2A	transcriptional repression
		H2B	transcriptional repression
6	Biotinylation	H2A	unknown
		H3	gene expression
		H4	DNA damage response

There exist many epigenetic modifications of N-terminal histones which govern structure and function of chromatin such as acetylation, methylation, phosphorylation, ubiquitylation, sumoylation and biotinylation (Figure 4 and Table 1). Each modification imposes different structural and functional changes in chromatin. These chemical modifications of histones are proposed to be stored in the form of 'histone code' which determines the structure and function of chromatin and ultimately govern the gene expression (Jenuwein and Allis 2001).

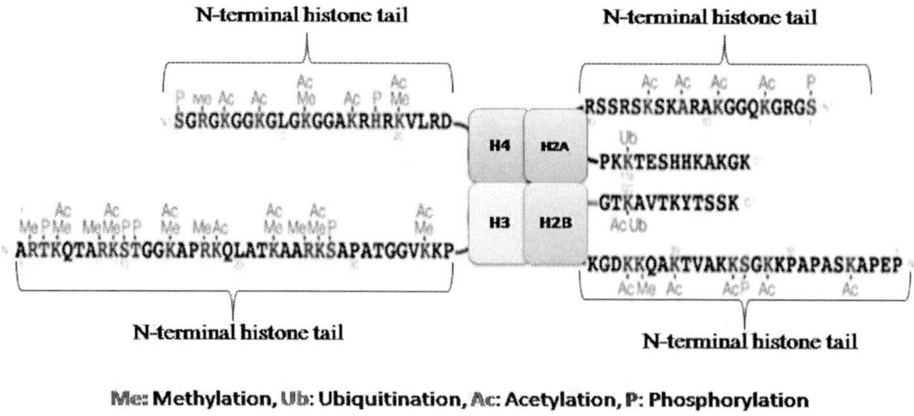

Figure 4. Major types of histone modifications.

Epi-Fingerprinting for Crop Improvement

In today's era of climate change and increasing global population, the greatest challenge is to enhance the food production. Enhancing crop production and crop quality are both necessary for sustained food production. Classical breeding approaches for crop improvement are good but introducing molecular approaches for crop breeding has become essential for sustainable food production. Marker-assisted introgressive hybridization is now a main stream process involved in crop improvement (Collard and Mackill 2008) however there are many beneficial traits which have arisen in crops spontaneously majorly through vegetative propagation. There were misconceptions that these phenotypic divergences have genetic rather than epigenetic origin probably due to use of conventional markers-assisted methods which are unable to differentiate between genetic and epigenetic consequences (Schmitz et al. 2013). That was the reason behind our incapability to fully predict the mechanism of phenotypic variability in many cases. Intensive researches carried out in recent years have extended our knowledge and there are many evidences that suggest significant role of epigenetic mechanisms in controlling plant development, phenotypic characters and plant management towards changing environment (Geyer et al. 2011;Brautigamet al. 2013). Applied epigenetics is emerging as a new way to enhance crop productivity. DNA methylation dynamics is linked with regulation of gene expression and it takes place through RNA-directed DNA methylation (RdDM) and hence it can be used as a potential tool for introduction of targeted DNA methylation in order to decrease methylation level of target gene (Wasseneggeret al. 1994). This process uses small interfering RNAs (siRNAs) and makes use of many enzymes such as DCL3 (DICER-like 3), AGO4 (Agronaute 4), DNA-dependent RNA Pol IV, Pol V and RNA-dependent polymerase RDR2 which finally direct specific DNA methyltransferase DRM2 for methyl group transfer (Naumannet al. 2011).Exogenous interventions are being applied to target one or more pathway enzymes in order to change the DNA methylation dynamics. Use of chemicals such as 5-azacytidine is being extensively used for inducing DNA demethylation at specific loci thereby, regulating gene expression both at genome-wide and locus-specific levels. The resulting 'epigenetic variants probably have desirably manipulated crop phenotype. In 2012, improved *Brassica napus* seedlings with higher seed protein content were produced through epigenetic approach using 5-azacytidine (Amoahet al. 2012). These chemicals are basically inhibitors of DNA methyltransferases.

Further, it has been shown through various studies that DNA methylation dynamics and associated gene expression is highly dependent on the environmental signals. Therefore, exploitation of this environment-mediated DNA methylation dynamics has been opted as a more direct approach for epigenetic intervention and crop improvement. There are many reports of RdDM-associated DNA methylation showing improved resistance against many abiotic and biotic stresses. In *Arabidopsis*, *P. syringae* exposure caused DNA demethylation that resulted in reduced rate of pathogen multiplication and vascular propagation (Yu et al. 2013). Epi-Fingerprinting is an emerging way for improved crop production and has application in plant production via both micro propagation and breeding (Rodriguez Lòpez and Wilkinson 2015). Micro propagation is an essential part of the improved crop production program. For successful production of genetically transformed micro propagated plant, it is necessary to maintain the genetic and epigenetic faithfulness of the parent plant. However, it has been observed that regenerated plants generally maintain only genetic faithfulness and show divergence in DNA methylation pattern which may cause phenotypic variations (Miguel and Marum 2011). These random phenotypic variations are one of the significant causes of economic loss due to random imposition of insignificant phenotypic traits. This also imposes syndromic symptoms such as mantled inflorescence syndrome is caused due to micro propagation-induced variation in total cytosine methylation in oil palm. Mantled inflorescence syndrome in oil palm causes drastic reduction in crop yield which ultimately is responsible for major industrial loss (Mattheset al. 2001). Although *in-vitro* culture is known to cause epigenetic variations in many plants such as *A. thaliana* (Bardiniet al. 2003), *Solanum tuberosum* (Joyce and Cassells 2002) and *Nicotiana tabacum* (Schmitt et al. 1997), their potential role in plant breeding programs cannot be denied. This is because some somaclonal variations are valuable and offer new and beneficial phenotypic traits (Henry 1998). Several authors have reported DNA methylation changes that are linked with yield related traits. For example hypermethylation of CYCLOIDEA gene in *Linaria vulgaris* changes bilaterally symmetrical flowers in to radial symmetry (Cubaset al. 1999). Another example of spontaneous somatic epi-mutation includes hypomethylation of FIE1 gene which causes flower-aberrant and dwarf phenotype in rice (Zhang et al. 2012). There are various epialles that have been found to be regulated by DNA methylation dynamics (Table 2).

Epi-fingerprinting is also being used as a reliable approach for development of stress resistant crops. Many studies conducted in last decade has now established the fact that abiotic or biotic stresses often induce global or locus-specific changes in the DNA methylation pattern and there exist a correlation between global cytosine methylation and onset/severity of particular stress (reviewed by Chinnusamy and Zhu 2009). Hence, cytosine-methylation fingerprinting is emerging as a new tool for early diagnosis of any particular abiotic or biotic stress in plants. Epi-fingerprinting based detection of stress has been implemented in many abiotic and biotic stresses such as drought (Trickeret al. 2013), low temperature (Pan et al. 2011), lower humidity (Trickeret al. 2012), heavy metals (Choi and Sano 2007), salinity (Choi and Sano 2007; Verhoevenet al. 2010) and many plant-pathogen interactions (Mason et al. 2008; Boyko and Kovalchuk 2011, Herrera and Bazaga 2013). Due to high degree correlation between DNA methylation pattern and plant's perspective towards growing environment, Rodriguze Lopez and Wilkinson (2015) have suggested use of C-methylation fingerprinting as an effective diagnostic tool for several aspects of crop quality.

Table 2. Some important epialleles regulated by DNA methylation

Epiallele locus	Governed trait	Reference
AGAMOUS	Flower development	Ito et al. (2012)
GT2-related proteins	Organ morphogenesis	Song et al. (2012)
MEA polycomb protein	Seed development	Grossniklauset al. (1998)
SUPERMAN	Floral whorl development	Jacobsen and Meyerowitz(1997)
NOR	Shelf life and quality regulator	Zhong et al. (2013)
PG2A	Shelf life and quality regulator	Zhong et al. (2013)
RIN	Shelf life and quality regulator	Zhonget al. (2013)
PSY	Fruit color regulator	Zhonget al. (2013)
PDS	Fruit color regulator	Zhonget al. (2013)
Kinesin-like protein	Determination of flowering time	Shan et al. (2013)
MADS box protein	Regulation of cell wall formation	Shan et al. (2013)
Pectinesterase-1	ABA signalling	Shan et al. 2013
SUVH8	Regulator of ripening	Berdascoet al. (2008)
GUN4	Regulation of chlorophyll biosynthesis	Li et al. (2014)
P1	Determination of grain color in Maize	Das and Messing (1994)
D1	Determination of plant height in rice	Miura et al. (2009)
CWF19	Cell cycle control protein	Jeon et al. (2015)
NtGPDL	Abiotic stress response	Choi and Sano (2007)

Epigenetic Silencing

Regulation of gene expression through epigenetic modifications is a natural phenomenon of plant defense and often silences the expression of transgene. For development of transgenic crops, their stability and segregation are the major concern. Many reports have been published regarding transgene instability due to silencing or inactivation of the transgene (Graham et al. 2011). There are two best explained molecular mechanisms for transgene silencing: one is transcriptional gene silencing (TGS) due to hypermethylation of promoters which in turn block the accessibility of transcriptional machinery and another is post-transcriptional gene silencing (PTGS) which involves degradation of transgene mRNA. Both TGS and PTGS are highly governed by epigenetic modifications such as DNA methylation and several histone modifications. Higher methylation level makes any silenced transgene less susceptible to endonucleases (Van Bloklandet al. 1997) hence it is predicted that TGS probably mimic heterochomatinization. Transcriptional gene silencing can be of two types: *cis* and *trans*. The position of transgene integration in plants is difficult to regulate hence they integrate at random positions. The copy number of transgene is inversely proportional to the expression of that gene and integration of multicopy transgene may result in gene silencing when it has been integrated in a *cis*-manner i.e., adjacent to any hyper methylated region of the host genome. Therefore this type of TGS is also termed as *cis*-inactivation (Mishibaet al. 2005).

Another method, where in spite of transcription there is no accumulation of mRNA, is termed as PTGS and causes degradation of specific RNA population using specific endo- and exonucleases (Metzlaffet al. 1997). There also exist mechanisms known as 'co-suppression' where both mRNA of both the transgene and endogenous gene are degraded. For example in *Petunia*, when CHS (chalcone synthase) gene was overexpressed, partial or complete loss of activity was observed for both transgene and endogenous gene (Napoli et al. 1990; Van der Krolet al. 1990). Further Van Eldik and co-workers (1998) analyzed the degradation product in tobacco, it was found that RNA was degraded by sequential action of endonuclease and then various exonucleases respectively (Table 3).

Table 3. List of few epigenetically silenced genes in plants

Gene	Plant	Remark	Reference
FLC (Flowering Locus C)	*Arabidopsis thaliana*	Epigenetic silencing made plants more adaptive/ responsive in spring season to floral promoting long day photoperiods.	Shindoet al. (2006)
DCL (Dicer-like)	*Nicotianabenthamiana*	Changed small RNA population and increased transgene expression	Katsarouet al. (2018)
ROS1 (a repressor of silencing gene)	*Nicotianatabacum*	Over expression of ROS1 in transgenic lines caused promoter demethylation of many flavonoid biosynthetic genes and genes involved in antioxidant defense mechanism.	Bharti et al. (2015)
CaMV35S promoter	*Nicotianatabacum*	De novo DNA methylation resulted in stable repressed epialleles.	Kˇrížová et al. (2013)
DBR2	*Artemisia annua*	Abiotic stress-mediated demethylation of DBR2-promoter led to its overexpression	Pandey and Pandey-Rai (2015)
AaPAL1	*Artemisia annua*	Abiotic stress-mediated demethylation of AaPAL1-promoter led to its overexpression	Pandey et al. (2018)
*npt*II	*Nicotianatabacum*	Locus-based PTGS of target gene was achieved	Khaitovaet al. (2011)
CHS	*Petunia*	Alteration in promoter methylation resulted in white-flowering phenotype	Kanazawa et al. (2007)
bar gene	*Oryzasativa*	Ubi1 promoter methylation resulted in *bar* gene silencing and development of bialaphos sensitivity in transgenic plants	Kumpatla and Hall (1998)

EPIGENETICS OF PLANT PATHOGEN INTERACTION AND ABIOTIC STRESS

Plants possess a well-organized defense system against wide variety of pathogens. In recent times, epigenetic modifications such as DNA methylation and various histone modifications have been found to have significant impact on plant's response towards both abiotic and biotic stresses. Clues for the involvement of epigenetics in plant pathogen interaction were given by initial studies on *Arabidopsis thaliana* infected by a bacterial pathogen *Pseudomonas syringae*. Infection by virulent strains of this bacterial pathogen resulted in alterations in DNA methylation pattern in all three sequence context i.e., CG, CHG and CHH, whereas avirulent strain caused DNA methylation alterations only at CG and CHG sites. Further these DNA methylation changes were found to be associated with thc overexpression of defense-related genes (Dowenet al. 2012). In tune with this study, Yu et al. (2013) also showed bacterial elicitor-mediated demethylation of transposable elements. Besides bacteria, few fungal pathogens such as *Botrytis cinerea* have been recently found to cause histone acetylation in *A. thaliana* and tomato (Crespo-Salvador et al. 2018). Nematode-infection in *A. thaliana* also caused massive changes in DNA methylation pattern which were probably associated with transcriptional activation of many defense-related genes (Heweziet al. 2017) (Table 4).

Table 4. List of few pathogens causing epigenetic changes in host genome

Pathogen	Host Plant	Reference
Trichoderma harzianum	Tomato	De Palma et al. (2019)
Botrytis cinerea	Arabidopsis Tomato	Crespo-salvador Ò et al. (2018)
Pseudomonas syringae	Tomato	Lopez-Galianoet al. (2018)
Uromyces appendiculatus (Bean Rust)	Common Bean	Melmaieeet al. (2015)
Alternaria brassicola	Arabidopsis	Luna et al. (2012)
H. arabidopsidis	Arabidopsis	Luna et al. (2012)
Xanthomonas oryzae (Blight Bacteria)	Rice	Li et al. (2013)
TMV (Tobacco Mosaic Virus)	Tobacco	Kathiriaet al. (2010)

In addition to plant pathogen interactions, epigenetic modifications are emerging as key regulators of abiotic stress tolerance in plants. Growing evidences suggest that epigenetic mechanisms such as DNA methylation and histone modification plays central role in abiotic stress adaptation in plants (Chinnusamy and Zhu 2009). These epigenetic mechanisms are known to have role in regulation of DNA replication, repair, transcription, translation and gene transposition. In recent years many studies were undertaken to find the epigenetic-basis of various abiotic stress responses in plants. So far, epigenetic regulation has been observed in abiotic stress like salinity, heavy metal stress, chilling, drought, high temperature and ultraviolet-B (UV-B) radiation (Chinnusamy and Zhu 2009). Our research group has found epigenetic regulation of UV-B stress in *Artemisia annua* plants. We have reported both global and locus-specific DNA demethylation at various transcription factor binding sites in promoters of flavonoid biosynthetic pathway gene and artemisinin biosynthetic pathway gene (Pandey and Pandey-Rai 2015 & Pandey et al. 2018).

In addition to the DNA methylation and histone modification, non-coding small RNAs have been shown to regulate abiotic stress mediated transcription and post-transcriptional modifications (Angers et al. 2010). Table 5 represents the list of miRNAs having role in epigenetic regulation in plants.

Table 5. List of miRNAs with roles in epigenetic regulation

miRNA	Sequence and Genome location	Target genes[a]	description[b]
miR396e-3p	Chr18:14914436-14914456	Potri.001G124300, Potri.018G081000, Potri.009G130600	Putative aspartate-arginine-rich mRNA binding protein mRNA, AAA-type ATPase family protein, Zinc finger (CCCH-type) family Protein
miR396e-5p	Chr18:14914544-14914565	Potri.001G215800	Similar to DNAJ heat shock N-terminal domain containing protein
miR166a	Chr01:6620351-6620372	-	-
Psi-miR1	Chr10:3868082-3868103	Potri.001G071100	Similar to 60S ribosomal protein L21
Ptc-miR6445a	TTCATTCCTCTTCCTAAAATGG/ Chr12:7706152-7706293	Potri.001G110700	Similar to dehydration responsive element binding protein 1 like protein(DREB9)
Ptc-miR156f	TGACAGAAGAGAGTGAGCAC/ Chr18:12448484-12448584	Potri.002G052800	Similar to cyclin 2
Ptc-miR472a	TTTTCCCTACTCCACCCATCCC/ Chr5:119630-119710	Potri.003G101000	NB-ARC domain-containing disease resistance protein
Psi-miR5	GTTCCATTTCTGATTCTAGGC/ Chr10:14521310-14521330	Potri.014G028200	Similar to bZIP transcription factor 6.(Pt-ABF2.1)
Psi-miR67	CTTAGAAGTAGAGACAGGATT/ Chr2:16497643-16497663	Potri.001G222600	Similar to CBL-INTERACTING PROTEIN KINASE 3

CONCLUSION

Understanding epigenetic regulation of plant metabolism and plant's responses towards changing environment is the burning area of plant science research. As various epigenetic modifications such as DNA methylation and many histone modifications are now known to be associated with the transcriptional switch, they can be better exploited for advanced breeding programs. Out of three major epigenetic regulators (DNA methylation, Histone modifications and small RNAs), DNA methylation followed by histone modifications are most studied. Majority of available literature explains the regulatory role of DNA methylation and how changes in methylation pattern alter accessibility of transcription factors to gene promoters. Role of histone acetylation in transcriptional regulation has also been observed in many plants. However roles of small RNA are yet to be dissected in depth. Some of the studies suggest significant roles of small RNAs in almost all kinds of plant metabolism and adaption, which need to be explored. Exploitation of only a fraction of information available on plant epigenome has resulted in revolutionary results in agriculture sector. Epi-Fingerprinting is emerging as a powerful tool for improved crop production and there are immense possibilities of using epigenomic tools for bringing 'epigreen' revolution in near future.

ACKNOWLEDGMENTS

All the authors are thankful to the Central Instrument Facility, Department of Botany, BHU, Varanasi for internet facilities. NP is thankful to Department of Botany, CMP College, University of Allahabad and AT is thankful to School of Biosciences, LPU Phagwada.

REFERENCES

Amoah, S., Kurup, S., Rodríguez López, C. M., Welham, S. J., Powers, S. J., Hopkins, C. J., Wilkinson, M. J., King, G. J. (2012) A hypomethylated population of *Brassica rapa*for forward and reverse epigenetics. *BMC Plant Biol*12:193-210.

Angers, B., Castonguay, E., Massicotte, R. (2010) Environmentally induced phenotypes and DNA methylation: how to deal with unpredictable conditions until the next generation and after. *Mol Ecol* 19:1283-1295.

Bartee, L., Malagnac, F. and Bender, J. (2001) Arabidopsis cmt3 chromomethylase mutations block non-CG methylation and silencing of an endogenous gene. *Genes Dev* 15: 1753-1758.

Bardini, M., Labra, M., Winfield, M., Sala, F. (2003) Antibiotic-induced DNA methylation changes in calluses of *Arabidopsis thaliana*. Plant. Cell. Tissue. *Organ Cult* 72: 157-162.

Berdasco, M., Alcazar, R., García-Ortiz M. V., Ballestar, E., Fernández, A. F., Teresa Roldán-Arjona, T., Tiburcio, A. F., Altabella, T., Buisine, N., Quesneville, H., Baudry, A. (2008) Promoter DNA hypermethylation and gene repression in undifferentiated *Arabidopsis* cells. *PLoS ONE*3:e3306.

Berger, S. L., Kouzarides, T., Shiekhattar, R., Shilatifard, A. (2009) An operational definition of epigenetics. *Genes Dev* 23: 781-783.

Bharti, P., Mahajan, M., Vishwakarma, A. K., Bhardwaj, J., Yadav, S. K. (2015) AtROS1 overexpression provides evidence for epigenetic regulation of genes encoding enzymes of flavonoid biosynthesis and antioxidant pathways during salt stress in transgenic tobacco. *J Exp Bot* 66: 5959-5969.

Boyko, A., Kovalchuk, I. (2011) Genetic and epigenetic effects of plant-pathogen interactions: an evolutionary perspective. *Mol Plant* 4: 1014-1023.

Bräutigam, K., Vining, K. J., Lafon-Placette, C., Fossdal, C. G., Mirouze, M., Gutiérrez Marcos, J. G., Fluch, S., Fraga, M. F., Guevara, M. Á., Abarca, D., Johnsen, Ø. (2013)

Epigenetic regulation of adaptive responses of forest tree species to the environment. *Ecol Evol* 3: 399-415.

Chan, S. W., Henderson, I. R., Jacobsen, S. E. (2005) Gardening the genome: DNA methylation in *Arabidopsis thaliana*. *Nat Rev Genet* 6: 351-360.

Chinnusamy, V., Zhu, J. K. (2009) Epigenetic regulation of stress responses in plants. *Curr Opin Plant Biol* 12:133-139.

Cao, X., Jacobsen, S. E. (2002) Locus-specific control of asymmetric and CpNpG methylation by the DRM and CMT3 methyltransferase genes. *Proc Natl Acad Sci USA* 99: 16491-16498.

Chen, T., Li, E. (2004) Structure and function of eukaryotic DNA methyltransferases. *Curr Top Dev Biol* 60: 55-89.

Choi, C. S., Sano, H. (2007) Abiotic-stress induces demethylation and transcriptional activation of a gene encoding a glycerol phosphodiesterase-like protein in tobacco plants. *Mol Genet Genomics* 277: 589-600.

Collard, B. C. Y., Mackill, D. J. (2008) Marker-assisted selection: an approach for precision plant breeding in the twenty-first century. *Philos Trans R Soc Lond B Biol Sci*363:557-572.

Crespo-Salvador, O., Escamilla-Aguilar, M., Lopez-Cruz, J., Lopez-Rodas, G., Gonzalez-Bosch, C. (2018) Determination of histone epigenetic marks in *Arabidopsis* and tomato genes in the early response to *Botrytis cinerea*. *Plant Cell Rep* 37: 153-166.

Cubas, P., Vincent, C., Coen, E. (1999) An epigenetic mutation responsible for natural variation in floral symmetry. *Nature* 401: 157-161.

Das, O. P., Messing, J. (1994) Variegated phenotype and developmental methylation changes of a maize allele originating from epimutation. *Genetics* 136: 1121-1141.

De Palma, M., Salzano, M., Villano, C., Aversano, R., Lorito, M., Ruocco, M., Docimo, T., Piccinelli, A. L., D'Agostino, N., Tucci, M. (2019) Transcriptome reprogramming, epigenetic modifications and alternative splicing orchestrate the tomato root response to the beneficial fungus *Trichodermaharzianum*. *Hortic Res* 6:5.

Dowen, R. H., Pelizzola, M., Schmitz, R. J., Lister, R., Dowen, J. M., Nery, J. R., Dixon, J. E., Ecker, J. R. (2012) Widespread dynamic DNA methylation in response to biotic stress. *Proc Nat Acad Sci USA* 109: E2183-E2191.

Dryden, D. T. F. (1999) Bacterial DNA methyltransferases. In: *S-Adenosylmethionine-dependent Methyltransferases: Structures and Functions* (Eds. Cheng X, Blumenthal RM), World Scientific Publishing, Singapore, pp. 283-340.

Geyer, K. K., Rodríguez López, C. M., Heald, J., Wilkinson, M. J., Hoffmann, K. F. (2011) Cytosine methylation regulates oviposition in the pathogenic blood fluke *Schistosomamansoni*. *Nat Commun* 9: 424-434.

Graham, M. W., Mudge, S. R., Sternes, P. R., Birch, R. G. (2011) Understanding and avoiding transgene silencing. In: *Plant Transformation Technologies* (Eds. Stewart C.

N., Touraev A., Citopvsky V.,Tzfira T.), Blackwell Publishing,West Sussex, pp. 169-196.

Grossniklaus, U., Vielle-Calzada, J. P., Hoeppner, M. A., Gagliano, W. B. (1998) Maternal control of embryogenesis by medea, a Polycomb group gene in *Arabidopsis*. *Science* 280: 446-450.

Gruenbaum, Y., Naveh-Many, T., Cedar, H., Razin, A. (1981) Sequence specificity of methylation in higher plant DNA. *Nature* 292: 860-862.

Hasbún, R., Valledor, L., Rodríguez, J. L., Santamaría, E., Ríos, D., Sánchez, M., Cañal, M. J., Rodríguez, R. (2008) HPCE quantification of 5-methyl-2'-deoxycytidine in genomic DNA: methodological optimization for chestnut and other woody species. *Plant Phys Biochem*46: 815-822.

Henry, R. J. (1998) Molecular and biochemical characterization of somaclonal variation. In: *Somaclonal Variation and Induced Mutations in Crop Improvement* (Eds. Jain SM, Brar DS, Ahloowalia BS), Academic Publisher, Dordrecht, Kluwer, pp. 485-499.

Herrera, C. M., Bazaga, P. (2013) Epigenetic correlates of plant phenotypic plasticity: DNA methylation differs between prickly and nonprickly leaves in heterophyllous*Ilex aquifolium* (Aquifoliaceae) trees. *Bot J Linnean Soc*171: 441-452.

Hewezi, T., Lane, T., Piya, S., Rambani, A., Rice, J. H., Staton, M. (2017) Cyst nematode parasitism induces dynamic changes in the root epigenome. *Plant Physiol* 174: 405-420.

Holliday, R. (1987) The inheritance of epigenetic defects. *Science* 238:163-170.

Ito, H. (2012) Small RNAs and transposon silencing in plants. *Dev Growth Diff* 54: 100-107.

Jacobsen, S. E., Meyerowitz, E. M. (1997) Hypermethylated SUPERMAN epigenetic alleles in *Arabidopsis*. *Science* 277: 1100-1103.

Jenuwein, T., Allis, C. D. (2001) Translating the histone code. *Science* 293: 1074-1080.

Jeon, J., Choi, J., Lee, G. W., Park, S. Y., Huh, A., Dean, R. A., Lee, Y. H. (2015) Genome-wide profiling of DNA methylation provides insights in to epigenetic regulation of fungal development in a plant pathogenic fungus, *Magnaportheoryzae*. *Sci Rep* 5:8567.

Joyce, S. M., Cassells AC (2002) Variation in potato microplant morphology *in vitro* and DNA methylation. *Plant. Cell. Tissue. Organ Cult* 70: 125-137.

Kˇrížová, K., Depicker, A., Kovarik, A. (2013) Epigenetic switches of tobacco transgenes associate with transient redistribution of histone marks in callus culture. *Epigenetics* 8: 666-676.

Kanazawa, A., O'Dell, M., Hellens, P. R. (2007) Epigenetic inactivation of chalcone synthase-a transgene transcription in petunia leads to a reversion of the post-transcriptional gene silencing phenotype. *Plant Cell Physiol* 48: 638-647.

Kathiria, P., Sidler, C., Golubov, A., Kalischuk, M., Kawchuk, L. M., Kovalchuk, I. (2010) Tobacco mosaic virus infection results in an increase in recombination frequency and resistance to viral, bacterial, and fungal pathogens in the progeny of infected tobacco plants. *Plant Physiol* 153: 1859-1870.

Katsarou, K., Mitta, E., Bardani, E., Oulas, A., Dadami, E., Kalantidis, K. (2018) DCL-suppressed Nicotianabenthamiana plants: valuable tools in research and biotechnology. *Mol Plant Pathol* doi: 10.1111/mpp.12761.

Khaitova, C. L., Fojtova, M., Krizova, K., Lunerova, J., Fulnecek, J., Depicker, A., Kovařík, A. (2011) Paramutation of tobacco transgenes by small RNA-mediated transcriptional gene silencing. *Epigenetics* 6:650-660.

Kumar, S., Cheng, X., Klimasauskas, S., Mi, S., Posfai, J., Roberts, R. J., Wilson, G. G. (1994) The DNA (cytosine-5) methyltransferases. *Nucleic Acids Res* 22:1-10.

Kumpatla, S. P., Hall, T. C. (1998) Recurrent onset of epigenetic silencing in rice harboring a multi-copy transgene. *Plant J* 14: 129-135.

Kurdistani, S. K., Grunstein, M. (2003) Histone acetylation and deacetylation in yeast. *Nat Rev Mol Cell Biol* 4: 276-284.

Li, R. Q., Huang, J. Z., Zhao, H. J., Fu, H. W., Li, Y. F., Liu, G. Z., Shu, Q. Y.(2014) A down-regulated epi-allele of the genomes uncoupled 4 gene generates a xantha marker trait in rice. *TheorAppl Genet* 127: 2491-2501.

Li, T., Chen, X., Zhong, X., Zhao, Y., Liu, X., Zhou, S., Cheng, S., Zhou, D. X. (2013) Jumonji C domain protein JMJ705-mediated removal of histone H3 lysine 27 trimethylation is involved in defense-related gene activation in rice. *Plant Cell* 25:4725-4736.

López-Galiano, M. J., González-Hernández, A. I., Crespo-Salvador, O., Rausell, C., Real, M. D., Escamilla, M., Camañes, G., García-Agustín, P., González-Bosch, C., García-Robles, I. (2018) Epigenetic regulation of the expression of WRKY75 transcription factor in response to biotic and abiotic stresses in Solanaceae plants. *Plant Cell Rep* 37:167-176.

Luger, K., Maeder, A. W., Richmond, R. K., Sargent, D. F., Richmond, T. J. (1997) X-ray structure of the nucleosome core particle at 2.8 Å resolution. *Nature* 389: 251-259.

Luna, E., Bruce, T. J., Roberts, M. R., Flors, V., Ton, J. (2012) Next-generation systemic acquired resistance. *Plant Physiol* 158: 844-853.

Matzke, M., Kanno, T., Claxinger, L., Huettel, B., Matzke, A. J. M. (2009) RNA-mediated chromatin based silencing in plants. *Curr Opin Cell Biol* 21: 367-376.

Mason, G., Noris, E., Lanteri, S., Acquadro, A., Accotto, G. P., Portis, E. (2008) Potentiality of methylation-sensitive amplification polymorphism (MSAP) in identifying genes involved in tomato response to tomato yellow leaf curl sardiniavirus. *Plant Mol Biol Rep* 26: 156-173.

Matthes, M., Singh, R., Cheah, S. C, Karp, A (2001) Variation in oil palm (*Elaeisguineensis* Jacq.) tissue culture-derived regenerants revealed by AFLPs with methylation-sensitive enzymes. *Theor Appl Genet* 102: 971-979.

Melmaiee, K., Kalavacharla, V. K., Brown, A., Todd, A., Thurston, Y., Elavarthi, S. (2015) Quantification and Gene Expression Analysis of Histone Deacetylases in Common Bean during Rust Fungal Inoculation. *Int J Genomics* 2015:153243.

Metzlaff, M., O'Dell, M., Cluster, P. D., Flavell, R. B. (1997) RNA-mediated RNA degradation and chalcone synthase A silencing in petunia.*Cell* 88:845-854.

Miguel, C., Marum, L. (2011) An epigenetic view of plant cells cultured *in vitro*: somaclonal variation and beyond. *J Exp Bot* 62: 3713-3725.

Mishiba K., Nishihara, M., Nakatsuka, T., Abe, Y., Hirano, H., Yokoi. T., Kikuchi, A., Yamamura, S. (2005) Consistent transcriptional silencing of 35S-driven transgenes in gentian. *Plant J* 44:541-556.

Miura, K., Agetsuma, M., Kitano, H., Yoshimura, A., Matsuoka, M., Jacobsen, S. E., Ashikari. M. (2009) A metastable DWARF1 epigenetic mutant affecting plant stature in rice. *Proc Natl Acad Sci USA* 106: 11218-11223.

Napoli, C., Lemieux, C., Jorgensen, R. (1990) Introduction of a chimeric chalcone synthase gene into petunia results in reversible co-suppression of homologous gene in trans. *Plant Cell* 2: 279-289.

Naumann, U., Daxinger, L., Kanno, T., Eun, C., Long, Q., Lorkovic, Z. J., Matzke, M., Matzke, A. J. (2011) Genetic evidence that DNA methyltransferase DRM2 has a direct catalytic role in RNA-directed DNA methylation in *Arabidopsis thaliana*. *Genetics* 187: 977-979.

Pan, Y., Wang, W, Zhao, X., Zhu, L., Fu, B., Li, Z. (2011) DNA methylation alterations of rice in response to cold stress. *Plant Omics J* 4: 364-369.

Pandey, N., Pandey-Rai, S. (2015) Deciphering UV-B-induced variation in DNA methylation pattern and its influence on regulation of DBR2 expression in *Artemisia annua* L. *Planta* 242: 869-879.

Pandey, N., Goswami, N., Tripathi, D., Rai, K., Rai, S. K., Singh, S., Pandey-Rai, S. (2018) Epigenetic control of UV-B induced flavonoid accumulation in *Artemisia annua* L. *Planta* 28: 1-8.

Posfai, J., Bhagwat, A. S., Posfai, G., Roberts, R. J. (1989) Predictive motifs derived from cytosine methyltransferases. *Nucleic Acids Res* 17: 2421-2435.

Reik, W. (2007) Stability and flexibility of epigenetic gene regulation in mammalian development. *Nature* 447: 425-432.

Rodriguez Lopez, C. M., Wilkinson, M. J. (2015) Epi-fingerprinting and epi-interventions for improved crop production and food quality. *Front Plant Sci* 6:397.

Saze, H., MittelstenScheid, O., Paszkowski, J., MittelstenScheid, O., Paszkowski, J., Paszkowski, J. (2003) Maintenance of CpG methylation is essential for epigenetic inheritance during plant gametogenesis. *Nat Genet* 34: 65-69.

Schmitt, F., Oakeley, E. J., Jost, J. P. (1997) Antibiotics induce genome-wide hypermethylation in cultured *Nicotianatabacum*plants. *J Biol Chem* 272: 1534-1540.

Schmitz, R. J., He, Y., Valdés-López, O., Khan, S. M., Joshi, T., Urich, M. A. (2013) Epigenome-wide inheritance of cytosine methylation variants in a recombinant inbred population. *Genome Res* 23: 1663-1674.

Shan, X., Wang, X., Yang, G., Wu, Y., Su, S., Li, S., Liu, H., Yuan, Y. (2013) Analysis of the DNA methylation of maize (*Zea mays* L.) in response to cold stress based on methylation-sensitive amplified polymorphisms. *J Plant Biol* 56: 32-38.

Shindo, C., Lister, C., Crevillen, P., Nordborg, M., Dean, C. (2006) Variation in the epigenetic silencing of FLC contributes to natural variation in *Arabidopsis* vernalization response. *Genes Dev* 20: 3079-3083.

Song, Y., Ma, K., Bo, W., Zhang, Z., Zhang, D. (2012) Sex-specific DNA methylation and gene expression in andromonoecious poplar. *Plant Cell Rep* 31:1393-1405.

Tricker, P., Gibbings, G., Rodríguez López, C. M., Hadley, P., Wilkinson, M. J. (2012) Low relative humidity triggers RNA-directed de novo DNA methylation and suppression of genes controlling stomatal development. *J Exp Bot* 63: 3799-3813.

Tricker, P., Rodríguez López, C. M., Hadley, P., Wagstaff, C., Wilkinson, M. J. (2013) Pre-conditioning the epigenetic response to high vapour pressure deficit increases the drought tolerance in *Arabidopsis thaliana*. *Plant Signal Behav* 8:e25974.

Van Blokland, R., Ten Lohuis, M., Meyer, P. (1997) Condensation of chromatin in transcriptional regions of an inactivated plant transgene: evidence for an active role of transcription in gene silencing. *Mol Genet Genomics* 257: 1-13.

Van der Krol, A. R., Mur L. A., Beld, M., Mol, J. N., Stuitje, A. R. (1990) Flavonoid genes in petunia: addition of a limited number of genes copies may lead to a suppression of gene expression. *Plant Cell* 2: 291-299.

Van Eldik, G. J, Litiere, K., Jacobs, J. J., VanMontagu, M., Cornelissen, M. (1998) Silencing of beta-1,3-glucanase genes into bacco correlates with an increased abundance of RNA degradation intermediates. *Nucleic Acids Res* 26: 5176-5181.

Verhoeven, K. J. F., Jansen, J. J., van Dijk, P. J., Biere, A. (2010) Stress-induced DNA methylation changes and their heritability in asexual dandelions. *New Phytol* 185: 1108-1118.

Vilkaitis, G., Merkiene, E., Serva, S., Weinhold, E., Klimasauskas, S. (2001) The mechanism of DNA cytosine-5 methylation. Kinetic and mutational dissection of Hhaimethyl-transferase. *J Biol Chem* 276: 20924-20934.

Waddington, C. H. (1942) Canalization of development and the inheritance of acquired characters. *Nature* 3811:563-565.

Wassenegger, M., Heimes, S., Riedel, L., Sanger, H. (1994) RNA-directed de-novo methylation of genomic sequences in plants. *Cell* 76: 567-576.

Yu, A., Lepere, G., Jay, F., Wang, J., Bapaume, L., Wang, Y., Abraham, A. L., Penterman, J., Fischer, R. L., Voinnet, O. (2013) Dynamics and biological relevance of DNA demethylation in Arabidopsis antibacterial defense. *Proc Nat Acad Sci USA* 110: 2389-2394.

Zemach, A., McDaniel, I. E., Silva, P., Zilberman, D. (2010) Genome-wide evolutionary analysis of eukaryotic DNA methylation. *Science* 328: 916-919.

Zhang, L., Cheng, Z., Qin, R., Qiu, Y., Wang, J. L., Cui, X. (2012) Identification and characterization of an Epi-Allele of FIE1 reveals a regulatory linkage between two epigenetic marks in rice. *Plant Cell* 24: 4407-4421.

Zhong, S. L., Fei, Z. J., Chen, Y. R., Zheng, Y., Huang, M. Y., Vrebalov, J., McQuinn, R., Gapper, N., Liu, B., Xiang, J., Shao, Y. (2013) Single-base resolution methylomes of tomato fruit development reveal epigenome modifications associated with ripening. *Nat Biotechnol* 31: 154-159.

In: Innovations in Life Science Research ISBN: 978-1-53615-868-7
Editors: R. P. Sinha, S. Pandey-Rai and N. Ghoshal ©2019 Nova Science Publishers, Inc.

Chapter 8

GENETIC ENGINEERING: AN EMERGING APPROACH FOR FOSTERING BIOREMEDIATION, CURRENT STATUS, CHALLENGES AND OUTLOOK

Shivani Uniyal, Indrajeet Kumar, Prince Kumar Singh and Rajesh Kumar Sharma*

Department of Botany, Banaras Hindu University
Varanasi, India

ABSTRACT

The pervasive industrialization and relentless use of pesticides during unscientific agricultural practices have escalated the release of toxic and carcinogenic contaminants in the environment. Conventional methods for pollutant removal are time-consuming and affected by environmental conditions. In this regard, a major breakthrough in molecular biology has accelerated the advancement of genetic engineering as a potential alternative to traditional bioremediation approaches. This chapter emphasizes the computational tools and enzyme engineering approaches being used for the optimization of catalytic pathways and acceleration of bioremediation processes, respectively. The chapter also enfolds the applications and evolution of genetic engineering in the field of bioremediation. Finally, existing challenges and future aspects of the application of genetic engineering technology in bioremediation methods were briefly discussed.

Keywords: genetic engineering, bioremediation, computational tools, enzyme engineering, catalytic pathway

* Corresponding Author's Email: rajeshbhu78@gmail.com.

INTRODUCTION

Continuous use of chemicals, fertilizers, pesticides and other types of components in industrial process and agricultural applications has resulted in the inception of a wide range of toxic contaminants such as pesticides, heavy metals, etc. which could adversely affect the environment and human health. These contaminants are transported mainly from the soil and water to the human and other living organisms and affect their health directly or indirectly. Therefore, there is a growing need for some eco-friendly strategies to remove these contaminants from our environment. Bioremediation is one of the best approaches for the removal of contaminants and defined as the use of living organisms to break down or remove the harmful environmental contaminants. Bioremediation can be achieved by utilizing fungi (mycoremediation), bacteria (bacterial bioremediation) or plants (phytoremediation) (Shukla and Vanker 2014).

The mycoremediation of contaminants from the environment has been evaluated by the application of yeast, microfungi, and macrofungi. Fungi possess different biochemical mechanisms to degrade organic contaminants either by modification in their chemical structure or by affecting their bioavailability (Prakash 2017). It has been reported that fungal mycelia secrete different types of extracellular enzymes which help in the breakdown of the toxic materials into simple and non-toxic compounds. There are also a group of fungi which have the capability to accumulate or absorb and concentrate heavy metals in their mycelia. Fungi such as *Phanaerochaete chrysosporium* and *Polyporus* sp. were considered as promising candidates for bioremediation due to their ability to degrade a wide range of persistent contaminants such as polychlorinated biphenyls (PCBs), polycyclic aromatic hydrocarbons (PAHs), and organochlorine pesticides, etc. (Favero et al. 1991, Abraham 2001, Xueheng et al. 2006, Wu and Yu 2007, Baiand, Ayu et al. 2011).

In addition, bacterial degradation is also considered as a major bioremediation strategy for the treatment of the contaminated environment. There are several reports on bacterial strains belonging to the different taxonomic group having an enormous bioremediation potential (Uniyal et al. 2016a; Uniyal et al. 2016b). For instance; Cycon et al. (2009) revealed remarkable diazinon degradation potential of bacterial strains *Serratia* sp., *Pseudomonas* sp. with a rate constant of 0.032-0.085 d^{-1}. In a Similar study, Oves et al. (2013) indicated that *Bacillus thuringiensis* showed remarkable biosorption capacity for Ni (94%), Cu (91.8%) and Cd (87%). Moreover, traditional phytoremediation methods were also used for pollutant removal in contaminated regions. Phytoremediation techniques have been considered as eco-friendly, efficient, cost-effective and aesthetically pleasing. It also sustains the biological processes, physical structure and fertility of the soil and is one of the best alternatives to physicochemical treatment technologies for remediation of the affected or contaminated ecosystem. The overall mechanism of phytoremediation of contaminants (for instance; heavy metals) have five major aspects: (1) Mobilization of soil containing heavy metals (2) Heavy metal ions uptake by plant roots (3) Movement towards aerial tissues (4) Metal ions sequestration in plant tissues and (5) Heavy metals tolerance to stress condition. Some families present in the environment which contains more capacity for accumulation or resisting contaminants in their biomass such as Asteraceae, Brassicaceae, Caryophyllaceae, Fabaceae, Poaceae, Violaceae, Euphorbiaceae families, etc. (Bolan et al. 2014).

Although eco-friendly, there are several drawbacks associated with the process of bioremediation. The rate of bioremediation in natural conditions is sometimes very low because microbes prefer the easily degradable carbon sources in comparison to recalcitrant pollutants. In addition, the mechanism of remediation of these contaminants is dependent on some fixed environmental conditions, therefore naturally occurring organisms are unable to survive and degrade the contaminants at some particular environmental condition due to inability to synthesize certain enzymes for degradation. The constraints can be overcome by the application of genetically engineered organisms, as they can specifically degrade specific pollutants in shorter time duration due to their advanced genetic makeup. Thus, an evolution of novel scientific researches i.e., genetic engineering technology combined with an interdisciplinary research strategy could allow the development of this idea into a promising, cost-effective, feasible, and eco-friendly technology.

Genetic Engineering: Advancement in Bioremediation Strategy

Evolution in molecular biology, biotechnology, and enzymology turned out to be the driving forces towards the development of engineer-improved stress tolerant organisms to achieve efficient bioremediation even in extreme environmental conditions. Genetic engineering, introduced around 1970 consists principally of nucleotide sequence alterations into DNA molecules (Arber 2010). The approaches used for developing engineered organism include certain critical steps (Dvorak et al. 2017). For instance; selection of an appropriate catabolic pathway for the compound of interest is the key step in the bioremediation process by genetically modified organisms. Various computational tools and databases are available, which provide useful information about the primary building units and catabolic pathways of the target compound. The University of Minnesota Biocatalysis/Biodegradation Database and Pathway Prediction System (UM-BBD/PPS) and MetaCyc and BioCyc databases are the major computations tools, which have used for this purpose (Gao et al. 2010; Caspi et al. 2016). Furthermore, numerous reports are consistently being generated by various authorities on the genome sequences of the organisms. For instance, *Schizosaccharomyces pombe* and *Neurospora crassa* were anticipated to contain 4,940 protein-coding genes and ten thousand genes encoding wood decaying enzymes respectively (Wood et al. 2002; Galagan et al. 2003). These sequence databases can further extend the dimensions of futuristic bioremediation of organopollutants. Once primary units and enzymes involved in the catalytic pathway are detected, these can be further quantified by mass spectroscopy, SDS-PAGE and Western blot analysis (Kurumbang et al. 2014). Following this, a suitable host is selected by deriving information from regenerated genome-scale metabolic models, transcriptomic, proteomic, metabolomic, and flux analyses (Sohn et al. 2010). These techniques define the catabolic strategies used by the organism in the presence of a target compound. The proposed catabolic pathway may be hampered by the over-expression of introduced genes (Lin et al. 2013; Silva et al. 2005; Sopko et al. 2006; Zha et al. 2014), redox imbalance engendered by the incompatible cofactor (Hong and Lee 2002; Kokura et al. 1999), accumulation of secondary toxic metabolites (Arevalo-Rodriguez et al. 2004; Choi and Lee 2013; McGary et al. 2013) or other bottlenecks (Ikeda 2006; Liu et al. 2002). Thus, the growth of engineered cells and the catabolic flux to target metabolites need to be optimized. The proposed catabolic pathway is therefore optimized in accordance with the host bacterial metabolism. It can be achieved by applying computational models like genome-scale metabolic models and experimental tools like balancing of gene expression. Further enzyme engineering can be performed to eliminate bottlenecks of the catalytic pathway. Enzyme engineering utilizes the modification in the basic amino acid structure of the enzyme for improvization of its activity, stress tolerance, temperature, pH, etc. (Rayu et al. 2012). Enzyme engineering necessitates the implementation of recombinant DNA technology and site-directed mutagenesis to introduce the required changes in amino acid sequences of an enzyme (Singh et al. 2013). Recombinant enzymes could be produced by recombinant DNA technology, which offers several advantages over native strains such as cost-effectiveness, easier purification, increased stability and activity (Alcalde et al. 2006). Amino acid sequences of an enzyme are used to determine by its fundamental structural and functional unit. Nonetheless, site-directed mutagenesis could aid in substitution of amino acids in the active site of an enzyme (Madigan et al. 2012).

Basic Tools for Genetic Engineering

The genetic engineering technology has been applied to escalate the efficiency of bioremediation processes by isolating and cloning the target genes, into another expression host (Gupta and Shukla 2015). The strategy of genetic engineering principally includes the application of restriction enzymes, electrophoresis, nucleic acid hybridization and molecular cloning (Madigan et al. 2012). Restriction enzymes are one of the most important tools in molecular genetics research and considered as key elements for *in vitro* DNA manipulation. These belong to the class of the enzymes, present in all cells that are able to chemically modify DNA. These enzymes are composed of two identical polypeptide subunits. Each polypeptide subunit of the restriction enzymes bind to the DNA at their specific recognition sequences and cut the DNA on one of its two strands within this sequences. The bits of DNA generated can be segregated from each other by gel-electrophoresis. Electrophoresis is a procedure by which a mixture of charged molecules are segregated on the basis of their corresponding size by migration in an electrical field and then visualized by numerous gel staining techniques such as ethidium bromide, SYBR® green, SYBR® safe, GelStar®, acridine orange, methylene blue, etc.(Robert and Farrell 2010). Following this, the DNA fragments are put through to denaturation in the gel, fragments collected at the end of the column (Barlaan et al. 2005) and then moved from the gel to a membrane (Merker and Arber 2010). One of the mostly used electrophoresis procedure, denaturing gradient gel electrophoresis (DGGE), is a fingerprinting technique which relies on the separation of PCR amplicons on the basis of their sequences. Nucleic acid hybridization is then performed with a labeled probe complementary to the target sequence. The signal derived from the labeled probes is detected and identified. Nucleic acid hybridization allows the detection of cloned gene expression in the host cell. Molecular cloning forms the foundation of genetic engineering and allows genetic manipulation in a number of ways. Molecular cloning thus includes isolation of the target gene with the help of restriction enzyme, its insertion into a cloning vector and introduction into a host organism.

PROGRESS IN GENETIC ENGINEERING

Bacteria

Bacteria possess the ability to adapt in the polluted environment by generating different kinds of defense mechanisms against pollutant (Pieper and Reineke 2000; Furukawa 2003). Traditional approaches used for contaminant removal may generate unknown toxic intermediates (Dua et al. 2002). However, GE bacteria may overcome the disadvantages associated with the conventional approaches due to their unique characteristics to remediate with no production of toxic intermediates (Singh et al. 2011). Genetic engineering further enhances this adaptation ability in the presence of stressors by enabling (i) amplification, disruption, and/or modification of the genes encoding enzymes involved in the metabolic pathways, (ii) reducing the pathway bottlenecks, (iii) encouraging redox and energy generation, and (iv) proving new properties derived from heterologous genes (Liu et al. 2006; Shimizu 2002; Timmis and Piper 1999). Genetic modification first observed in a bacterium, *Streptococcus pneumoniae*, involves the insertion of exogenous DNA sequences to induce an alteration in the genome of a target organism (Lorenz and Wackernagel 1997; Y. Al-Salloum et al. 2017). Further, various genetic refinements have been instigated for upgrading existing competencies and generating remarkable traits in bacteria involved in bioremediation (Table 1). For instance, overexpression of enzymes involved in bioremediation can be achieved by transferring the genes encoding the desired enzymes into another expression host (Gupta and Shukla 2015). Moreover, bacterial strains can be genetically modified to express a high amount of bacterial metallothionein in order to significantly reduce metal bioavailability. Bioaccumulation ability can be highly increased by expression of metallothioneins on the cell surface by fusion with cell surface proteins. Genetic engineering can be further extended to express high levels of metal-binding ligands, such as carboxyl, hydroxyl, sulfate, phosphate and, amine. These metal-binding ligands can increase the biosorption of contaminants like heavy metals either on the cell surface or microbial products such as EPS and biosurfactants (Naik and Dubey 2013). In addition to gene alteration, balancing of gene expression is also necessary and has been evolved as a powerful mechanism for harvesting maximum beneficial properties from genetically modified bacteria.

There has been enormous progress in the development of GE bacteria and is advantageous over physicochemical methods due to its high selectivity for pollutant transformation (Singh et al. 2011). In a study, *Ralstonia eutropha* was genetically modified by incorporating a mouse gene, encoding metallothionein (Valls et al. 2000). The genetically modified strain showed higher ability to sequester heavy metals than native strain. Nevertheless, the efficiency of gene expression can be influenced by environmental conditions such as matrix effect. For instance; *Rhodopseudomonas palustris* is known to consist of an *arsM* gene encoding mammalian Cyt19 As (III) S-adenosyl methionine methyltransferase. The target gene was introduced in *Sphingomonas desiccabilis* and *Bacillus idriensis* (Liu et al. 2011). Results showed that the rate of arsenic removal was more pronounced in the aqueous system (~ 10 folds increase in methylated arsenic gas) than in soil system (~ 2.2%-4.5% of arsenic removal). In another study, bacterial strain *Caulobacter crescentus* was genetically engineered to overexpress hexahistidine peptide on the cell surface (Xu et al. 2012). Results indicated a significant increase in cadmium biosorption capacity in modified strain than the control strain.

Genetic variations in bacterial strains have also been performed to achieve bioremediation of pesticides. Genetic transformations in bacterial strains either functions by inducing higher degradation ability directly or remove the bottlenecks associated with the degradation process. In a study, DNA shuffling was used to produce mutants with 25-fold and 725-fold higher degradation potential for two recalcitrant pesticides, methyl parathion, and chlorpyrifos, respectively (Cho et al. 2002; Cho et al. 2004). In another study, a synergistic system comprising a lipoprotein and an outer membrane protein A was designed and further used to eliminate transport obstruction of the parathion, across the cell membrane. The resultant system led to enhance the parathion degradation about 7 times faster in *E. coli* (Richins et al. 1997). Radiation resistant bacteria have also been used genetically modified to be used in bioremediation of potential contaminants. For instance; *E. coli* strain BL308 is known to contain Hg (II) resistance gene (merA). This gene was introduced in radiation resistant *D. radiodurans* strain and the engineered cells were observed to transform Hg (II) to the less toxic volatile elemental mercury (Brim et al. 2006). Genetic engineering has also been performed to incorporate thermostability in the enzymes involved in biodegradation. For instance; Site-directed mutagenesis was performed on an enzyme named 2-hydroxy biphenyl-2-sulfinate sulfinolyase, which is known to involve in the catalytic cycle of an organosulfur compound, dibenzothiophene. The genetic modification resulted in improved enzymatic activity and thermostability (Ohshiro et. al. 2007).

Fungi

Genetically modified fungi are known to survive in adverse environmental conditions, secrete pollutant degrading enzymes, and therefore degrade more environmental contaminants than the native strains. There are various reports on genetic manipulation in fungi used in mycoremediation studies (Table 2). For instance, *Trichoderma reesei* is known to secrete cellulose degrading cellobiohydrolase I (CBHI) and cellobiohydrolase II (CBHII) enzymes (Nam et al. 2002). Genes encoding the cellulose binding domain (CBD) from *Trichoderma reesei* have been expressed on the cell surface of *Saccharomyces cerevisiae*. The engineered cells have shown a remarkable binding affinity to cellulose. Furthermore, *S. cerevisiae* cells were transformed with the expression of a plasmid containing rat CYP1A1 with rat CYP1A2 with Yeast NADPH P-450 reductase to produce a recombinant strain (Sakaki et al. 2002). The resultant strain showed higher degradation of DD, MCDDs, DCDDs, and Tri-CDDs as compare to the wild strains.

The wild strain of *Fusarium solani* present in the soil is poor degrader of DDT. A strain of this fungus was modified by parasexual hybridization of two complementary isolates *viz.*, isolate 1 (P-1, genetically $Kel^{+}BenRDBP^{-}Lin^{-}$) and 4 (P-2, $Kel^{-}BenrDBP^{+}Lin^{+}$) (Mitra et al. 2001). The resultant mutant with genotype $Kel^{+}BenRDBP^{+}Lin^{+}$ showed superior mycoremediation capacity for DDT. In another study, a genetically modified strain of thermotolerant methylotrophic yeast, *Hansenula polymorpha*is used for removal of chromate. The recombinant strain expressing FCb2 (Flavocytochrome b2) enzymes, showed six-fold greater chromate removal than the wild-type strain (Smutok et al. 2011).

Enzyme-engineering methods have also been applied to acquire efficient removal of pollutants like heavy metals, polyaromatic hydrocarbons and, radionuclides (Dhanya 2014). For instance; genetically modified immobilized *Phanerochaete chrysosporium* was shown to remove Zn^{2+} about 16 fold greater than the pure strain (Iqbal and Edyvan 2004). In another study, Juand Parales, (2006) utilized this approach to increase the oxidation rate of nitrobenzene to catechol and nitrite by modification in the nitrobenzene 1,2-dioxygenase enzyme. The engineered strain showed a 2.5 fold higher oxidation of nitrobenzene against 2,6 dinitrotoluene.

Plants

Phytoremediation strength of a plant can be improved by inducing genetic modifications like the variations in primary or secondary metabolism via insertion of new genotypic or phenotypic characteristics. It involves genetic manipulation to cause overexpression of the gene of interest for uptake, transportation, metabolism activities, vacuolar sequestration of particular contaminants, triggering increased metal tolerance and more accumulation in plant biomass (Chen et al. 2013; Zhang et al. 2013) (Figure 1).

Table 1. A summary of genetically modified bacteria used in bioremediation

Modified enzyme/gene expression	Genetically modified bacteria	Pollutants	Reference
Metallothionein (MT)	*Ralstonia eutropha* CH34	Cd^{2+}	Valls et al. (2000)
Phytochelatin synthase (PCS)	*Mesorhizobium huakuii* B3	Cd^{2+} (From rice fields)	Sriprang et al. (2003)
Chromate reductase (ChrR)	*P. putida* strain	Cr (Bacterial cultures, as well as cell suspensions)	Ackerley et al. (2004)
PCS gene expression (SpPCS	*E.coli* strain	Cd^{2+} (Microbial sorbents for Cd removal)	Kang et al. (2007)
MerE protein encoded by the transposon Tn21 (broad Hg transporter)	*Pseudomonas strain* K-62	Hg (Across the bacterial membrane)	Kiyono et al. (2009)
Tetrahydrofuran monooxygenase (ThmD)	*E. coli*	Chlorinated solvents, tetrahydrofuran, Cyclic ethers 1,4-dioxane, etc.	Oppenheimer et al. (2010); Sales et al. (2013)
Overexpression of the arsM gene	*Sphingomonas desiccabilis* and *Bacillus Idriensis* strains	As (Laboratory conditions)	Liu et al. (2011)
Overexpression of hexahistidine peptide	*Caulobacter crescentus* strain JS4022/p723-6H	Cd^{2+}	Xu et al. (2012)
Mn peroxidase (Mn P)	*E. coli*	Phenols, amine containing aromatic compounds, dye, etc.	Chen et al. (2015)
Carboxylesterase	*E. coli* BL21	Pesticides, carbamates, chlorinecompound, etc.	Boonyuen et al. (2015)
Flavodoxin-like protein (Pst2)	*E. coli*	Quinine, phenolic component, etc.	Koch et al. (2017)

Metal binding proteins could be massively produced in the plant cell by genetic alteration. The occurrence of Expressed sequence gene (ESTs) for the gene encoding the metal binding protein in each progressive or developmental phase roughly shows the level of mRNA expression. Due to the availability of cDNA encoding metal binding protein, the recombinant metal binding protein could be easily generated *in vitro*. Figure 2 summarizes the steps involved in the process of generation of genetically improved metal hyper accumulator plant. The gene of interest can be isolated from the gene pool with the help of restriction enzymes and then can be transferred to plasmid by DNA ligase, which assists in the joining of cleaved DNA strands. This recombinant DNA could be easily transferred in the bacterial cell for enhanced production of these metal-binding proteins which is important for heavy metal accumulation in the plant cell. For instance. Gene *IIMt2a* encodes the metal lothionein producing enzyme and was characterized in a halophyte *Iris lactea* var. *chinensis* (Gu et al. 2014). The gene was introduced in the *Arabidopsis* genome and results showed increased resistance for cadmium and copper (Gu et al. 2014, 2015). In another study, Cu resistancc in *Ensifer medicae* was increased by triggering the expression of Cu resistance gene CopAB from *Pseudomonas* sp.,(Delgadillo et al. 2015). Followed by this, Patricia Perez-Palacios et al. (2017) generated a double symbiotic system for Cu resistance. One side of the operon comprised the metallothionein gene mt4a, expressed in *Medicago truncatula* from *Arabidopsis thaliana* and another side possessed Cu resistance gene copAB, overexpressed from *Pseudomonas fluorescence* derived by nodulation promotor if Hp. The developed system showed improved Cu resistance.

NEW INSIGHTS INTO PERFORMANCE IMPROVISATION OF GENETICALLY MODIFIED ORGANISMS

Modularization of Genetic Elements

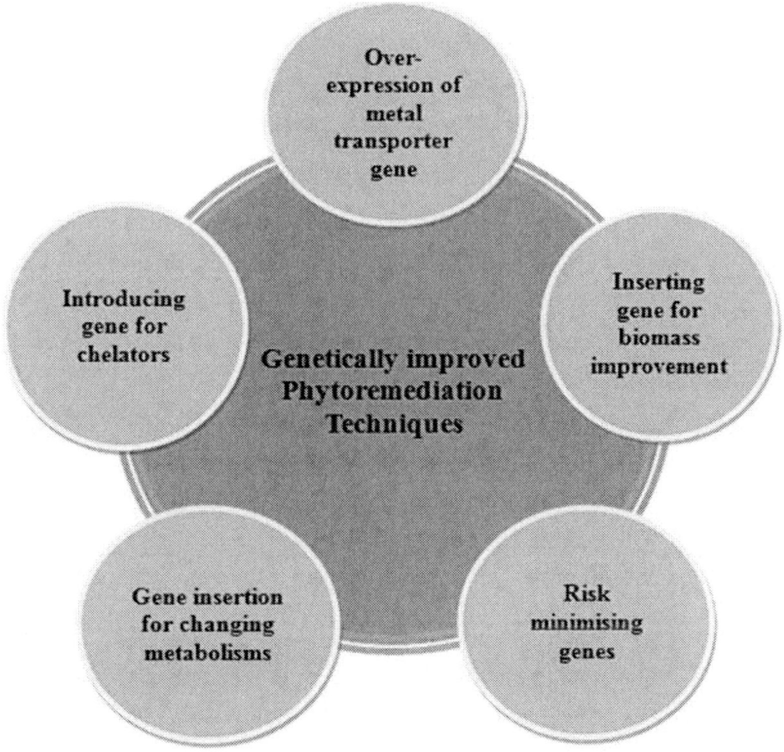

Figure 1. An integrated approach for genetically modified Phytoremediation techniques.

Modularization of genetic elements could pave the way for extending the application of genetically modified cells by improving the efficiency of genetically modified organisms with heterologous metabolic network and creating an artificial metabolic pathway (Qi et al. 2015). In this process, genetic elements are arranged into exchangeable modules by standardization of the genetic elements. Numerous methods and software are available which could modularize the promoter, terminator, coding sequence, and ribosome binding site for regulation of gene expression. These include optimized DNA assembly methods (Hatfield and Roth 2007) and Genome editing suite BioStudio (Dymond et al. 2011). Further, Computer assistant DNA assembly software (Chen et al. 2012) and J5 DNA assembly design automation software (Hillson et al. 2012) were also developed to extend the progress in regulation of gene expression.

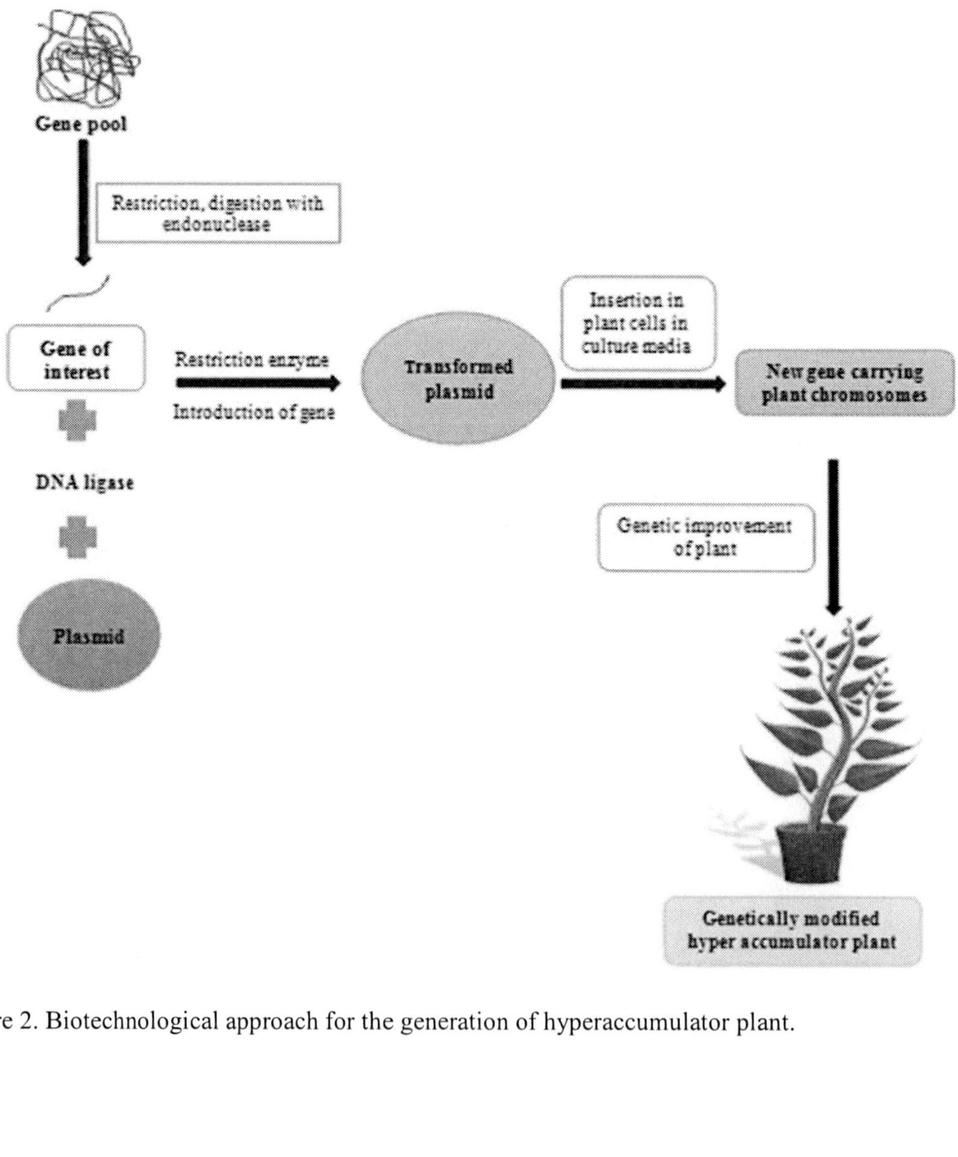

Figure 2. Biotechnological approach for the generation of hyperaccumulator plant.

Table 2. A summary of genetically modified plants used in bioremediation

Gene of interest	Genomic output	Gene origin	Host fauna	Effect	Reference
GR	Glutathione reductase	*E.coli*	Indian mustard	Improved Cd tolerance (100 mM $CdSO_4$)	Pilon-Smits et al. (2000)
TaPCS1	Phytochelatin synthase	*Triticum aestivum*	Tobacco	Increased resistance to Pb (1 mM) and Cd (50 mM)	Gisbert et al. (2003)
OASTL	Cysteine synthase	*A. thaliana*	Arabidopsis	Increased Cd build up	Dominguez-Solis et al. (2004)
AtPCS1	Phytochelatin synthase	*A. thaliana*	Arabidopsis	Increased resistance to As.	Li et al. (2004)
gshI, gshII and, APS1	c,-Glu-Cyssynthetase, glutathione synthetase, and ATP sulfurylase	*E. coli* *E. coli* *A. thaliana*	Indian Mustard	Increased resistance to Se	Banuelos et al. (2005)
mer B	Organomercuriallyase (MerB)	Bacteria	Tobacco	Increased resistance to methyl Hg	Nagata et al. (2010)
ScMTII	Metallothionein	*Saccharomyces cerevisiae*	*N.tabacum*	Enhanced Cd resistance	Daghan et al. (2013)
P450 CYP1A2	Cytochrome P450 monooxygenase	Human	*A.thaliana*	Increased resistance to phenylurea and chlortoluron	Kebeish et al. (2014)
nidA, nidB, NahAa and, NahAb	Naphthalene dioxygenase (nid A and nidB). Flavoproteinreductase and ferredoxin (NahAa and NahAb)	*Mycobacterium vanbaalenii PYR-1 Pseudomonas putida G7*	*A.thaliana*	Improved resistance for PAHs and faster accumulation of PAHs.	Peng et al. (2014)
AtHMA4	Export protein (Zn/Cd root to shoot translocation)	*A. thaliana*	*N.tabacum*	Reduced Cd uptake in roots and shoots due to enhanced cell wall lignification and H_2O_2 storage.	Siemianowski et al. (2014)
AtACBP1 and, AtACBP4	Acyl-CoA-binding proteins (ACBPs)	*A.thaliana*	*Brassica juncea*	Improved storage of Pb in roots.	Du et al. (2015)
IlMT2b	Metallothionein	*Iris lactea var. chinensis*	*A.thaliana*	Improved resistance for Cu metal and reduced H_2O_2.	Gu et al. (2015)
gshI	γ-Glutamylcysteinesynthetase	*Escherichia coli*	*Populustremula, P. alba*	Improved resistance for Cd in aerial part, An enhanced formation of glutathione (GSH) and oxidized GSH in shoot or root.	He et al. (2015)
CsGSTU1 and, CsGSTU2	Glutathione transferase	*Citrus sinensis*	*N.tabacum*	Resistance to Fluorodifen.	Lo et al. (2015)
dbphc	2,3-Dihydroxybiphenyl-1,2-dioxygenase	Soil metagenomic library	*Medicago sativa L. cv. Gongnong*	Improved resistance for PCBs and 2,4-DCP.	Wang et al. (2015)

Programmed Cell Death

The introduction of programmed cell death after detoxification can address the need for safe and efficient bioremediation. The principal of programmed cell death is based on the expression of two genes i.e., a stable killer toxin (protein) and a labile anti-killer toxin (antisense RNA or protein) (Zielenkiewicz and Ceglowski 2001). The engineered bacteria would encode the Gene(s) encoding the catabolic pathway of enzymes and anti-killer protein, either in plasmid or genome, which in turns regulated by a catabolic inducible bacterial promoter. Anti-killer genes are induced by the presence of contaminant initiating biodegradation. Once the contaminant will be completely degraded, synthesis of the anti-killer would disrupt, initiating the production of killer protein. The killer protein would exterminate the cell by destabilizing the bacterial cell membrane. In addition, horizontal gene transfer to other organisms would be averted by incorporating an opposite alignment, i.e., the killer genes would be encoded in the plasmid, thus instigating cell death in case of plasmid-mediated gene transfer. A recombinant strain of *Pseudomonas fluorescens* HK44 has been successfully assessed for field release in controlled field condition (Ripp et al. 2000). However, the upcoming execution of other GE bacteria in field condition still needs to be investigated.

Conclusion

Genetically modified organisms have great potential for the removal of different environmental contaminants. However, there are some limitations associated with the field release of genetically engineered organisms, such as invasion of mobile genetic material in non-target organism, reduced fitness and requirement of more energy driven by the presence of foreign genetic element in the cells, lack of sufficient information on the population dynamics of genetically engineered organisms in natural field condition and poor physiological control of target gene expression under stress condition (Saylor and Ripp 2000; Singh et al. 2011; Cases and Lorenzo 2005). In addition, social and administrative acceptance of GEMs application in the natural environment is still a challenge. Nevertheless, proper expression of introduced genes or enzymes requires integration within the regulatory and metabolic network (Cases and Lorenzo 2005). During the past few years, a number of sequence information for a broad spectrum of organisms has accumulated in databases. However, exploration of novel enzymes and regulatory mechanisms for efficient utilization of genetic modification is still a challenge (Qi et al. 2015). The currently available molecular toolbox can be applied to harbor more information on the metabolic pathways and bottlenecks of bioremediation (Stegmann 2001). Future research should be directed towards protein and pathway engineering to improve degradation efficiency. More future studies still remain such as exploring information from field work and risk assessment strategy, which would enhance the reliability or feasibility of genetic engineering.

Acknowledgments

Department of Science and Technology, New Delhi is gratefully acknowledged for providing the national postdoctoral fellowship to Dr. Shivani Uniyal. University Grant Commission, New Delhi for providing Junior Research Fellowship to Mr. Indrajeet Kumar and Mr. Prince Kumar Singh is also acknowledged.

References

Ackerley, D. F., Gonzalez, C. F., Keyhan, M., Blake, R. and Matin, A. (2004). Mechanism of chromate reduction by the *Escherichia coli* protein, NfsA, and the role of different chromate reductases in minimizing oxidative stress during chromate reduction. *Environ Microbiol*, 6: 851-860.

Alcalde, M., Ferrer, M., Plou, F. J. and Ballesteros, A. (2006). Environmental biocatalysis: from remediation with enzymes to novel green processes. *Trends Biotechnol*, 24: 281-287.

Al-Salloum, Y., Hadi, S., Abbas, H., Almusallam, T. and Moslem, M. A. (2017). Bio-induction and bioremediation of cementitious composites using microbial mineral precipitation-A review. *Constr Build Mater*, 154:857-876.

Arber, W. (2010). Genetic engineering compared to natural genetic variations. *New Biotechnol*, *27*:517-521.

Arevalo-Rodriguez, M., Pan, X., Boeke, J. D. and Heitman, J. (2004). FKBP12 controls aspartate pathway flux in *Saccharomyces cerevisiae* to prevent toxic intermediate accumulation. *Eukaryot Cell*, 3:1287-96.

Ayu, K. R., Tony, H., Tadashi, T., Yasuhiro, T. and Kazuhiro, M. (2011). Bioremediation of crude oil by white rot fungi Polyporus sp. S133. *J Microbiol Biotechn*, 21:995-1000.

Bai, S. R. and Abraham, T. E. (2001). Biosorption of Cr(VI) from aqueous solution by *Rhizopus nigricans*. *BioresourTechnol*, 79:73-81.

Banuelos, G., Terry, N., LeDuc, D. L., Pilon Smits, E. A. and Mackey, B. (2005). Field trial of transgenic Indian mustard plants shows enhanced phytoremediation of selenium-contaminated sediment. *Environ Sci Technol*, 39:1771-1777.

Barlaan, E. A., Sugimori, M., Furukawa, S. and Takeuchi, K. (2005). Profiling and monitoring of microbial populations by denaturing high-performance liquid chromatography. *J Microbiol Methods*, 61:399-412.

Bolan, N., Kunhikrishnan, A., Thangarajana, R., Kumpiene, J., Park, J., Makino, T., Kirkham, M. B. and Scheckel, K. (2014). Remediation of heavy metal(loid)s contaminated soils-to mobilize or to immobilize? *J Hazard Mater*, 26:141-166.

Boonyuen, U., Promnares, K., Junkree, S., Day, N. P. and Imwong, M. (2015). Efficient *in vitro* refolding and functional characterization of recombinant human liver carboxylesterase (CES1) expressed in E. coli. *Protein Expr Purif*, 107:68-75.

Cases, I. and de Lorenzo, V. (2005). Genetically modified organisms for the environment: stories of success and failure and what we have learned from them. *Int Microbiol*, 8:213-222.

Caspi, R., Billington, R., Ferrer, L., Foerster, H., Fulcher, C. A., Keseler, I. M., Kothari, A., Krummenacker, M., Latendresse, M., Mueller, L. A., Ong, Q., Paley, S., Subhraveti, P., Weaver, D. S. and Karp, P. D. (2016). The Meta Cyc database of metabolic pathways and enzymes and the BioCyc collection of pathway/genome databases. *Nucleic Acids Res*, 44: D471-480.

Chen, S., Liu, C., Peng, C., Liu, H., Hu, M. and Zhong, G. (2012). Biodegradation of chlorpyrifos and its hydrolysis product 3, 5, 6-Trichloro-2-pyridinol by a new fungal strain *Cladosporium cladosporioides* Hu-01. *Plos One*, 7: e47205.

Chen, W., Zheng, L., Jia, R. and Wang, N. (2015). Cloning and expression of new manganese peroxidase from *Irpexlacteus* F17 and its application in decolorization of reactive black 5. *Process Biochem*, 50:1748-1759.

Chen, Z., Fujii, Y., Yamaji, N., Masuda, S., Takemoto, Y., Kamiya, T. and Ueno, D. (2013). Mn tolerance in rice is mediated by MTP8 1, a member of the cation diffusion facilitator family. *J Exp Bot*, 64:4375-4387.

Cho, C. M. H., Mulchandani, A. and Chen, W. (2002). Bacterial cell surface display of organophosphorus hydrolase for selective screening of improved hydrolysis of organophosphate nerve agents. *Appl Environ Microbiol*, 68:2026-2030.

Cho, C. M. H., Mulchandani, A. and Chen, W. (2004). Altering the substrate specificity of organophosphorus hydrolase for enhanced hydrolysis of chlorpyrifos. *Appl Environ Microbiol*, 70:4681-4685.

Choi, Y. J. and Lee, S. Y. (2013). Microbial production of short-chain alkanes. *Nature*, 502:571-574.

Cicero, L. L., Madesis, P., Tsaftaris, A. and Piero, A. R. L. (2015). Tobacco plants over-expressing the sweet orange tau glutathione transferases (CsGSTUs) acquire tolerance to the diphenyl ether herbicide fluorodifen and to salt and drought stresses. *Phytochemistry*, 116:69-77.

Cycon, M., Wojcik, M. and Piotrowski-Seget, Z. (2009). Biodegradation of the organophosphorus insecticide diazinon by *Serratia* sp. and *Pseudomonas* sp. and their use in bioremediation of contaminated soil. *Chemosphere*, 76:494-501.

Daghan, H., Arslan, M., Uygur, V. and Koleli, N. (2013). Transformation of tobacco with ScMTII gene-enhanced cadmium and zinc accumulation. *Clean*, 41:503-509.

Delgadillo, J., Lafuente, A., Doukkali, B., Redondo-Gómez, S., Mateos-Naranjo, E., Caviedes, M. A., Pajuelo, E. and Rodríguez Llorente, I. D. (2015). Improving legume nodulation and Cu rhizostabilization using a genetically modified rhizobia. *Environ Technol*, 36:1237-1245.

Dhanya, M. S. (2014). Advances in microbial biodegradation of chlorpyrifos.*J Environ ResDevelop*, 9:232.

Dominguez Solis, J. R., Lopez Martin, M. C., Ager, F. J., Ynsa, M. D., Romero, L. C. and Gotor, C. (2004). Increased cysteine availability is essential for cadmium tolerance and accumulation in *Arabidopsis thaliana*. *Plant Biotechnol J*, 2:469-476.

Du, Z. Y., Chen, M. X., Chen, Q. F., Gu, J. D. and Chye, M. L. (2015). Expression of Arabidopsis acyl-CoA-binding proteins AtACBP1 and AtACBP4 confers Pb (II) accumulation in *Brassica juncea* roots. *Plant Cell Environ*, 38:101-117.

Dua, M., Singh, A., Sethunathan, N. and Johri, A. K. (2002). Biotechnology and bioremediation, successes and limitations. *Appl Microbiol Biotechnol*, 59:143-152.

Dvořáka, P., Nikelb, P. I., Damborský, J. and Lorenzoa, V. D. (2017). Bioremediation 3.0: Engineering pollutant-removing bacteria in the times of systemic biology. *Biotechnol Adv*, 35:845-866.

Dymond, J. S., Richardson, S. M., Coombes, C. E., Babatz, T., Muller, H. and Annaluru, N. (2011). Synthetic chromosome arms function in yeast and generate phenotypic diversity by design. *Nature*, 477:471-476.

Favero, N., Costa, P. and Massimino, M. L. (1991). *In vitro* uptake of cadmium by basidiomycete *Pleurotus ostreatus*. *Biotechnol Lett*, 10:701-704.

Furukawa, K. (2003). Superbugs for bioremediation.*Trends Biotech*, 21:187-190.

Galagan, J. E., Calvo, S. E., Borkovice, K. A., Selker, E. U., Read, N. D., Jaffe, D., FitzHugh, W., Ma, L. J., Smirnov, S., Purcell, S. and Rehman, B. (2003). The genome sequence of the filamentous fungus *Neurospora crassa*. *Nature*, 422:859-868.

Gao, J., Ellis, L. B. M. and Wackett, L. P. (2010). The University of Minnesota Biocatalysis/Biodegradation Database: improving public access. *Nucleic Acids Res*, 38: D488-491.

Gisbert, C., Ros, R., Haro, A. D., Walker, D. J., Bernal, M. P., Serrano, R. and Avino, J. N. A. (2003). Plant genetically modified that accumulates Pb is especially promising for phytoremediation. *Biochem Biophys Res Commun*, 303:440-445.

Gu, C. S., Liu, L. Q., Deng, Y. M., Zhu, X. D., Huang, S. Z. and Lu, X. Q. (2015). The heterologous expression of the *Iris lactea var. chinensis* type 2 metallothionein IlMT2b gene enhances copper tolerance in *Arabidopsis thaliana*. *Bull Environ Contam Toxicol*, 94:247-253.

Gu, C. S., Liu, L. Q., Deng, Y. M., Zhu, X. D., Huang, S. Z. and Lu, X. Q. (2015). The heterologous expression of the *Iris Lactea var. chinensis* type 2 metallothionein IlMT2b gene enhances copper tolerance in *Arabidopsis thaliana*. *Bull Environ Contam Toxicol*, 94:247-53.

Gu, C. S., Liu, L. Q., Zhao, Y. H., Deng, Y. M., Zhu, X. D. andHuang, S. Z. (2014). Overexpression of *Iris lactea var. chinensis* metallothionein IlMT2a enhances cadmium tolerance in *Arabidopsis thaliana*. *Ecotoxicol Environ Saf*, *105*:22-28.

Gupta, S. K. and Shukla, P. (2015). Advanced technologies for improved expression of recombinant proteins in bacteria: perspectives and applications. *Crit Rev Biotechnol*, *36*:1089-1098.

Hao, Q. i., Bing Zhi, L. i., Wen Qian, Zhang, Duo, Liu, and Ying, Jin Yuan. (2015). Modularization of genetic elements promotes synthetic metabolic engineering. *Biotechnol Adv*, *33*:1412-1419.

Hatfield, G. W. and Roth, D. A. (2007). Optimizing scale up yield for protein production: computationally optimized DNA assembly (CODA) and translation engineering. *Biotechnol Annu Rev*, *13*:27-42.

He, J., Li, H., Ma, C., Zhang, Y., Polle, A. and Rennenberg, H. (2015). Overexpression of bacterial γ-glutamyl cysteine synthetase mediates changes in cadmium influx, allocation and detoxification in poplar. *New Phytol*, *205*: 240-254.

Hillson, N. J., Rosengarten, R. D. and Keasling, J. D. (2012). J5 DNA assembly design automation software. *ACS Synth Biol*, *1*:14-21.

Hong, S. H. and Lee, S. Y. (2002). Importance of redox balance on the production of succinic acid by metabolically engineered *Escherichia coli*. *Appl Microbiol Biotechnol*, *58*:286-290.

Ikeda, M. (2006). Towards bacterial strains overproducing L-tryptophan and other aromatics by metabolic engineering. *Appl Microbiol Biotechnol*, *69*:615-626.

Iqbal, M. and Edyvan, R. G. J. (2004). Biosorption of lead, copper and zinc ions on loofa sponge immobilized biomass of *Phanerochaete chrysosporium*. *Miner Eng*, *17*:217-223.

Jafari, Morad., Danesh, Y. R. and Ghoosta, Y. (2013). Molecular techniques in fungal bioremediation. *In: Fungi as Bioremediators* (Eds. Goltapeh EM, Danesh, YR, Varma A), Springer, Berlin, Heidelberg, pp. 453-465.

Ju, K. S. and Parales, R. E. (2006). Control of substrate specificity by active-site residues in nitrobenzene dioxygenase. *Appl. Environ Microbiol*, 72:1817-1824.

Kang, S. H., Singh, S., Kim, J. Y., Lee, W., Mulchandani, A. and Chen, W. (2007). Bacteria metabolically engineered for enhanced phytochelatin production and cadmium accumulation. *Appl Environ Microbiol*, 73:6317-6320.

Kebeish, R., Azab, E., Peterhaensel, C. and El Basheer, R. (2014). Engineering the metabolism of the phenylurea herbicide chlortoluron in genetically modified *Arabidopsis thaliana* plants expressing the mammalian cytochrome P450 enzyme CYP1A2. *Environ Sci Pollut Res Int*, 21:8224-8232.

Kiyono, M., Sone, Y., Nakamura, R., Pan Hou, H. and Sakabe, K. (2009). The Mer E protein encoded by transposon Tn21 is a broad mercury transporter in *Escherichia coli*. *FEBS Lett*, 583:1127-1131.

Koch, K., Hromic, A., Sorokina, M., Strandback, E., Reisinger, M., Gruber K. and Macheroux, P. (2017). Structure, biochemical and kinetic properties of recombinant Pst2p from *Saccharomyces cerevisiae* an FMN-dependent NAD(P)H:quinine oxidoreductase. *Biochim Biophys Acta Protein Proteonomics*, 1865:1046-1056.

Kokura, S., Wolf, R. E., Yoshikawa, T., Granger, D. N. and Aw, T. Y. (1999). Molecular mechanisms of neutrophil-endothelial cell adhesion induced by redox imbalance.*Circ Res*, 84:516-524.

Kurumbang, N. P., Dvorak, P., Bendl, J., Brezovsky, J., Prokop, Z. and Damborsky, J. (2014). Computer-assisted engineering of the synthetic pathway for biodegradation of a toxic persistent pollutant. *ACS Synth Biol*, 3:172-181.

Li, Y., Dhankher, O. P., Carreira, L., Lee, D., Chen, A., Schroeder, J. I., Balish, R. S. and Meagher, R. B. (2004). Overexpression of phytochelatin synthase in *Arabidopsis* leads to enhanced arsenic tolerance and cadmium hypersensitivity. *Plant Cell Physiol*, 45:1787-1797.

Lin, F., Chen, Y., Levine, R., Lee, K., Yuan, Y. and Lin, X.N. (2013). Improving fatty acid availability for biohydrocarbon production in *Escherichia coli* by metabolic engineering. *PLoS One*, 8, e78595.

Liu, C. J., Blount, J. W., Steele, C. L. and Dixon, R. A. (2002). Bottlenecks for metabolic engineering of isoflavone glycoconjugates in Arabidopsis. *Proc Natl Acad Sci USA*, *99*:14578-14583.

Liu, S., Zhang, F., Chen, J. and Sun, G. X. (2011). Arsenic removal from contaminated soil via biovolatilization by genetically engineered bacteria under laboratory conditions. *J Environ Sci*, 23:1544-1550.

Liu, Z., Hong, Q., Xu, J. H., Jun, W. and Li, S. P. (2006). Construction of a genetically engineered microorganism for degrading organophosphate and carbamate pesticides. *Int Biodeter Biodegr*, 58:65-69.

Lorenz, M. G. and Wackernagel, W. (1994). Bacterial gene-transfer by natural genetic transformation in the environment. *Microbiol Rev*, 58:563-602.

Madigan, M. T., Martinko, J. M., Stahl, D. A. and Clark, D. P. (2012). Genetic Engineering. In: *Brock biology of microorganisms* (Eds. Michael TM, John MM, Jack P), 13th edition, Benjamin Cummings Publications, Sansome Street, San Francisco, CA, pp. 94111-94298.

McGary, K. L., Slot, J. C. and Rokas, A. (2013). Physical linkage of metabolic genes in fungi is an adaptation against the accumulation of toxic intermediate compounds. *Proc Natl Acad Sci USA*, 110:11481-11486.

Merker, J. D. and Arber, D. A. (2010). Molecular Diagnostics of Hematopoietic Disorders.In: *Molecular Diagnostics Techniques and Applications for the Clinical Laboratory* (Eds. Grody, W. W., Nakamura, R. M., Kiechle, F. L., Strom, C.), Academic Press, pp. 243-260.

Mitra, J., Mukherjee, P. K., Kale, S. P. and Murthy, N. B. K. (2001). Bioremediation of DDT in soil by genetically improved strain of soil fungi *Fusarium solani. Biodegradation*, 12:235-245.

Nagata, T., Morita, H., Akizawa, T. and Pan Hou, H. (2010). Development of a transgenic tobacco plant for phytoremediation of methyl mercury pollution. *App Microbiol Biotechnol*, 87:781-786.

Naik, M. M. and Dubey, S. K. (2013). Lead resistant bacteria: Lead resistance mechanisms, their applications in lead bioremediation and biomonitoring. *Ecotoxicol Environ Saf*, 98:1-7.

Nam, J. M., Fujita, Y., Arai, T., Kondo, A., Morikawa, Y., Okada, H., Ueda, M. and Tanaka, A. (2002). Construction of engineered yeast with the ability of binding to cellulose. *J Mol Catal B Enzym*, 17:197-202.

Ohshiro, T., Ohkita, R., Takikawa, T., Manabe, M., Lee, W. C., Tanokura, M. and Izumi, Y. (2007). Improvement of 20-Hydroxybiphenyl-2-sulfinate desulfinase, an enzyme involved in the dibenzothiophene desulfurization pathway from *Rhodococcusery-thropolis* KA2-5-1 by site-directed mutagenesis. *Biosci Biotech Bioch*, 71:2815-2821.

Oppenheimer, M., Pierce, B. S., Crawford, J. A., Ray, K., Helm, R. F. and Sobrado, P. (2010). Recombinant expression, purification, and characterization of ThmD, the oxidoreductase component of tetrahydrofuran monooxygenase. *Arch Biochem Biophys*, 496:123-131.

Oves, M., Khan, M. S. and Zaidi, A. (2013). Biosorption of heavy metals by *Bacillus thuringiensis* strain OSM29 originating from industrial effluent contaminated north Indian soil. *Saudi J Biol Sci*, 20:121-129.

Patricia Pérez, Palacios., Romero, Aguilar A., Delgadillo, J., Bouchra, Doukkali.,Caviedes, M. A., Rodríguez Llorente, I. D. and Eloísa, Pajuelo. (2017). Double genetically modified symbiotic system for improved Cu phytostabilization in legume roots. *Environ Sci Pollut Res*, 24:14910-14923.

Peng, R., Fu, X., Tian, Y., Zhao, W., Zhu, B. and Xu, J. (2014). Metabolic engineering of Arabidopsis for remediation of different polycyclic aromatic hydrocarbons using a hybrid bacterial dioxygenase complex. *Metab Eng*, 26:100-110.

Pieper, D. H. and Reineke, W. (2000). Engineering bacteria for bioremediation. *Curr Opin Bio technol*, 11:262-270.

Pilon Smits, E. A. H., Zhu, Y. L., Sears, T. and Terry, N. (2000). Overexpression of glutathione reductase in *Brassica juncea*: effects on cadmium accumulation and tolerance. *Physiol Plant*, 110:455-460.

Prakash, V. (2017). Mycoremediation of Environmental Pollutants. *Int j Chem Tech Res*, 10:149-155.

Rayu, S., Karpouzas, D. G. and Singh, B. K. (2012). Emerging technologies in bioremediation: constraints and opportunities. *Biodegradation*, 23:917-926.

Robert, E. and Farrell, Jr. (2010). Electrophoresis of RNA. In:*RNA Methodologies-Laboratory Guide for Isolation and Characterization* (Eds. Robert E, Farrell Jr), 4th Edition, Elsevier Inc., Academic press, pp. 179-219.

Richins, R., Kaneva, I., Mulchandani, A. and Chen, W. (1997). Biodegradation of organophosphorus pesticides by surface-expressed organophosphorus hydrolase. *Nat Biotechnol*, 15:984-987.

Sakaki, T., Shinkyo, R., Takita, T., Ohta, M. and Inouye, K. (2002).Biodegradation of polychlorinated dibenzo-p-dioxins by recombinant yeast expressing rat CYP1A subfamily. *Arch Biochem Biophys*, 401:91-98.

Saylor, G. S. and Ripp, S. (2000). Field applications of genetically engineered microorganisms for bioremediation processes. *Curr Opin Biotechnol*, 11:286-289.

Shimizu, H. (2002). Metabolic engineering-integrating methodologies of molecular breeding and bioprocess systems engineering. *J Biosci Bioeng*, 94:563-573.

Shukla, D. and Vankar, P. S. (2014). Role of Trichoderma Species in Bioremediation Process: Biosorption Studies on Hexavalent Chromium. Biotechnology and Biology of Trichoderma. pp. 405-412.

Siemianowski, O., Barabasz, A., Kendziorek, M., Ruszczynska, A., Bulska, E., Williams, L. E. and Antoniewicz, D. M. (2014). HMA4 expression in tobacco reduces Cd accumulation due to the induction of the apoplastic barrier. *J Exp Bot*, 65:1125-1139.

Silva, J. P., Shabalina, I. G., Dufour, E., Petrovic, N., Backlund, E. C., Hultenby, K., Wibom, R., Nedergaard, J., Cannon, B. and Larsson, N. G. (2005). SOD2 overexpression: enhanced mitochondrial tolerance but absence of effect on UCP activity. *EMBO J*, 24:4061-70.

Singh, J. S., Abhilash, P. C., Singh, H. B., Singh, R. P. and Singh, D. P. (2011). Genetically engineered bacteria: an emerging tool for environmental remediation and future research perspectives. *Gene*: 480:1-9.

Singh, R. K., Tiwari, M. K., Singh, R. and Lee, J. K. (2013). From protein engineering to immobilization: promising strategies for the upgrade of industrial enzymes. *Int J Mol Sci*, 14:1232-1277.

Smutok, O., Broda, D., Smutok, H., Dmytruk, K. and Gonchar, M. (2011). Chromate reducing activity of *Hansenula polymorpha* recombinant cell over producing flavocytochrome b2.*Chemosphere,* 83:449-454.

Sohn, S. B., Kim, T. Y., Park, J. M. and Lee, S. Y. (2010). In silico genome-scale metabolic analysis of *Pseudomonas putida* KT2440 for polyhydroxy alkanoate synthesis, degradation of aromatics and anaerobic survival. *Biotechnol J*, *5*:739-750.

Sopko, R., Huang, D., Preston, N., Chua, G., Papp, B., Kafadar, K., Snyder, M., Oliver, S. G., Cyert, M., Hughes, T. R. and Boone, C. (2006). Mapping pathways and phenotypes by systematic gene overexpression. *Mol Cell*, 21:319-330.

Sriprang, R., Hayashi, M., Ono, H., Takagi, M., Hirata, K. and Murooka, Y. (2003). Enhanced accumulation of Cd^{2+} by a *Mesorhizobium* sp. transformed with a gene from

Arabidopsis thaliana coding for phytochelatin synthase. *Appl Environ Microbiol*, 69:179-796.

Stegmann, R. (2001). Treatment of contaminated soil: fundamentals, analysis, applications. Berlin Springer Verlag.

Timmis, K. N.and Piper, D. H. (1999). Bacteria designed for bioremediation. *Trends Biotechnol*, 17:201-204.

Uniyal, S., Paliwal, R., Sharma, R. K. and Rai, J. P. N. (2016). Degradation of fipronil by *Stenotrophomonas acidaminiphila* isolated from rhizospheric soil of *Zea mays*. *3 Biotech*, 6:48.

Uniyal, S., Paliwal, R., Verma, M., Sharma, R. K. and Rai, J. P. N. (2016). Isolation and characterization of fipronil degrading *Acinetobacter calcoaceticus* and *Acinetobacter oleivorans* from rhizospheric zone of *Zea mays*. *Bull Environ ContamToxicol*, 96:833-838.

Valls, M., Atrian, S., De Lorenzo, V. and Fernandez, L. A. (2000). Engineering a mouse metallothionein on the cell surface of *Ralstonia eutropha* CH34 for immobilization of heavy metals in soil. *Nats Biotechnol*, 18:661-665.

Wang, Y., Ren, H., Pan, H., Liu, J. and Zhang, L. (2015). Enhanced tolerance and remediation to mixed contaminates of PCBs and 2, 4-DCP by transgenic alfalfa plants expressing the 2,3-dihydroxybiphenyl-1.,2-dioxygenase. *J Hazard Mater*, 286:269-75.

Wood, V., Gwilliam, R., Rajandream, M. A., Lyne, M., Lyne, R., Stewart, A., Sgouros, J., Peat, N., Hayles, J., Baker, S. and Basham, D. (2002). The genome sequence of *Schizosaccharomyces pombe*. *Nature*, 415:871-880.

Wu, J. and Yu, H. Q. (2007). Biosorption of 2, 4-dichlorophenol by immobilized white-rot fungus *Phanerochaete chrysosporiyum* from aqueous solutions. *Bioresour Technol*, 98:253-259.

Xu, Z., Lei, Y. and Patel, J. (2010). Bioremediation of soluble heavy metals with recombinant *Caulobacter crescentus*. *Bioeng Bugs*, 1:207-212.

Xueheng, Z., Hardin, I. R. and Huey-Min, H. (2006). Biodegradation of a model azodisperse dye by the white rot fungus *Pleurotus ostreatus*. *Int Biodeterior Biodegradation*, *57*: 1-6.

Zha, J., Shen, M., Hu, M., Song, H. and Yuan, Y. (2014). Enhanced expression of genes involved in initial xylose metabolism and the oxidative pentose phosphate pathway in the improved xylose-utilizing *Saccharomyces cerevisiae* through evolutionary engineering. *J Ind Microbiol Biotechnol*, 41:27-39.

Zhang, Y., Liu, J., Zhu, Y., Gong, T., Wang, J. and Ge, Y. (2013). Enhanced phytoremediation of mixed heavy metal (mercury)-organic pollutants (trichloroethylene) with transgenic alfalfa co-expressing glutathione S-transferase and human P450 2E1. *J Hazard Mater*, 260:1100-1107.

Zielenkiewicz, U. and Ceglowski, P. (2001). Mechanisms of plasmid stable maintenance with special focus on plasmid addiction systems. *Acta Biochim Polon,* 48:1003-1023.

In: Innovations in Life Science Research ISBN: 978-1-53615-868-7
Editors: R. P. Sinha, S. Pandey-Rai and N. Ghoshal ©2019 Nova Science Publishers, Inc.

Chapter 9

ENHANCING ABIOTIC STRESS TOLERANCE IN HYACINTH BEAN (*LABLAB PURPUREUS*. L)

***Krishna K. Rai*[1,2], *Nagendra Rai*[2,*] *and Shashi Pandey-Rai*[1]**

[1]Laboratory of Morphogenesis, Centre of Advanced Study in Botany, Institute of Science, Banaras Hindu University, Varanasi, India

[2]ICAR-Indian Institute of Vegetable Research, Varanasi, India

ABSTRACT

Legumes Play an important role in safeguarding global nutritional food security and also contribute to the improvement of soil fertility by fixing biological nitrogen in agriculture. However, the production of legumes is adversely affected by several abiotic stresses. Therefore, it has become imperative to understand molecular mechanisms of plant response to adverse environmental cues by discovering candidiate genes involved in stress tolerance. Recent developments in "omics" technology has witnessed advancements in legume genomics by providing reference genome of various crops such as chickpea, pigeonpea, cowpea, soybean etc. that has accelerated the genetic gains in these crops in order meet the supply demand gap of ever-increasing global populations. Hyacinth bean, an Indian origin crop of great economic value serving both as food crop and medicinal plant is still categorized as orphan crop at genomic level. With the availability of draft genomes of various important legume crops it has now become possible to identify structural variations at large scale. Advancement in genomic resources and with the low-cost ultra-throughput sequencing technology marker trait association studies and QTL mapping in hyacinth bean can be done with higher resolution and efficiency. More importantly, deployment of molecular breeding approaches can result in the development/identification of tolerant/resistant cultivar in hyacinth bean as it has been deployed for other legume crops such as chickpea, pigeonpea and cowpea. Once this crop is characterized at its genomic level, other realm of functional genomics such as proteomics, metabolomics and epigenomics can be used in the understanding hyacinth bean stress tolerance.

[*]Corresponding Author's Email:nrai1964@gmail.com.

Keywords: abiotic stress, epigenomics, functional genomics, phytohormones, stress tolerance

INTRODUCTION

Among various food and fibre crops, the crop legumes are the key sources of protein, minerals, vitamins, iron, zinc, calcium, and magnesium, as well as omega-3 fatty acids. The importance of these legumes is higher where a large section of the society depends on vegetarian food such as in India. The unique ability to fix atmospheric nitrogen by the legume crops play a crucial role in sustaining the farming system by making available the residual nitrogen to the non-legume crops. Legumes also serve as an excellent source of high quality and nutritious feed to livestock leading to ~20% increase in animal productivity (Tarawali and Ogunbile 1995). The human civilization has a long association with legume cultivation in the last 6000 years which has contributed significantly toward economical, nutritional, and improving the livelihood of subsistence farmers across the world.

Indian bean or hyacinth bean (*Dolichos lablab* L. or *Lablab purpureus* L.) is a twining herbaceous crop legume with 2n = 22 chromosomes belonging to family Fabaceae which is being maneuverer for various purposes such as a green vegetable, pulse, soil improvement and protection, forage and weed control (Rai et al. 2016). The wild forms of Indian bean are believed to have originated in India and were introduced into Africa from South East Asia during the eighth century. Outside India, hyacinth bean is prevalent in Africa, Zimbabwe, Cameroon, Ethiopia, Sudan, Uganda and Tanzania (Maass et al. 2010) and have been called by various common names such as poor man bean and tanage bean in England, rongaidolichos and lablab bean in Australia, betas in Philippines, lubiaafin in Sudan, hyacinth bean in Brazil and in India(Gibbs et al. 2008).

TAXONOMY

The genus lablab refers to the hyacinth bean, Bonavist bean, Sem, Waby Salad bean, Egyptian bean, Field bean, Australian pea, Indian butter bean belongs to the family leguminosae (Fabaceae), sub family papilionacae and varying chromosomes 2n = 20, 22, 24. In Flora Indica, Roxburgh (1832) exemplify "Dolichos" in *Dolichos lablab* as Greek word that symbolize long pod and "lablab" as Arabic or Egyptian word implying shallow vessel enclosing seed. He also distinguishes hyacinth bean into seven varieties, demarcating them into five cultivated and in two wild varieties. Later, Backer (1984) on the basis of shape, texture and angle of attachment of seeds to the suture of the pods categorised all the five cultivated varieties in two cross compatible botanical groups i.e., *Lablab purpureus* var. typicus and *Lablab purpureus* var. lignosus (Figure 1).

Lablab purpureus L. var. typicus (Syn. Lablab niger var. typicus) is widely cultivated as garden type bean with soft pods that has long axis in which seeds are arranged parallelly to the suture and *Lablab purpureus* var. lignosus (Syn. *Lablab niger* var. lignosus) where the seeds are arranged at right angle to the suture, is mainly cultivated as field bean and most commonly used as pulse containing more protein (20.9-29.2%) than *Lablab purpureus* var. typicus. Pods of *Lablab purpureus* var. lignosus is highly preferred by farmers and consumers as they radiate certain oily substances that gives pods their characteristic fragrance. This distinctive aroma of *Lablab purpureus* var. lignosus pods is due the presence of two dominant fatty acids i.e., trans-2-dodecenoic acid and trans-2-tetradecenoic acid (Uday Kumar et al. 2017).

BOTANY AND FLORAL BIOLOGY

Hyacinth bean plant is a tropical, dicotyledonous, herbaceous and temperature sensitive flowering plant with annual growth habit. Plants are robust twiner with woody stem, having large trifoliate heart shaped leaves along with broad ovate-rhomboid shape leaflets measuring 7 to 15 cm long and 4.5 -15 cm wide. The dorsal side of the leaf is smooth with the ventral side being hairy. Plant consists of tap root system and main root has several lateral roots distributed uniformly along the main axis. Flower is usually bisexual, hypogenous and pentamerous varies from white, pink to purple in colour with axillary, erect and raceme inflorescence (Figure 1). Stamens are usually arranged in 9+1 configuration, with uniform anther, having sessile ovary with incurved style and terminal glabrous stigma. Seeds are variable in colour such as white, ochreous with black's dots or black with white dots, uniformly brown or black.

The flowers generally open two days after anther dehiscence. The presence of superfluous floral nectarines at the bottom of the corolla attract bees, flies and ant, however, heavy insect pressure may supress the development of wing petals that could lead to stamen and stigma exertion. In hyacinth bean, flowers bloom between 6.30 hr. to 7.00 hr and are predominantly a self-pollinated crop (Kukade and Tidke 2014). However, insect mediated cross pollination occurs up to an extent of 6-13%. Anthesis in hyacinth bean occurs between 9.00-17.00 hr and anther dehiscence occurs between 5.00-14.00 hr. The optimum time for emasculation of hyacinth bean flower ranges between 16.30-17.30 hrs and for hand pollination 7.00-9.00 hrs, the next day in the morning. Pollens are fertile on the day of anthesis which remains viable for 42 hrs at room temperature of 28.5°C with 85-90% relative humidity. The pollens can remain viable for much longer duration i.e., up to 50-60 hrs if they are stored under cold environment in refrigerator.

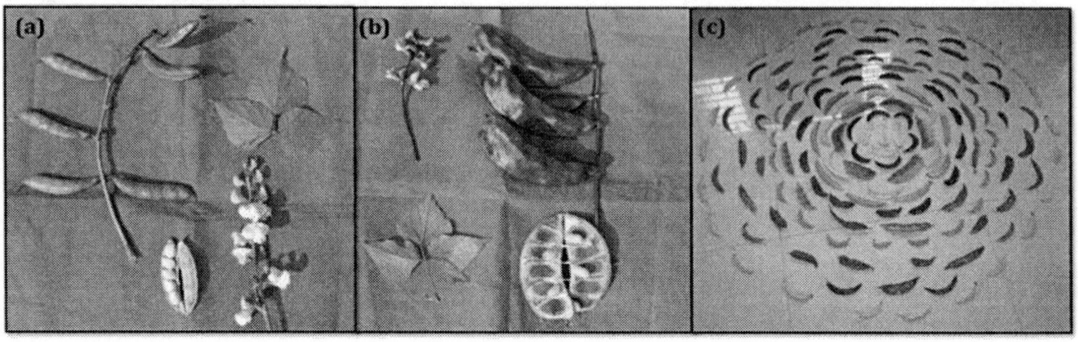

Figure 1. Phylogeny and pattern of floral biology in *Lablab purpureus* L. var. (a) typicus (b) lignosus (c) pod diversity.

NUTRITIVE AND MEDICINAL VALUE

Proteins are crucial macronutrient for growth as well as maintenance of human body and it is well known that a minimum protein intake of 0.8, 1.5 and 1.0 g protein/kg body weight/day is recommended for adults, children, and adolescents, respectively (Kafatos and Hatzis 2008). Among other crop legumes such as soybean, common bean, winged bean, cowpea, mung bean, pea, moth bean, urd bean, chickpea, pigeon pea and rice bean, hyacinth bean also serve as an excellent source of high quality and nutritious source of protein, minerals and vitamins and is richer than common bean in nutritive value. The nutritional composition of hyacinth bean pods varies greatly depending upon the genotypes and pod maturity stages. In general, 100 g of mature green pods contain 86.1 g moisture, 3.8 g protein, 0.7 g fat, 0.9 g minerals, 1.8 g fibres, 6.7 g carbohydrates, 34 mg magnesium, 68 mg phosphorus, 55.4 mg sodium, 0.13 mg copper, 210 mg calcium, 74 mg potassium, 1.7 mg iron, 40 mg sulphur, 0.06 mg riboflavin, 1 mg oxalic acid, 0.7 mg nicotinic acid, 0.1 mg thiamine, 312 I.U. vitamin A and 9 mg vitamin C.

Among the legumes, hyacinth bean constitutes an important source of therapeutic agents used in the modern as well as traditional systems of medicine (Morris 2003, 2009). The beans, including hyacinth bean, are also stack of flavonoids and isoflavones such as genistein, phaseollidin and isovestinol that can act both as signal molecule for pigments or co-pigments in mediating flower colour by modulating auxin distribution and as an armour for plants against UV damage and pathogenic microbes (Kan et al. 2016). Despite of its distinct performance as vegetables and fodder, it is also conceivable as medicinal legume and performs an indispensable role as therapeutic agents directly in the interception of discrete forms of cancer and various chronic diseases, most commonly sparring against breast cancer (Morris 2009). All these chemo-preventive proprietaries of hyacinth bean against various chronic and cancer diseases (Takagi et al. 2015) have interlineated it among important medicinal plants.

The Challenge of Multiple Environmental Stresses on Agriculture

Most of the agricultural crop plants thrive in substandard environments that preclude them from acquiring their full genetic capacity for their growth and reproduction thus imposing severe repercussion on their economic yield (Ray et al. 2013). For instance, in India, the most favourable season for hyacinth bean yield is November-January and in this period the yield can be up to 6-8 times high compared to their average yield under adverse environmental condition (www.lablablab.org).

Global Scenario

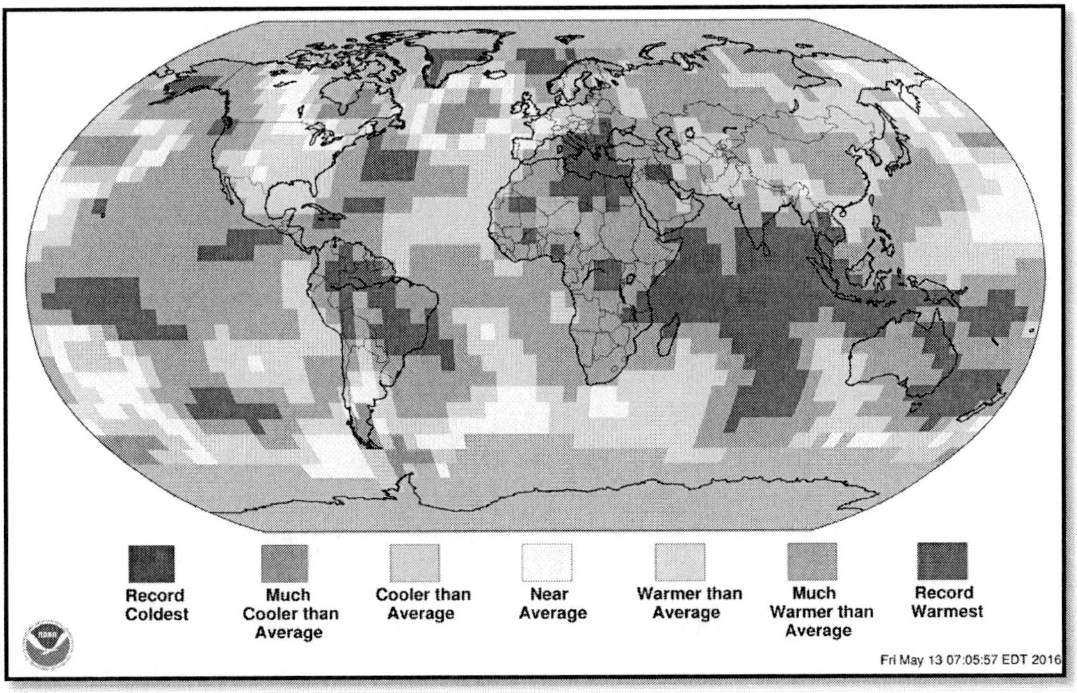

Figure 2. Global climate report of high temperature stress in August 2016 (Source: National Centre for Environmental Information, https://www.ncdc.noaa.gov/sotc/).

Current reports on climate change (Figure 2) simulates that mean surface temperatures will increase by 3-5°C in the next 50-100 years through reiteration of the heat waves, which will have drastic effect on global agriculture systems (Vereecken et al. 2016). Inconsistencies of temperature and in rainfall may impoverish nutritional quality of vegetables crops as well as can cause huge deflation in their yield and may also alter the inclination of insects and pests thereby expediting pathogen spread. Crop plants are more likely to meet a wide extent of biotic and environmental stresses, which when transpire, can have severe damaging effect on their usual corporal growth (Okazaki and Saito 2016). Several recent models for climate change have predicted that the probability of plants to stumble upon different biotic and abiotic stresses and their combinations is likely to be increased in the near future than previously anticipated. Climate change together with growing loads on global food security due to increasing population will compound the demand for stress-resistant/tolerant crop varieties which is still a milestone to cover.

Various diversified legumes such as soybean (*Glycine max*), pea (*Pisum Sativum*), chickpea (*Cicer arietinum*), cowpea (*Vigna unguiculata*), pigeon pea (*Cajanus cajan*), peanut (*Arachis hypogea*), faba bean (*Vicia faba*), lupin (*Lupinus luteus*) and mung bean (*Vigna radiata*) have indeed strengthen global food, nutrition security and had ensured environmental sustainability around the globe (Varshney et al. 2013). Recently, several underutilized tropical crop legumes essentially lima bean (*Phaseolus lunatus*), jack bean (*Canava liaensiformis*), winged bean (*Psophocarpus tetragonolobus*) and hyacinth bean (*Lablab purpureus* L.) are now being exploited as forage legumes in livestock production. However, most of these legume crops including hyacinth bean are cultivated in minimal environments, where plants are constantly threatened by various abiotic (heat, drought, salinity) and biotic stresses (white mould, anthracnose, angular leaf spot, yellow mosaic virus, thrips, bacterial blight), which cramp their growth and eventually their productivity (Rodziewicz et al. 2014).

Indian Scenario

In India, mainstream, production of beans, transpires in low level farms via using minimal agriculture by poor farmers which are much more prone to be attacked by abiotic stresses (high temperature, drought) and biotic stresses (insects and pests, diseases) as compared to high level farms as they have assets to fight such stresses by applying fertilizers, pesticides and proper irrigation (Ramalingam et al. 2015). Use of such protective measures, however, can reduce the profit and can have a negative impact on the environment, and majority of them are not even effectively controlled with chemicals. Thus, in agriculture, abiotic and biotic stresses are progressively becoming the major constraints on productivity and economic yield of hyacinth bean. Due to ever increasing global population, increase in global mean temperature (Figure 3) with decreasing land and water resources especially in India where people are largely dependent upon legume vegetables for their protein requirements, will lead to the protein unavailability in the near future, if the productivity of the legumes crops are not met with the protein requirement of the future population.

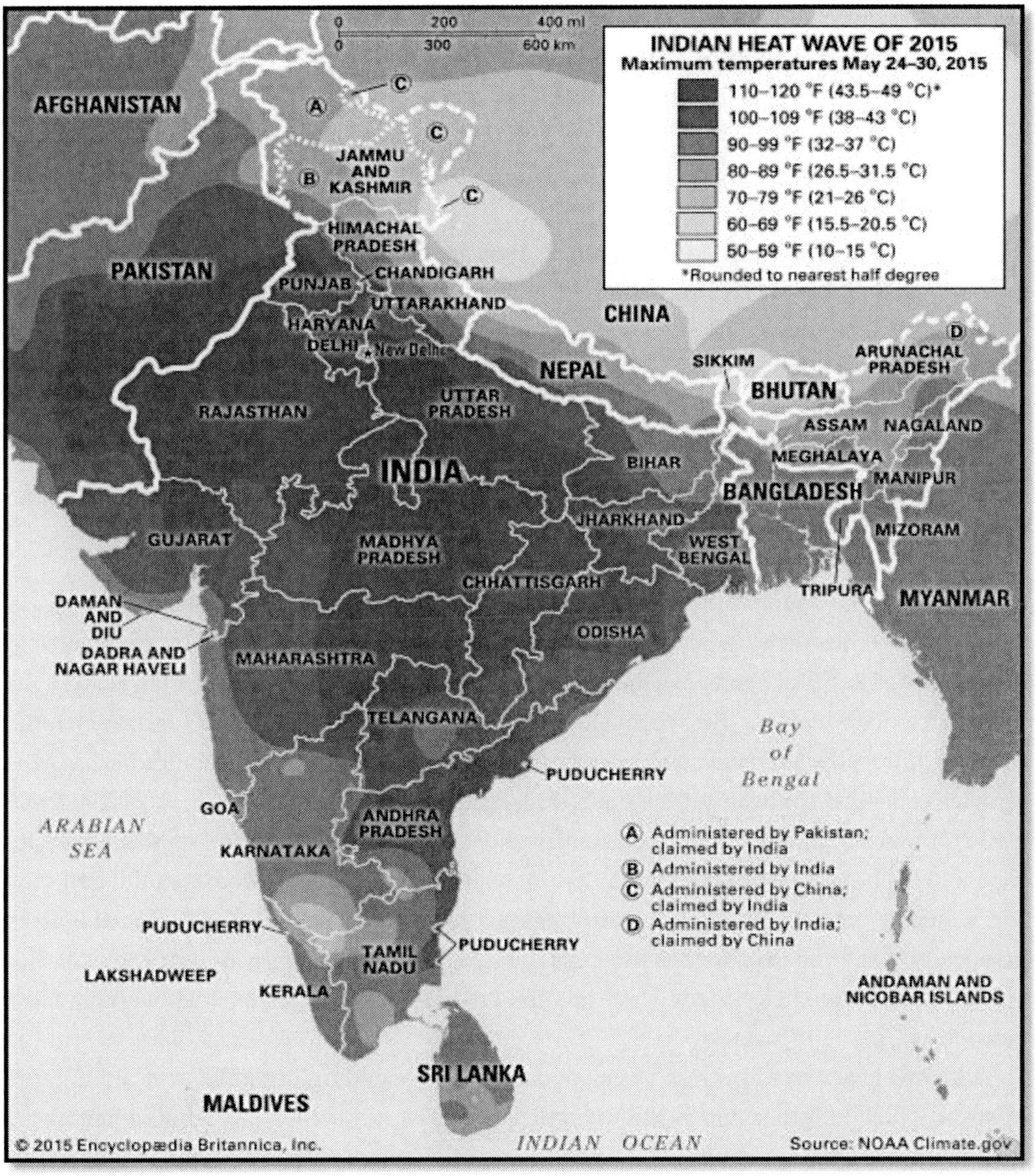

Figure 3. Geographical area under high temperature stress in May 2015 (Source: "India-Pakistan heat wave of 2015". Encyclopedia Britannica. Encyclopedia Britannica Online).

In a study Therefore, it has become imperative to unravel the issue of low productivity, stress tolerance and improved nutritional quality in some underutilized legumes perse., hyacinth bean, which is not happening at desired rate through conventional breeding approaches. Most of the elite crop varieties that we used today have been produced through traditional plant breeding approach. So, in order to comprehend the nature of multiple stress responses and to develop avenues to create varieties with enhanced stress tolerance together with high yields, an understanding of stress response mechanisms in legumes through genomics, transcriptomics, proteomics, metabolomics and recently epigenomic has become an indispensable biotechnological tool for their improvement.

MAJOR ABIOTIC STRESSES AFFECTING HYACINTH BEAN

Climate extremes such as high temperature, extreme frost, salinity and drought have massive impact on world agriculture that can cause deflation in the total yield of economically important crops by >50% for most of the commercial crops. This section deals with the major abiotic stresses that severely affect hyacinth bean growth and developmental processes.

High Temperature (HT) Stress

HT stress has symbolic repercussions on plant growth and reproductive processes which may limit their metabolism and productivity worldwide. HT stress increases metabolic toxicity in plants via generating excess reactive oxygen species (ROS) that stimulate oxidative stress (Siebers et al. 2015). Being sparsely susceptive, high temperature stress significantly alters various biological processes that occur in hyacinth beans. Most productive habitat for beans at the time of planting, flowering and fruiting is 18-29, 25-30, 25-32° C and temperature above can significantly affect their growth and reproduction (Rai et al. 2015). It has been documented that increase in temperature decreased the pod yield and reduced growth of *Lablab purpureus* L genotypes (Rai et al. 2015). HT stress drastically alters the expressions of stress induced genes, regulatory proteins, metabolites and antioxidant defence system which are involved in the re-establishment of cellular redox balance and ion homeostasis.

Plant response to HT stress varies dynamically depending upon the severity/duration of stress which in turn is dependent upon the plant types and other surrounding factors and till date none of the traits have been identified and confirmed that can confer tolerance to HT stress in plants. Scientists involved in the plant abiotic stress tolerance research are aspiring to discover the mechanisms behind the plant response that induces HT tolerance. Alteration in the expression of certain genes alters competency of antioxidant defence system, metabolites and proteins stability that triggers some of the physiological and biochemical processes in plants that may lead to fatal disruption of cellular organization and metabolic imbalance (Ahuja et al. 2010). Plant tolerance (a highly specific trait) to HT stress is predominantly interpreted as their ability to grow and reproduce economic yield. This ability varies significantly in closely related species and may even show distinct variation in tissues/organs of the same plant. To thrive under HT stress, plants have evolved several mechanisms such as enhanced functionating of ion transporters, expression of late embryogenesis abundant (LEA) proteins and upregulation of antioxidant defence related genes are one of the essential mechanisms adopted by plants to overcome stress effects.

Only recently, applications of exogenous protectants in the form of phyto-hormones ABA (abscisic acid), JA (jasmonic acids), SA (salicylic acid) etc, signalling molecules (NO: nitric oxide), osmo-protectants (proline, GB: glycine betaine etc.) have been found effective in mitigating HT stress-induced damage in plants (Shahzad et al. 2015). However, limited information is available on the use of phytohormones or plant growth regulators in the management of heat stress in hyacinth bean plants. Unfolding of new crop cultivars that are tolerant to high temperature stress is becoming a major constraint for plant scientists, who are continuously involved in research exclusively on high temperature stress are striving to conceive plant responses to heat tolerance and discretion used by the plants in HT environments. Overall, further insights on role of phytohormones in regulating defence networks in hyacinth bean plants can be a field worth further investigation, which can be uncovered by adopting an integrated approach of genomics, proteomics and metabolomics.

Drought Stress

Reciprocal and inconsistent climatic condition such as high temperature, salinity and drought are the main factors that limits plant growth and productivity. Due to ever-changing climatic conditions, drought is considered as one of the most devastating adverse environmental condition around the globe, and with the inflation in the CO_2 levels in the current atmosphere and aerial changes, drought prone areas will be contemplated as more stressful areas for agricultural crops in foreseeable future(Adams et al. 2016). In crop legumes per se, cowpea, soya bean and hyacinth bean, severe drought may reduce growth, photosynthesis and can significantly reduce their yield up to 40%. The period during anthesis and following anthesis are considered as one of the most pivotal periods for drought stress in legume plants. In general, on the basis of drought tolerance nature, the defensive response of plants to cope with drought stress can be grouped into following: drought escape, drought avoidance and drought tolerance (Basu et al. 2016).

Lablab purpureus L is believed to grow well under rain fed conditions is well adapted to drought and have better drought tolerance capacity than major legume crops such as common bean or cowpea (Maass et al. 2010). In the area experiencing heavy rainfall, hyacinth bean can be used as forage, once plant has attained its vegetative growth. In India, due to its nutritional and therapeutic property, hyacinth bean is exploited as pulse as well as a vegetable not only in the main season and this demand is expected to increase in the foreseeable future as therapeutic uses of legumes are now better understood. However, in the off-season, the growth of hyacinth bean is restrained recurrently in the month of April and May, especially in Uttar Pradesh which a drought prone region. Some of the early maturing varieties of hyacinth bean are sensitive to drought if the drought occurs between anthesis and reproductive phase. Therefore, understanding physiological and genetics of early maturing hyacinth bean lines is most important for the enhancement of growth, survival and drought tolerance for sustainable growth and yield under changing climates. So, a unifying approach including breeding and biotechnology is need to be integrated in breeding program so that the breeders can move from "chance breeding" to "knowledge-based breeding" for abiotic stress tolerance in hyacinth bean. In recent years, several drought related genes (*BhGRP1)* through subtraction subtractive hybridization (SSH) transcription factors (*R2R3-MYB*) through SSH library (Yao et al. 2016) and two miRNA genes (*miRNA 156* and *miRNA 172*) have been isolated and confirmed from hyacinth bean (Thilagavathy and Devaraj 2016). These advances made in hyacinth bean genomics will foster trait mapping, gene discovery and their functional characterization in the development of functional markers and their introgression through marker assisted selection (MAS).

Salinity Stress

Salinity stress is one of the major limiting factors for plant productivity. Based on the type and causes, salinity stress has been characterized as (1) Natural Salinity-It is also called as primary salinity stress that results from the accumulation of salts through natural process in the soil or groundwater. It is mainly caused by the weathering process that breaks down rocks that releases various mineral salts and to some extent sulphates and carbonates in to the soil. Salinity stress may also result from the periodic deposition of ocean salts that travel inland through wind and deposited by rainfall, and (2) Secondary salinization or human-induced salinization are caused by irrigation through salt-rich water or replacement of perennial crops by annual crops (Iseki et al. 2016).

Salinity stress causes rapid decrease in the osmotic potential of soil thereby causing hyper-ionic/hyperosmotic stress that reduces the availability of water to roots and this leads to reduction in their growth rate. The salinity factors like evaporation (increase salt concentration in soil and precipitation (decrease salt concentration in soil) are three times more severe as compared to other abiotic stresses as it adversely affects seed germination, photosynthetic pigment content, disturbed ion homeostasis, reduces water uptake, thereby limiting plant growth and productivity.

Most of the bean plants including hyacinth bean have shown remarkable flexibility/adaptivity to adverse environmental factors such as drought, high temperature under different belts and there are reports delineating about the salt tolerance nature of this crop (Rai et al. 2015). In hyacinth bean, with the advancement in the 'Omics' techniques considerable amount of expressed sequence tags (ESTs), miRNAs and transcription factors has been identified under various abiotic stresses (Table 1) including salt stress (Yao et al.

2016). However, till date no significant effort has been made to analyse hyacinth bean transcriptome/proteome level and to map QTLs for different abiotic stress tolerance in this plant. With the developments in genome sequencing technology, transcriptomics, and protein profiling together with metabolomics has emerge as a promising tool that can be successfully used to unravel different components concerning plant biotic and abiotic stress tolerance and simultaneously can identify functional which can be used to engineer plants with increase tolerance/resilience to adverse situations to meet the increasing demand of food for ever increasing populations.

ROS as Key Players in Plant Stress Signalling

Reactive oxygen species (ROS; $O_2^{\bullet-}$, H_2O_2, $OH^{\bullet}$, $^1O_2$) are moderately charged forms of atmospheric oxygen which are deliberately produced in photosynthetic organisms that causes cellular damage by reacting with biomolecules and simultaneously also function in cells as signalling molecules as well as in responses to adverse environmental conditions (Mittler 2017; Rai et al. 2017). In plants, ROSs (Figure 4) are mainly produced as unavoidable toxic by-products of aerobic metabolism that take place in different cellular compartments. A large body of literature data supports that under normal physiological conditions these reactive compounds are promote several basic biological processes including cellular growth and differentiation whereas under the situation of oxidative stress these reactive compounds are actively scavenged by distinctive enzymes antioxidant defence system. The equilibrium concerning generation and scavenging of free radicals varies greatly under adverse environmental conditions such as high temperature, drought and salinity that may result in the inflation of the intracellular levels of free radicals. Among ROSs, superoxide anion radical ($O_2{\bullet}$-), hydrogen peroxide (H_2O_2), and per-hydroxyl radical (HO_2), singlet oxygen ($^1O_2$), ozone (O_3), peroxides and alcoxides are non-radicals. Some free radicals (like hydroxyl radical, $HO^{\bullet}$ and the hydrogen radical, $H^{\bullet}$) are extremely reactive, whereas some free radicals, like ions of transition metals exhibit weak reactivity.

Recently, different studies conducted on plants have confirmed that reactive oxygen species are good for stimulating normal growth and physiological processes contrary to having toxic effect on their survival. High production of ROS in chloroplast and peroxisome during photosynthesis and photorespiration may cause cell death as a result of oxidative stress, which is now contemplated as ROS triggered physiological response that lead to cell death. Furthermore, the findings have also depicted that generation of ROSs below a threshold level may suppress cellular growth, differentiation and immunity (Diebold and Chandel 2016), which clearly pinpointed that maintaining basal level of ROS i.e., above cytostatic level, but below cytotoxic level is essential for normal life. Thus, ROS level either too low or too high can impair plant growth and development, whereas when ROS are maintained in adequate level stimulate plant health and when the level ROS break the boundary between cytostatic and cytotoxic levels, impairs the normal functioning of plants.

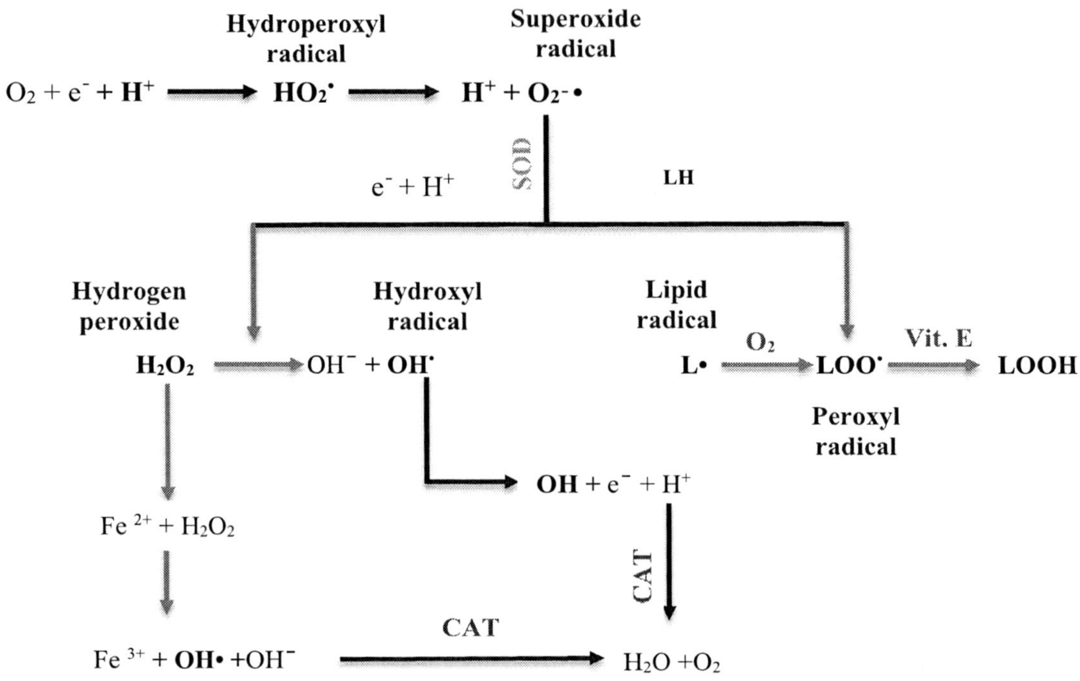

Figure 4. Overview of the reactions involved in generation of ROS. Purple arrows represent lipid peroxidation. Blue arrows represent the Haber-Weiss reactions and the red arrows represent the Fenton reactions. The bold letters represent reactive oxygen species (ROS). SOD and CAT refers to the enzyme superoxide dismutase and catalase.

Enzymatic and Non-Enzymatic Defence System

Plants must protect themselves from unfavourable environmental conditions such as temperature extremes, salinity and drought *via* adaptation or acclimation process in order to foster under different climate adversities. In crop plants, enhanced tolerance to oxidative stress is often associated with improved functioning of enzymatic and non-enzymatic antioxidants. Plants that shows the tendencies to over-come damaging effects of ROS induced oxidative stress are often associated with the enhanced production of ROS scavenging and detoxification enzymes which are described below.

Enzymatic Antioxidants

ROS scavenging enzymatic antioxidants in plants include catalase (CAT), superoxide dismutase (SOD), ascorbate peroxidase (APX), guaicol peroxidase (GPX) and glutathione reductase (GR). The metalloenzymatic SODs is the highly effective omnipotent antioxidant enzyme that act as the first line of defence and is found in all aerobic organisms that dismutase superoxide anion to H_2O_2 (Figure 5). Catalase, one of the largest heme containing tetrameric enzymes has maximum turnover rates among all the enzymes i.e., one catalase molecule can transform up to 6 million molecules of H_2O_2 to H_2O and O_2 per minute and is vital for scavenging of free radicals under stressed conditions. Apart from catalases, APX is believed to play a crucial role in scavenging of H_2O_2 thereby protecting cells from oxidative damage via ascorbate-glutathione (ASH-GSH) cycles that utilizes ASH as the electron donor. APX has high affinity for H_2O_2 as compared to CAT and POX. Catalytic action of APX involve oxidation of ascorbate to Malone dialdehyde (MDA), that result in the formation of monodehydroascorbate-reductase (MDHAR) and dehydroascorbatereductase (DHAR), using NADPH as reducing equivalents and by transforming reduced glutathione (GSH) to oxidized glutathione (GSSG) (Xin et al. 2014). Finally, a flavo-protein oxidoreductase enzyme *viz.,* glutathione reductase (GR), NADPH dependent regeneration of GSH from GSSG, plays an imperative role in strengthening of defence mechanism by scavenging free radicals and by maintaining glutathione in its reduced form. Target oriented roles for different antioxidant enzymes have been explored in many plant species using transgenic approaches (Shafi et al. 2015).

Non-Enzymatic Antioxidants

The cellular osmolyte and compounds like proline, flavonoids alkaloids tocopherol, carotenoids, ascorbate and glutathione are some the major redox buffers that are collectively characterized under the heading of non-enzymatic antioxidants (Figure 5). Commonly, ascorbic acid and tocopherols are known by their generic names i.e., Vitamins C and E that play an imperative role in the scavenging of reactive oxygen species (Boubakri et al. 2016). Ascorbic acid is one of the most generous, influential and water-soluble antioxidants that comprehends L-ascorbic acid and L-dehydroascorbic acid, two compounds with antioxidant activity. The concentration of ascorbic acid has been reported to maximum in mature leaves with fully developed chloroplast. Vitamin E (tocopherol) is a lipid soluble antioxidant with four isoforms *viz.,* α-tocopherol, β-tocopherol, γ-tocopherol and δ-tocopherol and tocotrienols per se., α-tocotrienol, β-tocotrienol, γ-tocotrienol. Vitamin E is one of the most exuberant antioxidants found in bio-membranes which is actively involved in the arrest of lipid peroxidation by contributing phenolic hydrogen to the hydroxyl radicals and converting it to tocopheroxyl radicals which is an unreactive radical which is unable to initiate oxidative chain reaction. It is the only major chain breaking lipid-soluble antioxidant present in RBCs, plasma and tissues that protect the integrity of lipid structures and primary membranes.

Flavonoids, an antioxidant metabolite commonly occur in leaves, floral parts and pollens are composed of isoflavonoids, flavanones and flavones flavonols, flavanols and anthocyanins that share the same diphenylpropane ($C_6C_3C_6$) skeleton (Rai et al. 2015).

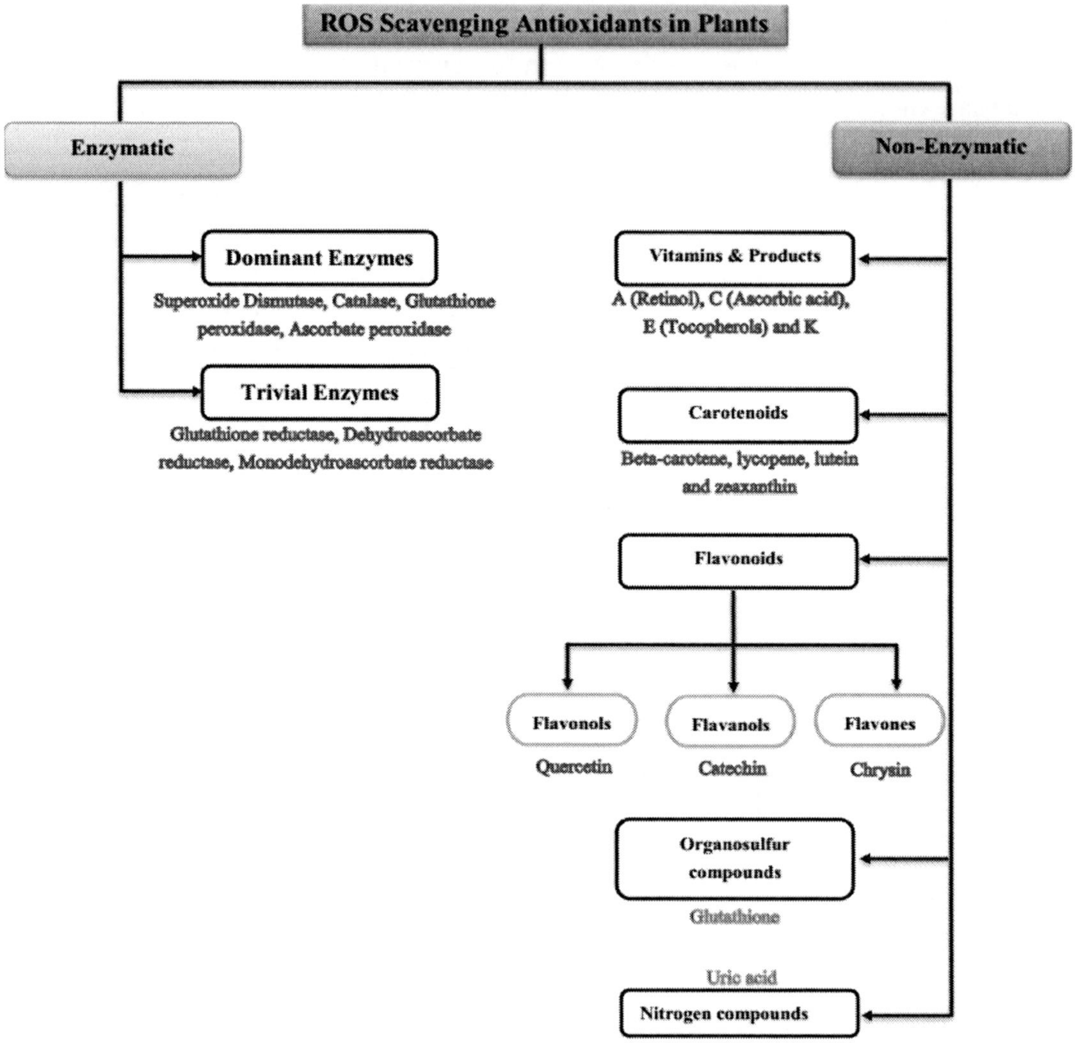

Figure 5. Natural antioxidants separated in classes.

Flavonoids are reported to be involved in myriad of function such as flower/fruits development, protection against UV damage and phyto-pathogens and also act as signalling molecule. Flavonoid and anthocyanin also function as hydrogen donators, reducing agents, singlet oxygen quenchers, superoxide radical scavengers and even as metal chelators. They also activate enzyme of halliwel-asada pathway, reduce a-tocopherol radicals (tocopheroxyls), absorb UV light thereby inhibiting oxidases and mitigating oxidative stress. Some of the most important flavonoids are catechin, catechin-gallate, quercetin and kaempferol. Carotenoids are a group of natural photosynthetic pigments which are synthesized by plants and microorganisms which serve as an important protective role. They are characterized into two broad categories: the carotenes containing lycopene, b-carotene; and the oxygenated xanthophyll, like zeaxanthin, lutein (Zou et al. 2015). The carotenoids are mainly involved in the quenching of singlet oxygen and triplet sensitizer (Chl^3) thereby directing PSI assembly and the stability of light harvesting complex proteins along with thylakoid membrane stability, thus returning to the unexcited state and allowing them to quench more radical species.

USE OF EXOGENOUS PROTECTANTS IN MITIGATING ABIOTIC STRESSES

Plant growth regulators serve dual function in plant that offers one of the ways to cope with climatic adversities and are also involve in the regulation of plant growth, developmental and signalling processes. In recent decades, exogenous applications of various plant growth regulators (SA; salicylic acid, ABA; abscisic acid and JA; Jasmonic acid), osmo-protectants (Pro; proline and GB; glycine betaine), signalling molecules (SNP; sodium nitroprusside) have shown to have constructive effect on plant growth as these molecules play significant role in mitigating abiotic stress induced oxidative damages in plants (Khan et al. 2013).

Salicylic Acid (SA)

Salicylic acid, a naturally occurring seven-carbon containing phenolic compound plays significant role in the modulation of plant growth and reproduction and is also believed to be effective in conferring abiotic stress tolerance by decreasing the content thiobarbituric acid reactive substances (TBARS), protecting photosynthetic machinery and stimulating antioxidant enzymes and metabolites (Zhang et al. 2011). Several researchers have pinpointed toward the involvement of SA in plants defence against major abiotic stress particularly drought, HT stress and salinity either through seed soaking, foliar spraying, or supplementing it with nutrient solution. In plants, SA exert these functions in a dose dependent manner, where induced or inhibited plant functions has been observed with high/low doses of SA, respectively.

In a study, foliar application of SA @ 0.1 and 0.5 mM have been shown to enhanced photosynthesis and growth, however an inhibited growth was evidenced with 2.0 mM of SA in *vigna radiata* (Nazar et al. 2011). Apart from dose, age, duration and nutritional stage of treated plant species also regulate the SA effects in plants. Wang and Li (2006) reported that exogenous application of 0.1 mM of SA has shown ameliorative effect by decreasing thiobarbituric acid reactive substance (TBARS) and increasing water content in young grape leaves subjected to heat stress, thus mitigating heat stress induced oxidative stress in grape vines. Treatment of plants with SA have also shown to mediate proline accumulation under abiotic stress. Foliar application of SA (at 0.5 mM) significantly increased the activity of proline biosynthesis enzyme (pyrroline-5-caboxylate reductase) under salinity stress that resulted in the increased proline content in *Lens esculenta*, (Misra and Saxena 2009). Recently, molecular studies have confirmed that SA can modulate the plant function at gene level, by regulating genes involve in secondary metabolites production, antioxidants, and heat shock proteins, thereby refining plant-abiotic stress tolerance.

Abscisic Acid (ABA)

Abscisic acid is considered as a signalling molecule involved in many physiological processes and play an important role in regulating plant defence against diverse environmental stresses such as heat, drought, salinity and flooding (Ding et al. 2010). An unprecedented increase in the endogenous level of the ABA under different stress conditions, has been associated with the enhancement of thermotolerance, suggesting its involvement in the initiation and regulation of defence related genes and signalling pathways which are essential for the survival of plants under climate extremes (Ding et al. 2010). Various studies have confirmed that foliar application of ABA is effective in ameliorating abiotic stress induced oxidative damage in crop plants by interacting and stimulating with other signalling molecules like nitric oxide (NO). In a study, ABA (2.5 μM) stimulated the growth of chickpea seedlings at 40-45°C by stimulating of endogenous level of ABA with osmolytes and by inducing the regulation of several heat shock proteins and heat shock transcription factors (Rojas et al. 1999) and this effect of ABA in conferring heat tolerance has been extensively studied in maize (Gong et al. 1998). Even though, a large body of literature have confirmed defensive role of ABA in conferring abiotic stress tolerance in plants, till date comparably less is known about physiological and biochemical mechanisms of ABA underlying HT stress tolerance in plants. Furthermore, the cooperative involvement of ABA with other stress induced phytohormones and osmolyte need to observed in plants under different stress condition.

Sodium Nitroprusside (SNP)

Nitric oxide (NO) is a highly diffusible bioactive molecule that plays a significant role in several key metabolic processes both as a signalling as well as protective molecule, that play a vital role in imparting abiotic stress tolerance in plants (Hasanuzzaman et al. 2012). Sodium nitroprusside (SNP), one of the most influential NO donar, supposed to have a wide range of function in plant growth and developmental processes such as osmotic balance, fruit ripening, germination, floral development, apoptosis due to which it is now being considered as a new member of phytohormones. Several researchers have shown that foliar application of nitric oxide improves the growth and development of various plants under the changing environmental conditions such as salinity, drought, UV as well as HT stress (Gill et al. 2015). In plants, exogenous application of NO donor SNP has been well documented to confer tolerance to severe drought and heat stress conditions by increasing water retention capacity and lowering membrane damage (Hasanuzzaman et al. 2010). SNP treated heat stress seedlings showed upregulation in the activities antioxidant defence systems of catalase, ascorbate peroxidase (APX), monodehydroascorbate reductase (MDHAR), dehydroascorbate reductase (DHAR), glutathione reductase thus protecting seedlings from heat stress induced oxidative damage (Hasanuzzaman et al. 2013). Results of several findings have indicated that exogenous application of SNP is effective in regulating plant growth and development processes under stress conditions via modulating antioxidant defence system. However, till date, not much study has been conducted regarding the effect of exogenous application NO on antioxidant metabolism and detoxification system of plants under stress conditions.

Proline (PRO)

Proline are osmolytes/compatible solutes found in a wide range of organisms from bacteria to plants and animals (Alcazar et al. 2006). They also serve as one of the main adaptive mechanism by which plants protect themselves against abiotic stresses via adjusting their osmotic level. Proline perform myriad of functions from detoxification of excess ROS, protects biological membranes to stabilizing enzymes/proteins thus providing protection to plants from damaging effects of various stresses (Hossain et al. 2014). Iqbal et al. (2014) in their study reported that proline provide protection against heat stress by regulating cellular homeostasis. Exogenous application of proline (10-20 μM) are able to maintain thermostability of thylakoid membranes thereby reducing protein oxidation, lipid peroxidation and increasing photosynthetic efficiency (Banu and Hoque 2010). Several studies have shown that proline has the ability to enhance antioxidant defence system by regulating enzymatic and non-enzymatic antioxidants thus improving stress tolerance in plants (Hossain et al. 2014). A study conducted by Okuma et al. (2004) concluded that exogenous application of proline stimulates intracellular proline content that prevent oxidative cell death in tobacco cultured cells via regulating antioxidant defence systems. Furthermore, researchers have shown that exogenous application of proline is effective in improving abiotic stress tolerance in plants by improving ion homeostasis and regulating the activities of antioxidant enzymes. However, in many of agriculturally important crops per se., hyacinth bean, the protective effect of exogenous application of proline in response to abiotic stress remain to be elucidated.

Biotechnological Tools for Empowering Abiotic Stress Tolerance

Major abiotic stresses influencing cool season food legumes has been a matter comprehensive discussion (Singh and Saxena 1993). Numerous stresses are known to affect hyacinth bean such as cold, drought, heat salinity, nutrition deficiency/toxicity, of these heat is considered as the most significant stress worldwide. Susceptibility of hyacinth bean to high temperatures has a major constrain on its productivity in tropical and temperate areas of the world. Though, pervasive germplasms are available that can withstand/tolerate high temperature stress that provide us the opportunity to breed summer hardy cultivars that can planted in late winter with a practical probability ofgrowth and reproduction the summer (Yao et al. 2016). Breeding for abiotic stress tolerance in vegetable crops is mainly incarcerated due to, (i) variation in time, strength and period of abiotic stress tolerance; (ii) dissemination of lethal genes along with desirable traits (iii) limited transfer of suitablealleles due to reproductive barricade from diverse genetic resources (Varshney et al. 2013). Integrated 'omics' techniques have emerged as indispensable tools for improving genotypes/lines that can enroot growth and development of plants under astringent conditions, especially for (ii) and (iii) above. Deployment of molecular breeding approaches can help in the identification of superior genes and alleles that can be transferred into elite genotypes. Furthermore, with the emerging breeder friendly genomics along with decision support tools will play a critical role in breeding programs in developing countries.

Advanced Genomic Tools Applicable to Hyacinth Bean Genomics

Decoding of plant genome sequence has enabled excavation of entire genomic sequences in model plants per se., in *Arabidopsis*, *Lotus japonicus* and *Medicago truncatula*. Although, the same approach does not seem to be applicable to other crops species including hyacinth bean not only due to its large genomic size compared to model species but also due to limitation in financial and human resources. With the advancement in modern genomic tools such issues can be avoided in an efficient way. By the virtue of sequencing-based trait mappings techniques has facilitated trait mapping at sequence level that has enabled the selection of plants at nucleotide levels. Two of these techniques *viz.,* genome wide association studies (GWAS) and genotyping by sequencing (GBS) has led to the identification of large-scale genome-wide SNPs that can facilitate high resolution trait mapping in plants of diverse mapping populations. GBS allows quick identification of vast number of SNPs that can be used for trait mapping, GWAS has been found promising in several crops (He et al. 2014). Both of these techniques i.e., GBS and GWAS has received immense recognition as they do not require any prior genome information, low cost genotyping, provide high-density genotyping data, and provide advantage over other sequencing based-methods as it allows substantial degree of false positive detection.

In legumes, with the advent of next generation sequencing technologies (NGS) several new genotyping platforms such as GBS and GWAS are currently available and have been deployed extensively for construction of genetic maps, genetic diversity/purity analysis and marker assisted selection (Varshney et al. 2014). The utilization of both GBS and GWAS for the improvement of horticultural crops has been extensively reviewed by Kim et al. (2016). Similarly, in hyacinth bean, GBS approach may emerge as most powerful tool as compared to other sequencing techniques for targeting candidate genes from diverse germplasm. It can also facilitate the understanding of the role of candidate genes for targeted traits by sequencing of protein coding regions and in the identification of SNPs. Similar techniques have also been used in rice for the identification of candidate genes against biotic stress.

Transcriptomics for Genome Annotation and Marker Development

Discovery of genes and characterization of their regulatory pathways that are stimulated under various abiotic stresses could play an important role in improving the productivity of legumes under adverse environmental conditions. In this context, transcriptome sequencing has emerged as an indispensable tool for trait specific gene sequencing.

In legumes, with the help of transcriptome sequencing various efforts has been made on the development of cDNA libraries, gene expression analysis, *in-silico* mining of expressed sequence tags (ESTs) before the genome sequence of crop became available (Morozova and Marra 2008). Significant amount of work has been done in legume transcriptomics and a considerable number of ESTs and transcription factors (Table 1) have been developed in several legume plants challenged by different stresses such as in soybean (1.5 million ESTs), *Medicago truncatula* (280,000 ESTs), *Lotus japonicus* (242,000 ESTs), chickpea (20,162 ESTs) and pigeon pea (9,888 ESTs) has been generated through Sanger sequencing approach (Varshney et al. 2015). So, in hyacinth bean, deployment of transcriptomics techniques needs to be focused on the generation of cDNA libraries, expressed sequence tags (ESTs), gene expression analysis together with in silico mining of ESTs data sets to obtain functional annotation and discovery of novel genes to understand the molecular basis of gene function under climate extremes.

Recently, numerous sequencing platforms are readily available at low cost that can be used for the generation of wide range of stress responsive transcript reads in different crop legumes. In the case of peanut, differentially expressed genes and mRNA transcripts, have been identified under drought stress and well-watered condition by using differential

Table 1. List of different transcription factors with a potential role in the abiotic stress signalling pathways and regulation gene expression in plants

Stress condition	Transcription factors	Sequence	Mode of action	Gene regulated	References
Drought	bZIP	PyACGTGGC	Promotion of salicylic acid biosynthesis by regulating drought-responsive genes	*Em, RAB16, Osem, RD29B*	Zhu (2002)
Drought, Cold	ERF/AP2	TGCCACCGG	Positively regulate dehydration responsive element binding protein (DREB/CBF) and stimulate freezing/drought tolerance pathways	*HVA1, RD29A, Cor15A, BN115*	Lee et al. (2004)
Drought	MYB	TGGTTAG	Regulates ABA-and JA biosynthetic genes overexpression which confers resistance to cold, drought, salt, ozone stresses	*RD22*	Schenke et al. (2011)
Drought	bHLH	CACATG	Activation of ABA and JA signalling response	*RD22*	Abe et al. (2003)
Drought	NAC	ACACGCA-TGT	An inducer of abiotic stress responses	*ERD1*	Seo et al. (2010)
Drought	ZFHD	CACTAAA-TTG	When over-expressed confers tolerance to both biotic and abiotic stresses	*ERD1*	Tran et al. (2004)
Salinity	ERF	GCCGCC	Regulation of tolerance to abiotic stresses	*BIERF1-4*	Park et al. (2001)
Salinity	AP2/ ERFBP	AGCCGCC	Overexpression enhances tolerance salt stress.	*OPBP1*	Guo et al. (2004)
Drought, salinity	ABF	PyACGTG-G/TC	ABA-mediated abiotic stress tolerance	*AREB1*	Orellana et al. (2010)
Heat, cold, salinity, drought	WRKY	TTGACCT	Regulation of abiotic stress response inducing ABA-responsive abiotic stress tolerance genes.	*WRKY45/82*	Peng et al. (2011)
Heat, salinity	Zinc-finger	CCCH/AAAG	Regulation of genes involved in heat and oxidative stress tolerance	*ZAT7/12*	Davletova et al. (2005)
Heat	HSF	AGAATTCT	Regulation of antioxidant defence related genes under abiotic stresses and maintains growth	*HSFA1b, HSFA2*	Mullineaux et al. (2011)
Drought, heat, salinity	DREB	GCCGAC	Mediate positive and negative regulation of ABA and sugar responses	*RD/COR*	Lee et al. (2004)

display reverse transcriptase PCR (DDRT-PCR) (Qiao et al. 2012). In addition, ESTs libraries has also been developed using suppression subtractive hybridization (SSH) techniques which has led to the development of 480 abiotic stress responsive ESTs in chickpea. Furthermore, SSH approach has also been used to strengthen the EST libraries in in chickpea and pigeon pea by generating 3,062 and 182 ESTs in response to drought and salinity stress (Qiao et al. 2012). Hyacinth bean, taxonomically one of the closest crops to the model legume *Medicago truncatula* is the most important food legume of semi-arid tropics and. In hyacinth bean, both sensitive and tolerant varieties have been characterized under different abiotic stress conditions, although very little is known about the genes involved in these responses. Though, transcriptome assemblies of genes involved in the differential behaviour and profiling of regulatory genes from developed transcriptome can facilitate in generating highly dense genetic map with well distributed markers and Hiz et al. (2014), identified 441 transcription factors from salt-stressedcommon bean and classified them under 59 different transcription factor families.

Proteomics Analysis

Proteomic analyses of plants under adverse environmental cues can provide valuable insight on proteins involved in the abiotic stress response which can lead to the identification cellular and molecular mechanisms that are linked between different abiotic stresses. Proteomics involve high-throughput analysis of (1) total cellular proteomes of a species that are expressed in specific cells/tissues/organs at distinct developmental stages (2) several post transcriptional modifications to find out how proteins get modified (3) protein-protein interactions to know how proteins interact with other biomolecules (4) deciphering protein structure and functional group attached to proteins (Jorrin-Novo et al. 2015). Advancement in proteomics approach has made complex biological questions to answer more efficiently due to the availability of wide range of proteome coverage that have facilitated the interpretation of protein structure and modulation/modifications of stress induced enzymatic and regulatory functions of specific proteins (Table 2) encoded by specific genes (Katam et al. 2010). However, proteomics in crop research still have to travel a milestone to study response mechanism of several stress specific traits as obtaining good sample quality is a bit challenging owing to the fact that plant tissues are often enriched with several proteases and oxidases which make it quite difficult to obtain stable protein mixtures.

Several lines of experiment have been conducted on model legumes *per se.,* Medicago, Lotus and in soybean (Hossain and Komastu 2014) and substantial load of protein reference map providing detailed proteomic information have made available which can be exploited in other legume crops to better understand abiotic and biotic stress responses. Earlier studies on hyacinth bean have shown that limited information is available on proteomics-based datasets which can be enhanced by increasing the availability of transcriptome data sets and legume genome sequences (Thilagavathy and Devaraj 2016). Such techniques have been successfully employed in other crop legumes such as chickpea, pea and groundnut under biotic and abiotic stresses that has facilitated the protein research in these crops by enhancing the availability of protein reference maps (Katam et al. 2010). Comparative proteomics together with differential expression studies can significantly contribute toward understanding of stress specific protein response in crop legumes such as response of dehydrin proteins in drought/osmotic stress, salt stress (Kottapalli et al. 2008). Therefore, in hyacinth bean, future investigation on these areas can improve our understanding by augmenting the use of comparative proteomics approach from other legume crops.

Metabolic Fingerprinting in Legumes

Metabolic fingerprinting (Table 2) allow contemporaneous analysis of all or a set of metabolites which have diverse functions and are manufacture at relatively large scale (Genovese et al. 2015). The advantage of metabolic fingerprinting is that it identifies certain metabolic signatures per se., specific pattern associated with particular stress response without prior information about the metabolites involved. Mass spectrometry (MS) and nuclear magnetic resonance (NMR) have facilitated the identification of large number distinct metabolites in plants (Sumner et al. 2003). Different analytical techniques such as liquid chromatography-MS (LC-MS), gas chromatography-MS (GC-MS) and GC-time of flight MS (GC-TOF-MS) may also be used for metabolic fingerprinting (Johnson et al. 2003). On the contrary, GC-MS is more advantageous over NMR spectroscopy because of its low resolving power, as GC-MS allow quantification of large number of primary metabolites. Furthermore, GC-TOF-MS has also emerged as more rapid and robust technique for identification of secondary metabolites that are expressed at certain developmental stages of plants and have better resolution and provide better peak compared to GC-MS (Fernie and Schauer 2009).

However, metabolic profiling technique has not been extensively used in crop legume but this approach has been successfully used in model legumes. As for example, in a study in drought tolerant Lotus genotypes, comparative metabolomic approach coupled with GC-TOF-MS lead to the profiling of rhizobia Nod factors to visualize metabolic variation among different symbionts. Similarly, using GC coupled to electron impact ionization (EI)-TOF-MS (GC-EI-TOF-MS), a comprehensive and advanced understanding of drought induce metabolic re-programming of metabolites were made in response to water stress (Sanchez et al. 2012). However, limited metabolic fingerprinting studies has been implicated in other forage legumes. For example, profiling of chickpea in response to biotic stress, pea in response to abiotic stress and in common bean under heavy metal stress

(Pandey et al. 2016). In addition, metabolite profiling in Lotus have also been done to analyse flavonoids under abiotic stress (Zhang et al. 2012). Hyacinth bean one of the rich source of flavonoids which characterize it as medicinal therapeutic plant (Rai et al. 2015), deployment of metabolite profiling techniques can facilitate the identification of candidate metabolites that are expressed in response to abiotic and biotic stresses.

Table 2. Application of proteomics/metabolomics for the identification of key protein/metabolites in model and crop legumes under abiotic stresses conditions

Legume	Stress conditions	Techniques used	Differentially expressed proteins and metabolites	References
Model legume				
Medicago	• Salt • Drought	• ESI-LC-MS/MS • 2D-GE • MALDI-TOF-MS • LC-MS/MS • GC/MS • LC-ESI-TOF-MS • HPLC	• Regulation of photosynthesis • Protein metabolism and chaperones metabolism	Staudinger et al. (2012)
Lotus	• Salt • Drought	• GC-EI-TOF-MS • GC-MS • LC-MS	• Storage proteins, • Sucrose synthase • Regulation of nod genes	Sanchez et al. (2011, 2012)
Crop legume				
Pea	• Salt • Drought • Osmotic	• 2D-GE • MALDI-TOF-MS • LC-ESI-TOF • ESI-TOF-MS • NMR	• Regulation of glycolysis, Flavonoid and sulphur metabolism • Antioxidant metabolites	Brosowska-Arendt et al. (2014)
Soybean	• Salt • Drought • Osmotic • Cold • Heat	• 2D-GE • MALDI-TOF-MS • LC-MS/MS • ESI-TOF-MS • NMR • UPLC-QTOF-MS • HPLC-UV-ESI-MS • GC-MS • CE-TOF-MS	• Regulation of defence related genes and proteins • Apoptotic proteins • Photosynthetic proteins • Secondary metabolites	Swigonska and Weidner (2013)
Chickpea	• Drought • Cold	• 2D-GE • MALDI-TOF-MS • LC-MS/MS • ESI-TOF-MS	• Regulation of defence related metabolites • Molecular chaperones • Cell wall modification proteins	Jaiswal et al. (2014)
Common bean	• Drought	• 2D-DIGE • LC-MS/MS	• Regulation of photosynthesis • Defence related genes • Protein synthesis and metabolism	Zadraznik et al. (2013)

Epigenetic Changes in Response to Abiotic Stress

Alteration in histone modification and DNA methylation due to regulation of stress-responsive genes have been shown to be involve plant growth and development thereby increasing the ability of them to adapt under stringent abiotic stress condition (Figure 6). Several histone modifications (Table 3) such as phosphorylation, methylation, acetylation and SUMOylation (small ubiquitin-related modifier) are known to influence abiotic stress responses in plants (Luo et al. 2012). In angiosperms, histone modifiers are well conserved as for example, histone deacetylase, histone methyltransferase, histone acetyltransferase, histone demethylases have been identified and phylogenetically classified in many of the agriculturally important crops such as tomato, rice, Arabidopsis (Cigliano et al. 2013). In drought stress, transcriptional fortitude of drought responsive genes shows strong association with DNA methylation and histone modification where To and Kim (2014) reported that modification of H3K4me3 and H3K9ac upregulated the expression of *RD20* and *RD29A* genes in severe drought condition as compared to moderate drought stress.

A body of literatures have suggested that stress responsive genes uses histone modification as one of the tools to memorize abiotic stress response and in this context H3K4me3 act as one of the potential markers (Table 4) for gene activation and stress memory (Ding et al. 2012). Acetylation of histone is mainly associated with activation of stress responsive genes which is under tight regulation of histone acetyltransferase and histone deacetylase. In *Arabidopsis*, gene ADA2b, which is known to regulate function of histone acetyltransferase modifier, have shown to activated in response to salt stress (Kaldis et al. 2011), thereby suggesting the defensive role of histone acetyltransferase modifier in imparting salinity tolerance. Several researchers in their study have confirmed positive connection between changes in histone modification and DNA methylation with changes in gene expression. In an interesting study was conducted by Sani et al. (2013) with the aim to decipher changes in DNA methylation in response to salinity stress. They reported that genes viz., H3K27me3 and HKT1 that show high affinity for $K(^{+})$ transporter was induce under salinity stress, thus playinga major role in salinity tolerance.

SUMOylation i.e., small ubiquitin-related modifier is also a type of histone modification that is predominantly found under thermal stress response. SUMOylation was previously identified as reversible post-translational modification with the ability to regulate protein-protein interaction in eukaryotes (Lu et al. 2009). Study conducted in model plant *Arabidopsis* have confirm that SUMOylation of core histone in particular H2A and H3 histone play significant role in the assembly of centromeric chromosomes, DNA repair pathways and is also able to regulate gene expression under heat stress (Coleman-Derr and Zilberman 2012). The researchers have showed that both *H4K16ac* and *H3K23ac* showed distinct expression patterns in association with particular developmental stages and also contribute to stress memory in plants (Chinnusamy and Zhu 2009). The techniques *viz.,* NGS-based whole genome bisulphite sequencing (WGBS) and chromatin immunoprecipitation-sequencing (ChIP-Seq) are being currently deployed to comprehend DNA methylation mechanism and heterosis in model plants under stress conditions. These techniques can also be used in other legume crops for understanding DNA methylation and histone modifications to study these epigenetic changes underbiotic and abiotic stresses for crop improvement and management (Pandey et al. 2016).

Table 3. List of various histone post-translational modifications and their relationship with gene expression in plants under abiotic stress conditions

Histone modification	Modified histone residue	Mode of action	Major enzyme involved	Regulation of gene expression	References
Phosphory-lation	Hydroxyl groups of serine and threonine in H2A, H2B, H3, and H4	Protein kinases of aurora family are phosphorylated	Phosphatase 1 (PP1) family enzymes	Transcription activation	Houben et al. (2007)
Acetylation	H2A (K 4, 5, 7) H2B (K 5, 11, 12, 15, 16, 20) H3 (K 4, 9, 14, 18, 23, 27) H4 (K 5, 8, 12, 16)	Acetyl group from histone acetyltransferases (HATs) is transfer to the ε-amino group of lysine	Histone deacetylases (HDACs)	Activation of transcription through acetylation	Peterson and Laniel, (2004)
Methylation	H3 (K4, 9, 27, 36) H4 K20 H3 R3	Methyl group from S-adenosyl-L-methionine is transfer to ε-amino group of lysine and guanidino group of arginine	Lysine-specific demethylase1 (LSD1/ KDM1) and Jumonji C (JmjC) domain-containing proteins	Methylation of H3K4 and H3K36 activate geneexpression. Methylation of H3K9, H3K27 and H4R3 repress gene expression.	Couture and Trievel (2006)
Ubiquiti-nation	H2A H2B (monoubiquitination)	Ubiquitination of Brefeldin A-sensitivity protein1, histone monoubiquitination1, and ubiquitin carrier protein1	Ubiquitin-specific thiol proteases, Utubain-like deubiquitinase (OTLD1)	Ubiquitination of H3K4me3 activate transcription	Deal and Henikoff, (2011)
SUMO-ylation	H2A.Z H2B H4	Small ubiquitin-like modifier (SUMO) E3-ligases	SUMO-specific proteases	SUMOylation of Histone deacetylases represses transcription	Garcia-Dominguez and Reyes (2009)

Table 4. Effects of different abiotic stresses and exogenous protectants on histone modification and DNA methylation in plants. The table depicts genes regulated, methylation status, duration of stress and techniques used to decipher epigenetic marks (ChIP-Seq: Chromatin immunoprecipitation sequencing, RT-qPCR: Real time quantitative PCR, RNA-Seq: RNA sequencing, Western blot: Western blotting)

Stress condition	Duration	Plant species	Histone modification	Genes regulated	Methylation status	Techniques used	References
Drought	1, 2, 4, 5h	*Arabidopsis*	H3K9ac, H3K1ac, H3K23ac, H3K4me3, H3K4me2, H3K4me1	*RD29A, RD29B, RD20, RAP2.4, AtGOLS2, ProDH*	Methylation	ChIP-Seq, RT-qPCR	Fang et al. (2014)
	35 h	*Oryzasativa*	H3K18ac, H3K27ac, H4K5ac, H3K9ac, H3K4me3	*OsHAC703, OsHAG703, OsHAM701, OsHAF701*	Methylation	RT-qPCR, Western blot	
Salinity	5, 20 days	*Arabidopsis*	H3K9K14ac, H3K4me3, H3K9me2	*ABI1, ABI2, RD29A, RD29B, KAT1, KAT2, DREB2A, ERF4*	Methylation	ChIP-Seq, RT-qPCR	Zhao et al. (2016)
	5-14 days	*Oryzasativa*	H3K18ac, H3K27ac, H4K5ac, H3K9ac, H3K4me3	*OsGA20ox2, OsGA20ox3, OsGA3ox1*	Methylation	RT-qPCR, Western blot	
Cold	6 days	*Arabidopsis*	H3K27me3	*COR15A, ATGOLS3*	Methylation	ChIP-qPCR	Roy et al. (2014)
	16 h	*Oryzasativa*	H3K9ac, H3K9K14ac, H3K27ac	*OsDREB1*	Methylation	ChIP-Seq, RNA-Seq Microarray	
	3-6 weeks	*Zea mays*	H3K9ac, H4K5ac, H4K4ac	*ZmDREB1, ZmICE1, ZmCOR413*	Methylation/ De-methylation	ChIP-qPCR, RT-qPCR, Western blot	Hu et al. (2012)
Heat	8 days	*Arabidopsis*	H3K27me3, H3K4me2, H3K9ac	*APX2, HSP18.2, HSP22.0, HSP70, WRKY53, FRK1, NHL10*	Methylation/ De-methylation	ChIP-qPCR, RT-qPCR, Western blot, RNA-Seq	Gan et al. (2014)
Submergence	24 h	*Oryzasativa*	H3K4me3, H3K4me2, H3K9/14ac	OsADH1, OsPDC1	Methylation	ChIP-qPCR, RT-qPCR	Tsuji et al. (2006)

Stress condition	Duration	Plant species	Histone modification	Genes regulated	Methylation status	Techniques used	References
Phytohormone priming							
Salicylic acid + Drought	100 μM	*Hordeumvulgare*	-	*HvHDAC2-1, HvHDAC2-2*	Methylation	ChIP-qPCR, RT-qPCR, Western blot,	Demetriou et al. (2009)
Abscisic acid + drought + salt	0.5-2.0 μM	*Arabidopsis*	H3K9K14ac, H3K4me3, H3K9me2	*ABI1, ABI2, RD29A, RD29B, KAT1, KAT2, DREB2A*	Methylation	ChIP-Seq, RT-qPCR	Chen et al. (2010)
	100 μM	*Hordeumvulgare*	-	*HvHDAC2-1, HvHDAC2-2*	Methylation	RT-qPCR	
Jasmonic acid + drought	100 μM	*Hordeumvulgare*	-	*HvHDAC2-1, HvHDAC2-2*	Methylation	ChIP-Seq, RT-qPCR	Demetriou et al. (2009)

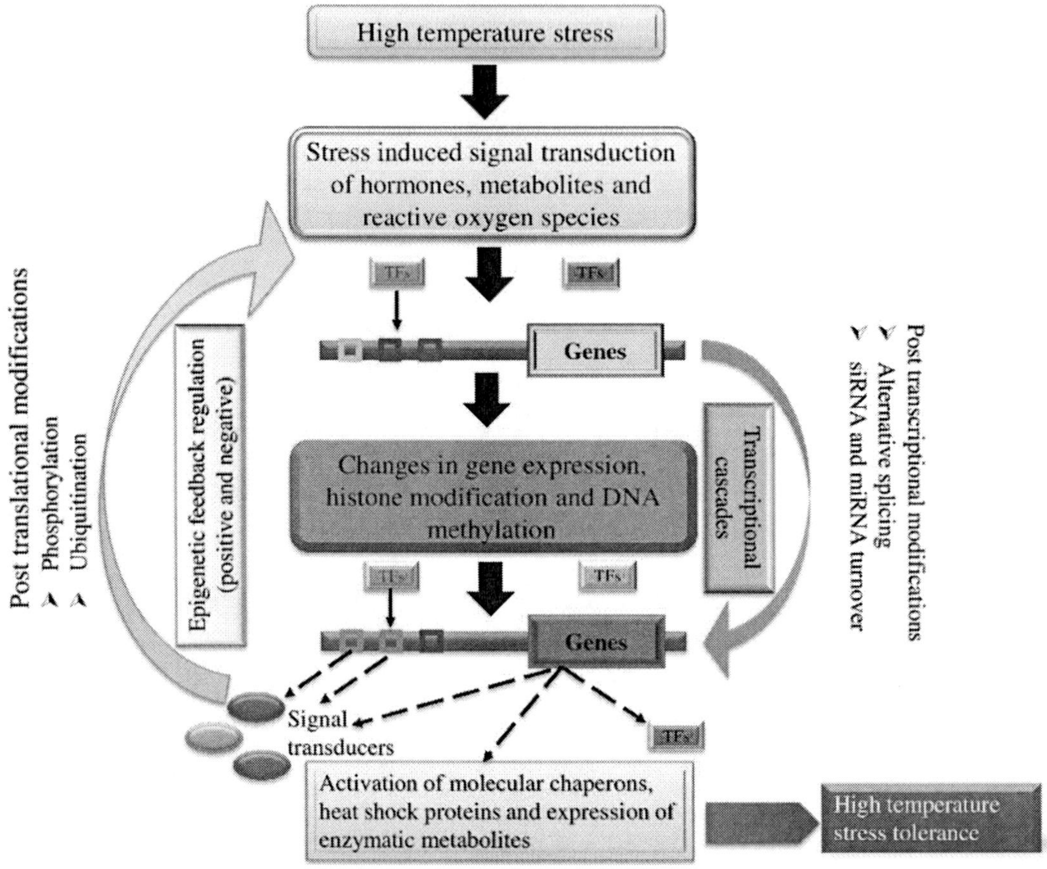

Figure 6. Overall model for high temperature stress tolerance in terms of post-transcriptional, post translational and epigenetic regulation. Plant cell perceive and transduce stress signal via various signalling pathway, plant hormones and secondary metabolites which in turn regulate stress inducible genes and transcription factors (TFs) thereby expressing functional metabolites and proteins that helps the plants in combating stress condition.

Conclusion and Future Outlook

The era of "omics" technology has empowered the development in the area of transcriptome, proteome and metabolome that has enhanced our possibility of studying regulation of plant growth and development at all the three levels. The effect of stress and their response are not straightforward as they involve series of physiological, biochemical and molecular changes that instigate severe repercussion on plants. For instance, to decipher quantitative relation between different proteins and metabolites that are regulated under various abiotic stress conditions, identification of bio-markers is an utmost obligation. With the advancement of "omics" techniques, several genomics and protein databases has been made available in model legume crops such as Medicago and Lotus and in crop legumes as well such as soybean and common bean, which can facilitate high throughput identification and validation of stress responsive proteins and metabolites in other crop legumes *per se*., hyacinth bean.

Legumes including hyacinth bean are cultivated widely in the tropical and sub-tropical part of the world where plants are affected not by one stress but also a combination of stress at a same point under natural field condition that severely effect their growth and productivity. The only option by which legume production can be sustained under these variable environments is to develop high yielding/resistant/tolerant cultivars by incorporating integrated "omics" technology into breeding program that can pinpoint those genes that responsible for desired traits and thus facilitate the selection plants on the basis of their genotyping information. Few of these techniques have been successfully employed in some legume crops such as chickpea, soybean and pea that has significantly contributed towards the understanding of stress related traits in these crops leading to implication of effective strategy for achieving higher yield. With the ever-increasing global population, the demand-supply gap for the legumes is widening day by day. Therefore, in order to meet that gap in near future, hyacinth bean need to be advanced at all the three "omics" levels for better understanding of physiological, biochemical and molecular pathways involved in stress response to sustain its cultivation under adverse environmental cues.

ACKNOWLEDGMENTS

The authors are highly grateful to DBT (Department of Biotechnology), Govt. of India, for the financial support (Grant No. BT/PR10067/AGR/02/554/2007). The authors are also thankful to DST PURSE (DST/BHU/PURSE 2017-18) and FIST program (DST/BHU/FIST 2016-17) for financial support and central internet facility, Department of Botany BHU, Varanasi, to carry out literature survey.

REFERENCES

Abe, H., Urao, T., Ito, T., Seki, M., Shinozaki, K., Yamaguchi Shinozaki, K. (2003) Arabidopsis AtMYC2 (bHLH) and AtMYB2 (MYB) function as transcriptional activators in abscisic acid signaling. *Plant Cell* 15: 63-78.

Adams, M. A., Turnbull, T. L., Sprent, J. I., Buchmann, N. (2016) Legumes are different: Leaf nitrogen, photosynthesis, and water use efficiency. *Proc Natl Acad Sci USA* 113: 4098-4103.

Ahuja, I., de Vos, R. C., Bones, A. M., Hall, R. D. (2010) Plant molecular stress responses face climate change. *Trends Plant Sci* 15: 664-674.

Alcazar, R., Altabella, T., Marco, F., Bortolotti, C., Reymond, M., Koncz, C., Carrasco, P., Tiburcio, A. F. (2010) Polyamines: Molecules with regulatory functions in plant abiotic stress tolerance. *Planta* 231: 1237-1249.

Baker, R. J. (1984) Quantitative genetic principles in plant breeding. *In Gene manipulation in plant improvement* (Ed. Gustafson, J. P.), NY Plenum Press, New York, pp. 147-176.

Banu, M. N. A., Hoque, M. A., Watanabe-Sugimoto, M., Islam, M. M., Uraji, M., Matsuoka, K., Nakamura, Y., Murata, Y. (2010) Proline and glycinebetaine ameliorated NaCl stress via scavenging of hydrogen peroxide and methylglyoxal but not superoxide or nitric oxide in tobacco cultured cells. *Biosci Biotechnol Biochem* 74: 2043-2049.

Basu, S., Ramegowda, V., Kumar, A., Pereira, A. (2016) Plant adaptation to drought stress. *F1000 Research* 5: 1554.

Boubakri, H., Gargouri, M., Mliki, A., Brini, F., Chong, J., Jbara, M. (2016) Vitamins for enhancing plant resistance. *Planta* 244: 529-543.

Brosowska-Arendt, W., Gallardo, K., Sommerer, N., Weidner, S. (2014) Changes in the proteome of pea (*Pisumsativum* L.) seeds germinating under optimal and osmotic stress conditions and subjected to post-stress recovery. *Acta Physiol Plant* 36: 795-807.

Chen, L. T., Luo, M., Wang, Y. Y., Wu, K. Q. (2010) Involvement of Arabidopsis histone deacetylase HDA6 in ABA and salt stress response. *J Exp Bot* 61:345-3353.

Chinnusamy, V., Zhu, J. K. (2009) Epigenetic regulation of stress responses in plants. *CurrOpin Plant Biol* 12: 133-139.

Cigliano, R. A., Sanseverino, W., Cremona, G., Ercolano, M. R., Conicella, C., Consiglio, F. M. (2013) Genome-wide analysis of histone modifiers in tomato: Gaining an insight into their developmental roles. *BMC genomics* 14: 57.

Conrad, M., Angeli, J. P. F., Vandenabeele, P., Stockwell, B. R. (2016) Regulated necrosis: Disease relevance and therapeutic opportunities. *Nat Rev Drug Discov* 15: 348.

Couture, J. F., Trievel, R. C. (2006) Histone-modifying enzymes, encrypting an enigmatic epigenetic code. *Curr Opin Struct Biol* 16: 753-760.

Davletova, S., Schlauch, K., Coutu, J., Mittler, R. (2005) The zinc-fingerprotein Zat12 plays a central role in reactive oxygen and abiotic stresssignaling in Arabidopsis. *Plant Physiol* 139: 847-856.

Deal, R. B., Henikoff, S. (2011) Histone variants and modifications in plant gene regulation. *Curr Opin Plant Biol* 14: 116-122.

Demetriou, K., Kapazoglou, A., Tondelli, A., Francia, E., Stanca, M. A., Bladenopoulos, K., Tsaftaris, A. S. (2009) Epigenetic chromatin modifiers in barley: I. Cloning, mapping and expression analysis of the plant specific HD2 family of histone deacetylases from barley, during seed development and after hormonal treatment. *Physiol Plant* 136: 358-368.

Diebold, L., Chandel, N. S. (2016) Mitochondrial ROS regulation of proliferating cells. *Free Radic Biol Medi* 100: 86-93.

Ding, J., Shen, J., Mao, H., Xie, W., Li, X., Zhang, Q. (2012) RNA-directed DNA methylation is involved in regulating photoperiod-sensitive male sterility in rice. *Mol plant* 5: 1210-1216.

Ding, W., Song, L., Wang, X., Bi, Y. (2010) Effect of abscisic acid on heat stress tolerance in the calli from two ecotypes of *Phragmitescommunis*. *BiolPlantarum* 54: 607-613.

Fang, H., Liu, X., Thorn, G., Duan, J., Tian, L. (2014) Expression analysis of histone acetyltransferases in rice under drought stress. *Biochem Biophys Res Commun* 443: 400-405.

Fernie, A. R., Schauer, N. (2009) Metabolomics-assisted breeding: A viable option for crop improvement. *Trends Genet* 25: 39-48.

Gan, E. S., Xu, Y., Wong, J. Y., Geraldine, Goh, J., Sun, B., Wee, W. Y., Huang, J., Ito, T., (2014) Jumonjidemethylases moderate precocious flowering at elevated temperature via regulation of FLC in Arabidopsis. *Nat Commun* 5: 5098.

Garcia-Dominguez, M., Reyes, J. C. (2009) SUMO association with repressor complexes, emerging routes for transcriptional control. *Biochim Biophys Acta* 1789: 45-459.

Genovese, S., Fiorito, S., Epifano, F., Taddeo, V. A. (2015) A novel class of emerging anticancer compounds: Oxyprenylated secondary metabolites from plants and fungi. *Curr Med Chem* 22: 3426-3433.

Gibbs, A. J, Mackenzie, K. J., Gibbs, M. J. (2008) Thepotyvirus of Australia. *Arch Virol* 153:1411-1420.

Gill, S. S., Anjum, N. A., Gill, R., Yadav, S., Hasanuzzaman, M., Fujita, M., Mishra, P., Sabat, S. C., Tuteja, N. (2015) Superoxide dismutase-mentor of abiotic stress tolerance in crop plants. *Environ Sci Pollut Res* 22: 10375-10394.

Gong, M., Li, Y. J., Chen, S. Z. (1998) Abscisic acid-induced thermotolerance in maize seedlings is mediated by calcium and associated with antioxidant systems. *J Plant Physiol* 153: 488-496.

Guo, Z. J., Chen, X. J., Wu, X. L., Ling, J. Q., Xu, P. (2004) Overexpression of the AP2/ EREBP transcription factor OPBP1 enhances diseaseresistance and salt tolerance in tobacco. *Plant Mol Biol* 55:607-618.

Hasanuzzaman, M., Ahamed, K. U., Rahmatullah, M., Akhter, N., Nahar, K., Rahman, M. L. (2010) Plant growth characters and productivity of wetland rice (*Oryzasativa* L.) as affected by application of different manures. *Emir J Food Agric* 46-58.

Hasanuzzaman, M., Alam, M., Rahman, A., Hasanuzzaman, M., Nahar, K., Fujita, M. (2014) Exogenous proline and glycine betaine mediated upregulation of antioxidant defenseand glyoxalase systems provides better protection against salt-induced oxidative stress in two rice (*Oryzasativa* L.) varieties. *Bio Med Res Int 2014*.

Hasanuzzaman, M., Hossain, M. A., Teixeira da Silva, J. A., Fujita, M. (2012) Plant responses and tolerance to abiotic oxidative stress: Antioxidant defense is a key factor.

In: *Crop Stress and its Management: Perspectives and Strategies* (Eds. Bandi V, Shanker A. K., Shanker C., Mandapaka, M.) Springer, Berlin, pp. 261-316.

Hasanuzzaman, M., Nahar, K., Alam, M. M., Roychowdhury, R., Fujita, M. (2013) Physiological, Biochemical, and Molecular Mechanisms of Heat Stress Tolerance in Plants. *Int J Mol Sci* 14: 9643-9684.

He, J., Zhao, X., Laroche, A., Lu, Z. X., Liu, H., Li, Z. (2014) Genotyping by-sequencing (GBS), an ultimate marker-assisted selection (MAS) tool to accelerate plant breeding. *Front Plant Sci* 5: 484.

Hiz M. C., Canher B., Niron H., Turet, M. (2014) Transcriptome analysis of salt tolerant common bean (*Phaseolus vulgaris* L.) under saline conditions. *Plo S ONE* 9: e92598.

Hossain, M. A., Hoque, M. A., Burritt, D. J., and Fujita, M. (2014) Proline protects plants against abiotic oxidative stress: Biochemical and molecular mechanisms. In: *Oxidative damage to plants* (Ed.Ahmad P), Elsevier Inc., Academic press, UK, pp. 477-522.

Hossain, Z., Komatsu, S. (2014) Soybean proteomics. *In: Plant Proteomics Methods* (Eds. Jorrin-Novo, J. V., Komatsu, S., Weckwerth, W., Wienkoop, S.) Mol Biol, Springer, New York, pp. 315-331.

Houben A, Demidov D, Caperta AD, Karimi R, Agueci F, Vlasenko L (2007) Phosphorylation of histone H3 in plants-a dynamic affair. *Biochim Biophys Acta* 1769: 308-315.

Hu, Y., Zhang, L. U., He, S., Huang, M. I., Tan, J., Zhao, L. I., Yan, S., Li, H. U., Zhou, K. U., Liang, Y., Li, L. (2012) Cold stress selectively unsilences tandem repeats in heterochromatin associated with accumulation of H3K9ac. *Plant Cell Environ* 35: 2130 -2142.

Iqbal, N., Umar, S., Khan, N. A., Khan, M. I. R. (2014) A new perspective of phytohormones in salinity tolerance: Regulation of proline metabolism. *Environ Exp Bot* 100: 34-42.

Iseki, K., Takahashi, Y., Muto, C., Naito, K., Tomooka, N. (2016) Diversity and evolution of salt tolerance in the genus Vigna. *Plo S One* 11: e0164711.

Johnson, H. E., Broadhurst, D., Goodacre, R., Smith, A. R. (2003) Metabolic fingerprinting of salt-stressed tomatoes. *Phytochemistry* 62: 919-928.

Jorrin-Novo, J. V., Pascual, J., Sánchez-Lucas, R., Romero-Rodríguez, M. C., Rodríguez-Ortega, M. J., Lenz, C., Valledor, L. (2015) Fourteen years of plant proteomics reflected in Proteomics: Moving from model species and 2DE-based approaches to orphan species and gel-free platforms. *Proteomics* 15: 1089-1112.

Kafatos, A., Hatzis, C. (2008) Clinical nutrition for medical students. Crete University of Crete.

Kaldis, A., Tsementzi, D., Tanriverdi, O., Vlachonasios, K. E. (2011) Arabidopsis thaliana transcriptional co-activators ADA2b and SGF29a are implicated in salt stress responses. *Planta* 233: 749-762.

Kan, L., Nie, S., Hu. J., Liu, Z., Xie, M. (2016) Antioxidant activities and anthocyanins composition of seed coats from twenty-six kidney bean cultivars. *J Funct Foods* 26: 622-631.

Katam, R., Basha, S. M., Suravajhala, P., Pechan, T. (2010) Analysis of peanut leaf proteome. *J Proteome Res* 9: 2236-2254.

Khan, M. I. R., Iqbal, N., Masood, A., Per, T. S., Khan, N. A. (2013) Salicylic acid alleviates adverse effects of heat stress on photosynthesis through changes in proline production and ethylene formation. *Plant Signal Behav* 8: e26374.

Kim, C., Guo, H., Kong, W., Chandnani, R., Shuang, L. S., Paterson, A. H. (2016) Application of genotyping by sequencing technology to a variety of crop breeding programs. *Plant Sci* 242: 14-22.

Kottapalli, K. R., Payton, P., Rakwal, R., Agrawal, G. K., Shibato, J., Burow, M., Puppala, N. (2008) Proteomics analysis of mature seed of four peanut cultivars using two-dimensional gel electrophoresis reveals distinct differential expression of storage, anti-nutritional, and allergenic proteins. *Plant Sci* 175: 321-329.

Kukade, S. A., Tidke, J. A. (2014) Reproductive biology of *Dolichos lablab* L. (Fabaceae). *Indian J Plant Sci* 3: 22-25.

Lee, J. H., Hong, J. P., Oh, S. K., Lee, S., Choi, D., Kim, W. T. (2004) Theethylene-responsive factor like protein 1 (*CaERFLP1*) of hot pepper*capsicum annuum* L.) interacts *in vitro* with both GCC and DRE/CRT sequences with different binding affinities: Possible biological roles of CaERFLP1 in response to pathogen infection and high salinity conditions in transgenic tobacco plants. *Plant molbiol* 55: 61-81.

Lu, C., Ward, P. S., Kapoor, G. S., Rohle D., Turcan, S., Abdel-Wahab, O., Edwards, C. R., Khanin, R., Figueroa, M. E., Melnick, A., Wellen, K. E. (2012) IDH mutation impairs histone demethylation and results in a block to cell differentiation. *Nature* 483: 474.

Luo, M., Wang, Y. Y., Liu, X., Yang, S., Lu, Q., Cui, Y., Wu, K. (2012) HD2C interacts with HDA6 and is involved in ABA and salt stress response in *Arabidopsis*. *J Exp Bot* 63: 3297-3306.

Maass, B. L, Knox, M. R., Venkatesha, S. C., Angessa, T. T., Ramme, S., Pengelly, B. C. (2010) *Lablab purpureus*-A crop lost for Africa? *Tropi plant biology* 3: 123-135.

Misra, N., Saxena, P. (2009) Effect of salicylic acid on proline metabolism in lentil grown under salinity stress. *Plant Sci* 177: 181-189.

Mittler, R. (2017) ROS are good. *Trends Plant Sci* 22: 11-19.

Morozova, O., Marra, M. A. (2008) Applications of next-generation sequencing technologies in functional genomics. *Genomics* 92: 255-264.

Morris, J. B. (2009) Morphological and reproductive characterization in hyacinth bean, *Lablab purpureus* (L.) Sweet germplasm with clinically proven nutraceutical and pharmaceutical traits for use as a medicinal food. *J Diet Suppl* 6: 263-279.

Mullineaux, P. M., Bechtold, U., Fryer, M. J., Lawson, T., Kim, Y. S., Hickman, R, Ott, S., Beynon, J. L., Buchanan-Wollaston, V., Baker, N. R., Friedrich, S. (2011) *Arabidopsis* heat shock transcription factor 1b is a major determinant of seed yield and constitutively regulates basal resistance to abioticand biotic stresses. *In: Society for Experimental Biology Annual Main Meeting,* Glasgow, UK.

Nazar, R., Iqbal, N., Syeed, S., Khan, N. A. (2011) Salicylic acid alleviates decreases in photosynthesis under salt stress by enhancing nitrogen and sulfur assimilation and antioxidant metabolism differentially in two mung bean cultivars. *J Plant Physiol* 168: 807-815.

Okazaki, Y., Saito, K. (2016) Integrated metabolomics and phytochemical genomics approaches for studies on rice. *Giga Science* 5: 11.

Okuma, E., Murakami, Y., Shimoishi, Y., Tada, M., Murata, Y. (2004) Effects of exogenous application of proline and betaine on the growth of tobacco cultured cells under saline conditions. *J Soil Sci Plant Nutr* 50: 1301-1305.

Orellana, S., Yanez, M., Espinoza, A., Verdugo, I., Gonzalez, E., Ruiz-Lara, S., Casaretto, J. A. (2010) The transcription factor SlAREB1confers drought, salt stress tolerance and regulates biotic and abioticstress-related genes in tomato. *Plant Cell Environ* 33:2191-2208.

Pandey, M. K., Roorkiwal, M., Singh, V. K., Ramalingam, A., Kudapa, H., Thudi, M., Chitikineni, A., Rathore, A., Varshney, R. K. (2016) Emerging genomic tools for legume breeding: Current status and future prospects. *Front Plant Sci* 7: 455.

Park, J. M., Park, C. J., Lee, S. B., Ham, B. K., Shin, R., Paek, K. H. (2001) Overexpression of the tobacco TSI1 gene encoding an EREBP/AP2-type transcription factor enhances resistance against pathogen attackand osmotic stress in tobacco. *Plant Cell* 13: 1035-1046.

Peng, X. X., Tang, X. K., Zhou, P. L., Hu, Y. J., Deng, X. B., He, Y., Wang, H. H.(2011) Isolation and expression patterns of rice WRKY82 transcriptionfactor gene responsive to both biotic and abiotic stresses. *Agric Sci China* 10: 893-901.

Peterson, C. L., Laniel, M. A. (2004) Histones and histone modifications. *Curr Biol* 14: R546-R551.

Qiao, G., Wen, X., Yu, L., Ji, X. (2012) Identification of differentially expressed genes preferably related to drought response in pigeon pea (*Cajanuscajan*) inoculated by arbuscular mycorrhizae fungi (AMF). *Acta Physiol Plant* 34: 1711-1721.

Rai, K. K., Rai, N., Rai, S. P. (2017) Downregulation of γECS gene affects antioxidant activity and free radical scavenging system during pod development and maturation in *Lablab purpureus* L. *Biocatal Agric Biotechnol* 11: 192-200.

Rai, N., Rai, K. K., Tiwari, G., Singh, P. K. (2015) Changes in free radical generation, metabolites and antioxidant defense machinery in hyacinth bean (*Lablab purpureus* L.) in response to high temperature stress. *Acta Physiol Plant* 37:46.

Rai, N., Rai, K. K., Venkataravanappa, V., Saha, S. (2016). Molecular Approach Coupled with Biochemical Attributes to Elucidate the Presence of DYMV in Leaf Samples of *Lablab purpureus*. L Genotypes. *Applbio chem biotechnol* 178: 876-890.

Ramalingam, A., Kudapa, H., Pazhamala, L. T., Weckwerth, W., Varshney, R. K. (2015) Proteomics and Metabolomics: Two Emerging Areas for Legume Improvement. *Front Plant Sci* 6: 1116.

Ray, D. K., Mueller, N. D., West, P. C., Foley, J. A. (2013) Yield trends are insufficient to double global crop production by 2050. *Plo S One* 8: 66428.

Rodziewicz, P., Swarcewicz, B., Chmielewska, K., Wojakowska, A., Stobiecki, M. (2014) Influence of abiotic stresses on plant proteome and metabolome changes. *Acta Physiol Plant* 36: 1-19.

Rojas, A., Almoguera, C., Jordano, J. (1999) Transcriptional activation of a heat shock gene promoter in sunflower embryos: Synergism between ABI3 and heat shock factors. *Plant J* 20: 601-610.

Roxburgh, W. (1832) *Flora Indicavol* 3 W Thackery and Co Calcutta.

Roy, D., Paul, A., Roy, A., Ghosh, R., Ganguly, P., Chaudhuri, S. (2014) Differential acetylation of histone H3 at the regulatory region of OsDREB1b promoter facilitates chromatin remodelling and transcription activation during cold stress. *PLo S ONE* 9: e100343.

Sanchez, D. H., Pieckenstain, F. L., Escaray, F., Erban, A., Kraemer, U. T. E., Udvardi, M K., Kopka, J. (2011) Comparative ionomics and metabolomics in extremophile and glycophytic Lotus species under salt stress challenge the metabolic pre-adaptation hypothesis. *Plant Cell Environ* 34: 605-617.

Sanchez, D. H., Schwabe, F., Erban, A., Udvardi, M. K., Kopka, J. (2012) Comparative metabolomics of drought acclimation in model and forage legumes. *Plant Cell Environ* 35: 136-149.

Sanchez, D. H., Schwabe, F., Erban, A., Udvardi, M. K., Kopka, J. (2012) Comparative metabolomics of drought acclimation in model and forage legumes. *Plant Cell Environ* 35: 136-149.

Sani, E., Herzyk, P., Perrella, G., Colot, V., Amtmann, A. (2013) Hyperosmotic priming of Arabidopsis seedlings establishes a long-term somatic memory accompanied by specific changes of the epigenome. *Genome biol* 14: R59.

Schenke, D., Bottcher, C., Scheel, D. (2011) Crosstalk between abioticultraviolet-B stress and biotic (flg22) stress signalling in Arabidopsis prevents flavonol accumulation in favor of pathogen defence compound production. *Plant Cell Environ* 34: 1849-1864.

Seo, P. J., Kim, M. J., Park, J. Y., Kim, S. Y., Jeon, J., Lee, Y. H., Kim, J., Park, C. M. (2010) Cold activation of a plasma membrane-tethered NAC transcription factor induces a pathogen resistance response in Arabidopsis. *Plant J* 61: 661-671.

Shafi, A.,Chauhan, R., Gill, T., Swarnkar, M. K., Sreenivasulu, Y., Kumar, S., Kumar, N., Shankar, R., Ahuja, P. S., Singh, A. K. (2015) Expression of SOD and APX genes positively regulates secondary cell wall biosynthesis and promotes plant growth and yield in Arabidopsis under salt stress. *Plant Mol Biol* 87: 615-631.

Shahzad, R., Waqas, M., Khan, A. L., Hamayun, M., Kang, S. M., Lee, I. J. (2015) Foliar application of methyl jasmonate induced physio-hormonal changes in *Pisumsativum* under diverse temperature regimes. *Plant Physiol Biochem* 96: 406-416.

Siebers, M. H.,Yendrek, C. R., Drag, D., Locke, A. M., Rios Acosta, L., Leakey, A. D., Ainsworth, E. A., Bernacchi, C. J., Ort, D. R. (2015) Heat waves imposed during early pod development in soybean (*Glycine max*) cause significant yield loss despite a rapid recovery from oxidative stress. *Global Change Biol* 21: 3114-3125.

Singh, K. B., Saxena, M. C. (1993) *Breeding for stress tolerance in cool-season food legumes* (No. 635.65 S52). Chichester, U.K., Wiley.

Staudinger, C., Mehmeti, V., Turetschek, R., Lyon, D., Egelhofer, V., Wienkoop, S. (2012) Possible role of nutritional priming for early salt and drought stress responses in *Medicago truncatula*. *Front Plant Sci* 3: 285.

Sumner, L. W., Mendes, P., Dixon, R. A. (2003) Plant metabolomics: Large-scale phytochemistry in the functional genomics era. *Phytochemistry* 62: 817-836.

Swigonska, S., Weidner, S. (2013) Proteomic analysis of response to long-term continuous stress in roots of germinating soybean seeds. *J Plant Physiol* 170: 470-479.

Takagi, A., Kano, M., Kaga, C. (2015) Possibility of breast cancer prevention: Use of soy isoflavones and fermented soy beverage produced using probiotics. *Int J Mol Sci* 16: 10907-10920.

Tarawali, G., Ogunbile, O. A. (1995) Legumes for sustainable food production in semi-arid savannahs. *ILEIA newsletter* 11: 18-23.

Thilagavathy, A., Devaraj, V. R. (2016) Evaluation of appropriate reference gene for normalization of micro RNA expression by real-time PCR in *Lablab purpureus* under abiotic stress conditions. *Biologia* 71: 660-668.

Tran, L. S. P. (2004) Functional analysis of Arabidopsis NACtranscription factors controlling expression of erd1 gene underdrought stress. *Plant Cell* 16: 2482-2498.

Tsuji, H., Saika, H., Tsutsumi, N., Hirai, N., Nakazono, M. (2006) Dynamic and reversible changes in histone H3-Lys4 methylation and H3 acetylation occurring at submergence-inducible genes in rice. *Plant Cell Physiol* 47: 995-1003.

Uday Kumar, H. R., Byre Gowda, M., Ramesh, S., Vasundhara, M. (2017) Characterization and Identification of Dolichos Bean (*Lablab purpureus* L. sweet) Recombinant Inbred Lines (RIL) with High Pod Yield and High Pod Fragrance. *Int J Pure App Biosci* 5: 428-436.

Varshney, R. K., Mohan, S. M., Gaur, P. M., Gangarao, N. V., Pandey, M. K., Bohra, A., Sawargaonkar, S. L., Chitikineni, A., Kimurto, P. K., Janila, P., Saxena, K. B. (2013) Achievements and prospects of genomics-assisted breeding in three legume crops of the semi-arid tropics. *Biotechnol Advances* 31: 1120-1134.

Varshney, R. K., Thudi, M., Nayak, S. N., Gaur, P. M., Kashiwagi, J., Krishnamurthy, L., Jaganathan, D., Koppolu, J., Bohra, A., Tripathi, S., Rathore, A. (2014) Genetic dissection of drought tolerance in chickpea (*Cicerarietinum* L.) *Theor Appl Genet* 127: 445-462.

Varshney, R. K., Singh, V. K., Hickey, J., Xun, X., Marshall, D. F., Wang, J., Edwards, D., Ribaut, J. M. (2015) Analytical and decision support tools for genomics-assisted breeding. *Trends Plant Sci* 21: 354-363.

Vereecken, H., Schnepf, A., Hopmans, J. W., Javaux, M., Or, D, Roose, T., Vanderborght, J., Young M. H., Amelung, W., Aitkenhead, M., Allison, S. D. (2016) Modeling soil processes: Review, key challenges, and new perspectives. *Vadose Zone J* 15.

Wang, L. J., Li, S. H. (2006) Salicylic acid-induced heat or cold tolerance in relation to Ca2+ homeostasis and antioxidant systems in young grape plants. *Plant Sci* 170: 685-694.

Xin, X., Tian, Q., Yin, G., Chen, X., Zhang, J., Ng, S., Lu, X. (2014) Reduced mitochondrial and ascorbate-glutathione activity after artificial ageing in soybean seed. *J Plant Physiol* 171: 140-147.

Yao, L., Jiang, Y., Lu, X., Wang, B., Zhou, P., Wu, T. (2016) A R2R3-MYB transcription factor from *Lablabpurpureus* induced by drought increases tolerance to abiotic stress in Arabidopsis. *Plant Mol Biol Report* 43: 1089-1100.

Zadraznik, T., Hollung, K., Egge-Jacobsen, W., Me glic, V., SuStar-Vozlic, J. (2013) Differential proteomic analysis of drought stress response in leaves of common bean (*Phaseolus vulgaris* L.). *J Proteomics* 78: 254-272.

Zhang, N., Venkateshwaran, M., Boersma, M., Harms, A., Howes-Podoll, M., Os, D., Ané, J. M., Sussman, M. R. (2012) Metabolomic profiling reveals suppression of oxylipin biosynthesis during the early stages of legume-rhizobia symbiosis. *FEBS Letter* 586: 3150-3158.

Zhang, X. W., Zhang, M., Wang, Q. H., Qiu, X. K., Hu, G. Q., Dong, Y. J. (2011) Effect of exogenous nitric oxide on physiological characteristic of peanut under iron deficient stress. *J Plant Nutr Fert* 17: 665-673.

Zhao, J., Li, M., Gu D., Liu X., Zhang J., Wu K., Zhang X., Teixeira da Silva, J. A., Duan, J. (2016) Involvement of rice histone deacetylase HDA705 in seed germination and in response to ABA and abiotic stresses. *Biochem Biophys Res Commun* 470: 439-444.

Zhu, J. K. (2002) Salt and drought stress signal transduction in plants. *Annu Rev Plant Biol* 53: 247-273.

Zou, C. M., Wang, Y. Q., Cao, W. D., Liu, Y, Zhang, X. H., Tang, S. (2015) Response of photosynthesis and growth to weak light regime in different Adzuki bean (*Vigna angularis*) varieties. *J Appl Ecol* 26: 3687-3692.

In: Innovations in Life Science Research ISBN: 978-1-53615-868-7
Editors: R. P. Sinha, S. Pandey-Rai and N. Ghoshal ©2019 Nova Science Publishers, Inc.

Chapter 10

PAPAYA RINGSPOT DISEASE AND ITS MANAGEMENT STRATEGIES

Rajesh Kumar[1,*], Vishal Srivashtav[1] and Rajeshwar P. Sinha[2]
[1]Plant Biotechnology Laboratory, Department of Genetics and Plant Breeding, RGSC, Banaras Hindu University, Mirzapur, Uttar Pradesh, India
[2]Laboratory of Photobiology and Molecular Microbiology, Centre of Advanced Study in Botany, Institute of Science, Banaras Hindu University, Varanasi, India

ABSTRACT

Papaya plant is widely grown in tropics and sub-tropics and possesses a great economic importance for its worldwide consumption and medicinal values. The leaf-extract of papaya has been recognised for its medicinal value against dengue, a serious health hazard in tropical and the sub-tropical regions. However, the plant is severely affected by various pathogens like insect-pests, fungi, bacteria and viruses in every corner of growing areas causing significant loss in its fruit production. Among all the pathogens that infect papaya, *Papaya Ringspot Virus* (PRSV) is the key pathogen responsible for damage of plant and significant loss of yield. Vigour of trees, fruit-set, flavour and quality is usually deteriorated depending on the age of plants when get infected. Various management strategies have been employed to control and minimise the effect of PRSV in agricultural field; however, transgenic approaches has been successfully implemented in controlling the effect of PRSV on papaya.

Keywords: papaya, *Papaya Ringspot Virus*, transgenic papaya

[*]Corresponding Author's Email: rajitrc@gmail.com.

INTRODUCTION

Papaya plant *(Carica papaya* L.), known to have originated in America, is herbaceous, perennial, and evergreen dicot with a hollow soft wooded trunk of 1-10 m in height. It also adopts itself to diverse soil and climatic conditions which make it easy to grow and sustain. The plant produces fruits throughout the year and gives quick returns. Globally, India was the largest producer of papaya fruit in 2016 with 5.69 million tonnes followed by Brazil with1.42 million tonnes (FAOSTAT).

Traditionally, leaf, seed, fruit and latex parts of papaya are being used from ancient times for medicinal and nutritional purposes. It is a good source of vitamin A and C as well as mineral elements like Ca, Mg, Na and K (Chukwuka et al. 2013). The plant parts show medicinal properties such as anti-inflammatory and immunomodulatory effects (Pandey et al. 2016), wound healing (Nayak et al. 2007), anti-tumour (Otsuki et al. 2010), antioxidant (Maisarah et al. 2012), antifungal (Singh and Ali 2011), antimicrobial (Yahaya et al. 2017), hypoglycaemic (Juárez-Rojop et al. 2012) and antifertility effect (Udoh et al. 2005). Researchers also demonstrated its beneficial roles in the treatment of cardiovascular diseases (Ravikant et al. 2012), dengue fever (Ahmad et al. 2011), malaria (Kovendan et al. 2012), gastrointestinal tract disorders (Muss et al. 2013), intestinal parasite infections (Okeniyi et al. 2007) and diuretic patients (Adam et al. 2013). The digestive enzyme papain, isolated from papaya, is used as an ingredient in brewing, meat tenderizing as well as in cosmetic and leather industry (Amri and Mamboya 2012).

However, a major problem globally faced by papaya fruit industry is significant pre and post-harvest losses along the marketing chain. Various factors such as fungal diseases, physiological disorders, mechanical damages, or a combination of these are the leading causes of these losses. However, a large amount of losses are contributed by various microbial infections on plant. Among virus pathogens, the major contributor is *Papaya Ringspot Virus* (PRSV) which belongs to genus *Potyvirus* of family Potyviridae (Wylie et al. 2017). On the basis of infected host, PRSV is further classified into two pathotypes: PRSV-P, infects the papaya family (Caricaceae) as well as cucurbits (Gonsalves 1998) and PRSV-W, primarily infects watermelons and other cucurbits (Tennant et al. 2007). In this chapter, the disease symptoms, transmission, structural and molecular organization of PRSV- P strain and its detection and disease management strategies for papaya plants have been discussed.

DISEASE SYMPTOMS AND TRANSMISSION OF PRSV

The occurrence of papaya ringspot disease was first noted in 1945 in Hawaii, USA (Lindner et al. 1945). Since then, this virus has been recognized in nearly every region where papaya plant is grown. The disease limits the production of papaya worldwide particularly in Hawaii (USA), Thailand, Taiwan, India, Mexico, Brazil, Bangladesh, Philippines, and southern region of China. In India, the disease has been reported in all papaya growing regions. The mixed infections of PRSV with other viruses such as *Papaya mosaic virus* (PapMV), *Papaya leaf distortion mosaic virus* (PLDMV), *Papaya leaf curl virus* (PaLCuV) have also been reported in papaya plants in Mexico, Philippines, Taiwan, China and India (Noa-Carrazana et al. 2006; Bau et al. 2008; Cruz et al. 2009; Usharani et al. 2013; Shen et al. 2015). Recently, a mixed infection of PRSV with *Lettuce chlorosis virus* and *Tomato yellow leaf curl virus-IL* was detected on papaya plants in Texas, USA (Olufemi et al. 2017).

Papaya ringspot disease is characterized by various symptoms such as severe leaf mosaic, reduction of leaf canopy during disease progression (Figure 1A), stunted growth of plant, water-soaked oily streaks on the petiole and upper part of the trunk, and the most important "ringed spots" on the surface of the fruits from which the disease name "ringspot" was derived. At initial stage, disease symptoms can be seen as yellowing and vein-clearing of the young leaves followed by yellow mottling of the leaves and sometimes severe blistering and leaf distortion. In later stages, characteristics 'shoe-string' formation in the leaves is clearly observed (Figure 1A and 1B). The dark-green streaks and rings may also appear in the leaf-stalks and stems. These consist of concentric rings and spots or C-shaped markings, a darker green than the background-green fruit colour. The symptoms also persist on the ripe fruit as darker orange-brown rings. The symptoms in the mixed infections with other viruses multiple patterns can be observed (Noa-Carrazana et al. 2006; Olufemi et al. 2017) and the order of the virus infections also determines whether it may be antagonistic or synergistic interaction which, in turn, results in symptoms severity accordingly (Chávez-Calvillo et al. 2016). The papaya plants when get infected with virus lose vigor and become stunted. Fruit-set is usually reduced and quality of fruit particularly flavour is adversely affected. Plants get infected with virus before flowering usually does not produce good quality of marketable fruits and thus, in turn, causing a significant impact in the economic trade of papaya.

The PRSV is naturally spread in the field primarily via several species of aphid vectors in a non-persistent manner i.e., virus doesn't replicate itself inside aphids but physically transmitted when aphids feed on plant-sap (Purcifull et al. 1984). Hence, aphid population plays a significant role in the incidence and spread of virus in papaya. More than 20 aphid species including *Myzus persicae, Aphis gossypii, A. craccivora A. citricola, A. rumicis* and *Rhopalosiphum maidis* are known to transmit this virus (Kalleshwaraswamy and Krishna Kumar 2008; Krishna Kumar et al. 2010; Singh and Singh 2010). The virus fails to survive on dead plant material as well as in soil; however, it is transmitted through the sap. Hence, it is transmitted by mechanical activities like pruning and grafting.

Figure 1. Disease symptoms of PRSV infection on papaya plant: (A) Infected plant; (B) Leaves showing various symptoms.

Structural and Genomic Organization of PRSV

PRSV is a flexuous rod shaped virus, without envelope (non-enveloped) with length of about 700 nm - 800 nm and a diameter of about 12 nm -13 nm (Figure 2) (Yeh and Gonsalves 1984a; Tripathi et al. 2008; Muske et al. 2014). The virus particle contains about 94.5% protein and about 5.5% nucleic acid by weight in its composition (Gonsalves et al. 2010). The nucleocapsid contains approximately 2000 copies of capsid protein arranged in a helical symmetry with pitch of 3.4 nm (Tennant et al. 2007).

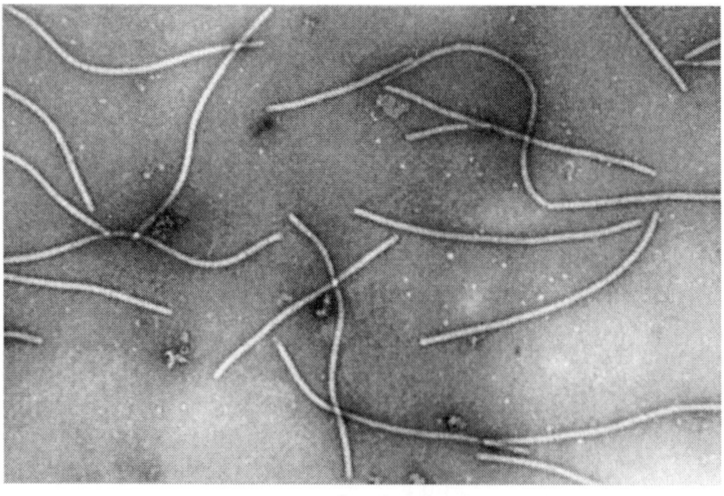

Figure 2. Electron micrograph of Papaya ringspot virus, PRSV-P (© Jorge A.M. Rezende), (CABI, The Plantwise Blog, https://blog.plantwise.org/2017/12/18/update-new-pest-disease-records-15-dec-17.

Table 1. List of reported full genome sequences of PRSV strains infecting papaya plants

S.N.	Virus isolate	GenBank Accession no.	Locality	Genome size (nucleotide bases)
1	PRSV P isolate HN-1	HQ424465	China	10332
2	PRSV isolate PRSV-HNVb	KF791028	China	10326
3	PRSV P isolate HN	EF183499	China	10323
4	PRSV strain Hainan HN-2	KF734962	China	10332
5	PRSV P isolate PRSV-LM	KT633943	China	10325
6	PRSV P isolate DEL	EF017707	India	10317
7	PRSV isolate HYD	KP743981	India	10341
8	PRSV isolate Meghalaya	MF356497	India	10343
9	PRSV isolate Mex-VrPO	AY231130	Mexico	10320
10	PRSV isolate P-5-19	EU882728	Taiwan	10320
11	PRSV strain Leaf-mottling	DQ340771	Taiwan	10326
12	PRSV strain Leaf-mottling and necrosis	DQ340770	Taiwan	10326
13	PRSV strain Leaf-deformation	DQ340769	Taiwan	10326
14	PRSV P isolate prT3-AX-N-Ad6	JX448369	Taiwan	10326
15	PRSV P isolate prT3-AX-N-Ph	JX448370	Taiwan	10326
16	PRSV P isolate prT3-AX-M-Ph	JX448371	Taiwan	10326
17	PRSV P isolate pfT3-AX-D-Ph	JX448372	Taiwan	10326
18	PRSV P isolate pFT3-NP	JX448373	Taiwan	10326
19	PRSV isolate YK	X97251	Taiwan	10326
20	PRSV P, Thailand isolate	AY162218	Thailand	10323
21	PRSV isolate -HA	X67673	USA	10326
22	PRSV isolate PG	EU126128	USA	10320
23	PRSV, HA, Hawaii isolate	S46722	USA	10326
24	PRSV FEV polyprotein gene	KP462721	Brazil	10349
25	PRSV isolate E2	KC345609	France	10343

The genetic material of virus is a single stranded, positive sense RNA molecule contains nucleotide bases varies from 10,317- 10,349 with 3' poly-A tail. It contains single open reading frame encoding for a single polypeptide ranged from 3,341- 3,344 amino acids precursor. A list of reported complete genome of PRSV isolates originating from different countries has been given in Table 1. The polyprotein is further proteolytically processed by virus-encoded proteinases into seven smaller proteins (Figure 3): P1, helper component (HC-Pro), P3, cylindrical inclusion (CI), nuclear inclusion A (NIa), nuclear inclusion B (NIb), coat protein (CP) and two small putative proteins known as 6K1 and 6K2. The possible functions of above encoded proteins have been reported to be involved in virus infection and multiplication in the host plant (Gonsalves et al. 2010). Protein P1

(~63 kDa) is a serine protease involved in cell-to-cell movement. HC-Pro (~52 kDa) is a helper component proteinase involved in vector transmission and also acts as a viral RNA silencing suppressor. The P3 (~41 kDa) involved in virus replication and appears to be significant in symptom development. CI (~72 kDa) is an RNA helicase with ATPase activity and involved in membrane attachment. NIa (48 kDa) is further auto catalytically cleaved into a NIa-Pro (~27 kDa) and VPg (~21 kDa) proteins. The role of VPg is predicted in priming of RNA synthesis during replication of viral genome and possibly packaging of virion. NIa-Pro is a cysteine protease responsible for cleavage of most sites in the polyprotein. NIb (~59 kDa) is an RNA-dependent RNA polymerase. The function of 6K1 protein (6 kDa) is not clearly known. 6K2 protein (6 kDa) having single transmembrane domain is accumulated in host cellular membranes and thought to play a role in forming the replication vesicles of the virus.

The CP is the most important protein of virus and the gene sequence ranged in size from 840 to 873 nucleotide bases which encodes for about 30-35 kDa protein (Bateson et al. 2002; Wei et al. 2007). The CP shows a high genetic diversity region among PRSV isolates (Wei et al. 2007; Tripathi et al. 2008) and plays an important role in RNA encapsidation, vector transmission, pathogenicity and cell-to-cell movement (Gonsalves et al. 2010). The continuous alteration in the capsid structure is main cause of PRSV pathogenicity. The highly mutable capsid protein allows virus to fool the host plant machinery and allows the proceeding of virus infection.

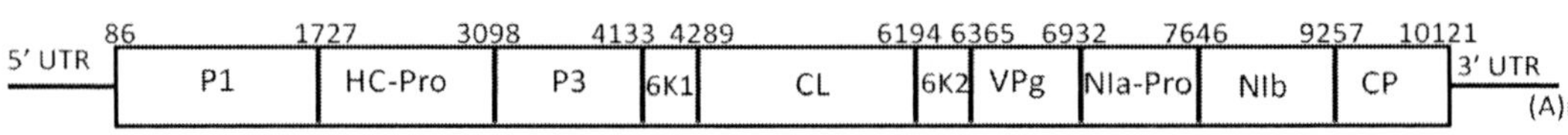

Figure 3. PRSV genome map of Meghalaya isolate (GenBank accession number MF356497). Numbers at the top indicate nucleic acid position. The 5′ and 3′ untranslated regions (UTRs) are marked.

MOLECULAR DETECTION OF PRSV

PRSV is primarily diagnosed by assessment based on appearance of symptoms. The visual diagnosis is, however, quick but it is also unreliable. Because in mixed infection with other viruses such as PaLCuV and PapMV, it may cause indistinguishable symptoms pattern in papaya such as mosaic, yellow-green leaf discoloration and distortion on leaves which makes it very difficult to identify and differentiate without further molecular testing (Noa-Carrazana et al. 2006; Shen et al. 2015; Chávez-Calvillo et al. 2016; Olufemi et al. 2017)

The presence of PRSV can be confirmed by using different molecular diagnostic techniques such as Enzyme-linked Immunosorbent Assay (ELISA), Western blotting, Reverse Transcriptase- Polymerase Chain Reaction (RT-PCR), Real-time RT-PCR, and RT- Loop-mediated Isothermal Amplification (RT-LAMP). Several ELISA test has been reported for rapid detection of PRSV as a quick and reliable technique (Tennant et al. 1994; Kuan et al. 1999; Muske et al. 2014; Shen et al. 2015). Very sensitive and reliable commercial ELISA kits are also available in the market for detection of PRSV in papaya (Agdia, USA; Bioreba, Switzerland). Northern blot technique (Tennant et al. 1994; Noa-Carrazana et al. 2006) and Western blot technique (Tennant et al. 1994; Kertbundit et al. 2007) have also been utilized for the detection of PRSV in papaya. RT-PCR has also extensively been used for sensitive detection of PRSV targeting genomic mRNA especially CP gene (Ruiz-Castro and Silva-Rosales 1997; Chiang et al. 2001; Noa-Carrazana et al. 2006; Kertbundit et al. 2007; Kalleshwaraswamy and Krishna Kumar 2008; Cruz et al. 2009; Usharani et al. 2013; Singh et al. 2017; Wu et al. 2018) and P3 gene (Tuo et al. 2014). Real-time RT-PCR technique involves real-time machine which provides more sensitive detection results for PRSV. Huo et al. (2015) developed real-time RT-PCR based detection of PRSV targeting CI region.

Recently, a novel diagnostic tool RT-LAMP was developed for rapid detection of PRSV (Shen et al. 2014). This method is a one-step, single-tube, reverse transcription and amplification reaction that detects a target RNA sequence with high sensitivity and specificity within 60 min. The detection of amplification product can be observed by the naked eye for turbidity or color change. A novel isothermal amplification method RT-recombinase polymerase amplification (RT-RPA) may be developed and utilised for rapid detection of PRSV within 30 min (Silva et al. 2018). Another method, the dipstick strip for on-site rapid detection of virus can also be developed with suitable specificity to limit the spread of disease.

Management of PRSV Infection

PRSV is causing serious loss to plants in papaya growing countries. To curtail this, different strategies are being used to control as well as minimize the effects of PRSV. These approaches are such as plantation in partially isolated areas and rouging; plantation under screen-houses; use of insecticides and fertilizers; development of resistance in papaya either through conventional approaches or creating virus-resistant transgenic plants.

Rouging and Netting

It is a phytosanitary method to combat the virus infections. Rouging involves removal of the infected plants from cultivated areas. It is a preventive step to avoid the spread of infection. As there is no cure for PRSV disease after the infection of virus on plants, they should be destroyed by cutting them as close to the ground as possible if any symptom appears, and further immediately apply suitable preventive chemicals. This prevents the contamination to other plants. It involves clinical assessment and morphological screening.

However, it is very difficult to control the spreading of PRSV through rouging because it is spread very quickly and effectively by aphids. But PRSV may be controlled by other preventive practices that reduce or delay the spread of the virus within the orchard. These practices include the use of virus-free seedlings to start new crops, planting in partially isolated areas, avoidance of cucurbit plants in or near the orchard, and systematic weed control to reduce the aphid population (Gonsalves 1994).

Papaya plant is not the preferred host of vector aphids. Vector visits and probes papaya plants but do not colonize papaya orchard. Hence, the disease management by controlling the aphids within the papaya orchards is very difficult. However, plantation under screen-house (net-house) can prevent it from landing of aphid vectors on plants and it is very much suitable for small scale producers. This technique was first tried in Taiwan and has produced satisfactory results in managing this disease (Rezende and Costa 1995). Sheen et al. (1998) also reported that growing papaya in net-house to protect the vectors transmitting PRSV was extremely effective.

Use of Insecticides and Antiviral Agents

Another principle management technique to control this disease is the eradication of the vectors. In case of PRSV, the primary vectors or the carriers of the virus are aphids. Removal of aphids from the environment using insecticide can be considered as an effective management strategy. These systemic insecticides viz. carbosulphan, imidacloprid, acetamiprid, *etc.* should be used in low concentrations at nursery stage and at regular concentrations at field stage (Takada 1995). In order to implement successful vector control, frequent preventive sprays of insecticides are required. But, the vectors cannot be successfully controlled by application of insecticide alone.

Some antiviral substances against virus are also currently available in the market such as Inhibitovir (produces by Proveedora Fitozoosanitaria, Mexico), Virus Stop (Fagro S.A., Mexico), Q 2000 VI (Quimcasa, Mexico), and Ekologik, the preventive natural antiviral (marketed by Bioaga Cellular Biology Lab, USA). Hernández et al. (2017) observed the effectiveness of viral inhibitor (Inhibitovir) as a preventive and curative treatment to control PRSV under field conditions. The Inhibitovir prevented the damages caused by PRSV allowing increased yield of 1.5 - 3 times more.

Time of Plantation and Border Cropping

It is essential to analyse and calculate the time of planting papaya to prevent it from biotic losses and for a good yielded harvest. In case of PRSV, planting time in the field can be used as a preventive measure to manage the disease to minimum. Aphids play a significant role in the incidence and spread of PRSV in papaya. Few test bites by an infected aphid is sufficient to spread virus in healthy plants. This spread can be curtailed by adopting cultural management approach like "host escape". It involves growing of plants in the season when there is a minimal population of aphid vectors. If planting can be done from February to April during which population of aphids remain low, the incidence of PRSV will be significantly less as compared to those planted from September to January (Chandrashekar et al. 2015). Chavan et al. (2010) reported that plantation of papaya during comparatively virus free period (July or April) reduces the chances of severe infection of PRSV during the active growth periods of the plants. Plantation during peak vector activity such as December-January should be avoided to decrease the spread of PRSV in the field.

The disease incidence can also be significantly lowered when plantation is done with maize as border crop which may serve as a barrier for alighting winged aphids on papaya (Chandrashekar et al. 2015). It would be beneficial to grow papaya in months having lower aphid vector population such as March, and protecting papaya using maize as border crop.

Use of Fertilizer

The use of balanced fertilizer during transplanting papaya can be helpful in reducing the incidence and severity of PRSV. The best useful fertilization consists of a combination of farmyard manure, castor cake, nitrogen, K_2O and P_2O_5 (Kudada and Prasad 1999; Ray et al. 1999). Effective microorganisms, a mixture of beneficial microoganisms including lactic acid bacteria, yeasts, actinomycetes and photosynthetic bacteria, have also been reported to reduce disease incidence in Taiwan (Tsai 1998). Application of boron through soil and foliar spray at bimonthly intervals along with recommended dose of fertilizers shows significant effect on growth, yield, quality and incidence of PRSV.

Resistance against PRSV

All above methods have limited success to manage PRSV. None of these methods have proven effective completely against PRSV in the field. The developing resistant in the papaya plants against PRSV is effective approach which includes: conventional breeding approach, cross-protection, systemic induced resistance and transgenic approach.

Conventional Breeding Approach

Natural resistance in papaya for PRSV-P has not been reported. However, resistance to PRSV exists in some *Vasconcellea* species which belongs to the same family as papaya. The resistant species are *V. cauliflora*, *V. pubescens, V. quercifolia* and *V. stipulate* (Conover 1964), but these species show varying degree of PRSV-P tolerance in different parts of the world. Intergeneric hybridization in papaya was carried out in many countries by conventional means as well as by embryo rescue and *in vitro* methods. Introgression of tolerant trait from *Vasconcellea quercifolia* to papaya against PRSV has been successfully made by wide crosses (Alamery and Drew 2011; Siar et al. 2011). Intergeneric crosses were also carried out between PRSV resistant wild species *Vasconcellea cauliflora* and papaya variety resulted in the PRSV resistant intergeneric hybrid progenies (Dinesh et al. 2013; Chalak and Hasbanis 2017; Yanthan et al. 2017). However, not much progress has been done in developing fully PRSV resistant papaya variety which can be grown widely.

Cross-Protection

Cross-protection has been used for controlling the papaya ringspot disease in different countries with varying degrees of success. Cross protection is a natural form of pathogen-derived resistance and involves deliberate inoculation of papaya seedlings at the nursery stage with a mild strain of PRSV; the seedlings develop immunity against severe strain of PRSV in the field. For cross-protection, availability of a protective virus strain is essential. Two mild PRSV strains HA 5-1 and HA 6-1 were carefully evaluated for their cross-protection effectiveness under greenhouse and field conditions (Yeh and Gonsalves 1984b). In Hawaii and Taiwan, papaya was infected with mild strain PRSV HA 5-1 which provides protection against some severe local strains, but not found successful for strains of other countries (Yeh and Gonsalves 1984b; Yeh et al. 1988; Tennant et al. 1994; Yeh and Gonsalves 1994).

Sheen et al. (1998) reported a high degree of protection with mild PRSV strain HA 6-1 in papaya against severe strain of PRSV in Taiwan; however, it was unable to completely control PRSV completely. Wang et al. (1987) tested two mild PRSV strains (HA 5-1 and HA 6-1) under greenhouse and field conditions in Taiwan and observed a high degree of protection in papaya against severe effects of two prevalent PRSV strains. McMillan and Gonsalves (1987) performed field trial of papaya protected with mild PRSV strain HA 5-1 in Florida and observed a high degree of protection against severe strain of PRSV.

Despite successful employment of cross-protection in papaya in some countries, the use of this technique is limited due to some reasons such as strain-specific local protection varieties, limited availability of a protective virus and inconsistent results. Resistance in some papaya varieties primarily resulting in a delayed progression of the disease, but in many cases this resistance was too weak or the mild that virus still cause certain disease symptoms.

Systemic Induced Resistance

In a different approach to PRSV resistance, the induction of systemic resistance in papaya plant is also an alternative to suppress and prevent the PRSV disease. Induced systemic resistance (ISR) leads to systemic resistance in susceptible plants, induced by rhizobacteria and higher plant proteins. Systemic resistance-inducing proteins from non-host plants such as CIP-29, CAP-34 and BD-30 from *Clerodendrum inerme*, *C. aculeatum*, and *Boerhaavia diffusa* respectively, have been characterised against PRSV in papaya (Verma and Awasthi 1979; Prasad et al. 1995; Verma et al. 1996; Van Loon et al. 1998; Prasad and Srivastava 2001; Srivastava et al. 2009, Singh et al. 2011; Srivastava et al. 2015). These proteins when sprayed on leaves before virus inoculation may be easily absorbed into the leaves and translocated into the whole plant to induce the production or synthesis of some proteins that are actually antiviral and protect the plant against infection.

Transgenic Approach

The conventional methods of management are not much successful in controlling the PRSV. However, transgenic approach is an efficient and alternate management method which is successfully being utilized to minimize the effect of PRSV. Virus resistant transgenic papaya mainly developed using strategy based on pathogen-derived resistance (PDR). PDR in papaya is utilized in two ways. First, it is based on RNA-mediated resistance in which gene expression is regulated by post-transcriptional gene silencing (PTGS), a host defense response to foreign RNA. Second, the expressed transgene interfere with normal functioning and/or assembly of viral enzyme complexes which leads to inhibit the replication of virus in the cells of transgenic plants. PTGS-mediated transgenic resistance depends on the sequence homology between a transgene and the corresponding viral genome. In PDR, genes from various sources are used to interfere with the replication process of a targeted virus in the papaya plant. So far, PRSV-rcsistant transgenic papaya has been successfully developed mainly through CP gene with PTGS technology.

1. CP-Mediated Virus Resistance

CP-mediated resistance approach to generate PRSV resistant papaya was initiated in late 1980s. However, commercialization of transgenic papaya with CP gene of PRSV was initiated in Hawaii in 1998. 'SunUp' and 'Rainbow' two important cultivars were developed and observed excellent resistance and horticultural qualities, and these cultivars were commercially released in Hawaii in 1998. Transgenic resistance in papaya depends on growth stage, doses of transgene, and transgene homology (Tennant et al. 1994). 'SunUp' shows broader resistance than 'Rainbow', in that 'SunUp' is homozygous for the inserted coat protein gene, while Rainbow is hemizygous (or has half the gene dosage of SunUp).

CP-transgenic resistance of papaya works in a nucleotide-sequence-homology-dependent manner (Kung et al. 2015) which expresses varying levels of resistance against PRSV isolates from other geographical regions. It is generally considered that sequence identity greater than 90% is sufficient for inducing resistance (Moreno et al. 1998; Tripathi et al. 2008). Resistant was successfully achieved at 89% similarity (Bau et al. 2003) and 97 % similarity (Jia et al. 2017). Transgenic papaya cultivar 'SunUp' containing CP gene of mild strain HA 5-1 of PRSV showed resistance to severe strain of PRSV isolated from Hawaii, but not to PRSV isolated from Thailand and other countries (Tennant et al. 1994).

The CP resistant papaya cultivar, 'Rainbow,' is also susceptible to PRSV isolates from Taiwan (Kung et al. 2009). It leads to suggestion that resistance levels depend on the variability among CP genes of PRSV isolates which is highly variable part (Gonsalves 1998; Bau et al. 2003; Davis and Ying 2004). The high levels of genetic divergence in PRSV isolates from Hainan is likely to be the cause of the failure of transgenic papaya lines that targets specific viral CP genes (Zhao et al. 2016). The resistance to PRSV is observed in isolate specific manner, therefore, it must be targeted to all strains from the same geographical region.

Further, an untranslatable CP gene was transferred for the development of a PRSV-resistant papaya variety which showed resistance to homologous PRSV isolates from Hawaii, Australia, Taiwan, Mexico, Jamaica, the Bahamas, and Brazil (Gonsalves et al. 1998). Similarly, researchers in Australia used an untranslatable CP gene region to create two transgenic papaya cultivars (local variety) which shows immunity to PRSV isolates from the local region (Lines et al. 2002). The researchers from Florida also developed transgenic papaya breeding lines using non-translatable CP gene against PRSV isolate H1K (Davis and Ying 2004).

Further, some attempts were tried focusing on broad-spectrum resistance. CP-transgenic papaya lines was developed using CP gene of PRSV YK, a severe mosaic strain from Taiwan and these lines were having great potential for use in Taiwan and other geographical areas to control PRSV (Bau et al. 2003; Bau et al. 2004). Jia et al. (2017) developed transgenic papaya plants using conserved CP region of the PRSV to develop broad spectrum resistance and this approach has a useful appliance against PRSV in the major papaya growing area of Hainan, China. However, breakdown of CP-transgenic PRSV resistance is a major concern facing in papaya cultivation, since genetically distinct strains of PRSV have been identified throughout the world (Gonsalves 2006).

2. Replicase Gene (RP)-Mediated Virus Resistance

The replicase protein gene is the second most widely used transgene against virus and this strategy is known as replicase protein-mediated resistance (RPMR). It is reviewed that expression of replicase-derived sequences at the protein level interferes with normal functioning and/or assembly of viral replicase enzyme complexes which leads to inhibit the replication of virus in the transgenic plants. Transgenic papaya plants with mutated replicase (RP) gene of PRSV was developed which showed high resistance or immunity against PRSV in the field (Wei et al. 2007), and not a single loss of resistance was observed in the transgenic papaya plants about 5 to 6 years with RP gene (Tecson Mendoza et al. 2008). 'Huanong No. 1', a PRSV resistant transgenic papaya cultivar expressing a replicase gene is used for commercial production from southern China, instead of the CP gene (Azad et al. 2014; Guo et al. 2009). However, an example of resistance breakdown in 'Huanong No. 1' has been observed in recent years (Zhao et al. 2015).

Apart from above two genes, another gene HC-Pro has also been tried in transgenic development. In 2015, transgenic papaya was developed in Taiwan using untranslatable constructs of HC-Pro from a highly virulent PRSV strain. The plants were resistant to the newly emerging, more virulent virus strains and recombinant strains known to break CP-mediated resistance in transgenic papaya (Kung et al. 2015).

Further, a novel CRISPR/Cas9 technology can also expedite the development of virus-resistant papaya either by targeting the virus directly, or editing onc or more host genes essential for virus replication. CRISPR/Cas9-mediated mutations of host plants can potentially confer resistance against many invading viruses (Green and Hu 2017).

Eventually, the development of PRSV-resistant lines is generally considered to be the best strategy for efficient PRSV disease control in papaya for long-term protection. So far, complete resistance to PRSV strains has not been observed in any known papaya cultivars. Hence, papaya cultivars with resistance to a broad-spectrum of PRSV strains and focusing individually for various papaya growing regions are urgently needed by the papaya industry. The attempt can also be tried by considering transgenic papaya as an important component of Integrated Pest Management (IPM) which might improve its potential of resistance against PRSV.

CONCLUSION

PRSV is the most significant virus limiting the papaya production worldwide. Some newly emerged virulent PRSV strains or the increased virulence of existing strains has caused severe disease outbreaks among resistant cultivars in many papaya-growing areas. Several conventional strategies have been adopted to minimise the effect of this disease which has been proved to be successful to some extent. But developing PRSV-resistant transgenic plants are still the most effective means of controlling this disease which shows comparatively higher efficacy. Transgenic resistance may also be combined with appropriate elements of an IPM program to maximise the disease combating effect. In future, CRISPR/Cas9 technology may lead to the development of durable, broad-spectrum, PRSV-resistant transgenic cultivars.

REFERENCES

Adam, Y., Nasaruddin, A. A., Zuraini, A., Arifah, A. K., Omar Fauzee, M. S., Zakaria, Z. A., Somchit, M. N. (2013) Diuretic activity of roots from *Carica papaya* L. and *Ananas comosus* L. *Int J Pharm Sci Rev Res* 23: 163-167.

Ahmad, N., Fazal, H., Ayaz, M., Abbasi, B. H., Mohammad, I., Fazal, L. (2011) Dengue fever treatment with *Carica papaya* leave extract. *Asian Pac J Trop Biomed* 1: 330-333.

Alamery, S., Drew, R. A. (2011) Studies on the genetics of PRSV-P resistance genes in intergeneric hybrids between *Carica papaya* and *Vasconcellea quercifolia*. International Society of Horticultural Sciences; Third International Symposium on Papaya, Chiang Mai, Thailand, pp. 16.

Amri, E., Mamboya, F. (2012) Papain, a plant enzyme of biological importance: a review. *Am J Biochem Biotechnol* 8: 94-104.

Azad, M. A., Amin, L, Sidik, N. M. (2014) Gene technology for *Papaya ringspot virus* disease management. *Sci World J* 2014: 1-11.

Bateson, M. F., Lines, R. E., Revill, P., Chaleeprom, W., Ha, C. V., Gibbs, A. J., Dale, J. L. (2002) On the evolution and molecular epidemiology of the potyvirus *Papaya ringspot virus*. *J Gen Virol* 83: 2575-2585.

Bau, H. J., Cheng, Y. H., Yu, T. A., Yang, J. S., Yeh, S. D. (2003) Broad-spectrum resistance to different geographic strains of *Papaya ringspot virus* in coat protein gene transgenic papaya. *Phytopathology* 93: 112-120.

Bau, H. J., Cheng, Y. H., Yu, T. A., Yang, J. S., Liou, P. C., Hsiao, C. H., Lin, C. Y., Yeh, S. D. (2004) Field evaluation of transgenic papaya lines carrying the coat protein gene of *Papaya ringspot virus* in Taiwan. *Plant Dis* 85: 594-599.

Bau, H. J., Kung, Y. J., Raja, J. A., Chan, S. J., Chen, K. C., Chen, Y. K., Wu, H. W., Yeh, S. D. (2008) Potential threat of a new pathotype of *Papaya leaf distortion mosaic virus* infecting transgenic papaya resistant to *Papaya ringspot virus*. *Phytopathology* 98: 848-856.

Chandrashekar, K., Chavan, V. M, Sharma, S. K. and Bhosle, A. B. (2015) Management of PRSV-P in papaya through time of planting and border cropping. *Indian J Hort*. 72:423-425.

Chávez-Calvillo, G., Contreras-Paredes, C. A., Mora-Macias, J., Noa-Carrazana, J. C., Serrano-Rubio, A. A., Dinkova, T. D., Carrillo-Tripp, M., Silva-Rosales, L. (2016) Antagonism or synergism between *Papaya ringspot virus* and *Papaya mosaic virus* in *Carica papaya* is determined by their order of infection. *Virology* 489: 179-191.

Chalak, S. U., Hasbnis, S. N. (2017) Study on resistance to *Papaya ringspot virus* (PRSV) in intergeneric hybrid population of papaya C V Washington (*Carica papaya* L.) and *Vasconcellea cauliflora*. *J Pharmacogn Phytochem* 6: 653-657.

Chavan, V. M., Tomar, S. P. S., Dhale, M. G. (2010) Management of *Papaya Ring Spot Virus* (PRSV-P) of papaya under Pune conditions. *Acta Hort* 851: 447-452.

Chiang, C. H., Wang, J. J., Jan, F. J., Yeh, S. D., Gonsalves, D. (2001) Comparative reactions of recombinant papaya ringspot viruses with chimeric coat protein (CP) genes and wild-type viruses on CP-transgenic papaya. *J Gen Virol* 82: 2827-2836.

Chukwuka, K. S., Iwuagwu, M., Uka, U. N. (2013) Evaluation of nutritional components of *Carica papaya* L. at different stages of ripening. *IOSR-JPBS* 6: 13-16.

Conover, R. A. (1964) Distortion ringspot, a severe virus disease of papaya in Florida. *Proc Fla State Hort Soc* 89: 440-444.

Cruz, F. C. S., Tanada, J. M., Elvira, P. R. V., Dolores, L. M., Magdalita, P. M., Hautea, D. M., Hautea, R. A. (2009) Detection of mixed virus infection with *Papaya ringspot virus* (PRSV) in papaya (*Carica papaya* L.) grown in Luzon, Philippines. *Philipp J Crop Sci* 34: 62-74.

Davis, M. J., Ying, Z. (2004) Development of papaya breeding lines with transgenic resistance to *Papaya ringspot virus*. *Plant Dis* 88: 352-358.

Dinesh, M. R., Veena, G. L., Vasugi, C., Krishna, M. R., Ravishankara, K. V. (2013) Intergeneric hybridization in papaya for 'PRSV' tolerance. *Sci Hort* 161: 357-360.

FAOSTAT Website (http://www.fao.org/faostat/en/#rankings/countries_by_commodity, accessed on 04 August 2018)

Gonsalves, D. (1994) *Papaya ringspot virus*. *In: Compendium of Tropical Fruit Diseases* (Eds. Ploetz, R. C., Zentmyer, G. A., Nishijima, W. T., Rohrbach, K. G., Ohr, H. D.), APS Press, Minnesota, USA, pp. 67-68.

Gonsalves, D. (1998) Control of *Papaya ringspot virus* in papaya: A case study. *Annu Rev of Phytopathol* 36: 415-437.

Gonsalves, C., Cai, W., Tennant, P. F., Gonsalves, D. (1998) Effective development of *Papaya ringspot virus* resistant papaya with untranslatable coat protein gene using a modified microprojectile transformation method. *Acta Hort* 461: 311-319.

Gonsalves, D., (2006) Transgenic papaya: development, release, impact and challenges. *Adv Virus Res* 67: 317-354.

Gonsalves, D., Tripathi, S., Carr, J. B., Suzuki, J. Y. (2010) *Papaya Ringspot virus*. *The Plant Health Instructor*. DOI: 10.1094/PHI-I-2010-1004-01.

Green, J. C., Hu, S. J. (2017) Editing plants for virus resistance using CRISPR-Cas. *Acta Virol* 61: 138-142.

Guo, J., Yang, L., Liu, X., Guan, X., Jiang, L., Zhang, D. (2009) Characterization of the exogenous insert and development of event-specific PCR detection methods for genetically modified Huanong No. 1 papaya. *J Agric Food Chem* 57: 7205-7212.

Hernández, R., Guillen, D., Pérez, M., Casanova, E. (2017) Viral inhibitors to control the *Papaya ringspot virus* on *Carica papaya*. *Cien Inv Agr* 44: 312-319.

Huo, P., Shen, W. T., Yan, P., Tuo, D. C., Li, X. Y., Zhou, P. (2015) Simultaneous detection of *Papaya ringspot virus*, *Papaya leaf distortion mosaic virus*, and *Papaya mosaic virus* by multiplex real-time reverse transcription PCR. *Acta Virol* 59: 380-388.

Jia, R., Zhao, H., Huang, J., Kong, H., Zhang, Y., Guo, J., Huang, Q., Guo, Y., Wei, Q., Zuo, J., Zhu, Y. J., Peng, M., Guo, A. (2017) Use of RNAi technology to develop a PRSV-resistant transgenic papaya. *Sci Rep* 7: 12636.

Juárez-Rojop, I., Díaz-Zagoya, J. C., Ble-Castillo, J. L., Miranda-Osorio, P., Castell-Rodríguez, A. E., Tovilla-Zárate, C. A., Rodríguez-Hernández, A., Aguilar-Mariscal, H., Ramón-Frías, T., Bermúdez-Ocaña, D. Y. (2012) Hypoglycemic effect of *Carica papaya* leaves in streptozotocin-induced diabetic rats. *BMC Comp Alt Med* 12: 236.

Kalleshwaraswamy, C. M., Krishna Kumar, N. K. (2008) Transmission efficiency of *Papaya ring spot virus* by three aphid species. *Phytopathology* 98: 541-546.

Kertbundit, S., Pongtanom, N., Ruanjan, P., Chantasingh, D., Tanwanchai, A., Panyim, S., Juříček, M. (2007) Resistance of transgenic papaya plants to *Papaya ringspot virus*. *Biol Plantarum* 51: 333-339.

Kovendan, K., Murugan, K., Panneerselvam, C., Aarthi, N., Kumar, P., Subramaniam, J., Amerasan, D., Kalimuthu, K., Vincent, S. (2012) Anti-malarial activity of *Carica papaya* (Family: Caricaceae) leaf extract against *Plasmodium falciparum*. *Asian Pacific J Trop Dis* 2: 306-311.

Krishna Kumar, N. K., Singh, H. S., Kalleshwaraswamy, C. M. (2010) Aphid (Aphididae: Homoptera) Vector of *Papaya ring spot virus* (PRSV), bionomics, transmission efficiency and factors contributing to epidemiology. *Acta Hort* 851: 431-43.

Kuan, C. P., Chen, K. H., Su, H. J. (1999) Serological characterization of *Papaya ringspot virus* isolates in Taiwan. *Bot Bull Acad Sin* 40: 231-236.

Kudada, N., Prasad, S. M. (1999) Effect of planting time on the incidence of *Papaya ringspot virus* disease and yield of fruits. *Indian Phytopathol* 52: 224-227.

Kung, Y. J., Bau, H. J., Wu, Y. L., Huang, C. H., Chen, T. M., Yeh, S. D. (2009) Generation of transgenic Papaya with double resistance to *Papaya ringspot virus* and *Papaya leaf-distortion mosaic virus*. *Phytopathology* 99: 1312-1320.

Kung, Y. J., You, B. J., Raja, J. A. J., Chen, K. C., Huang, C. H., Bau, H. J. (2015) Nucleotide sequence-homology-independent breakdown of transgenic resistance by more virulent virus strains and a potential solution. *Sci Rep* 5: 9804.

Lindner, R. C., Jensen, D. D., Ikeda, W. (1945) Ringspot: new papaya plunderer. Hawaii Farm and Home. 8: 10-14.

Lines, R., Persley, D., Dale, J., Drew, R., Bateson, M. (2002) Genetically engineered immunity to *Papaya ringspot virus* in Australian papaya cultivars. *Mol Breed* 10: 119-129.

Maisarah, A. M., Nurul, Amira, B., Asmah, R., Fauziah, O. (2012) Anti-oxidant analysis of different parts of *Carica papaya*. *Int Food Res J* 20: 1043-1048.

McMillan Jr, R. T., Gonsalves, D. (1987) Effectiveness of cross protection by a mild mutant of *Papaya ringspot virus* for control of ringspot disease of papaya in Florida. *Proc Fla State Hort Soc* 100: 294-296.

Moreno, M., Bernal, J. J., Jimenez, I., Rodriguez-Cerezo, E. (1998) Resistance in plants transformed with the P1 or P3 gene of tobacco vein mottling potyvirus. J Gen Virol 79:2819-2827

Muske, D. N., Peter, A, Swetha, P. S., Jingade, P., Satish., K. K. (2014) Molecular and serological detection of *Papaya ringspot virus* infecting papaya (*Carica papaya*). *J Pl Dis Sci* 9: 8-15.

Muss, C., Mosgoeller, W., Endler, T. (2013) Papaya preparation (Caricol ®) in digestive disorders. *Neuro Endocrinol Lett* 34: 38- 46.

Nayak, S. B., Pinto Pereira, L., Maharaj, D. (2007) Wound healing activity of *Carica papaya* L. in experimentally induced diabetic rats. *Indian J Exp Biol* 45: 739-743.

Noa-Carrazana, J. C., González-de-León, D., Ruiz-Castro, B. S., Piñero, D., Silva-Rosales, L. (2006) Distribution of *Papaya ringspot virus* and *Papaya mosaic virus* in papaya plants (*Carica papaya*) in Mexico. *Plant Dis* 90: 1004-1011.

Okeniyi, J. A., Ogunlesi, T. A., Oyelami, O. A., Adeyemi, L.A. (2007) Effectiveness of dried *Carica papaya* seeds against human intestinal parasitosis: A pilot study. *J Med Food* 10: 194-196.

Olufemi, J. A., Al Rwahnih, M., Jifon, J. L., Sétamou, M., Brown J. K., Gregg, L., Park, J. W. (2017) A mixed infection of *Lettuce chlorosis virus*, *Papaya ringspot virus*, and *Tomato yellow leaf curl virus*-IL detected in a Texas papaya orchard affected by a virus-like disease outbreak. *Plant Dis* 101: 1094-1102.

Otsuki, N., Dang, N. H., Kumagai, E., Kondo, A., Iwata, S., Morimoto, C. (2010) Aqueous extract of *Carica papaya* leaves exhibits anti-tumor activity and immunomodulatory effects. *J Ethnopharmacol* 127: 760-767.

Pandey, S., Peter, J., Cabot, P., Nicholas, S., Amitha, K. H. (2016) Anti-inflammatory and immunomodulatory properties of *Carica papaya*. *J Immunotoxicol* 13: 590-602.

Prasad, V., Srivastava, S. (2001) Inducible mechanisms of plant resistance to virus infection. *J Plant Biol* 28: 1-11.

Prasad, V., Srivastava, S., Varsha, Verma, H. N. (1995) Two basic proteins isolated from *Clerodendrum inerme* Gaertn. are inducers of systemic antiviral resistance in susceptible plants. *Plant Sci* 110: 73-82.

Purcifull, D. E., Edwardson, J. R., Hiebert, E., Gonsalves, D. (1984) *Papaya ringspot virus*. CMI/AAB Descriptions of plant viruses. 292: 8.

Ravikant, T., Nishant, G., Shashipal, S., Samriti, T., Kumar, T. R., Vikas, V., Dishant, S. (2012) Antihypertensive effect of ethanolic extract of Indian *Carica papaya* L. root bark (Caricaceae) in renal artery occluded hypertensive rats. *Int J Pharm Clin Res* 4: 20-23.

Ray, P. K., Yadav, J. P., Kumar, A. (1999) Effect of transplanting dates and mineral nutrition on yield and susceptibility of papaya to ringspot virus. *Hort J* 12: 15-26.

Rezende, J. A. M., Costa, A. S. (1995) Alternatives for integrated control of *Papaya ringspot virus*. *Acta Hort* 370: 129-132.

Ruiz Castro, S., Silva, R. L. (1997) Use of RT-PCR for Papaya ringspot detection in papaya plants from Veracruz, Tabasco and Chiapas. *Rev Mex Fitopatol* 15: 86-90.

Sheen, T. F., Wang, H. L., Wang, D. N. (1998) Control of *Papaya ringspot virus* by cross protection and cultivation techniques. *J Japan Soc Hort Sci* 67: 1232-1235.

Shen, W., Tuo, D., Yan, P., Yang, Y., Li, X., Zhou, P. (2014) Reverse transcription loop-mediated isothermal amplification assay for rapid detection of *Papaya ringspot virus*. *J Virol Methods* 204: 93-100.

Shen, W., Tuo, D., Yang, Y., Yan, P., Li, X., Zhou, P. (2015) First report of mixed infection of *Papaya ringspot virus* and *Papaya leaf distortion mosaic virus* on *Carica papaya* L. *J Plant Pathol* 96: 113-131.

Siar, S. V., Beligan, G. A,. Sajise, A. J. C., Villegas, V. N., Drew, R. A. (2011) *Papaya ringspot virus* resistance in *Carica papaya* via introgression from *Vasconcellea quercifolia*. *Euphytica* 181: 159-168.

Silva, G., Oyekanmi, J., Nkere, C. K., Bömer, M., Kumar, P. L., Seal, S. E. (2018) Rapid detection of potyviruses from crude plant extracts. Anal Biochem 546:17–22.

Singh, O., Ali, M. (2011) Phytochemical and antifungal profiles of the seeds of *Carica papaya* L. *Indian J Pharm Sci* 73: 447-451.

Singh, S., Awasthi, L. P., Singh, R. K. (2011) Induction of systemic resistance through antiviral agents of plant origin against papaya ring spot disease (*Carica papaya* L.). *Arch Phytopathol Plant Protect* 44: 1676-1682.

Singh, S., Jangre, A., Kumar, P., Awasthi, L. P. (2017) Studies on the molecular variability in Indian isolates of *Papaya ringspot virus*. *Virol Res J* 1: 10-16.

Srivastava, A., Srivastava, S., Prasad, V. (2015) Systemic antiviral resistance induced in papaya by CAP-34, a resistance inducing protein from *Clerodendrum aculeatum*, is associated with a proteinaceous virus inhibitory activity. *J Plant Pathol* 97: 45-54.

Srivastava, A., Trivedi, S., Krishna, S. K., Verma, H. N., Prasad, V. (2009) Suppression of *Papaya ringspot virus* infection in *Carica papaya* with CAP-34, a systemic antiviral resistance inducing protein from *Clerodendrum aculeatum*. *Eur J Plant Pathol* 123: 241-246.

Singh, V., Singh, D. (2010) Studies on natural transmission of *Papaya ringspot virus* disease in eastern Uttar Pradesh. *Ann Pl Protec Sci* 18: 188-192.

Takada, H. (1995) IPM of vector aphids. Japan (JP): Kyoto Prefectural University.

Tennant, P. F., Fermin, G. A., Roye, R. E. (2007) Viruses infecting papaya (*Carica papaya* L.): etiology, pathogenesis, and molecular biology. *Plant Viruses* 1: 178-188.

Tennant, P. F., Gonsalves, C., Ling, K. S., Fitch, M., Manshardt, R., Slightom, J. L., Gonsalves, D. (1994) Differential protection against *Papaya ringspot virus* isolates in coat protein gene transgenic papaya and classically cross-protected papaya. *Phytopathology* 84: 1359-1366.

Tecson Mendoza, E. M., Laurena, A. C., Botella, J. R. (2008) Recent advances in the development of transgenic papaya technology. *Biotechnol Annu Rev* 14: 423-462.

Tripathi, S., Suzuki, J. Y., Ferreira, S., Gonsalves, D. (2008) Papaya ringspot virus P: characteristics, pathogenicity, sequence variability and control. *Mol Plant Pathol* 9: 269-280.

Tsai, C. T. (1998) E. M. application protected papaya culture. *J Chinese Soc Hort Sci* 44: 93-101.

Tuo, D., Shen, W., Yang, Y., Yan, P., Li, X., Zhou, P. (2014) Development and validation of a multiplex reverse transcription PCR assay for simultaneous detection of three papaya viruses. *Viruses* 6: 3893-3906.

Udoh, P., Essien, I., Udoh, F. (2005) Effects of *Carica papaya* (paw paw) seeds extract on the morphology of pituitary-gonadal axis of male Wistar rats. *Phytother Res* 19: 1065-1068.

Usharani, T. R., Laxmi, V., Jalali, S., Krishnareddy, M. (2013) Duplex PCR to detect both *Papaya ring spot virus* and *Papaya leaf curl virus* simultaneously from naturally infected papaya (*Carica papaya* L.). *Indian J Biotechnol* 12: 269-272.

Van Loon, L. C., Bakker, P. A. H. M., Pieterse, C. M. J. (1998) Systemic resistance induced by rhizophere bacteria. *Annu Rev Phytopathol* 36: 453-483.

Verma, H. N., Awasthi, L. P. (1979) Antiviral activity of *Boerhaavia diffusa* root extract and the physical properties of the virus inhibitor. *Canad J Bot* 57: 926-932.

Verma, H.N., Srivastava, S., Varsha Kumar, D. (1996) Induction of systemic resistance in plants against viruses by a basic protein from *Clerodendrum aculeatum* leaves. *Phytopathology* 86: 485-492.

Wang, H. L., Yeh, S. D., Chiu, R. J., Gonsalves, D. (1987) Effectiveness of cross-protection by mild mutants of *Papaya ringspot virus* for control of ringspot disease of papaya in Taiwan. *Plant Dis* 71: 491-497.

Wei, J. Y., Liu, D. B., Li, X. Y., Zhou, P. (2007) Variation in the coat protein gene of *Papaya ringspot virus* isolates from multiple locations of China. *J Integr Plant Biol* 49: 1062-1069.

Wei, X., Lan, C., Lu, Z., Ye, C. (2007) Analysis on virus resistance and fruit quality for T4 generation of transgenic papaya. *Front Biol China* 2: 284-290.

Wu, Z., Mo, C., Zhang, S., Li, H. (2018) Characterization of *Papaya ringspot virus* isolates infecting transgenic papaya 'Huanong No.1' in South China. *Sci Rep* 8: 8206.

Wylie, S. J., Adams, M., Chalam, C., Kreuze, J., López-Moya, J. J., Ohshima, K., Praveen, S., Rabenstein, F., Stenger, D., Wang, A., Zerbini, F. M. (2017) ICTV Report Consortium, ICTV Virus Taxonomy Profile: Potyviridae. *J Gen Virol* 98: 352-354.

Yahaya A, Ali M, Idris A (2017) Antibacterial activity and phytochemical screening of *Carica papaya* on some enteric bacterial isolates of public health importance. *Greener J Biol Sci* 7: 001-007.

Yanthan, J. L., Vasugi, C., Dinesh, M. R., Reddy, M. K., Das, R. (2017) Evaluation of F6 intergeneric population of papaya (*Carica papaya* L) for resistance to *Papaya ring spot virus* (PRSV). *Int J Curr Microbiol App Sci* 6: 289-298.

Yeh, S. D., Gonsalves, D. (1984a) Purification and immunological analysis of cylindrical inclusion protein induced by *Papaya ringspot virus* and *Watermelon mosaic virus*-1. *Phytopathology* 74: 1273-1278.

Yeh, S. D., Gonsalves, D. (1984b) Evaluation of induced mutants of *Papaya ring spot virus* for control by cross protection. *Phytopathology* 74: 1086-1091.

Yeh, S. D., Gonsalves, D. (1994) Practices and perspective of control of *Papaya ringspot virus* by cross protection. In: *Advances in Disease Vector Research* (Eds. Harris KF), Springer, New York, USA, pp. 237-257.

Yeh SD, Gonsalves D, Wang H.W, Namba R, Chiu RJ (1988) Control of *Papaya ringspot virus* by cross protection. *Plant Dis* 72: 375-380.

Zhao, H., Jia, R. Z., Zhang, Y. L., Zhu, Y. J., Zeng, H. C., Kong, H., McCafferty, H., Guo, A. P., Peng, M. (2016) Geographical and genetic divergence among *Papaya ringspot virus* populations within Hainan province, China. *Phytopathology* 106: 937-944.

Zhao, G., Yan, P., Shen, W., Tuo, D., Li, X., Zhou, P. (2015) Complete genome sequence of *Papaya ringspot virus* isolated from genetically modified papaya in Hainan island, China. *Genome Announc* 3: 01056-15.

In: Innovations in Life Science Research ISBN: 978-1-53615-868-7
Editors: R. P. Sinha, S. Pandey-Rai and N. Ghoshal ©2019 Nova Science Publishers, Inc.

Chapter 11

WITHANIA COAGULANS: AN OVERVIEW ON THERAPEUTIC POTENTIAL AND USE OF RECENT BIOTECHNOLOGICAL APPROACHES FOR CONSERVATION AND ENHANCEMENT OF SECONDARY METABOLITES

Deepika Tripathi and Shashi Pandey-Rai*
Laboratory of Morphogenesis, Centre of Advanced Study in Botany,
Institute of Science, Banaras Hindu University,
Varanasi, Uttar Pradesh, India

ABSTRACT

Natural herbal products represent goldmine specific therapeutic bio-molecules containing pharmaceutical agents to prevent various human health problems.Presently worldwide ~ 80% global population utilizes aromatic plants for their therapeutic uses. In the last few decades, *Withania coagulans* has attained much popularity for seeing its importance as pharmaceutical product. It is well reputed as hypoglycemic agent and distributed throughout the temperate and tropical regions of Asia. It has many other pharmaceutical properties like anticancer, antilipidemic, immunosuppressive, etc. Chemically *W. coagulans* saturated with unique type of triterpenes chemical group which are known as withanolides. However, indiscriminate use of natural resources overtime and habitat fragmentation may pose a serious threat to their survival. Therefore, currently it is found much attention through scientific communities for sustainable availability in nature, higher production of withanolides content and enhance the efficiency of herbal product through biotechnology as well as nanotechnology. This chapter is focusing on botanical features, bioactive compounds and therapeutic potential of *W. coagulans* with the future point of view.

* Corresponding Author's Email: shashi.bhubotany@gmail.com.

Keywords: *Withania coagulans*, withanolides, hypoglycemic, endangered plant

Introduction

At the present scenario, the plants covered approximately 30% land area of earth at global level. But about centenary years ago this percentage was found higher than now and peoples were highly dependent upon plant systems for several benefits like- food, fodder, and medicines. From the last several years, indigenous Ayurvedic literatures were found fully saturated with plants benefit as pharmaceutical products and life saving sources for humans. In literatures, *Withania coagulans* Dunal (family Solanaceae) is well known important medicinal plant with very high hypoglycemic potential. It is the only herb of genus *Withania*, having milk coagulating property due to the presence of an enzyme 'withanin' and therefore known as 'Punirbundh' or 'vegetable rennet'. It occurs in dry to hot climatic conditions and stony places up to 1,700 feet height. In phyto-chemical screening of *W. coagulans*, withanolides (steroidal lactones) are present as main bioactive compounds and related to its important pharmaceutical properties. Other than withanolides, flavonoides, alkaloids, and fatty acids (linoleic, palmitic, stearic acid, etc.) have been also mentioned in chemical profiling. Coagulinolides are specific chemical groups which contain 14, 20-epoxide bridge, only present in *W. coagulans*. With the high potential as anti-diabetic/hypoglycemic agent, *W. coagulans* also work very strongly against hypercholesterolemia, cancer, various kinds of microbial infections, inflammations and different other health problems (Budhirajaet al. 1977; Mauryaet al. 2008). It has been mentioned in traditional literatures as folk medicine for the cure of tooth ache, bronchitis, ulcer, rheumatism and degenerative diseases. It is presently listed under critically endangered plants category due to the over exploitation for medicinal purposes and as fodder for animals, etc. (Thomas and Hoshino 2016). Destruction of natural habitat, low seed germination rate, reproductive failure, unisexual flowers and poor seed germination are major problems in its mass propagation and also in enhancement of bioactive compounds (Tripathiet al. 2018; Rathore and Kheni 2017). Therefore, in the last few decades, scientists have been focused on various biotechnological approaches for conservation and sustainable production of *W. coagulans* plant. To achieve the goal of demand, several counterstrikes were taken for last few decades to develop tissue culture systems of *W. coagulans* using nodes, leaf discs and inter-nodal segments, shoot tips as explants, seedling derived direct and indirect organogenesis. This chapter is based upon botanical features, bioactive compounds and pharmaceutical activities with the recent advancement of biotechnological approaches related to sustainable conservation and production of *W. caogulans*.

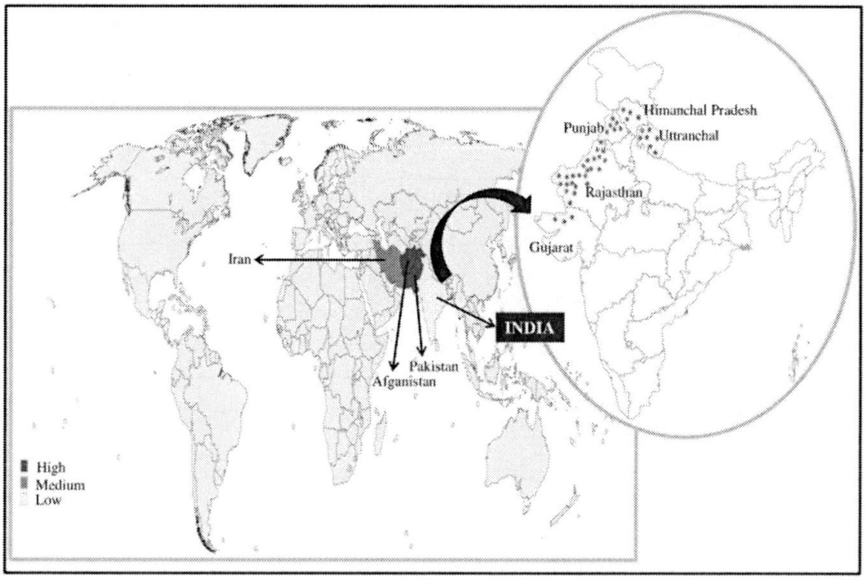

Figure 1. Distribution of *Withania coagulans* Dunal.

Origin and Distribution

In the nature among 23 known species of *Withania*, especially two species *Withania somnifera* and *Withania coagulans* are medicinally/economically important and used since ancient times for medicinal purposes. Although *W. somnifera* is widely distributed plant, while *W. coagulans* is available very less frequently or in low abundance. *W. coagulans* is a native of the Asia temperate and Asia tropical regions; and majorly distributed in Iran, Afghanistan, Pakistan and India. On the Indian map it mainly presents in North West India, Shimla, Punjab, Gujarat, Garhwal and Kumaon region (Figure 1).

Scientific Classification and Local Names

Kingdom	**Plantae**
Division	Angiosperms
Class	Dicotyledon
Order	Tubiflorae
Family	Solanaceae
Genus	*Withania*
Species	*coagulans*

Most commonly *W. coagulans* is known as 'vegetable rennet', because its fruits have milk coagulating property. It is also known by various other names in different languages (Table 1).

Table 1. Various vernacular names of *Withania coagulans* Dunal

Languages	Synonyms
English	Indian cheese maker, Vegetable rennet
Hindi	Akri, punir, paneerdoda
Urdu	Paneer band, Tukme-Hayath
Sanskrit	Rishyagandha
Bengali	Ashwagandha
Punjabi	Spin bajja, panir, Khamjira
Telugu	Panneru-gadda
Gwalior	Asgandha
Arabic	Javzulmizaja, Kaknajehindi
Sindhi	Punirjafota
Chinese	Ninggushuiqie
Tamil	Ammukkura
Latin	*Withania coagulans* Dunal

BOTANICAL FEATURES AND DESCRIPTION

W. coagulans is a perennial under shrub that grows from 24 to 36 inches in height. Stem is rigid gray and leaves are lanceolate- oblong, obtuse, thick and acute. The flowers are yellowish dioecious in axillary cluster. The type of fruit is berry, which is 6-8 mm in size, globose, red and enclosed by leathery calyx. The seeds are 2.5 - 3 mm dark brown in color, ear shaped, glabrous with fast fruity odour (Figure 2). Other morphological features are listed under Table 2.

Figure 2. *Withania coagulans* plant, leaf and fruits.

Table 2. Morphological features of *Withania coagulans* Dunal

Plant parts	Characters
Vegetative Characteristics:	
Plant type	Small shrub, 24- 36 inch approximate height
Stem	Rigid, grayish yellow, stellate tomentose
Leaves	Lanceolate oblong, obtuse, entire, base acute, 1-2.2 cm in size, petiolate
Floral Characteristics:	
Flower	Dioecious, yellow, pedicle 0-6mm long, pubescent
Inflorescence	Axillary cluster
Calyx	Campanulate, gamosepalous with 5 sepals, clothed with fine stellate grey tomentum
Corolla	Campanulate, greenish yellow petals
Androecium	Stamens 5, filament thin and straight, 0.4- 0.5cm long
Gynoecium	Ovary ovoid, style glabrous, stigma Mushroom shaped, 2 lamellate
Fruit	Barries, globose, red in color, enclosed in leathery calyx, 6-8 mm in diameter
Seeds	Dark brown, ear shaped, having nauseous fruity odour
Flowering period	November - March

ETHANOMEDICINAL APPLICATIONS

In traditional indigenous medicinal system *W. coagulans* is either used as single plant product or with a combination of some other herbal products. Different parts of this plantare used in various pharmaceutical purposes, which are mention in Table 3. Local and regional peoples of India use stem and twig for tooth cleaning and their smoke in relief against tooth ache since ancient times. *W. coagulans* has been used as folk medicine against small pox in widely occurring zone. Fruits are widely used for respiration problems, asthma; seeds used in the reduction of piles inflammation and kidney problems. Fruits/berries also alternatively used for blood purification. Traditional healers used it to cure of hyperglycemic and hyper-lipidemic conditions. Leaf extract with some other herbal drugs used as a tonic and some local regional tribes used it against snake venom.

Table 3. Different plant parts of *Withania coagulans* and their medicinal importance

Plant Parts	Uses
Twig	Tooth cleaning and toothache
Leaves	Antifungal and antibacterial activity
Roots	Antidiabetic
Flower	Antidiabetic
Fruits (Berries)	Blood purification, Hepatoprotective, anti-inflammatory and antidiabetic
Seeds	Milk coagulative property, anti-diabetic and anti-lipidemic

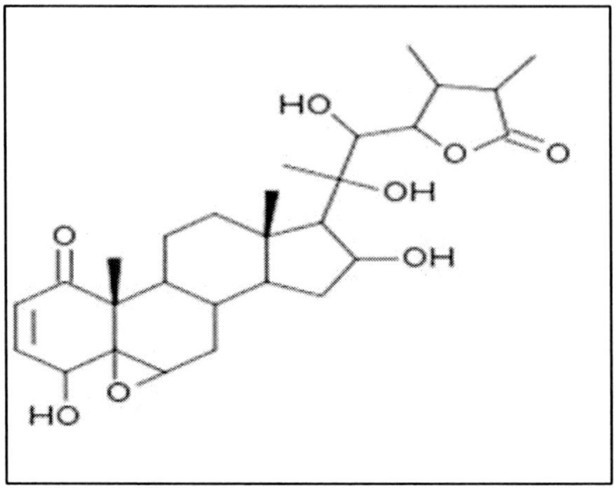

Figure 3. The Common skeleton of all withanolides.

The Bioactive Compounds: Withanolides

W. coagulans have more than 12 alkaloids, 40 withanolides and several sitoindosites (Withanolides chemical compound with glucose molecule at carbon 27) but the main bioactive compound of *W. coagulans* is withanolides. Withanolides are a group of chemical compounds with C_{28}steroides containing an ergostane skeleton in which C_{26} and C_{22} or C_{26} and C_{23} are oxidized in order to form a δ or γ lactone. The term "Withanolide'' made by two words 'Withan' and 'olide', in which Withan comes from *Withania* genus and olide used for lactone. All Withanolides are showing common basic Withanolide skeleton (Figure 3).

All withanolide has a common characteristic skeleton that has one C_8 or C_9 side chain with a lactone or lactol (six- or five- member) ring. In *W. coagulans*, a specific withanolide having 14, 20-epoxide bridge is also reported. Withanolides are present in various forms due to diversityin steroid nucleus structure and the side chain. In whole plant kingdom withanolides naturally present in mainly solanaceous (Nightshade family) plants (approximately 22 genera) including *Withania*. But it is not limited only in Solonaceae family, besides it is also found in Leguminosae, Labiatae, Myrtaceae and Taccaceae family. Withanolides containing plants have an extraordinary capability to introduce oxygen, which is functionally active at almost every position of the carboxylic skeleton and side chain. Withanolide are synthesized via the mevalonate pathway of terpenoids formation and arise from the initial cyclization of 3S-squalene-2,3epoxide. The whole plant contains withacoagulin A- I, withacoagulin, bispicropodophyllinglucoside, coagulins A-S and coagulin U, 14, 15 β- epoxywithanolide I, 17β- hydroxywithanolide K and co-agulinolide.

OTHER CHEMICAL CONSTITUENTS

Despite of main bioactive compound withanolides, *W. coagulans* having various fatty acids, free amino acids and esterases. Seeds contain specific enzyme 'withanin' which is responsible for coagulative property (Hemalathaet al. 2008). In mineral contents, Mg and Ca are found very high, but K found in low amounts, as well as energy percentage value and protein contents were also reported very less in *W. coagulans* (Ullahet al. 2013). The petroleum ether seeds extract contains a number of different fatty acids mainly, oleic acid, palmitic acid, stearic acid and arachidic acid. The unsaponifiable part of the seeds have dihydrostigmosterol and β- sitosterol. Berries have various alkaloids and free amino acids like; proline, hydroxyproline, valine, tyrosine, aspartic acid, glycine, cysteine, asparagines and glutamic acid. Ali et al. 2015 isolate two new fatty acids: (1) n-dotriacont-21- enoic acid, and (2) n- tetratriacontanoic acid, aromatic ester and monoterpenic benzyl glucoside from the methanolic extract of *W. coagulans* fruit (Table 4).

THERAPEUTIC POTENTIAL OF *WITHANIA COAGULANS*

Natural product used as a rich source of chemical entities that control various health problems. *W. coagulans* has been highlighted in Ayurveda for the treatment of diabetes, but this plant is also goodfor its effectiveness against other health problems like; hypolipidemic, hepatoprotective, anti-inflammatory, anticancer, free radical scavenging, central nervous system related disorders, etc. (Table 5).

Table 4. Identified chemical constituents in *Withania coagulans*

Constituent Name	Molecular formula	Molecular weight
Methyl hexadecanoate	$C_{17}H_{34}O_2$	270
n-Hexadecanoic acid	$C_{16}H_{32}O_2$	256
Methyl octadeca-9,12-dionate	$C_{19}H_{34}O_2$	294
Methyl octadecanoate	$C_{19}H_{38}O_2$	298
Octadeca-9,12-dionic acid	$C_{18}H_{32}O_2$	280
n-Ocadecanoic acid	$C_{18}H_{36}O_2$	284
Ergost-5-en-3β-ol	$C_{28}H_{48}O$	400
Stigmasta-4,22-dien-3β-ol	$C_{29}H_{48}O$	412
Stigmast-5-en-3β-ol	$C_{29}H_{50}O$	414
Olean-12-en-3β-ol	$C_{30}H_{50}O$	426
Lup-20(29)-en-3β-ol	$C_{30}H_{50}O$	426
Linalool	$C_{10}H_{18}O$	154
Octanoic acid	$C_8H_{16}O_2$	144
1-Dodecene	$C_{12}H_{24}$	168
Capric acid	$C_{10}H_{20}O_2$	172

Table 4. (Continued)

Constituent Name	Molecular formula	Molecular weight
1-Tetradecene	$C_{14}H_{28}$	196
n-Tetradecane	$C_{14}H_{30}$	198
Lauric acid	$C_{12}H_{24}O_2$	200
1-Heptadecene	$C_{17}H_{34}$	238
Myristic acid	$C_{14}H_{28}O_2$	228
Thiogeraniol	$C_{10}H_{18}S$	170
Pentadecanoic acid	$C_{15}H_{30}O_2$	242
1-Nonadecene	$C_{19}H_{38}$	266
Phthalic acid	$C_{12}H_{14}O_4$	222
Palmitoleic acid	$C_{16}H_{30}O_2$	254
Palmitic acid	$C_{17}H_{34}O_2$	270
Linoleic acid	$C_{19}H_{34}O_2$	294
n-Heneicosane	$C_{21}H_{44}$	296
Stearic acid	$C_{19}H_{38}O_2$	298
1-Eicosanol	$C_{20}H_{42}O$	298
11-Eicosenoic acid	$C_{21}H_{40}O_2$	324
Ethanol, 2-(9,12-octadecadienyloxy)	$C_{20}H_{38}O_2$	310
Arachidic acid	$C_{20}H_{40}O_2$	312
Oleic acid	$C_{19}H_{36}O_2$	296
Docosanoic acid	$C_{23}H_{46}O_2$	354
3β-Stigmasta-5,24(28)-dien-3-ol	$C_{29}H_{48}O$	412
Lupeo	$C_{30}H_{50}O$	426
24-Methyl-9,19-cyclolanost-25-en-3-ol	$C_{31}H_{52}O$	440
β-Sitosterol	$C_{29}H_{50}O$	414
n-Tetracontane	$C_{40}H_{82}$	562
Cyclohexane	C_6H_{12}	84
Borane carbonyl	CH_3BO	42
3-Methyl, hexane	C_7H_{16}	100
Heptane	C_7H_{16}	100
Hexanoic acid	$C_6H_{12}O_2$	116
Nonanoic acid	$C_9H_{18}O_2$	158
Longifolene	$C_{15}H_{24}$	204
Allene	C_3H_4	40
d-Cadinene	$C_{15}H_2$	204
Caryophyllene	$C_{15}H_{24}$	204
4-*tert*-Butyltoluene	$C_{11}H_{16}$	148
3-Carene	$C_{10}H_{16}$	136
Methyl ester of hexadecanoic acid	$C_{17}H_{34}O_2$	270
Methyl ester of nonadecanoic acid	$C_{20}H_{40}O_2$	312
Ar-Tumerone	$C_{15}H_{20}O$	216
Methyl ester 8,11-octadecadienoic acid	$C_{19}H_{34}O_2$	294
Ethyl ester of linoleic acid	$C_{20}H_{36}O$	308
18 Methyl ester 9-octadecenoic acid	$C_{19}H_{36}O_2$	296
Acetaldehyde	C_2H_4O	44
5-Dimethyl 2-Undecene-2	$C_{13}H_{26}$	182

HYPOGLYCEMIC ACTIVITY

In previous literatures, *W. coagulans* was specially mentioned for their hypoglycemic potential and described as an effective medicine for the cure of diabetes (Hemalathaet al. 2008) (Figure 4). Within the group of withanolides: coagulin C, 17β-hydroxywithanolide K, withanolide F, coagulanolide and coagulin L are well reported as potential bioactive chemical compounds which actively work against the hyperglycemic condition. During the balance of sugar content in human body various kinds of minerals play very crucial role, such as higher concentration of Mg, Ca and lower concentration of K performed very important role in the regulation of diabetes and *W. coagulans* fruits extract has been well known for the presence of higher concentrations of both Mg and Ca (Lopez-Ridauraet al. 2004, Jaiswalet al. 2009). On this basis antidiabetic potential of *W. coagulans* correlated with higher concentration of Mg along with Ca because during prevention of the hyperglycemic condition if Ca^{++} concentration increases in the body it quickly activates insulin gene expression by calcium responsive element binding (CREB) protein (Figure 5) (Veigaet al. 2006).

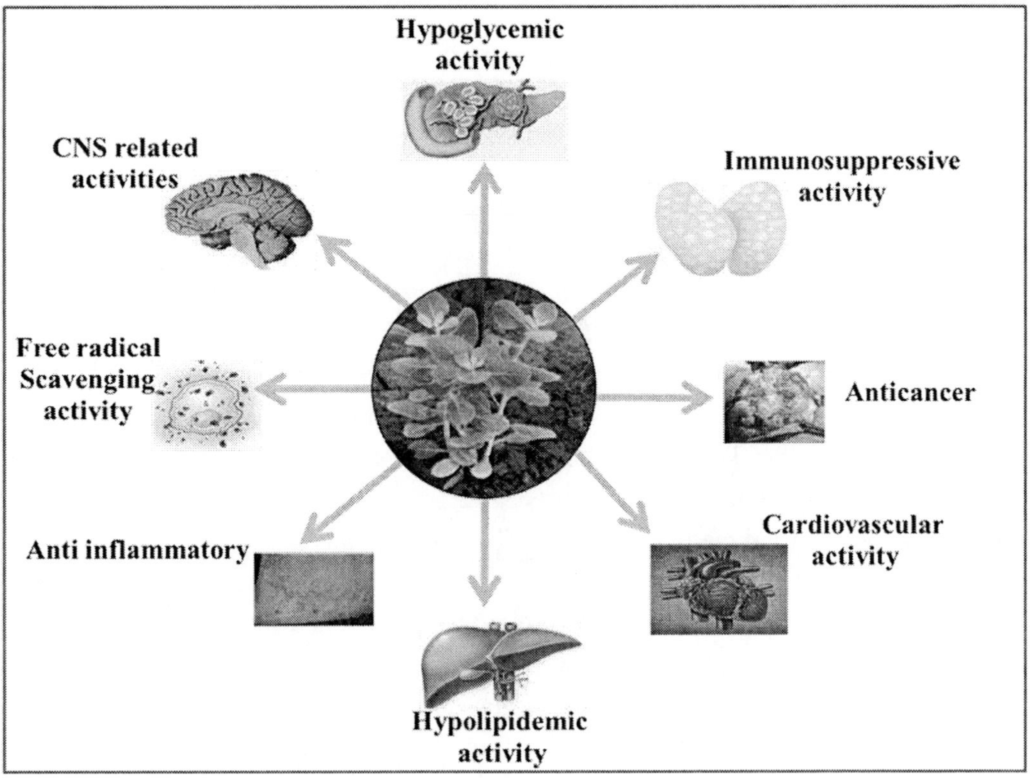

Figure 4. Pharmceutical activities of *Withania coagulans* Dunal.

Anti-Cancer

W. coagulans have many effective chemical compounds which help against cancer through inhibition of cancerous cell proliferation, cell cycle arrest and final cell death. For the inhibition of proliferation withanolides induce apoptosis in HL-60 leukamia cells *via* mitochondria with Cytochrome C and caspase activation (Senthilet al. 2007). Withaferin A is one of the important member of withanolides having strong anticancer potential, which reduce cell viability in cell lines and activate anti proliferative activity against head and neck squamous carcinoma (Subramanian and Sethi 1969). Along with withaferin A different other withanolides like withanocoagulin A, withacoagulin C, withacoagulin D, withacoagulin E, withanolide L, withanolide J, withanolide F and (22)R- 14α15α17β20β tetrahydroxy- 1oxowitha 2,5,24triene 2,6,22-olide are also reported as effective chemical constituent for the inhibition of cancer cells proliferation. Coagulin L and withaferin A sequentially induce S phase and G2/M phase cell cycle arrest through P27 and CDKs, coagulin L mainly down regulates PPARy, C/EBPα. *W. coagulans* aqueous and methanolic fruit extract are showing the cytotoxic effect on swiss albino mice against cancerous cells, which prevent the formation of tumors and decreases the rate of tumor yield. It has been also proven that methanolic extract of fruits are more effective than aqueous extract as anti cancerous herbal formulation against human breast cancer (Ahmad et al. 2017).

Anti-Inflammatory

Traditionally the fruits and leaves extract of *W. coagulans* are applied to treat inflammation. Cyclooxygenase-1 (cox-1) and cyclooxygenase-2 (cox-2) enzyme are responsible for the conversion of arachidonic acid into prostaglandins, thromoboxane and prostacyclin, which are involved in inflammatory responses of the body. After an inflammatory stimulus, cox-2 expressed and releases various metabolites which induce pain and inflammation. Therefore, nonselective nonsteroidal anti-inflammatory drugs (NSAIDs) and COX-2 inhibitors bind to COX-2, which provides relieffrom the pain of inflammation. Anti-inflammatory activity of withanolides has been assessed under *in vitro* conditions using leukotriene formation, COX-1 and COX-2 inhibitory assays (Wubeet al. 2008). Anti inflammatory activity of *W. coagulans* is related with withanolides like, 3β-hydroxy-2,3-dihydro withanolide F has a maximum or significant anti-inflammatory capacity. The alcoholic extract of *W. coagulans* showed anti-inflammatory effects during acute inflammation. The hydro alcoholic extract of *W. coagulans* berry also performed anti-inflammatory activity in rat paw oedema model induced with carragenin (Rajurkaret al. 2001). The aqueous extract of *W. coagulans* fruits has also significant anti-inflammatory activity in sub-acute models of inflammation, such as granuloma formation and formalin induced arthritis in rats (Budhirajaet al. 1984). Coagulin H, has a powerful inhibitory effect on inflammation, lymphocyte proliferation and Th-1 cytokine production (Mesaiket al.2006). Withanosides and withanolideglycansare shown perform better interaction with COX-2 than COX-1, therefore in upcoming time withanolides can be used as substitutes of NSAIDs (Nithyaet al. 2010).

Immunosuppressive Effect

In various types of withanolides, withaferin A and withanolide E are mainly responsible for immunosuppressive activity. These chemical compounds show their specific immunosuppressive affects on human B and T lymphocytes as well as on mice thymocytes (Bahr and Hansel 1982). Coagulin H also works as immunosuppressive compound through effect on lymphocyte proliferation, inter leukin-2 (IL-2) cytokine expression with the inhibition of activated T-cell proliferation and phytohaemagglutinin (PHA) (Mesaiket al.2006). Huang et al. 2009 isolated withacoagulins from areal parts of *W. coagulans* and reported its inhibitory effect on T and B lymphocytes proliferation in murine spleen cells. 5, 20α(R)-dihydroxy-6α, 7α-epoxy-1-oxo (5α)-witha-2, 24-dienolide is also perform immunosuppressive activity in spleen cell culture.

Anti-Fungal and Anti-Bacterial

W. coagulans has powerful antimicrobial activity and it is used to control microbial contamination especially for pathogenic bacterial infection. Antibacterial activity of *W. coagulans* fruits extracted volatile oil against *Micrococcus pyogenes*, *Staphylococcus aureus* and *Vibrio cholera* have been well reported. Fruits aqueous extract has also been reported for inhibition of *Salmonella* infection. Methanolic and ethanolic extract of plant leaves has also antibacterial activity against a spectrum of bacteria (Budhiraja 1967). In various withanolides, 14,15β- epoxywithanolide I [(20S, 22R) 17β,20β-dihydroxy-14β,15β-epoxy-1-oxo-witha-3,5,24trienolide], 17β-hydroxywithanolide K [(20S,22R)14α, 17β,20β-trihydroxy-1-oxo-witha-2,5,24 trien-olide], withanolide F, Withaferin A show very high inhibitory activity against bacterial and fungal contaminations. In 1983 Vandeet al. obtain 17β-hydroxywithanolide K by partial derivatization of withanolide F and 17β-hydroxyWithanolide K has also found to be highly useful for inhibition of pathogenic fungal growth (Choudharyet al. 1995). Withanolide F has great potential to inhibit growth and development of highly pathogenic fungal isolates *Nigrospora oryzae*, *Aspergillus niger*, *Curvularia lunata*, *Pleuretuso streatus*, *Stachy botrysatra*, *Allescheria boydii*, *Drechsleraro strata*, *Microsporum canis* and *Epidermophyton floccosum*. Withanolide F also represents high effectiveness against, gram positive bacteria. Another highly present withanolide in *W. coagulans*, withanolide D also retain antifungal activity against human pathogenic fungi. Withaferin A exhibited a significant antibacterial activity against gram positive bacteria, filamentous fungi and yeast group, but it is inactive in case of gram-negative bacteria and non filamentous fungi.

Anti-Lipidemic Activity

W. coagulans widely used for its anti-lipidemic activity which is attributed to coagulin L. Along with coagulin L, high percentage of β sitosterol and linoleic acid are also responsible for the hypo-cholesterolemic effect (Maurya et al. 2010). In 2008, Hemalatha et al. reported that the aqueous extract of *W. coagulans* fruits significantly decrease the amount of total lipid, serum cholesterol, triglycerides and the LPO levels in high fat diet induced hypercholesterolemic rats. The chloroform extracts of the fruits reduce triglycerides, total cholesterol, LDL and VLDL but increase the HDL level (Hoda et al. 2010).

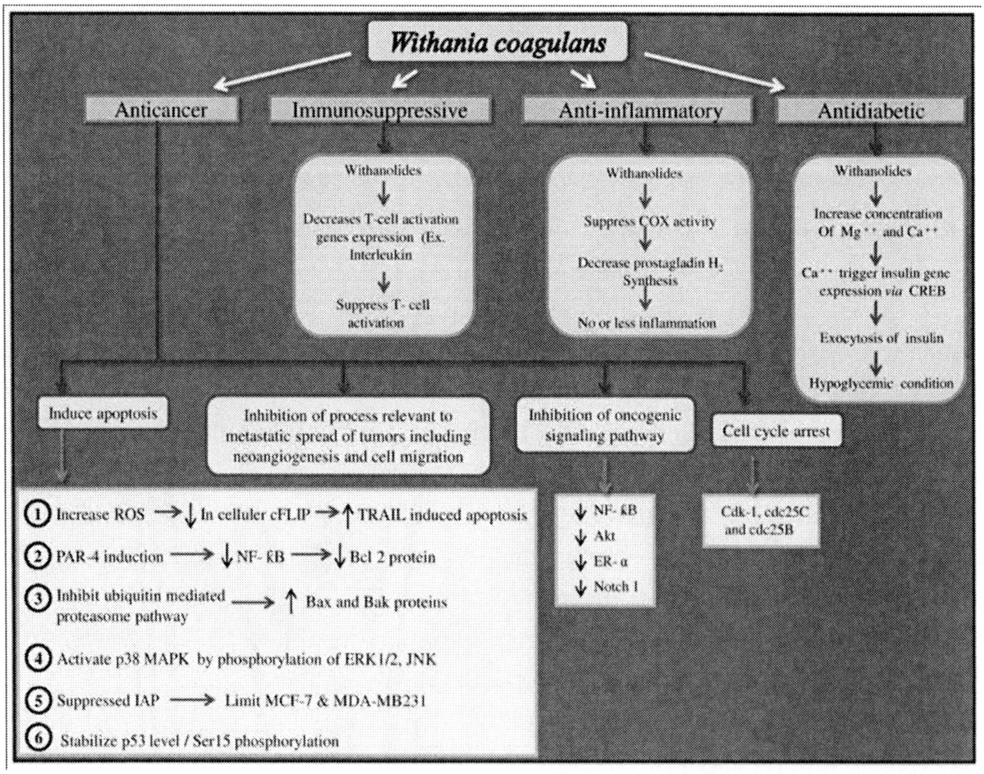

Figure 5. *Withania coagulans* hypothetical mode of action against various health problems (as cancer, diabetes, inflammations and immune degeneration diseases).

CNS Related

W. coagulans isolated bioactive compound withanolides and glycowithanolides having ability to alleviating many central nervous system disorders like: epilepsy, anxiety, depression, catalepsy and sleep. In mice, rabbits and dogs it has been observed that the total extract of fruits has central nervous system (CNS) depressant activity (Mauryaet al. 2010). The extract of its various plant part have capacity to modulate many neurotransmitters. The alcoholic extract, total alkaloids and aqueous extract of fruits exhibited CNS depression in albino rats characterized by sedation, reduced exploratory and hypothermia. 3β-hydroxy-2,3-dihydrowithanolide F has its CNS depressant activity (Budhirajaet al. 1987).

Cardiovascular Effects

It is well known that in the fruits of *W. coagulans* have some withanolides, which are very much structurally similar with aglycones and possessing a 6-membered unsaturated lactone ring attached to the steroidal ring at C_{20} instead of at $C_{17,}$ responsible for cardiovascular effects (Khodaeiet al. 2012). It is reported that 3β-hydroxy-2,3-dihydrowithanolide F effective against cardiac arrestin dogs and moderately decreases blood pressure (Budhirajaet al. 1983). *W. coagulans* fruits extracte having coagulin L that actively participate in significant drop of a fasting blood glucose profile and also improved the glucose tolerance of db/db mice (Mauryaet al. 2008). Another withanolide, withaferin A is also known for significant importance in the treatment of the inflammatory cascade of cardiovascular diseases.

Anti-Oxidative Activity

Anti-oxidative potential means inhibition of free radicals, which alters the physiological chemical reactions. The majority of withanolides in *W. coagulans* are known for antioxidative activities. The aqueous extract of *W. coagulans* having the potential for free radical scavenging and lipid peroxidation in diabetic rats that helped in providing protection against oxidative damage of beta cells of the pancreas (Hemalathaet al. 2004). The aqueous extract of *W. coagulans* exhibited antioxidant activity in an *in vitro* system using 1,1-diphenyl-2-picrylhydrazyl (DPPH). *W. coagulans* fruits extracts could be used against aging and artherosclerosis symptoms, which are caused by reactive oxygen species (Mathuret al. 2011).

Hepatoprotective Activity

W. coagulans aqueous fruit extractis beneficial for hepatoprotective health problems. In many of withanolides 3β-hydroxy-2,3-dihydrowithanolide F has significant importance as a hepatoprotective potential agent (Maurya et al. 2010). In root extract of *W. coagulans*, withanolide 3β-hydroxy-2,3-dihydrowithanolide F is present, which show hepato-protective activity against CCl_4 induced hepatotoxicity in both male and female adult albino rats (Budhirajaet al. 1986).

Table 5. Pharmaceutical properties of *Withania coagulans* Dunal and related withanolides with their medicinal potential

Activities	Bioactive withanolides	Plant parts	References
Anti cancerous	Withanocoagulin A Withanocoagulin C Withanocoagulin D Withanocoagulin E Withanolide J Withanolide L Withanolide F Coagulin L Withaferin A	Aerial parts Aerial parts Aerial parts Aerial parts Whole plant Aerial parts Aerial parts Whole plant Roots	Huang et al. (2009) Huang et al. (2009) Huang et al. (2009) Huang et al. (2009) Atta-ur-Rahman (1998) Huang et al. (2009) Huang et al. (2009) Atta-ur-Rahman et al. (1998) Subramanian and Sethi (1969)
Anti hyperglycemic	17β-hydroxywithanolide K Coagulin C Coagulin L Coagulanolide Withanolide F	Whole Plant and fruit Plant and fruit Fruits Fruits Fruits	Choudharyet al. (1995) Atta-ur-Rahman et al. (1998) Atta-ur-Rahman et al. (1998) Mauryaet al. (2008) Huang et al. (2009)
Anti lipidemic	Coagulin L	Whole plant	Atta-ur-Rahman et al. (1998)
Anti microbial	17β-hydroxywithanolide K Withaferin A 14,15β-epoxywithanolide I Withanolide F Withanolide D	Whole plant Root Whole plant Aerial parts Fruits	Choudharyet al. (1995) Subramanian and Sethi (1969) Choudhary et al. (1995) Huang et al. (2009) Budhirajaet al. (1983)
Anti inflammatory	3β-hydroxy-2,3-dihydrowithanolide F Coagulin H	Fruits Whole plant	Budhirajaet al. (1983) Atta-ur-Rahman et al. (1998)
Immunosuppressive	Withaferin A Withanolide E Coagulin H 5,20α(R)-dihydroxy-6α,7α-epoxy-1-oxo-(5α)-witha-2,24dienolide	Root Root Whole plant Leaves	Subramanian and Sethi (1969) Budhirajaet al. (1983) Atta-ur-Rahman et al. (1998) Subramanaianet al. (1971)
CNS Related	3β-hydroxy-2,3-dihydrowithanolide F	Fruits	Budhirajaet al. (1983)
Alzheimer	Withanolide A		
Cardiovascular	3β-hydroxy-2,3-dihydrowithanolide F Withaferin A	Fruits Root	Budhirajaet al. (1983) Subramanian and Sethi (1969)
Hepatoprotective	3β-hydroxy-2,3-dihydrowithanolide F	Fruits	Budhirajaet al. (1983)

Anti-Arthritic Activity

Rheumatoid arthritis is chronic inflammatory and auto immune disease. In *W. coagulans* fruits alkaloids and withanolides are major chemical constituents, which are responsible and participant for the treatment of rheumatoid arthritis. Methanolic extract and hydro alcoholic extract of *W. coagulans* both possess anti arthritic activity, but hydro alcoholic extract has more potential against arthritis than methanolic extract (Shendkar et al. 2014).

Alzheimer's Disease

Alzheimer's disease is extracellular deposits of amyloid beta (Aβ) protein with intracellular accumulation of neurofibrillary tangles (NFTs) and associated with mutation in AβPP genes (Mattson 2004). Among all withanolides, withanolide A highly useful in the treatment of Alzheimer *via* Aβ pathway. In Aβ pathway, BACE-1 (beta-site APP cleaving Enzyme 1) is the rate limiting enzyme, which is significantly down regulated by withanolide A (Khodaei et al. 2012).

Anti-Diuretic Potential

The anti diuretic potential of *W. coagulans* is related to the polar nature of withanolides. Dabheliyaet al. 2010, observed the diuretic potential of fruits aqueous extract of *W. coagulans*.

Biotechnological Approaches for *Withania coagulans* Conservation and Secondary Metabolites Production

Biotechnological approaches have well marked for improvement and conservation of plants (Figure 6).These approaches involve *in vitro* clonal propagation, synthetic seeds production, genetic transformation, cryopreservation, molecular markers development, protoplast fusion, development of transgenic, etc. to improve the quality and quantity of important biochemical constituents (bioactive compounds) with the large scale sustainable production and conservation. Recently, breeding practices in *W. coagulans* have been advanced by the application of biotechnological approaches, including *in vitro* propagation, molecular markers (identifying genetic diversity and selection of specific traits), and gene transfer (plants protection against abiotic and biotic stress). Due to the presence under endangered category and having large scale pharmaceutical properties, various tissue culture methods have been developed such as using nodes as explants (Valizadeh and Valizadeh 2011; Rathoreet al. 2012), leaf discs and inter-nodal segments (Valizadeh and Valizadeh 2009), shoot tips as explants, seedling derived direct and indirect organogenesis (Jat et al. 2014; Edalatifardet al. 2014), thin cell layer explants (Tripathiet al. 2018) as milestone to conserve this plant with improve level of its pharmaceutical potential. Although various reports have describe about conservation and propagation but still very limited studies have reported about molecular markers (such as RAPD and ISSR) based assessment of true-to-type *in vitro* clonal propagation related with *W. coagulans* (Tripathi et al. 2018). Other than conservation Mirjaliliet al. 2011 have been reported hairy root culture in *W. coagulans* with increased level of tritepenoids (withanolides) by the over expression of squalene synthase (*SQS*) and Mishra et al. 2013 genetic transformation of marker (*gusA*) and selector (*nptII*, *hptII*) genes with the help of *Agrobacterium tumefaciens*-mediated gene transfer. Synthetic seeds production has also been reported using shoot tip and nodal section of *W. coagulans* plant (Rathore and Kheni 2017).

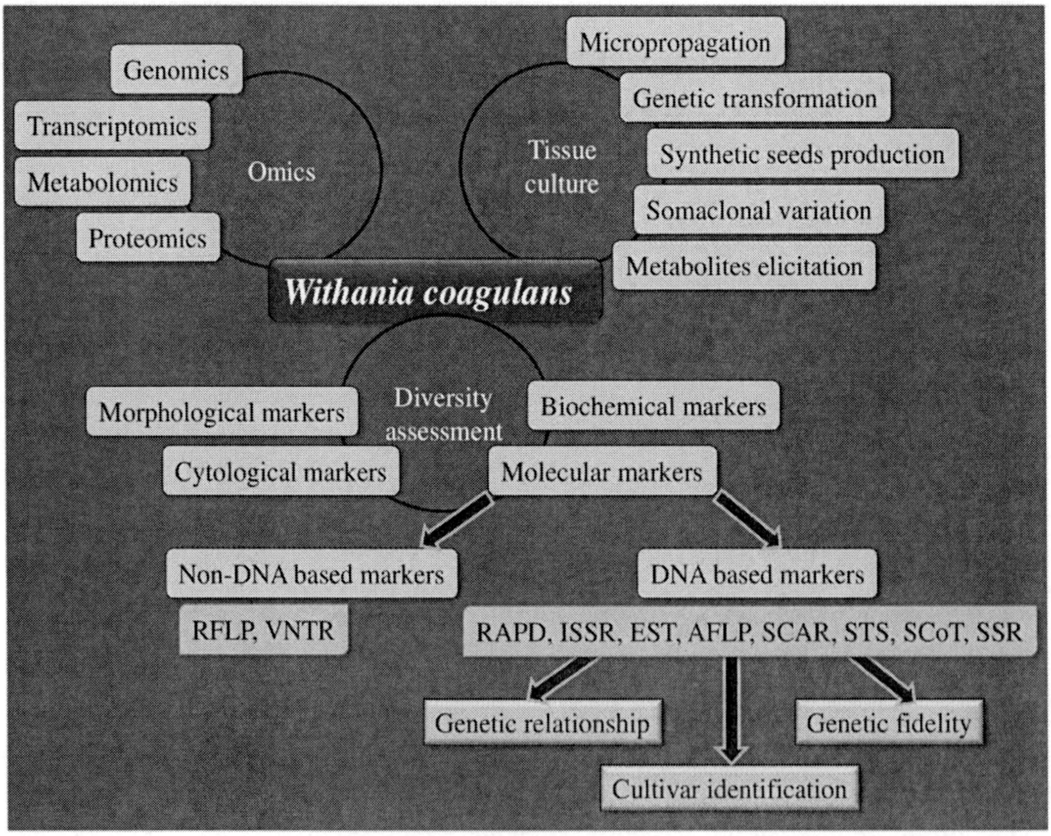

Figure 6. Various biotechnological approaches related to conservation and improvement of *Withania coagulans* Dunal.

Conclusion

The modern medications which are commonly using to treat different pharmacological problems having number of side effects. Therefore, phyto medicin edevelopment is emerging research area as safer and economic herbal medicine systems. *W. coagulans* is a traditional herbal plant with the presence of with anolides and various other bioactive compounds, which is helpful to offer relief in severe health problems such as cancer, hyperglycemia, microbial contamination, CNS problems and cardiovascular problems, etc. Now a days, there is enhancing research focus towards investigation of mechanisms of action related to the various with anolides. This study is emphasizing on the possible with anolides mode of actions, related to pharmacological problems. Although, several counterstrikes have been performed to improve the cultivation and production of *W. coagulans* with enhanced bioactive compounds, but still imperative demand of higher production of this useful plant with high medicinal efficiency is required, which can be achieved with the help of modern biotechnological tools and nanotechnology.

Acknowledgments

Authors are thankful to CAS- Botany (BHU), DST-PURSE and FIST programmes for providing funds to Department of Botany, Institute of Sciences, Banaras Hindu University, Varanasi. Deepika Tripathi is thankful to CSIR/UGC- New Delhi, India for providing financial support.

References

Ahmad, R., Fatima, A., Srivastava, A. N. & Khan, M. A. (2017). Evaluation of apoptotic activity of *Withania coagulans* methanolic extract against human breast cancer and Vero cell lines. *J Ayurveda Integr Med*, 8, 177-183.

Ali, A., Jameel, M. & Ali, M. (2015). New fatty acid, aromatic ester and monoterpenic benzyl glucoside from the fruits of *Withania coagulans* Dunal. *Nat Prod Res*, *29*, 1307-1314.

Atta-ur-Rahman, S. M., Dur-e-Shahwar, C. M., Voelter, W. & Hohnholz, D. (1998). New steroidal lactones from *Withania coagulans*. *Heterocycles*, *47*, 1005-1012.

Bähr, V. & Hänsel, R. (1982). Immunomodulating Properties of 5, 20α (R)-Dihydroxy-6α, 7α-epoxy-1-oxo-(5α)-witha-2, 24-dienolide and Solasodine. *Planta Medica*, *44*, 32-33.

Budhiraja, R. D., Bala, S. & Garg, K. N. (1977). Pharmacological investigations on fruits of *Withania coagulans* Dunal. *Planta Medica*, *32*, 154-157.

Budhiraja, R. D., Sudhir, S. & Garg, K. N. (1983). Cardiovascular effects of a withanolide from *Withania coagulans* dunal fruits. *Indian J Physiol Pharmacol*, *27*, 129-134.

Budhiraja, R. D., Sudhir, S. & Garg, K. N. (1984). Antiinflammatory activity of 3 β-Hydroxy-2, 3-dihydro-withanolide F. *Planta Medica*, *50*, 134-136.

Choudhary, M. I., Dur-e-Shahwar, N. A., Parveen, Z., Jabbar, A., Ali, I. & Atta-ur Rahman, A. (1995). Antifungal steroidal lactones from *Withania coagulance*. *Phytochemistry*, *40*, 1243-1246.

Dabheliya, J. A. N. A. K., Khan, S. A., Joshipura, M., Vasoya, M., Patel, S. & Vijaya, S. (2010). Diuretic potential of aqueous extract of fruits of *Withania coagulans* Dunal in experimental rats. *Int J Pharm Pharm Sci*, *2*, 51-53.

Edalatifard, L., Modarres-Sanavy, S. A. M. & Askari, H. (2014). The optimum condition under light and media for seed germination of *Withania coagulans*. *Int J Farm Allied Sci*, *3*, 722-728.

Hemalatha, S., Kumar, R. & Kumar, M. (2008).*Withania coagulans* Dunal: A review. *Pharmacogn Rev*, *2*, 351.

Hemalatha, S., Wahi, A. K., Singh, P. K. & Chansuria, J. P. (2004). Hypoglycemic activity of *Withania coagulans* Dunal. In streptozotocin induced diabetic rats. *J Ethanopharmacol*, *93*, 261.

Hemalatha, S., Wahi, A. K., Singh, P. N. & Chansouria, J. P. N. (2006). Hypolipidemic activity of aqueous extract of *Withania coagulans* Dunal in albino rats. *Phytother Res*, *20*, 614-617.

Hoda, Q., Ahmad, S., Akhtar, M., Najmi, A. K., Pillai, K. K. & Ahmad, S. J. (2010). Antihyperglycaemic and antihyperlipidaemic effect of poly-constituents, in aqueous and chloroform extracts, of *Withania coagulans* Dunal in experimental type 2 diabetes mellitus in rats. *Hum Exp Toxicol*, *29*, 653-658.

Huang, C. F., Ma, L., Sun, L. J., Ali, M., Arfan, M., Liu, J. W. & Hu, L. H. (2009). Immunosuppressive Withanollides from *Withania coagulans*. *Chem Biodivers*, *6*, 1415-1426.

Jaiswal, D., Rai, P. K. & Watal, G. (2009). Antidiabetic effect of *Withania coagulans* in experimental rats. *Indian J Clin Biochem*, *24*, 88-93.

Jat, B. L., Meena, G. P., Choudhary, C. R., Maheshwari, R. K. & Jeswani, G. (2014).*In vitro* propagation of *Withania coagulance* [sic] through seedling segment (epicotyledonary node). *Int J Chem Pharm Sci*, *2*, 979-989.

Khodaei, M., Jafari, M. & Noori, M. (2012). Remedial use of withanolides from *Withaniacoagolans* (Stocks) Dunal. *Int J Adv Life Sci*, *2*, 6-19.

Lopez-Ridaura, R., Willett, W. C., Rim, E. B., Liu, S., Stampfer, M. J., Manson, J. E. & Hu, F. B. (2004). Magnesium intake and risk of type 2 diabetes in men and women. *Diabetes Care*, *27*, 134-140.

Taxonomic Classification and Nomenclature

The plant's name *P. foetida* L. originated from the Greek word paederos meaning opals, because many species of this genus have translucent, opal-like drupes (Soni et. al. 2013) and the specific epithet foetida means stinking or bad smelling, because of the unpleasant odour released by the plant when broken or crushed (Osman et al. 2009). *P. foetida* is a member of family Rubiaceae belonging to order Gentianales of class Magnoliopsida (Table 1).

Table 1. Taxonomic classification of *Paederiafoetida* L.

Kingdom	**Plantae**	Harboring plants most of them are photosynthesizing eukaryotes with cellulosic cell wall
Subkingdom	Tracheobionta	Plants with true vascular bundles, both xylem and phloem present. Sporophytic phase of the plant having leaf, stem and roots
Superdivision	Spermatophyta	plants having true flowers containing ovary as Seed producing organ
Division	Magnoliophyta	Ovary is enclosed by carpels, triple fusion is prevalent and sieve elements with companion cells.
Class	Magnoliopsida	Seeds having 2 cotyledons (dicotyledons)
Subclass	Asteridae	Plants containing pentamerous flowers having fused floral organs with axile placentation
Order	Rubiales	Trees, shrub or herb. Length of the stamen is high up to corolla tube
Family	Rubiaceae	Madder family. Presence of interpetiolar stipules, bearing glands called colleters
Genus	*Paederia* L.	sewer vine
Species	*Paederia foetida* L.	Stinkvine

Table 2. Various vernacular names of *Paederia foetida* L.

Languages	**Synonyms**
Assamese	Bhedailota, Padurilota
Bangladesh	Gondhabadal
Bengali	Gandhabhaduliya, Gandhabhadule, Gandal
Chinese	Fever vine
English	Skunkvine, Stinkvine, Chicken excrement plant, Flatulent vine
Gujrati	Gandhana
Hindi	Gandhaprasarani or Pasaran
Latin	*Paederiafoetida*L.
Malaysian	Gandhali,Akarsekuntut
Marathi	Hiran- vel
Tamil	Pinarisangai
Telugu	Takkeda, Savirel

There are about 30 species of *P. foetida* reported worldwide, distributed from continental tropical Asia which is thought to be the origin of this plant, Malesia, Madagascar, East Africa, and tropical America. *P. foetida* used as type species of the genus *P.*, described and published by Linnaeus in 1767 and later lectotypified by Puff in 1991. *P. foetida* widely distributed and common in Asia and demonstrates the most variability in morphological characters. Many infraspecific and common names have been attributed to geographical and morphological variants (Table 2).

DISTRIBUTION

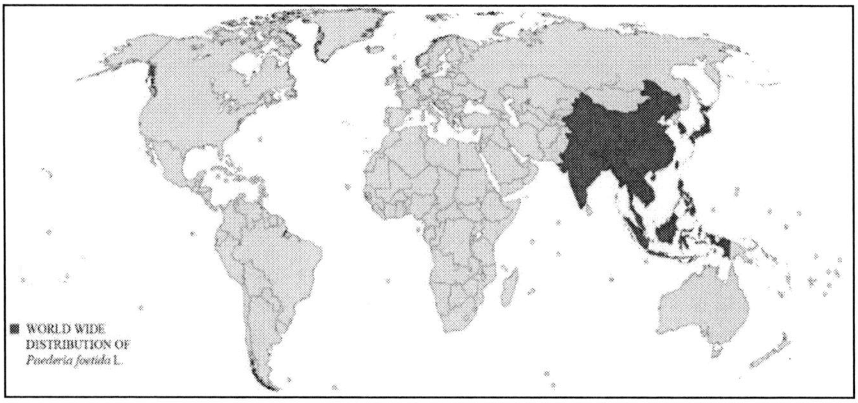

Figure 2. Map showing the worldwide distribution of *P. foetida* L.

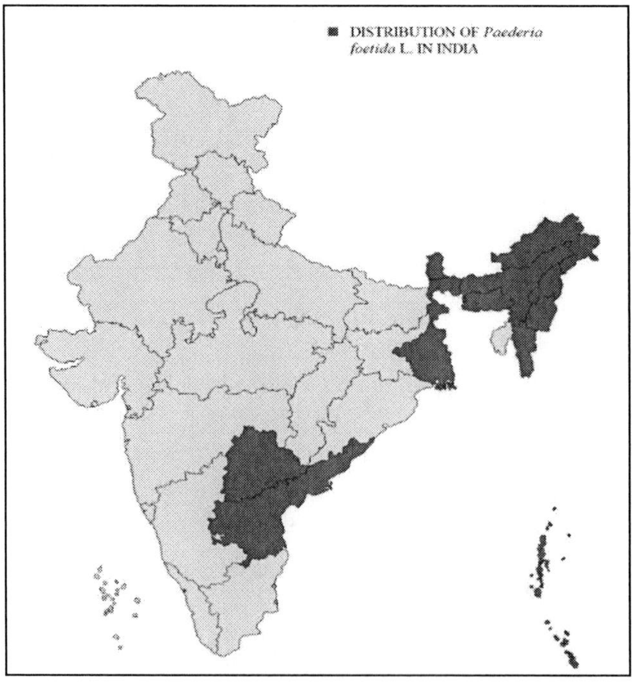

Figure 3. Map showing the distribution of *P. foetida*in India.

P. foetida is native to temperate and tropical East Asia and tropical South-East Asia, covering the area of Japan and Taiwan, parts of China and India, Hong Kong, Macau, Bangladesh, Bhutan, Cambodia, Laos, Myanmar, Thailand, Vietnam, Indonesia, Malaysia, Philippines, and Singapore (Figure 2).

In India, *P. foetida* is found in the Darjeeling Himalayan region (Borgohainet al. 2017), Western India, Bengal, Assam, Bihar, and Orissa (De et al. 1994). According to Puff 1991, *P. foetida* has also been introduced to the islands of Sri Lanka and New Guinea (Figure 3).

It has also been introduced to Mauritius and Reunion in the western Indian Ocean to the southwestern Pacific island of New Caledonia, the Hawaiian Islands of Oahu, Kauai, Hawaii and Maui, and the southern USA, including Florida and restricted locations in Texas, Louisiana, South Carolina, and North Carolina.

DESCRIPTION

P. foetida is a perennial, semi-woody, herbaceous, vine creeper (Figure 1). It crawls upward on the support of other plant species or their own older part. In some regions e.g., Florida, it is invasive in nature and covers the whole canopy, becomes a strong competitor of small herbs and shrubs which grow in sunlight near the ground. Some of the morphological features of the plant are described below (Figure 4) for easy identification:

The younger parts of the stem are light green in color whereas the older parts are dark green. The stem is slender, about 7 m long and about 5 mm in diameter, flexible, it is provided with short fine hairs. They climb upward by twining around other plants or around their own older stems; the portion of the plant which grows near the ground produces roots at the nodes.

Leaves are simple, opposite, lanceolate to elongated-ovate with a pointed tip, 2-5 cm wide, 3-14 cm long bases rounded, or heart-shaped. Leaves are medium to dark green on the upper surface and sparsely scattered with, multicellular hairs with pointed tips. The lower surface of the leaves is lighter in color, veins of the leaves are lined with the transparent, bristly multicellular hairs; bent at the tip. The margin is entire, fringed with closely appressed (strigose) hairs. Leaves are joined to the stem by the petioles. Petioles are green, 2-5 cm long, slender, with fine, short hairs in lines. The stipules are interpetiolar and broad wedge-shaped.

The inflorescence is double scorpioid or lax scorpioid (cymes). Flowers are stalked, arranged on double scorpioid inflorescences; terminal and axillary position; sepals are connate, calyx lobes are deltate, corolla funnel-shaped, 4-5 lobes, margins crinkled, 7-11 mm long, white to cream, dark purplish red within, outer surface pubescent. It has 5 stamens that are inserted in the middle of the tube which includes 2-2.5mm long anthers. The 2-celled and 2 ovuled ovary is inferior with a small disk and 4-15mm long style. The stigmas joined the style up to 2mm of its length. The 2 stigma branches are thread-like and irregularly twisted. Flowers appear during October-January.

Figure 4. (a) *P. Foetida* Plant; (b) Stem node with emerging roots; (c) Leaves dorsal view; (d) Leaves ventral view; (e) Inflorescence; (f) Flower; (g) Fruits; (h) Seeds.

Fruit are shiny brown, capsule, nearly globose, 4-6 mm diameter, with persistent calyx. They are often dotted with white raphides and contain one or two seeds. They appear during November-April.

Seeds are ellipsoidal, plano-convex or convexo-concave, without conspicuous wings, black, often covered with raphides. The whole plant is malodorous; the dominant foul-smelling aroma component is believed to be the sulfur-containing compound dimethyl disulfide which has been identified from volatile oils isolated by steam distillation from leaves, stems, and flowers (Wong and Tan 1994).

Puff (1991), describes the extreme polymorphism in *P. foetida*, especially in leaf shape, leaf size, the indumentum or hairiness of leaf blades, flower size, and inflorescence structure. Some of the variables correlated with the different geographical distribution and some variation show a random pattern. Narrow-leafed variants were reported from Japan, China, and Thailand, yet they also occur elsewhere. Hairy variants or those with particular hair characteristics have been published, as have types with entirely glabrous leaf blades although none has been correlated with geographical distribution, populations of hairy and glabrous plants do not grow intermingled. The longest corolla tube can be up to three times the length of the shortest, but there is no correlation between corolla tube length and native geographical distribution. Corollas with short, wide, nearly campanulate tubes are also known. Inflorescences can appear leafy and difficult to separate from stem foliage or they may lack lower leaf-like bracts to appear distinct.

HABITAT TYPES AND PLANT COMMUNITIES

In its native Asian range, *P. foetida* occurs in a wide variety of habitats and plant communities but is most common in weedy, sunny, disturbed sites that include "waste" areas, roadsides, and fence lines. Skunkvine is also possible in the openings and at the edges of wet to dry forests and woodlands, in open montane vegetation, on sandy, rocky seacoasts, and in the cracks of rocks. Based on the notes from herbarium records in Japan and Taiwan, skunkvine occupies secondary forests, shady forests, gaps in primary forests, grassy hillsides, river banks, canal banks, thickets, and roadsides. Based on an extensive and concise array of field observations and herbarium material, Puff (1991) describes the ecology of *P. foetida* in its native range thus: "in gaps and at edges of wet, evergreen, semi-deciduous or dry, deciduous forests; in woodland and open vegetation types with scattered bushes and trees; in montane communities, on steep wooded slopes, sometimes growing in cracks of rocks; on sandy or rocky sea coasts, sometimes in salt spray. Most commonly, however, "weedy" and found in sunny, disturbed places and secondary vegetation such as in hedges, thickets or in the scrub, in waste places, roadsides, on fences, in cracks of rock walls, etc. It occurs in disturbed areas in both rural and urban areas (Liu and Pemberton 2008). In evergreen and deciduous forests in India, China, Malaysia, Indonesia, and Japan it is common at altitudes of up to 1800 m (Wong and Tan 1994; Sriantaet al. 2012). The species can be locally common in disturbed mesic forest, coastal sites, dry forest, and subalpine woodland, from sea level to 1830 m.

P. FOETIDA AS A WEED

P. foetida has been and remains a principal weed of sugarcane on the island of Mauritius. Changes over time in cultural practices, such as green cane trash blanketing, have favored its more recent proliferation of the sugarcane crops (Seeruttunet al. 2005). It is also well entrenched as a weed and contaminant in the Hawaiian nursery industry for ornamental foliage plants (Pemberton and Pratt 2002).

P. foetida becomes a serious horticultural weed when it moves into lawns and yards in suburban Florida where little can be done to control it. Green, thin and pliable, its fast-growing stems creep into lawn grass safely below the circulating blade of a lawnmower. The stems readily anchor themselves by rooting at the nodes that are continuously in contact with the ground. Stems can run from lawns into the shrubbery, from where, after taking many twining turns around basal branches, the vines twist their way to the top of bushes and hedges, blanketing their surfaces and moving onwards to find the next vertical support.

GENETICS AND REPRODUCTIVE BIOLOGY OF *P. FOETIDA*

Similar to the other members of subfamily Rubiaceae, *P. foetida* has a base chromosome number of n=11. In nature *P. foetida* present as tetraploid, hexaploid and sometimes aneuploid. Puff (1991) reported that tetraploid seeds from Asian plants had a higher germination rate (70%) than seeds collected from hexaploid plants (≤40%).

P. foetida can reproduce both sexually and asexually. Sexual reproduction in *P. foetida* takes place by formation of seeds. Its flowers are perfect but the development of flower is protandrous (anthers matures before the stigma becomes receptive), so it requires pollinating agents for pollination and fruit set (Liu et al. 2006). In *P. foetida* flowering occurs in the late growing season i.e., it has a long reproductive cycle. Puff (1991) reported that in extratropical latitudes the late development of inflorescence in the growing season is caused by the climate which results in depauperate and few-flowered inflorescences. Fruits may overwinter for a short period on the dormant vine, during which the capsules become dry and brittle, usually releasing their one or two seeds.

Minimal information is available on the seed viability and the seed bank dynamics of *P. foetida* recorded a low (10%) germination rate from seeds collected from vines during mid-winter in central Florida. Field research conducted in Japan demonstrated that the freshly matured seeds, collected in autumn, will not germinate until subjected to a period of overwintering (up to ten months), indicative of innate seed dormancy. Similar research in Florida suggested that *P. foetida* creates a short-term persistent seed bank even though most of the seeds germinate or die during their first year of outdoor storage, especially those stored in open habitats. Viable seeds remained in the seedbank (up to 38% were stored in a shady, forest floor) for the second season, although they quickly lost their viability within two years (Liu and Pemberton 2008). A previous study (Puff 1991) had demonstrated continuous viability forupto two years, though these seeds were likely stored indoors during the study period and, regardless, germination rates declined markedly afterward. In summary, seeds of *P. foetida* require a dormancy period prior to germination which, in theory, allows for dispersal and short-term seed bank development over a few years.

Asexual reproduction through the fragmentation of stems bearing nodal roots cannot be discounted, especially during flood events and when human activities and equipment are involved.

CHEMICAL CONSTITUTION OF *P. FOETIDA*

To investigate the chemical composition of the essential oil of *P. scandens* (Yang et al. 2012) aerial parts of *P. foetida* L. were taken and steam distillation using Clavenger apparatus was performed in order to obtain the volatile oils. Gas chromatography/mass spectrometric (GC/MS) analyses of the essential oil showed the following composition (Table 3).

Table 3. Showing relative chemical components in leaf, stem, and flower of *P. foetida*

Sl. No.	Name of chemical compound	Content of chemical compound (%)		
		Leaf	Stem	Flower
1.	Butanedione	0	0.6	0
2.	Pentan-2-one	0.6	0	0
3.	Pentanal	0	0	6.3
4.	2-Methylbut-3-en-2-ol	0	14.8	0
5.	S-Methyl thioacetate	0.8	0	0
6.	Dimethyl disulphide	19.5	1.9	1.1
7.	Hexanal	0.1	0.2	6.9
8.	β-Pinene	0	0.3	0
9.	Dicthyl carbonate	0	0	0.04
10.	(E)-Pent-3-en-2-one	0.2	0.2	0
11.	Butan-1-ol	0	0	0.9
12.	Pent-I-en-3-one	0.1	0	0
13.	Myrcene	0.03	0	0.03
14.	Pen t-3-en-2-ol	0	0	3.3
15.	Pyridine	0	0.1	0
16.	Heptanal	0	0.03	0.04
17.	Limonene	0.1	0.5	0.03
18.	Dodccanc	0.04	0	0
19.	β-Phellandrene	0.02	0.2	0
20.	3-Methylbutan-1-ol	0.2	0.3	0
21.	(E)-Hex-2-enal	0.04	0.04	0
22.	Thiazole	0	0.03	0
23.	Styrene	0	0.02	0.03
24.	Pentan- 1-ol	0.1	0	0.04
25.	(E)-p-Ocimene	0.03	0	1.9
26.	Methyl propyl sulphide	0.04	0.1	0
27.	3-Hy drox ybutan-2-one	0	0.03	0
28.	3-Methylbut-2-en-I –ol	0.3	7.1	0
29.	(E)-Hept-2-enal	0.02	0	0
30.	Hexan-I –ol	0.5	0.1	0
31.	4-Hydroxy-4-methylpentan-2-one	0	0.04	1.8
32.	(E)-Hex-3-en-l-ol	0.1	0	0
33.	Dimethyl trisulphide	0.4	0.4	0
34.	(Z)-Hex-3-en-l-ol	2.4	0.1	0.04
35.	Nonanal	0	0	0.6

Table 3. (Continued)

Sl. No.	Name of chemical compound	Content of chemical compound (%)		
		Leaf	Stem	Flower
36.	(E)-Hex-2-en-l-ol	1	0	0.04
37.	2-Methoxy-3-(1 -methylethyl)pyrazine	0	0.05	0.05
38.	cis-Linalol oxide (furanoid)	0.04	0.8	0
39.	S,S-Dimethyl dithiocarbonate"	0.8	0	0.04
40.	3-(Methythio)propanaI	0.04	0.4	0.03
41.	2-Furancarboxaldehyde	0.8	3.5	0
42.	2-Ethylhexan-1 –ol	0	0	0.6
43.	Pentadecane	0.04	0.04	0
44.	2-Acetylfuran	0	0.04	0
45.	Pyrrole	0	0.03	0
46.	Benzaldehyde	0.4	0.3	0.03
47.	2-Methoxy-3-(2-methylpropyl)pyrazine	0	0.2	0
48.	Linalol Octan-1-ol	38.8	21.5	25.2
49.	2-Methylbutanoic acid	0.03	0.1	0
50.	Hexadecane	0	0.1	0
51.	Terpinen-4-ol	0.3	0	0
52.	α,4-Dimethyl-3-cyclohexene-1 -aldehyde"	0	0.2	0
53.	y-Butyrolactone	0	0.2	0
54.	Phenylacetaldehyde	0.1	0	0.03
55.	2-Furanmethanol	0.8	9.4	0.04
56.	Benzofuran	0.5	0.8	0.3
57.	α-Terpineol 1 ,2-Dimethoxybenzene	0.6	0.9	4.5
58.	3-Methyl-2(5H)-furanone	7.8	4.8	6
59.	3-Ethyl-4-methyl-2,5-furandione	0	0.3	0.5
60.	Methyl salicylate	0.1	0	0
61.	Nerol	0.4	0	0
62.	β-Damascenone	0.4	2.5	2.6
63.	1 –Phenylethanol	2.3	1.5	2
64.	Geraniol	0.5	0.1	0
65.	Benzyl alcohol	0	0	0.7
66.	2-Methoxyphenol	7.3	5	7.2
67.	2-Phenylethanol	0.1	1.2	0
68.	2,6-Di-t-butyl-4-mcthylphenol	0	0	9.2
69.	3,5,5-Trimethylhexanoic acid"	0	0.8	2
70.	Eugenol	0.5	0	0
71.	p-Vinylguaicol	0	0.7	0
72.	Carvacrol	0	0.8	1.4
73.	3-Ethyl-4-methylmaleimide	4.9	9.2	3
74.	2.3-Dihydrobenzofuran	0	0.2	0
75.	Indole	0.2	0	0

High Medium Low

*Above table taken from, Wong and tan 1994.

PHARMACOLOGICAL ACTIVITIES

P. foetida is popularly used in traditional ethnic remedies (Table 2). This plant contains many useful properties to cure human diseases; some of which are described in brief (Figure 5)-

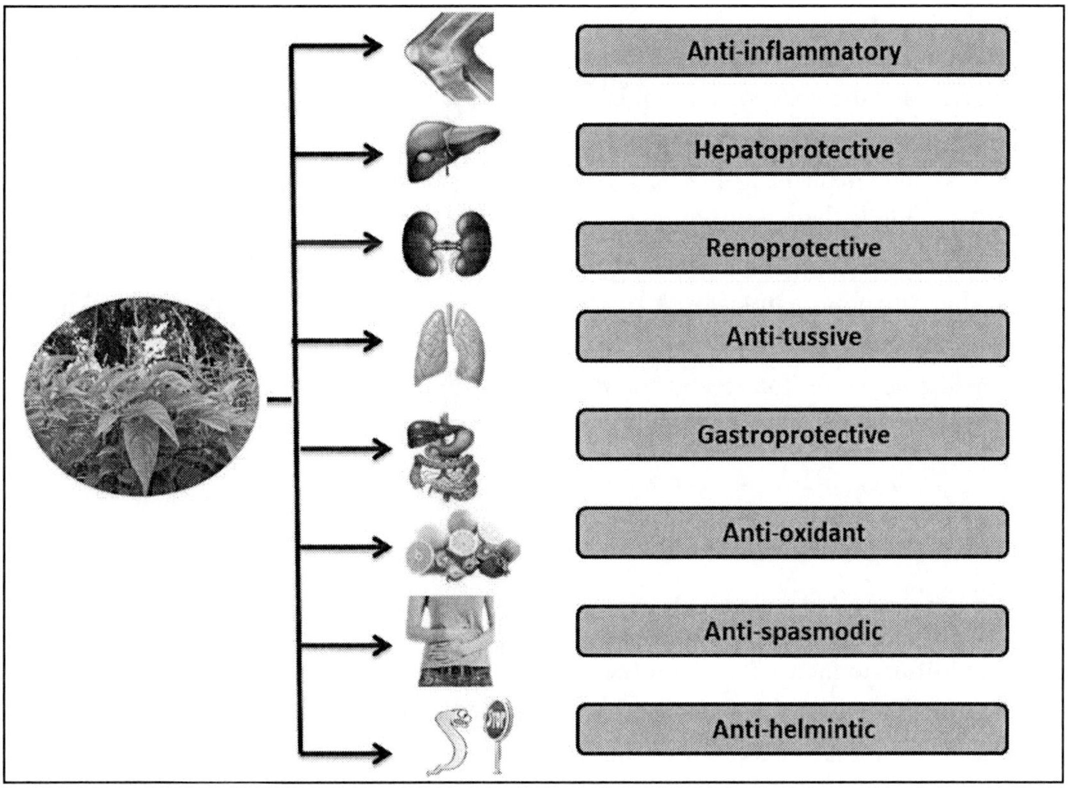

Figure 5. Different pharmacological properties of *P. foetida*.

Anti-Inflammatory and Antioxidant Activity

Butanol fraction of a methanol extract of the leaves of *P. foetida* was shown as an anti-inflammatory by De et al. 1994. In the albino rat, induced colitis is treated successfully by the ethanolic extract of leaves and concludes that this happens because of its anti-inflammatory and antioxidant property (Das et al. 2013). The extract of *P. foetida* is reported to have anti-inflammatory property and is a promising drug against arthritis as experimented on rats in anti-gouty arthritis by Ma et al. 2009 and shows the effect of EPS on the induction of inflammatory cytokines TNF-α and IL-1β, and the activity of NF-κB in GA to investigate its anti-inflammatory mechanism in GA rats.

Methanolic extract of *P. foetida* reported enhancing the activity of enzymatic and non-enzymatic antioxidant. Leaf extract showed the antioxidant property (Osman et al. 2009).

The antioxidant activity of the plant extract might be due to inhibition of lipid peroxidation and increase in the SOD, GPx, and CAT (Kumar et al. 2014).

Hepatoprotective and Gastroprotective Activity

The hepatoprotective activity of this plant was reported long back in 1993 by De et al. then after that, there are several reports present which confirms its hepatoprotective nature. It is reported that the possible mechanisms of those plant-derived. Hepatoprotective agents might be correlated to their significant antioxidant and anti-inflammatory effects (Ravikumar and Gnanadesigan 2012). Iridoid glycosides present in *P.* species are reported to be effective in hepatic injury or inflammation (Peng et al. 2015).

P. foetida can be used in traditional medicine, as an antidiarrhoeal agent. In animal models, the ethanolic extract of *P. foetida* showed antidiarrhoeal activity by inhibiting gastrointestinal motility (Afroz et al. 2006). *P. foetida* emerge as gastroprotective agent for use as medicine in the future and the gastroprotective activity might be due to the Nrf2 mediated antioxidant and anti-secretory effects (Chanda et al. 2015)

Renoprotective Potential

Uric acid nephropathy (UAN) is caused by excessive uric acid formation, the deposition of urate microcrystals in the kidneys causes inflammation and the condition is known as nephropathy. Uric acid/urate microcrystal provokes an inflammatory response which leads to fibrosis, loss of nephrons, and eventually to chronic renal failure (Shimizu and Hori 2009). Iridoid glycosides present in these plant plays a major role in inhibition of inflammation in the kidney through blockage of NFκB, a key molecule in inflammation (Zhu et al. 2012).

Antitussive, Antispasmodic and Antihelmintic Activity

Antitussive activity of ethanolic extract of *P.foetida* in conscious cats is similar to that of commonly used non-narcotic antitussive dropropizine, suggesting its possible therapeutic use. However, further studies need to be done to evaluate its efficacy in clinical conditions. The study showed the ethanolic extract of *P. foetida* had a cough suppressant effect, with a decrease in cough intensity and frequency. The effect was less than codeine but similar to non-narcotic dropropizine. The effect could be related to its anti-inflammatory activity. (Soni et al. 2013)

Table 4. Showing the traditional use of *P. foetida*

Plant part	Region	Use	References
Aerial green parts (Leaf)	Assam	Allergy, in gastralgia, post-natal pain, and bleeding, diarrhoea and dysentery, as vegetable, abdominal pain, Leaves are used as Curries, young leaf and shoot are used kidney trouble, diarrhoea and for children and women after child birth. Decoction of leaves increases appetite, good remedy for rheumatic pain	Choudhury et al.(2012); Upadhyaya(2013); Saikiaet al.(2006)
	Tripura	Tribal of Tripura uses the leaves with dry fish. Traditionally, the prime use of the plant in different digestive problems like gastric trouble, to clean stomach, against stomach swelling, gastritis, in loose motion, diarrhoea, in indigestion, ulceration etc.	Chanda et al. (2014)
	Telangana, Andhra Pradesh	As vegetable, skin diseases, rheumatism, dysentery	Sreeramulu et al.(2013)
	Arunachal Pradesh	Urinary disorder, kidney stone and digestive problem also with rice as vegetables in indigestion, in gastric trouble, to clean stomach and against stomach swelling and diarrhoea, gastritis and loose motion	Kagyunget al.(2010); Srivastava and Singh (2010); Tangjang et al. (2011)
	Sikkim and Darjeeling	Anti-diabetic, applied to the abdomen in case of retention of urine, as a diuretic and to dissolve vesicle calculi	Chhetri et al. (2005)
	Meghalaya;	As antidote for snake bite; rheumatism and gout	Hynniewta and Kumar (2008); Singh et al.(2010)
	Orissa	Leaf paste is applied externally on rheumatic parts of the body until the complete cure. Leaves are cooked with rice and taken one or two days in the treatment of rheumatism and gout. Plant extract is also taken internally for 8-10 days in joint pain	Singh et al.(2010)
	Nagaland	The decoction of fresh crushed root 10 g in 100 ml of water is taken orally for the removal of worms especially roundworm. The decoction of leaves and shoots 30 ml is taken in the treatment of abdominal colic and gastric ulcer	Jamir et al. (1999)
	West Bengal	A curry is prepared with the leaf of this plant and fed to the patients of enteritis and dysentery for 10- 15 days.	Pattanayak et al.(2015)
	Bangladesh	Diarrhoea, to relieve distention and flatulence, antihyperglycemic	Khan et al.(2011)
	Malaysia	As vegetable eaten raw or steamed	Osman et al.(2009)
Root	India	Asthma , emetic, piles, inflammation of spleen, pain in chest and liver, seminal weakness.	Chopra et al. (1969)
Bark	Philippine	Emetic	
Fruit	India, Banglades, Philippine	Toothache	Khan et al. (2011)
Seeds	India, Bangladesh	Piles and leukoderma	Ghani A (2003)

The ethanolic extract also exhibited antispasmodic activity on isolated guinea-pig ileum. Juice of leaves showed a potent anthelmintic effect against bovine helminths– *Strongyloides* spp, *Trichostrongylus*, and *Haemonchus* spp. Methanolic extract of the leaves of *P.foetida* found to be effective against *Pheretimaposthuma* and *Tubifex tubifex* (Pal et al. 2011).

CHRONIC INFLAMMATION, ARTHRITIS, AND *P. FOETIDA*

Inflammation is the response of the immune system to infection and injury and has been implicated in the pathogeneses of arthritis, cancer, and stroke, as well as in neurodegenerative and cardiovascular disease. Inflammation itself is a beneficial event that leads to the removal of offending factors and restoration of tissue structure and physiological function. The acute phase of inflammation is characterized by the rapid influx of blood granulocytes, typically neutrophils followed swiftly by monocytes that mature into inflammatory macrophages that subsequently proliferate and thereby affect the functions of resident tissue macrophages. This process causes the cardinal signs of acute inflammation: redness, heat, swelling, and pain. Once the initiating noxious stimulus is removed via phagocytosis, the inflammatory reaction can decrease and resolve. During the resolution of inflammation, granulocytes are eliminated and macrophages and lymphocytes return to normal pre-inflammatory numbers and phenotypes. The usual outcome of the acute inflammatory program is a successful resolution and repair of tissue damage, rather than persistence and dysfunction of the inflammatory response, which can lead to scarring and loss of organ function. It may be anticipated, therefore, that failure of acute inflammation to resolve may predispose to auto-immunity, chronic dysplastic inflammation and excessive tissue damage. Clinically, several non-steroidal anti-inflammatory drugs (NSAIDs), TNF-α inhibitors, selective cyclooxygenase inhibitors (COX-2), immunosuppressive and antioxidant drugs are used to treat chronic inflammatory, pain, and bone destruction during arthritis. But above treatment having the limitation in the treatment of Rheumatoid arthritis (RA) due to their undesirable effects like less effectiveness with chronic use, myelosuppression, gastrointestinal disorder (mainly ulcer), neutropenia, cardiovascular disorder, hemorrhage and high in cost. With these complexities, the RA research field has been improvised towards plant-based therapies that are thought of safe and effective altogether elevating chronic pain related to arthritis. RA is that most often occurs arthritis, being classified treated with the antioxidant and anti-inflammatory drugs. At this moment *P.foetida* becomes the matter of interest because of its amazing anti-inflammatory property. There are several reports about the anti-inflammatory property and gives relieve in inflammatory diseases like RA and others. RA, the joint inflammatory disease caused by chronic inflammation in joints. There are different agents which play a major role in the chronic inflammatory response and eventually the disease (Figure 6).

There are two main paths from which inflammation can initiate, a. microorganism invasion and signaling pathway through Toll-Like Receptor (TLR), which is a transmembrane protein and it activates the cytosolic NF-κB. The NF-κB proteins are normally sequestered in the cytoplasm by a family of inhibitory proteins, including IκB family members, most important IκB family member is IκBα. There is again two ways of activation of the NF-κB pathway, canonical and noncanonical (or alternative). In canonical, the primary mechanism for canonical NF-κB activation is the inducible degradation of IκBα triggered through its site-specific phosphorylation by a multi-subunit IκB kinase (IKK) complex. IKK is composed of two catalytic subunits, IKKα, and IKKβ, and a regulatory subunit named NF-κB essential modulator IKKγ or NEMO. IKK can be activated by different stimuli, including cytokines, growth factors, mitogens, microbial components, and stress agents. Upon activation, IKK phosphorylates IκBα at two N-terminal serines and, thereby, triggers ubiquitin-dependent IκBα degradation in the proteasome, resulting in rapid and transient nuclear translocation of canonical NF-κB members predominantly the p50 and p65 dimers. Being a transcription factor NF-κB activates different proinflammatory genes such as TNF-α and IL-1β hence eventually triggered the inflammation. Extract of *P. foetida* inhibit the expression of NF-κB (Borgohainet al. 2017) and hence inflammation as in Figure 6 above.

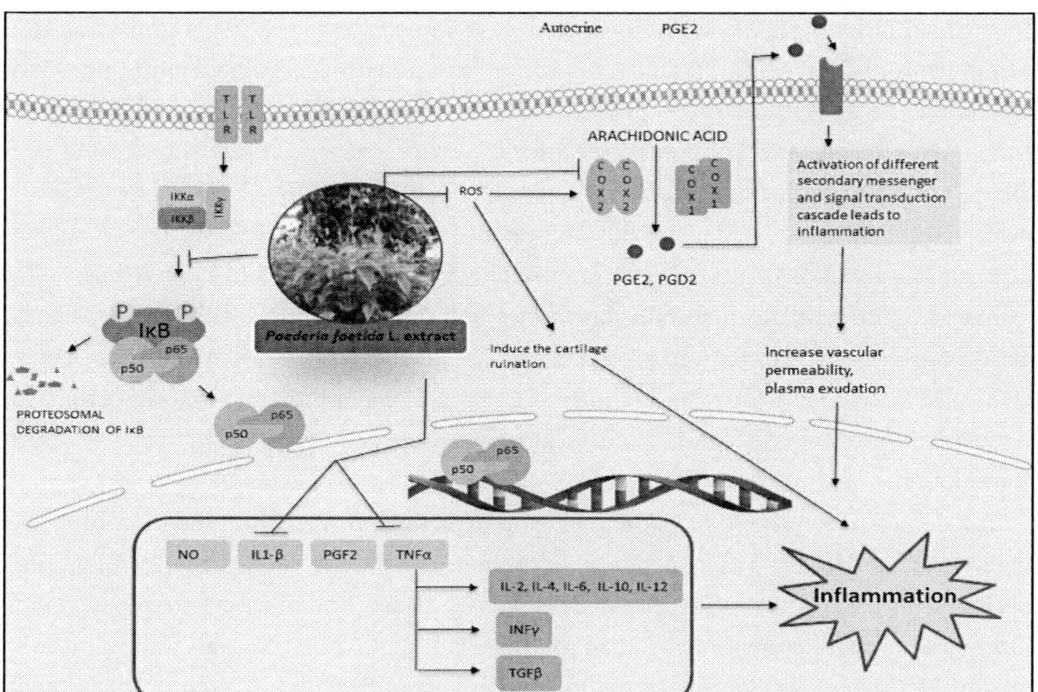

Figure 6. Molecular mechanism showing the inhibitory role of *P. foetida*against inflammation.

Again TNF-α and IL-1β production are inhibited by *P.* extract at their gene-level expression. Furthermore, these inhibitions might be correlated with the blockage of NF-κBactivation in synovial tissue, which suggests that the extract of P. *foetida* has a potential therapeutic role in Gouty Arthritis (Ma et al. 2009).

In addition, the noncanonical NF-κB activation does not involve IκBα degradation but rather relies on processing of the NF-κB2 precursor protein, p100. A central signaling molecule for this pathway is NF-κB-inducing kinase (NIK), which activates and functionally cooperates with IKKα to mediate p100 phosphorylation, which in turn induces p100 ubiquitination and processing. The processing of p100 involves the degradation of its C-terminal IκB-like structure, resulting in the generation of mature NF-κB2 p52 and nuclear translocation of the noncanonical NF-κB complex p52/RelB. Functionally, canonical NF-κB is involved in almost all aspects of immune responses, whereas the noncanonical NF-κB pathway appears to be evolved as a supplementary signaling axis that cooperates with canonical NF-κB pathway in the regulation of specific functions of the adaptive immune system. Not shown.

Another cause is Prostaglandins (PG), PGs are lipid autacoids derived from arachidonic acid. They both sustain homeostatic functions and mediate pathogenic mechanisms, including the inflammatory response. They are generated from arachidonate by the action of cyclooxygenase (COX) isoenzymes and their biosynthesis is blocked by nonsteroidal anti-inflammatory drugs (NSAIDs), including those selective for inhibition of COX-2. Prostaglandins play a key role in the generation of the inflammatory response. Their biosynthesis is significantly increased in inflamed tissue and they contribute to the development of the cardinal signs of acute inflammation. Prostaglandin production depends on the activity of prostaglandin G/H synthases, colloquially known as COXs, bifunctional enzymes that contain both cyclooxygenase and peroxidase activity and which exist as distinct isoforms referred to as COX-1 and COX-2. COX-1, expressed constitutively in most cells, is the dominant source of prostanoids that subserve housekeeping functions, such as gastric epithelial cytoprotection and homeostasis. COX-2, induced by inflammatory stimuli, hormones and growth factors, is the more important source of prostanoid formation in inflammation and in proliferative diseases, such as cancer. However, both enzymes contribute to the generation of autoregulatory and homeostatic prostanoids, and both can contribute to prostanoid release during inflammation.

In the case of RA, another pathogenesis is free radical formation, especially reactive oxygen species (ROS) which actively participate to induce the cartilage ruination, in addition with other stress they cause conversion of arachidonic acid to prostaglandins consequently the inflammation. And again, the Plant *P. foetida* plays an important role in quenching ROS and thus inhibit the pathogenic effects of ROS in RA. According to reports, it is found that extract of this plant is able to enhance the production of GSH (glutathione reductase), a potent ROS scavenging non-enzymatic molecule in the human body.

Conclusion and Future Prospective

P. foetida has immense importance in the field of herbal drugs. This plant has therapeutic values due to the presence of many pharmacologically bioactive compounds like asperuloside, paederosidic acid, phenolic compounds, alkaloid, volatile oil, sitosterols, stigmasterol, campesterol, ellagic acid, lignans, iridoids, methylenedioxy compound, tannins, triterpenoids, urosil acid, and epifriedelinol. Chemical formulations from this plant are made for the treatment of ulcers, diarrhoea, diabetes, and stomach diseases. Due to the presence of a number of medicinal properties in *P. foetida* it is a utilized ethnic by many tribes and villages but their commercial production is even now not very much successful, only some ayurvedic companies produce its packaged product in a small level. To make this plant of commercial use there is a need to do some more serious research and formulation of new drugs of low cost and without any side effects.

Acknowledgments

Authors are thankful to CAS-Botany (BHU) and DST-FIST for providing funds and internet facilities to the Department of Botany, Institute of Sciences, Banaras Hindu University. Niraj Kumar Goswami is thankful to CSIR-UGC New Delhi, India for providing financial support in the form of fellowship.

References

Afroz, S., Alamgir, M., Khan, M. T. H., Jabbar, S., Nahar, N., Choudhuri, M. S. K. (2006) Antidiarrhoeal activity of the ethanol extract of *Paederia foetida* Linn (Rubiaceae). *J Ethnopharmacol* 105: 125-130.

Borgohain, M. P., Chowdhury, L., Ahmed, S., Bolshette, N., Devasani, K., Das, T. J., Mohapatra, A., Lahkar, M. (2017) Renoprotective and antioxidative effects of methanolic *Paederia foetida* leaf extract on experimental diabetic nephropathy in rats. *J Ethnopharmacol* 198: 451-459.

Chanda, S., Deb, L., Tiwari, R. K., Singh, K., Ahmad, S. (2015) Gastroprotective mechanism of *Paederia foetida* Linn (Rubiaceae)-a popular edible plant used by the tribal community of North-East India. *BMC complement Altern Med* 15: 304.

Chhetri, D. R., Parajuli, P., Subba, G. C. (2005) Antidiabetic plants used by Sikkim and Darjeeling Himalayan tribes, India. *J Ethnopharmacol* 99: 199-202.

Chopra, R. N., Chopra, R. N. (1969) *Supplement to glossary of Indian medicinal plants.*

Choudhury, K. D., Choudhury, M. D., Paul, S. B., Baruah, M. K. *Bioactivities of some ethno-medicinal Rubiaceous plants available from Assam–a review.*

Das, S., Kanodia, L., Mukherjee, A., Hakim, A. (2013) Effect of ethanolic extract of leaves of *Paederiafoetida* Linn on acetic acid induced colitis in albino rats. *Indian J Pharmacol* 45: 453.

De, S., Ravishankar, B., Bhavsar, G. C. (1994) Investigation of the anti-inflammatory effects of *Paederia foetida*. *J Ethnopharmacol* 43: 31-38.

Ghani, A. (2003) *Medicinal Plants of Bangladesh, the Asiatic Society of Bangladesh,* 2nd Revised Edn.Dhaka, Bangladesh 603.

Hou, S. X., Zhu, W. J., Pang, M. Q., Jeffry, J., Zhou, L. L. (2014) Protective effect of iridoid glycosides from *Paederiascandens* (LOUR.) MERRILL (Rubiaceae) on uric acid nephropathy rats induced by yeast and potassium oxonate. *Food Chem Toxicol* 64: 57-64.

Hynniewta, S. R., Kumar, Y. (2008) *Herbal remedies among the Khasi traditional healers and village folks in Meghalaya.*

Jamir, T. T., Sharma, H. K., Dolui, A. K. (1999) Folklore medicinal plants of Nagaland, India. *Fitoterapia* 70: 395-401.

Kagyung, R., Gajurel, P. R., Rethy, P., Singh, B. (2010). Ethnomedicinal plants used for gastro-intestinal diseases by Adi tribes of Dehang-Debang Biosphere Reserve in Arunachal Pradesh.

Kapadia, G. J., Sharma, S. C., Tokuda, H., Nishino, H., Ueda, S. (1996) Inhibitory effect of iridoids on Epstein-Barr virus activation by a short-term in vitro assay for anti-tumor promoters. *Cancer Lett* 102: 223-226.

Khan, S., Zahan, D., Das, R., Nasrin, D., Ahsan, S., Ahmed, R., Sadat, A. F., Bashar, A. B., Nahar, N., Rahmatullah, M. (2011) Antihyperglycemic activity studies with methanol extract of *Madhucaindica* JF Gmel leaves and *Paederiafoetida* L stems in mice. *Adv Nat Appl Sci* 5: 122-126.

Kumar, V., Anwar, F., Ahmed, D., Verma, A., Ahmed, A., Damanhouri, Z. A., Mishra, V., Ramteke, P. W., Bhatt, P. C., Mujeeb, M. (2014) *Paederia foetida* Linn leaf extract: an antihyperlipidemic, antihyperglycaemic and antioxidant activity. *BMC complement Altern Med* 14: 76.

Liu, H., Pemberton, R. W. (2008) Differential soil seed bank longevity of *Paederia foetida* L. an invasive woody vine, across three habitats in Florida. *J Torrey Bot Soc* 135: 491-496.

Liu, H., Pemberton, R. W., Stiling, P. (2006) Native and introduced pollinators promote a self-incompatible invasive woody vine (*Paederiafoetida* L) in Florida. *J Torrey Bot Soc* 133: 304-311.

Ma, Y., Zhou, L. L., Yan, H. Y., Liu, M. (2009) Effects of extracts from *Paederiascandens* (LOUR.) MERRILL (Rubiaceae) on MSU crystal-induced rats gouty arthritis. *Am J Chin Med* 37: 669-683.

Nosáľová, G., Mokrý, J., Ather, A., Khan, M. T. H. (2007) Antitussive activity of the ethanolic extract of *Paederiafoetida* (Rubiaceae family) in non-anaesthetized cats. *Acta Vet Brno* 76: 27-33.

Osman, H., Rahim, A. A., Isa, N. M., Bakhir, N. M. (2009) Antioxidant activity and phenolic content of *Paederia foetida* and *Syzygiumaqueum*. *Molecules*14: 970-978.

Pal, M. K., Dey, Y. N. (2011) Evaluation of anthelmintic activity of leaves of *Paederia foetida*. *Int J Pharma Biosci* 2: 227-231.

Pattanayak, S., Mandal, T. K., Bandyopadhyay, S. K. (2015) A Study on Use of Plants to Cure Enteritis and Dysentery in Three Southern Districts of West Bengal, India. *J Med Plants* 3: 277-283.

Peng, W., Qiu, X. Q., Shu, Z. H., Liu, Q. C., Hu, M. B., Han, T., Rahman, K., Qin, L. P., Zheng, C. J. (2015) Hepatoprotective activity of total iridoid glycosides isolated from *Paederia scandens* (lour.) Merr var. tomentosa. *J Ethnopharmacol* 174: 317-321.

Puff, C. (1991) Revision of the genus *Paederia* L. (Rubiaceae: Paederieae) in Asia. *Op Bot Belg* 3: 207-289.

Ravikumar, S., Gnanadesigan, M. (2012) Hepatoprotective and antioxidant properties of *rhizophoramucronata* mangrove plant in CCl4 intoxicated rats. *J Exp Clin Med* 4: 66-72.

Saikia, A. P., Ryakala V. K., Sharma, P., Goswami, P., Bora, U. (2006)Ethnobotany of medicinal plants used by Assamese people for various skin ailments and cosmetics. *J Ethnopharmacol* 106: 149-157.

Seeruttun, S., Barbe, C., Gaungoo, A. (2005) Vine weeds in sugarcane: fluroxypyr provides cost-effective post-emergence control in Mauritius. *Sugarcane Int* 23: 3-5.

Shimizu, T., Hori, H. (2009) The prevalence of nephrolithiasis in patients with primary gout: a cross-sectional study using helical computed tomography. *J Rheumatoljrheum-*081128.

Singh, H., Krishna, G., Baske, P. K. (2010) Plants used in the treatment of joint diseases (rheumatism, arthritis, gout and lumbago) in Mayurbhanj district of Odisha, India. *Rep opin* 2: 22-26.

Soni, R. K., Irchhaiya, R., Dixit, V., Alok, S. (2013) *Paederiafoetida* Linn: phytochemistry, pharmacological and traditional uses. *Int J Pharm Sci Res* 4: 4525.

Sreeramulu, N., Suthari, S., Ragan, A., Raju, V. S. (2013) Ethno-botanico-medicine for common human ailments in Nalgonda and Warangal districts of Telangana Andhra Pradesh India. *Ann Plant Sci* 2: 220-229.

Srianta, I., Patria, H. D., Arisasmita, J. H., Epriliati, I. (2012) Ethnobotany, nutritional composition and DPPH radical scavenging of leafy vegetables of wild *Paederiafoetida* and Erechtiteshieracifolia. *Int Food Res J* 19: 245-250.

Srivastava, R. C., Singh, R. K., Mukherjee, T. K. (2010) *Indigenous biodiversity of Apatani plateau:* Learning on biocultural knowledge of Apatani tribe of Arunachal Pradesh for sustainable livelihoods.

Tangjang, S., Namsa, N. D., Aran, C., Litin, A. (2011) Anethnobotanical survey of medicinal plants in the Eastern Himalayan zone of Arunachal Pradesh, India. *J Ethnopharmacol* 134: 18-25.

Upadhyaya, S. (2013) Screening of phytochemicals, nutritional status, antioxidant and antimicrobial activity of Paederiafoetida Linn from different localities of Assam India. *J Pharm Res* 7: 139-141.

Wong, K. C., Tan, G. L. (1994) Steam volatile constituents of the aerial parts of *Paederiafoetida* L. *Flavour Fragr J* 9: 25-28.

Yang, K., Liu, Q. Z., Liu, Z. L., Du, S. S. (2012) GC-MS analysis of insecticidal essential oil of aerial parts of *Paederiascandens* (Lour) Merrill (Rubiaceae). *Trop J Pharm Res* 11: 461-467.

Zhu, W., Pang, M., Dong, L., Huang, X., Wang, S., Zhou, L. (2012) Anti-inflammatory and immunomodulatory effects of iridoid glycosides from *Paederiascandens* (LOUR.) MERRILL (Rubiaceae) on uric acid nephropathy rats. *Life Sciences* 91: 369-376.

In: Innovations in Life Science Research ISBN: 978-1-53615-868-7
Editors: R. P. Sinha, S. Pandey-Rai and N. Ghoshal ©2019 Nova Science Publishers, Inc.

Chapter 13

CARICA PAPAYA: PHARMACOLOGICAL, NUTRITIONAL AND ITS THERAPEUTIC USES

Bipin Maurya[1], Lakee Sharma[1], Apoorva[1], Shashi Pandey-Rai[1] and Sanjay Kumar Rai[2,*]

[1]Laboratory of Morphogenesis, Centre of Advance Study in Botany, Institute of Science, Banaras Hindu University, Varanasi, India

[2]Department of Horticulture, Dr. Rajendra Prasad Central Agricultural University, Pusa, Samastipur, India

ABSTRACT

Carica papaya is a highly medicinal fruit crop plant and it belongs to family Caricaceae. The plant contains various pharmacologically active compounds with high medicinal importance. Due to presence of these compounds, this plant is used for the treatment of many ailments like arthritis, cancer, dengue, ulcer and many more. Although all parts of the plant harbour medicinally important compounds. In this chapter, we are mainly concerned with the importance of papaya leaf extract and mainly to the cure of dengue. Dengue can be cured by oral consumption of *Carica papaya* leaf extract (CPLe). Leaf extract of papaya reduces the symptoms of dengue by prevention of replication and spread of dengue virus (DV) and by helping in the increase of platelet count (prevention of thrombocytopenia). Prevention of viral replication by papaya is possible only due to presence of a flavonoid component named quercetin which inhibits the serine proteases NS2B and NS3 (precursor of polyprotein required for viral assembly). Platelet count is increased by megakaryocyte activation and aggregation. Papaya contains cysteine endopeptidases like chymopapain, glycylendopeptidases and papain. Papaya also have antidiabetic, anti-inflammatory, cardioprotector (reduces blood pressure), antioxidant, antimicrobial activities.

* Cooresponding Author's Email: sanjayhortipusa@gmail.com.

Keywords: *Carica papaya*, Dengue virus, thrombocytopenia, quercetin

INTRODUCTION

Papaya (*Carica papaya* Linn.) is a tropical and subtropical fruit, which is now frequently used in traditional medicine for many ailments of human beings (Vij and Prashar 2015). It is a perennial plant of family Caricaceae having great medicinal properties. Not only the fruits but whole plant parts like roots, bark, seeds, and pulp are known to contain biologically active compounds like Papain, chymopapain, quercetin, pectin, carposide, carpaine, pseudocarpine, dehydrocarpines and carotenoids (Saran and Chaudhary 2013). The papaya can be regarded as neutraceutical and vitalizer. C*arica papaya* leaf extract (CPLe) have been demonstrated to have many medicinal properties like antioxidative, anti-inflammatory, anti-bacterial, nephroprotective, anticancer, hepatoprotective, hypoglycaemic effects (Lim 2012). Papaya contains chymopapain and papain which are biologically active compounds which are used for the treatment of arthritis (Saran and Chaudhary 2013; Zakaria et al. 2006). Besides papain and chymopapain, the main concern towards papaya is due to the presence of other biologically active compounds like cyanogenic glycoside and flavonoid quercetin. Efficacy of CPLe to cure dengue fever (DF) is perhaps due to the presence of these compounds because it acts as anti-thrombocytopenic and anti-viral agent (Subenthiran et al. 2013). Thrombocytopenia is a pathological condition in which the platelet count is less than 150×105/ L. Papaya is available throughout year and it is a rich source of vitamins, minerals, fibre, polysaccharides, alkaloids, glycosides, fats, oil, lectins, saponins, flavonoids, phytosterol, phenolic compounds, terpenoids, tannins, antioxidants etc (Lim 2012). It improves the cardiovascular system, remedy for heart diseases, and colon cancer (Krishna et al. 2008). The fruit of papaya is a rich source of β-carotene which prevents the damages caused by free radicals and especially it is very useful in the abdominal and digestive disorders. Papaya controls the high cholesterol level and also a good source of fibre. Papaya is the rich source of proteolytic enzymes which are helpful in the digestion of proteins thus, also used to cure digestive ailments. Ripe fruits of papaya are the best medicine for habitual constipation (Shenekar et al. 2014). The papaya also has antiaging property. The skin of papaya used as a medicine for wounds and worts (Starley et al. 1999).

ORIGIN AND DISTRIBUTION

Mexico and Northern South America are native countries of *Carica papaya* (figure 1). The luxuriant growth of the papaya plant is found in tropical and subtropical countries between 32°North and South latitudes but the best growth is found between 23°N and S latitudes (Nakasone and Paull 1998). Later it was introduced in Asian countries like India, Sri Lanka, South Africa and other parts of the word by various methods of seed dispersal. The distribution was further continued throughout Asia and Pacific regions. In eighteenth-century its seeds were dispersed from the Caribbean to Malacca and India by botanists.

MORPHOLOGY

Carica papaya L. is 6 to 12 m tall delicate tree belonging to Class Magnoliopsida, order Brassicales and family Caricaceae. In this plant branching rarely takes place and when branching is found it is profuse. Stem of the plant is hollow from inside. Leaves are arranged spirally on top of the trunk.

Leaves are simple, palmately arranged and lobed. Petiole of the leaf is very large and hollow which is directly attached to the main stem. Leaves of papaya have "kranz anatomy" and thus plant performs "Hatch and Slack"/C4 cycle during triose phosphate formation.

Papaya is polygamous and dichasial flowers which may be monoecious or dioecious that is it may be male, female or hermaphrodite (figure 2). The flower is pentamerous with 5 sepals and 5 petals.

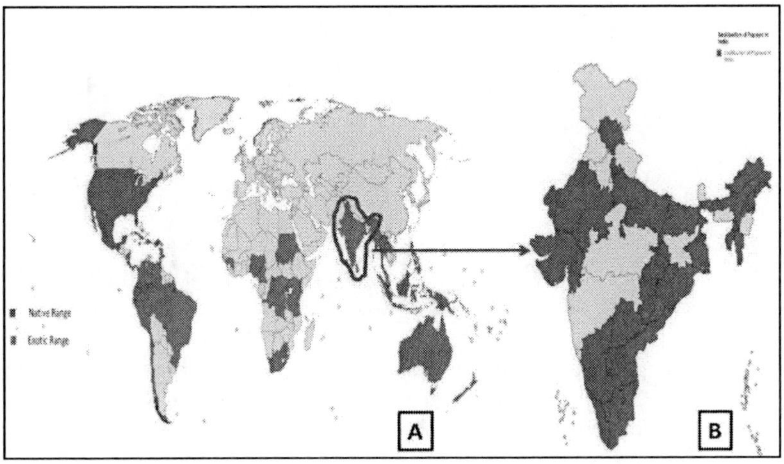

Figure 1. Distribution of Papaya (A) distribution worldwide (B) and distribution in India.

Male flowers have epipetalous stamen only and female have pistil only while hermaphrodite has both. In both, pistillate and hermaphrodite flowers, ovary is superior.

Fruits of papaya are berry type, green (unripe) and yellow (ripe) in colour and oval or rounded in shape. Since this plant can have both pistillate as well as hermaphrodite (perfect) flowers, thus the shape of the fruit also varies according to its origin. That is fruits from pistillate are smaller in size and oval shaped while from hermaphrodite flowers are larger and club-shaped. Pulp of fruit is very sweet in taste with specific aroma. The pulp contains proteins, carbohydrates, vitamins and minerals having antibacterial, anti-diabetic and anti-implantation activities. It is used in the treatment of diarrhoea, digestive disorders, and psoriasis.

Parietal placentation is found in papaya. Seeds are numerous in number, rounded in shape and white in colour when fruit is unripe but turns to black when fruit ripens. Seeds of papaya have been used since China and Japan as natural medicine, and to strengthen the liver. Oleic and politic acids are contained within seeds which can help protect the body from cancer. It contains vitamin C and A. Seeds have antiparasitic activity by which it protects from intestinal infections and worms.

It has antibacterial activity and it effectively works on *E.coli, Salmonella* and *Staphylococcus*. Seeds have a negative impact on fertility so it is a serious health issue. The paste of seeds is used commonly for the treatment of psoriasis and for removal of ringworms.

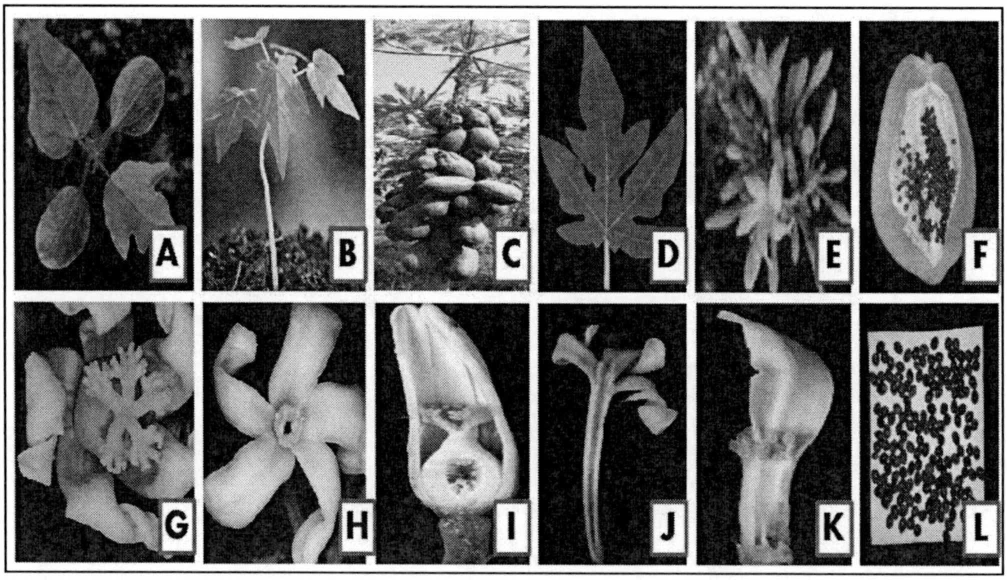

Figure 2. Morphology of Papaya A- Seedling of Papaya, B-Plantlet of Papaya, C-Papaya fruits, D-Papaya Leaf, E- Inflorescence of Papaya, F -Seed and flesh sliced fruit, G- Hermaphrodite flower, H-Male flower, I- L.S of female Flower, J- L.S of male flower, K- L.S of hermaphrodite Flower, L- Seeds of Papaya.

ECOLOGY AND CULTIVATION OF *CARICA PAPAYA*

Warm areas with 21 - 30°C temperature and adequate rainfall (near about 1200mm) are suitable for papaya cultivation. Fruit ripening and taste generally depend upon weather and other climatic conditions like over cool and frost decreases the quality of fruit and also delays the ripening. However, in warm conditions, proper rainfall and elevation around 1000m like in East Africa, high yield of papaya with better fruit quality is obtained.

Besides the above conditions, cultivation of papaya can be done by sufficient irrigation and water conservation practices but waterlogged condition (even for short time duration) eventually kills the plant. For better yield, the soil in which cultivation of plant is to be done should be well drained and rich in essential organic constituents and slightly acidic i.e., pH 6.0 - 6.5.

Plantation should be done in somehow shady place and surrounded by windbreaks because strong wind flow is detrimental for the plant. Under favourable conditions like appropriate oxygen availability, aeration, temperature, heat intensity, humidity, seeds of papaya get germinated. Generally, the germination is epigeal and takes place after 2-3 weeks of sowing. The cotyledonary leaves and first leaves have rounded margins but lobes arise after the emergence of second leaves and further all the leaves have lobed margins.

The development of papaya is relatively quite fast and after 3-9 months of germination, flowering takes place and harvesting is done after 1.0-1.5 years.

CULTIVARS AND DIFFERENT VARIETIES

Besides the wild type of papaya, there are different types of true breeding varieties. Nowadays, systematic breeding programmes, as well as many biotechnological strategies, are used by plant breeders and researchers respectively to produce new varieties (figure 3).

With the help of biotechnological tools different transgenic varieties are developed which are high yielding, disease resistant, having great fruit quality, and enhanced amount of pharmacologically active compounds.

Figure 3. Different varieties of Carica Papaya A-Coorg Honey Dew, B-Pusa Dwarf, C- Pusa Giant, D-Wainamalo Papaya, E- Pusa Majestry, F-Thai Red, G-Chinmay-1, H- PusaNanha, I-Golden Papaya, J-Solo Sunrise.

Papaya Nutritional Value

Nutritional index of papaya fruit is very high so it is listed amongst the important nutritious plant. Fruit of papaya contains vitamins, minerals, fibres, proteins, essential amino acids, fatty acids, carbohydrates (figure 4). Amongst these compounds, the largest fraction is covered by carbohydrates alone i.e., 80 – 90%, protein content is 5%, fat content is 3% (when water is not considered).

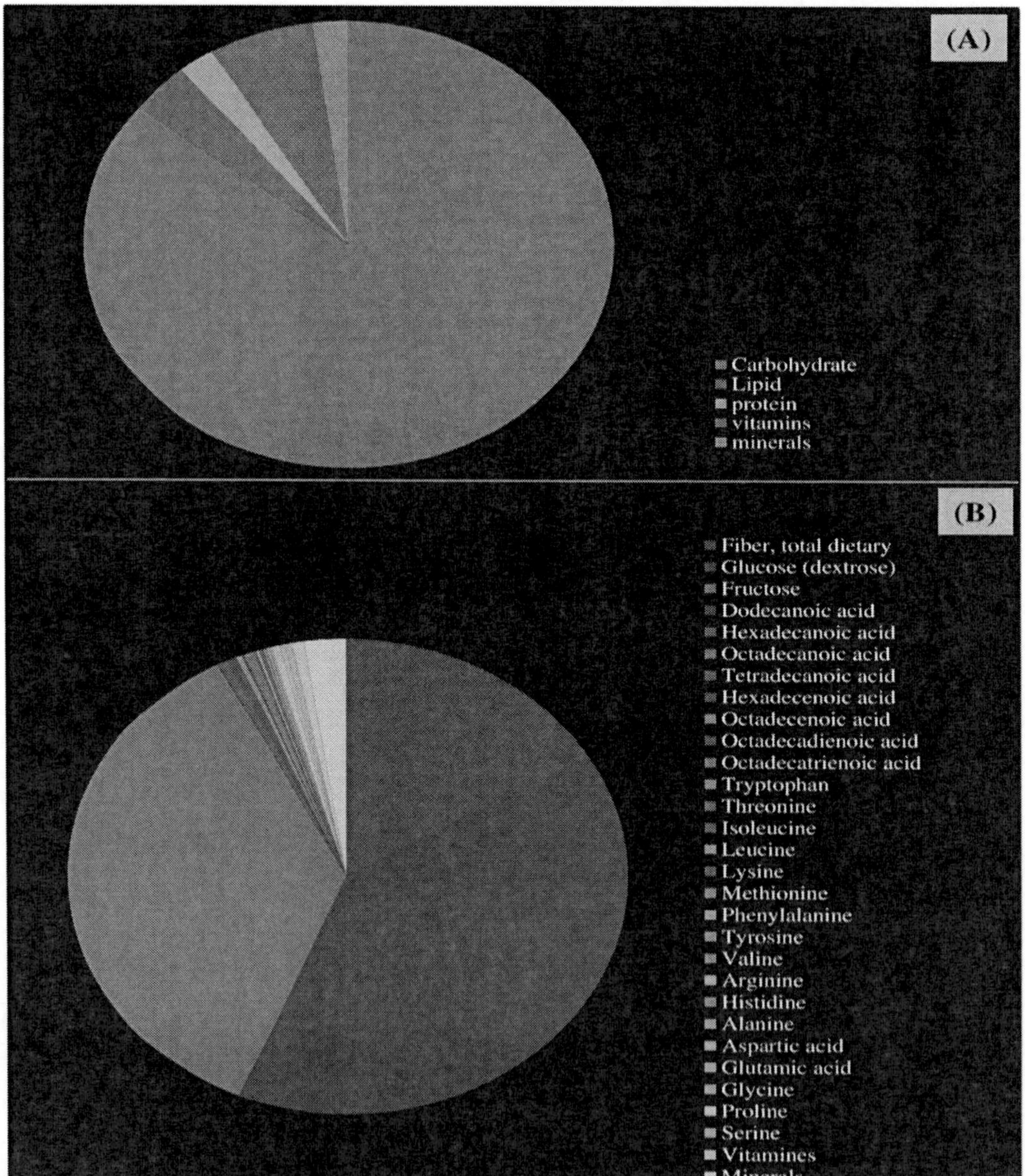

Figure 4. Papaya nutritional value pie chart A- Simple main nutritional components of papaya B- Detailed nutritional components of papaya.

MEDICINAL AND PHARMACOLOGICAL ACTIVITIES OF PAPAYA

Carica papaya is a highly valuable medicinal plant, extensively growing in tropical and subtropical countries. Each part of the plant (root, stem, bark, leaves, flower, fruits and seeds) is known to harbour pharmacologically active compounds (table 1). Main pharmaceutical compounds found in papaya are papain, chymopapain, flavonoids (quercetin), phytosterol, terpenoids etc.

Carica papaya as a Potent Dengue Virus Inhibitor- CPLe in Cure of Dengue and Increased Platelets Count

Table 1. Medicinal properties of different plant parts of Papaya and their constituents (Krishna et al. 2008)

S. No	Plant parts	Constituents	Medicinal uses
1.	Fruits	Amino acids, carbohydrates, carotene, citric acid, fat, fibre, malic acid, niacin, phosphorous, protein, riboflavin, thiamine, various minerals such as calcium, iron and vitamin C.	Abortifacient, antibacterial, anti-implantation, carminative, diuretic, expectorant, laxative, and sedative. Used in the treatment of chronic diarrhoea, dysentery (and other digestive ailments), obesity, skin diseases like psoriasis. Dried fruit reduces the enlarged liver and spleen.
2.	Juice	Linolenic acid, myristic, N-butyric, N-hexanoic, N-octanoic acid, oleic acid, steariclinoleic.	Used in the treatment of bleeding piles, ringworm, snake bite, wound of urinary tract. It is flatulent and diuretic.
3.	Seed	Benzylglucosinolate, benzylisothiocynate, benzylthiourea, carpaine, crude fibre, crude protein, glucotropacolin, hentriacontalinmyrosin, sitosterol.	Abortifacient, anti-fertility, carminative, counter-irritant, vermifuge. Paste used in the treatment of bleeding piles, enlarged liver, ringworm and psoriasis.
4.	Root	Carposide and enzyme myrosin	Abortifacient, antifungal, checking irregular bleeding from the uterus and diuretic used for relieving toothache.
5.	Leaves	Carpain, carposide, choline, dehyrocarpaine I and II, quertecin, pseudocarpain, papain and vitamine C and E.	Antibacterial, anticancerous, anti-diabetic, anti-inflammatory leaf extract is used in the treatment of dengue, gonorrhoea, jaundice (fine paste), ulcers, and urinary complications.
6.	Bark	Fructose, galactose, glucose, sitosterol, sucrose and xylitol.	Anti-fungal, anti-haemolytic activity, anti-malarial. Dried powder and paste is used in jaundice, STD (sexually transmitted diseases) and toothache.
7.	Latex	Chemopapain, chymopapain A, B, and C, cyclortransferase, glutamine, lysozyme, papain, and peptidase A and B.	Anathematic, bleeding haemorrhoids it cures burn, diarrhoea, dyspepsia, and whooping cough.

After many medical trials and research works, it is now confirmed that CPLe does have beneficial properties in dengue. It is found that platelet count gets increases after 24 to 48 hours in patients having dengue fever (DF) and dengue hemorrhagic fever (DHF), after the consumption of extract (Gowda et al. 2015).

Dengue virus (DV) causes thrombocytopenia and increased vascular permeability and plasma leakage (Swati et al. 2013; Sarala and Paknikar 2014). Thus, the platelets get frazil and their destruction takes place.

These platelets also react with leucocytes and endothelial cells, leading to their destruction. Thrombocytopenia may occur by four main ways i.e., Inhibition in platelet formation, inhibition of platelet activation, inhibition in platelet aggregation or platelet destruction.

DV attacks on platelets and promotes mediators like TNF-α and interleukin-1β which causes thrombocytopenia (Nascimento et al. 2014).

Molecular Aspect of Dengue Cure by CPLe-

Since in dengue fever and dengue hemorrhagic fever, the virus performs lytic cycle and it attacks the wall of macrophages and causes the apoptosis of the cell, loosens the immunity of host and reduces the platelet count etc. Researchers had tried hard to make a drug which can antagonist the symptoms caused by DV. Now, it is confirmed that oral consumption of CPLe reverses the above symptoms and drastically enhances the platelet count within 1 or 2 consecutive days. Moreover, there may be 2 aspects to cure dengue i.e., either by an increase in the platelet count or prevention of DV to replicate inside the host body (Figure 5). Here, below both, the aspect is taken into account.

Increase in Platelet Count

It is known earlier that *ALOX12* (arachidonate 12-lipoxygenase) gene is responsible for the megakaryocytes production as well as differentiation. This gene has a promoter sequence for the transcription factor named *RUNX1* and it is platelet-specific (Subenthiran et al. 2013). Megakaryocytes later get differentiated into preplatelets and finally into platelets after the formation of 12- HETE (12-hydroxyeicosatetraenoic acid). Platelet aggregation is an important process during the platelet formation. Aggregation is facilitated by another gene named PTAFR (platelet-activating factor receptor) (Subenthiran et al. 2013). Expression of PTAFR takes place in megakaryocytes.

Since the genes ALOX12 and PTAFR play a crucial role in the platelet production, aggregation and activation as well, researchers found that after the oral consumption of CPLe in dengue patients, there was a magical 12-20 fold increase in the expression of these two genes. Thus, CPLe is treated as a remedy for the enhancement of platelet count.

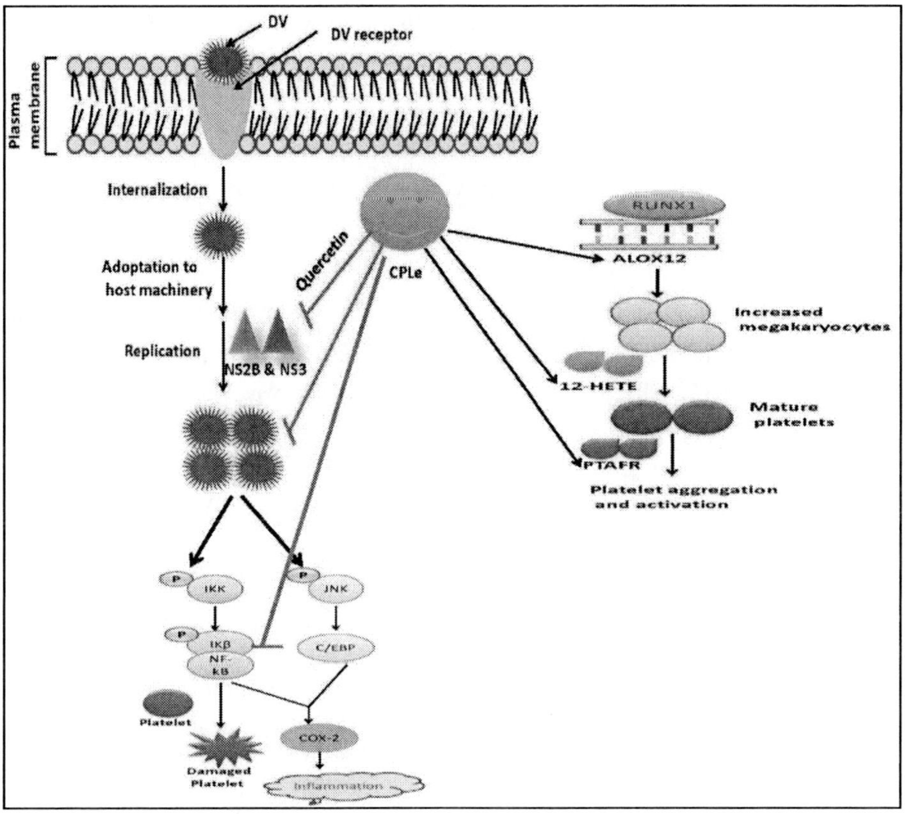

Figure 5. Molecular mechanism of CPLe in dengue cure.

Prevention of DV Replication

Two components are important for DV to replicate inside the host body after infection. These are NS2B and NS3, the part of viral serine protease (Senthilvel et al. 2013). These two components play as pivotal role during the assembly step of the viral packaging and replication. NS2B and NS3 provide polyprotein precursors to the DV before the assembly. Researchers found that the flavonoid components of CPLe, quercetin has significant inhibitory activity against NS2B-NS3 serine protease, particularly against Dengue virus serotype 2 (DVS2) and performs its antiviral activity by preventing viral assembly.

Other Medicinal Properties of *Carica papaya*

Carica papaya as Cardioprotector and Reduces the Blood Pressure

After many research works it is now confirmed that leaf extracts of the papaya plant contain an alkaloid named carpaine which can maintain the heart rate (by slowing down the rate) in humans thus reduces the blood pressure and protects the heart and reduces the chances of cardiac attack (Elgadir et al. 2014). Carpaine is found in leaves of plant having concentration of about 0.3% which does not depend upon temporal or spatial factors.

Herbal infusion of the leaf extract having carpaine is used against malaria because it also has anti-plasmodial activity.

Treatment in Ulcers and Wounds- Anti-Inflammatory Activity

Fruit and leaf of papaya contain many biologically active compounds which are antimicrobial and anti-inflammatory.

These compounds heal the inflamed skin cells and also restore the mucus membrane linings in the digestive and gastric tracts. Due to the presence of papain, chymopapain, carpaine, pectin, carposide, carotenoids, crypto glavine and antheraxanthin the plant parts are beneficial in dressing different types of inflammations, burns, cuts and wounds (Mahmood et al. 2005; Nayak et al. 2007; Ezike et al. 2009).

Besides curing the ulcers in the alimentary canal, papain also plays a role to diminish both acute as well as chronic digestive disorders by regulating the expression of replicasepolyproteins which is involved in bloating and chronic indigestion.

Anti-Microbial Activities

CPLe shows antimicrobial activities against fungi (*Aspergillus* and *Candida*) as well as gram-positive bacteria like species of *Bacillus, Streptococcus, Staphylococcus* and gram-negative like *E. coli* and *Salmonella)*. However, CPLe shows better antibacterial activity than antifungal activity. The methanolic leaf extracts were found to harbour potent antimicrobial activity (Mohammad 2016).

Anti-Diabetic Activity

Researchers found that aqueous leaf extract of papaya show hypoglycaemic effect on diabetic mice because aqueous leaf extract reduces the blood glucose level in diabetic mice and also reduces the blood cholesterol (Juarez-Rojop et al. 2012). In diabetic mice, the aqueous extract was found to help in the regeneration of cells of pancreatic islands as well as prevented the hepatocytes.

Anti-Oxidant Activity

Research works related to papaya have revealed that it contains antioxidant compounds, such as vitamin C, betacarotene, lycopene and vitamin E (Zhou et al. 2011). Besides these, other pharmacologically active compounds like quercetin and β-sitosterol are also found to be novel antioxidant substances in papaya leaf extracts. These all antioxidants have efficacy to reduce the oxidative stress. It is also emphasised that CPLe is responsible for a significant increase in the antioxidative enzymes like glutathione reductase, glutathione peroxidase and glucose-6-phosphate dehydrogenase. These foundations again throw the light towards *Carica papaya* to have potent antioxidative potential.

GENETICS AND GENOMICS OF *CARICA PAPAYA*

Knowledge of genomic constitution and genetics of a species is very essential and helpful to develop its conservation strategies and also crop improvement (Jimenez et al. 2014). In this regard, papaya has been proven to be an interesting organism system in genetic and genomic studies and one of the pioneer plant species to have its genome sequenced (initially by traditional Sanger method using whole genome shotgun approach) (Ming et al. 2008). The genome of papaya is diploid and very small with a size of only 372 Mb. In these references, papaya is again an interesting species for biologists because of presence of the primitive sex-chromosome system. The genome of papaya harbours a large percentage of repetitive sequences i.e., more than 52%. In these repeat sequences there is presence of mainly long terminal retrotransposons. In total genome, there is about 65% euchromatin and 35% heterochromatin and 90% of total euchromatic region is covered contain 92% unigenes and 92.4% of the genetic markers (Wie and Wing 2008).

Recent studies are focused on DNA sequencing by next generation methods which lead to breeding programmes and crop improvement strategies in tropical fruit crops like papaya (Litz and Padilla 2012). Further in this field, the tools like *de novo* genome sequencing, whole genome resequencing, use of restriction enzymes, analysis of transcriptomes, study of epigenetic changes like hypo or hypermethylation of promoter regions of genes have proved to play very important role in crop improvement, domestication studies and other breeding programmes (Guerra-Garcia and Pinero 2017).

The generation time of papaya is 1 year (±3 months) and it can be easily transformed. It is known as the first transgenic cultivar which was developed to mitigate the papaya ringspot virus, in 1998 (USA) it was commercialized (Tecson Mendoza et al. 2008; Manshardt 2014). Whole Genome sequencing of papaya plant has open the field with the advances in molecular markers, gene mapping (physical and linkage mapping) which help in the study of genetic diversity studies (Sharma et al. 2014; Perez et al. 2006; Eustice et al. 2008). Several molecular markers have been developed till now (mainly SSRs and AFLPs from majority of cultivated varieties) with the help of which, several studies on genetic diversity and conservation strategies have been developed and still it is a broad criteria for future studies.

CONCLUSION

This chapter mainly includes botanical, pharmacological as well as nutritional properties of papaya. *Carica papaya* is known as important neutraceutical plant due to the presence of papain, chymopapain, flavonoids (quercetin), phytosterol, terpenoids etc. Different plant parts of papaya involve in various medical problems such as cancer, hyperglycemia, inflammation, microbial infection etc. The leaf extract of papaya has high efficiency to cure dengue fever. It is a potent agent for recovery of blood platelets count and prevents the replication of dengue virus. Therefore, in near future, *Carica papaya* plant can be efficiently utilized in the formation of herbal drugs to prevent dengue fever. Various efforts are going on in screening, evaluation and improvement of varieties in India by AICRP-Fruits-ICAR Programme.

ACKNOWLEDGMENT

The corresponding author is thankful to AICRP-Fruits-ICAR Programme for collection of data. We are thankful to DST-PURSE, DST-FIST and CAS (Botany) for providing funds and internet facility to Department of Botany, Banaras Hindu University, Varanasi, India.

REFERENCES

Elgadir, M. A., Salama, M., Adam, A. (2014) *Carica papaya* as a Source of Natural Medicine and Its Utilization in Selected Pharmaceutical Applications. *Int. J. Pharm. Pharm. Sci.*, 6: 880-884.

Eustice, M., Yu, Q., Lai, C. W., Hou, S., Thimmapuram, J., Liu, L., Alam, M., Moore, P. H., Presting, G. G., Ming, R. (2008) Development and application of microsatellite markers for genomic analysis of papaya. *Tree Genet. Genomes,*4: 333-341.

Ezike, A. C., Akah, P. A., Okoli, C. O., Ezeuchenne, N. A., Ezeugwu, S. (2009) *Carica papaya* (Paw-Paw) unripe fruit may be beneficial in ulcer. *J. Med. Food,* 12: 1268-1273.

Gowda, A. C., Kumar, N. B., Kasture, P. N., Nagabhushan, K. H. (2015) A pilot study to evaluate the effectiveness of *Carica papaya* Leaf extract in increasing the platelet count in cases of dengue with thrombocytopaenia. *Ind. Med. Gaz.,* 149: 109-116.

Guerra-García, A., Piñero, D. (2017) Current approaches and methods in plant domestication studies. *Bot. Sci.,* 95: 345-362.

Jiménez, V. M., Mora-Newcomer, E., Gutiérrez-Soto, M. V. (2014) Biology of the papaya plant. *In: Genetics and Genomics of Papaya* (Eds. Ming, R., Moore, P. H.), Springer New York, pp. 17-33.

Juárez-Rojop, I. E., Díaz-Zagoya, J. C., Ble-Castillo, J. L., Miranda-Osorio, P. H., Castell-Rodríguez, A. E., Tovilla-Zárate, C. A., Rodríguez-Hernández, A., Aguilar-Mariscal, H., Ramón-Frías, T., Bermúdez-Ocaña, D. Y. (2012) Hypoglycemic effect of *Carica papaya* leaves in streptozotocin-induced diabetic rats. *BMC Complement Altern. Med.,*12: 236.

Krishna, K. L., Paridhavi, M., Patel, J. A. (2008) Review on nutritional, medicinal and pharmacological properties of Papaya (*Carica papaya* Linn.).

Lim, T. K. (2012) Carica papaya. *In Edible Medicinal and Non-Medicinal Plants* (Ed. TK Lim), Springer, Netherlands, pp. 693-717.

Litz, R. E., Padilla, G. (2012) Genetic transformation of fruit trees. *In*: *Genomics of tree crops* (Eds. Priyadarshan, P. M., Schnell, R. J.), Springer, Berlin, pp. 117-153.

Mahmood, A. A., Sidik, K., Salmah, I. (2005) Wound healing activity of *Carica papaya* L. aqueous leaf extract in rats. *Int. J. Mol. Med. Adv. Sci.,* 1: 398-401.

Ming, R., Ho, U. S., Feng, Y., Yu, Q., Dionne-Laporte, A., Saw, J. H., Senin, P., Wang, W., Ly, B. V., Lewis, K. L., Salzberg, S. L. (2008) The draft genome of the transgenic tropical fruit tree papaya (*Carica papaya* Linnaeus). *Nature, 452*: 991.

Mohammad, Z. (2016) Antimicrobial activity of papaya leaves extract. Doctoral dissertation, Faculty of Agro-Based Industry.

Nakasone, H. Y., Paull, R. E. (1998) *Tropical fruits*. Cab International.

Nascimento, E. J., Hottz, E. D., Garcia-Bates, T. M., Bozza, F., Marques, Jr. E. T., Barratt-Boyes, S. M. (2014) Emerging concepts in dengue pathogenesis: interplay between plasmablasts, platelets, and complement in triggering vasculopathy. *Crit. Rev. Immunol., 34*.

Nayak, B. S., Pereira, L. P., Maharaj, D. (2007) Wound healing activity of *Carica papaya* L. in experimentally induced diabetic rats.

PÉrez, J. O., Dambier, D., Ollitrault, P., D'Eeckenbrugge, G. C., Brottier, P., Froelicher, Y., Risterucci, A. (2006) Microsatellite markers in *Carica papaya* L.: isolation, characterization and transferability to Vasconcellea species. *Mol. Ecol. Resour.,* 6: 212-217.

Sarala, N., Paknikar, S. (2014) Papaya extract to treat dengue: A novel therapeutic option? *Ann. Med. Health Sci. Res.,* 4: 320-324.

Saran, P. L., Choudhary, R. (2013) Drug bioavailability and traditional medicaments of commercially available papaya: A review. *Afr. J. Agric. Res.* 8: 3216-3223.

Senthilvel, P., Lavanya, P., Kumar, K. M., Swetha, R., Anitha, P., Bag, S., Sarveswari, S., Vijayakumar, V., Ramaiah, S., Anbarasu, A. (2013) Flavonoid from *Carica papaya* inhibits NS2B-NS3 protease and prevents Dengue 2 viral assembly. *Bioinformation,* 9: 889.

Sharma, S. K., Tripathi, S. (2014) Papaya ringspot virus-P: overcoming limitations of resistance breeding in *Carica papaya* L. *In:Plant Virus-Host Interaction* (Eds. Gaur, K., Hohn, T., Sharma, P.), Elsevier Inc., Academic press, pp. 177-194.

Shenekar, P. N., Ukirade, P. S., Salunkhe, S. D., Sutar, S. T., Magdum, C. S., Mohite, S. K., Lokapure, S. G., Metri, S. M. (2014) In vitro evaluation of sun protection factor of fruit extract of *Carica papaya* L. as a lotion formulation. *Euro J. Exp. Bio.,* 4: 44-47.

Starley, I. F., Mohammed, P., Schneider, G., Bickler, S. W. (1999). The treatment of paediatric burns using topical papaya. *Burns,* 25: 636-639.

Subenthiran, S., Choon, T. C., Cheong, K. C., Thayan, R., Teck, M. B., Muniandy, P. K., Afzan, A., Abdullah, N. R., Ismail, Z. (2013) *Carica papaya* leaves juice significantly accelerates the rate of increase in platelet count among patients with dengue fever and dengue haemorrhagic fever. *Evid. Based Complement Alternat. Med.,* 2013.

Swati, P., Supritha, S., Rama, B., Shridhar, N. (2013) Evaluation of platelet augmentation activity of *Carica papaya* Leafaqueous extract in rats. *J. Pharmacogn. Phytochem.,* 1: 57-60.

Vij, T., Prashar, Y. (2015) A review on medicinal properties of *Carica papaya* Linn. *Asian. Pac. J. Trop., Dis* 5: 1-6.

Wei, F., Wing, R. A. (2008) A fruitful outcome to the papaya genome project. *Genome boil.*, 9: 227.

Zakaria, Z. A., Mat Jais, A. M., Sulaiman, M. R., Mohamed Isa, S. S. P., Riffin, S. (2006) The *In vitro* Antibacterial activity of methanol and ethanol extracts of *Carica papaya* flowers and *Mangiferaindica* leaves. *J. Pharmacol. Toxicol.,* 1: 278-283.

Zhou, K., Wang, H., Mei, W., Li, X., Luo, Y., Dai, H. (2011) Antioxidant activity of papaya seed extracts. *Molecules*, 16: 6179-6192.

In: Innovations in Life Science Research ISBN: 978-1-53615-868-7
Editors: R. P. Sinha, S. Pandey-Rai and N. Ghoshal ©2019 Nova Science Publishers, Inc.

Chapter 14

LAND USE, LAND USE CHANGE AND SUSTAINABLE LAND MANAGEMENT

Sunil Singh[1,2,*], Mahesh Kumar Singh[2], Chandra Mohan Kumar[2] and Nandita Ghoshal[2]

[1]Department of Botany, Kutir P.G. College, Chakkey, Jaunpur, India
[2]Centre of Advanced Study in Botany, Institute of Science, Banaras Hindu University, Varanasi, India

ABSTRACT

Land is one of the most important natural resources used by human since time immemorial for various purposes like agriculture, human settlements, protected areas, plantations, etc. These land uses are broadly classified by FAO into three land use classes i.e., forest, other woodland and other land. The ways that historical as well as present people used and managed land has emerged as the primary cause of land use change world-wide. Land use change is the result of direct and indirect human actions to secure essential resources. Population growth is the most active driver of land use changes as growing population depends on the native lands and vegetation mainly for food and shelter. Forests and grasslands particularly have gone through large changes due to agricultural expansion, industrialization and urbanization. India is also facing similar land use changes as that of going on world-wide. Land use changes have larger impact on structure and function of ecosystems and lead to large environmental changes including modification of the global, regional and local climate system, biodiversity loss, modification of ecosystem services, alterations in earth's biogeochemical cycles, and changes in the distribution, and abundance of biological. Consequences of land use changes resulted in degradation of land of various magnitudes mainly due to changes in structure and function of ecosystems involving soil physicochemical properties, soil carbon dynamics, soil microbial biomass dynamics, soil aggregate size fractions, etc. An important solution to overcome the impacts of land use change may be the "sustainable land management". Sustainable land

* Corresponding Author's Email: sunilsingh.kutir@yahoo.com.

management of the land degraded due to land use change may be achieved by developing eco-friendly rehabilitation strategies involving Social forestry, plantation of tree outside forest (TOF) in rural and urban lands so as to ensure long term supply of food and other resources to the growing population.

Keywords: land, land use, land use change, sustainable land management, rehabilitation, social forestry, tree outside forest (TOF)

INTRODUCTION

LAND AND LAND USE

One of the most important natural resources, upon which all human activity is based since time immemorial, is land. Therefore, humans have changed the terrestrial and aquatic systems on the land for centuries to fulfill their basic needs, and this led to modifications in global land use, atmospheric gaseous concentrations as well as the soil capacity to sequester carbon in future (Dale et al. 2000). 'Land' has been described in various ways, the most accepted description of 'land' is by Food and Agricultural organization (FAO 1995), i.e., "Land is a delineable area of the earth's terrestrial surface, encompassing all attributes of the biosphere immediately above or below this surface including those of the near-surface climate the soil and terrain forms, the surface hydrology (including shallow lakes, rivers, marshes, and swamps), the near-surface sedimentary layers and associated groundwater reserve, the plant and animal populations, the human settlement patterns, and the physical outcome of the past and present human activity (terracing, water storage or drainage structures, roads, buildings, etc.)". Thus land refers not only to soil but also landforms, climate, hydrology, vegetation and fauna, together with land improvements such as terraces and drainage work. Land includes all elements of the physical and biological environment that influence the land use.

Land use is the management of land to meet human needs, i.e., the purpose to which land is set by the humans. Protected areas, forestry for timber products, plantations, row-crop agriculture, pastures, or human settlements, are some common land uses set by humans (Turner and Meyer 1994, Dale et al. 2000). Food and Agricultural Organization (FAO) identified three land use classes based on FAO forest definition (FAO 2013):

I. **Forest (Lands):** Land spanning more than 0.5 ha with trees higher than 5 m and a canopy cover of more than 10 percent, or trees able to reach these thresholds *in situ*. It does not include the land predominantly under agricultural or urban land use.
II. **Other Wooded Land:** Land not classified as Forest, spanning more than 0.5 ha; with trees higher than 5 m and a canopy cover of 5-10 percent, or trees able to

reach these thresholds *in situ*; or with a combined cover of shrubs, bushes and trees above10 per cent. It does not include land that is predominantly under agricultural or urban land use.

III. **Other Land:** All land that is not classified as Forest (land) or Other Wooded Land. It includes agricultural land, meadows and pastures, built-up areas (including settlements and infrastructure), barren land, land under permanent ice, other land with tree cover, and other land with trees outside forest (TOF).

Other land is further categorized into **three** sub categories:

(a) **Other Land with Tree Cover:** It is other land spanning more than 0.5 hectares with a canopy cover of more than 10 percent or trees able to reach a height of 5 m at maturity. The difference between Forest and Other land with tree cover is the land use criterion. It includes groups of trees and scattered trees in agricultural landscapes, parks, gardens and around buildings, provided that area, height and canopy cover criteria are met. It also includes tree stands in agricultural production systems e.g., fruit tree plantations and agroforestry systems when crops are grown under tree cover.

(b) **Other Land with TOF:** It is other Land spanning more than 0.05 ha with trees higher than 5 m and a canopy cover above 5 percent, or with trees able to reach these thresholds *in situ*; or with a combined cover of shrubs and trees above 10 percent. It includes land with trees or/and shrubs predominantly under agricultural or urban use. It also includes some land with trees or/and shrubs that is not predominantly under agricultural or urban use with an area spanning less than 0.5 ha or a width of less than 20 m but more than 3 m e.g., windbreaks, shelterbelts and corridors of trees and shrubs.

(c) **Other Land with No TOF:** It is other Land not classified as Other Land with TOF i.e., area is < 0.05 ha, canopy cover < 5 per cent if trees are present, or < 10 per cent if combined trees, bushes and shrubs, or for linear structures a width < 3 m or length < 25m. For example, inland water bodies, barren land, stone outcrops, snow caps and glaciers, deserts, peat bogs, meadows without trees, annual crops without trees, etc.

LAND USE CHANGE

It refers to the modification of earth's terrestrial surface through anthropogenic activities such as urbanization, agriculture, etc. (Kalnay and Cai 2003). The ways that historical as well as present people used and managed land, has emerged as the primary cause of land use change world-wide (Dale et al. 2000; Mitsuda and Ito 2011). Land use changes give rise to large environmental changes including modification of the global climate system, reductions in stratospheric ozone, alterations in the earth's biogeochemical cycles, changes in the distribution and abundance of biological resources, and decrease in water quality (Meyer and Turner 1994, IPCC 1996, Vitousek et al. 1997, Mahlman 1997, Dale et al. 2000; Foley et al. 2005). Thus land use and land management increasingly represent the primary source of changes in the global environment (Dale et al. 2000; Mitsuda and Ito 2011).

Since time immemorial, land use change has been taking place along with the rise and fall of human civilizations as well as the change in world population (Perlin 1989; Turner et al. 1990). Two significant trends are observed in land use change in the past centuries; (i) dramatic increase in land area dedicated to human uses, (ii) intensification of both use and control of the land for increasing production of goods and services (Richards 1990) Figure 1. Land use change may have firstly taken place for increasing the availability of land for wild games by clearing the area through burning, birth of agriculture further accelerated the land use change and at present industrialization further changing the land through urbanization.

Forests and grasslands particularly, have gone through vast changes due to land use change all over the world at different times (Houghton 1995; Dale et al. 2000). Around 75 per cent of the natural forested areas across the world, have either been cleared or dominated by human activity since the last ice age (Ellis and Ramankutty 2008). The global rate of forest loss is currently reported to be ~ 0.6% per year (Hansen et al. 2010). Forest degradation as a result of resources extraction, and conversion of forest areas to cropland, settlements (urban) and other land use types has led to forest fragmentation (Crooks et al. 2011), a decrease in productivity (Hijmans et al. 2005), an increase in forest isolation (McGarigal and Cushman 2002), and changes in community composition (Saunders et al. 1991). Studies have shown that, if not controlled, natural old-growth forests can be critically fragmented to a point at which, they can neither maintain the viable flora and fauna, nor maintain their ecological integrity (Kettunen et al. 2007). The rate of change has been the highest in the tropical regions of the globe for forests and grasslands (Houghton et al. 1991; FAO 1993; Richards and Flint 1994; FAO 1996).

Globally, the land use has changed considerably in the past decades - mostly reflecting the enormous rise in human population and their food need. The world population has doubled since 1960. The developing world accounts for about 95 per cent of the population growth with the tropical region as the world's fastest growing region. The growing population has many implications but most of all it requires an increase in agricultural production to meet food, fuel, fodder demand. This demand can be met by expansion of agricultural land or by intensification of the existing systems.

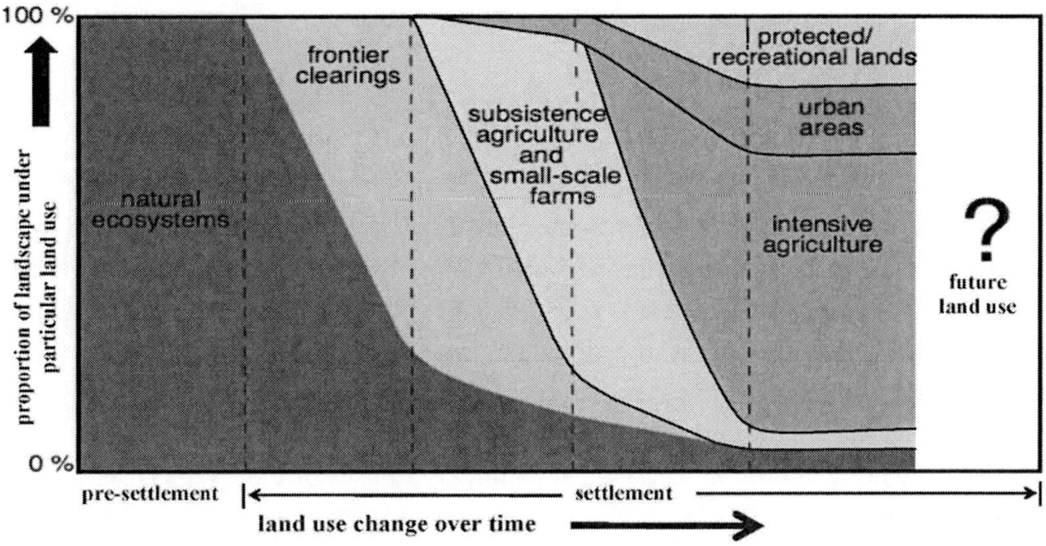

Figure 1. Land use changes. Changes in land use activities that may be followed within a given region over time. As with demographic and economic changes, societies also appear to follow a sequence of different land use regimes i.e., from pre-settlement natural vegetation land use regime to different settlement land use regimes. Furthermore, not all parts of the world move linearly through these land uses. Rather, some places remain in one stage of land use for a long period of time, while others move rapidly between different stages of land use (Adapted from Foley et al. 2005).

Land use change is the result of direct and indirect human actions to obtain useful and essential resources. Population growth is considered as the most active driver of land use changes (Dale et al. 2000) as growing population depends on the native lands and vegetation mainly for food and shelter which involves infrastructure development, technological advancement, political issues, policies, institutional development and socio-cultural factors (Kates et al. 1990, Liu et al. 1993, Turner et al. 1993, Riebsame et al. 1994, Diamond and Noonan 1996). The rate of land consumption and clearance of native land by the growing human population is continuously increasing worldwide (Lugo1991; Pathirana et al. 2014). Only 14 per cent of the world's population was urban in 1900 (Douglas 1994) but due to rapid urbanization ~45 per cent of the global population turned urban by 1995; that figure is projected to increase to 61 per cent by 2025 (WRI 1997). There are some natural drivers of land use change too such as fire, insect-pest outbreak, and the subsequent changes through successional processes. Land use changes significantly affect climate change through their impact on global energy balance and biogeochemical cycles, which in turn exert impact on the properties of land and also on various ecosystem services (Alkama and Cescatti 2016; Foley et al. 2005; Quéré et al. 2016; Turner et al. 2007). However, estimation of global land use change is not easy and quite challenging (Song et al. 2018).

LAND USE CHANGE: INDIAN SCENARIO

India is projected to have 50 percent of its population in urban areas in the next few decades which is presently ~ 30 per cent only (Manjula et al. 2016). India is the 7th largest country of the world that covers land area of 328.73 Mha, which is only 2.4 per cent of the world's total land area and harbors 16.7 per cent of the entire global population (Manjula et al. 2016). Urbanization in India has become an important and irreversible event, and an important determinant of national economic growth and poverty reduction. The urbanization is characterized by the dramatic increase in the number of large cities, although India may be said to be in the midst of transition from a predominantly rural to a quasi-urban society. At the current rate of growth, urban population in India might reach a staggering total of 575 million by 2030. According to Census of India (2011), as many as 52 cities in India had population of a million plus. Therefore, rapid growing population occupies forests and agriculturally suitable land for shelter (Mitsuda and Ito 2011). Various types of land use and its changes in India are shown in the Figure 2 and Table 1.

In the past five decades, India's human population expanded three times whereas, the total area of land under cultivation increased by 20.0 per cent i.e., from 118.75 Mha in 1951 to 140.13 Mha in 2014-15. In spite of the fast expansion of the agricultural land lesser agricultural land is available on per capita basis because of the population growth (State of Environment report 2009). Most of this expansion is the result of clearance of forest and grazing land. Therefore, agricultural expansion accelerated the land use change, and industrialization has further changed the land through urbanization.

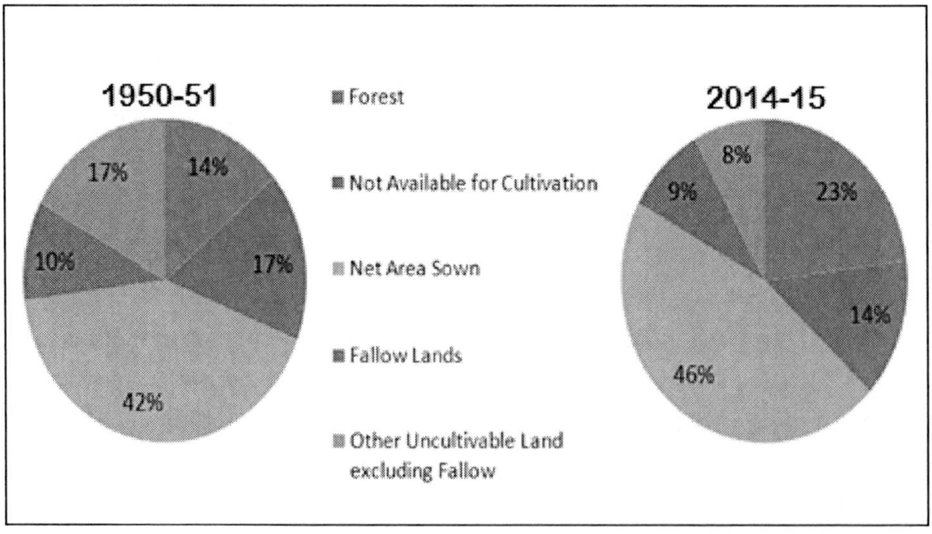

Source: Pocket book of Agricultural Statistics 2017, Ministry of Agriculture, Govt. of India

Figure 2. Land use Classification in India and land use change in over 60 years.

CONSEQUENCES OF LAND USE CHANGE

Land use changes have larger impact on structure and function of ecosystems and lead to large environmental changes including modification of the global, regional and local climate system, biodiversity loss, modification of ecosystem services, alterations in earth's biogeochemical cycles, and changes in the distribution, and abundance of biological resources (Dixon et al. 1994; Ojima et al. 1994; Meyer and Turner 1994; Dale et al. 2000; Pimm and Raven 2000; Sala et al. 2000; DeFries et al. 2004; Rindfuss et al. 2004; Foley et al. 2005; Millennium Ecosystem Assessment 2005; Satake and Iwasa 2006; Quetier et al. 2007; Yamaura et al. 2009; Mitsuda and Ito 2011). Land use changes may or may not result in land cover change that is if aselectively harvested forest is changed to protected forest, in the short term, it will not result in the observable cover change at the site, but if changed to agricultural land, it will bring about larger changes in the land cover. Land cover refers to the physical appearance and ecological state of the land surface like closed forests, open forests, or grasslands (Turner and Meyer 1994, Dale et al. 2000). Land cover change means the conversion of one type of cover to another, regardless of its use. Natural disturbances, for example, fire and insect-pest outbreaks, and subsequent changes through succession, and anthropogenic activities such as agriculture are responsible for land cover change.

Land use change is the result of direct and indirect human actions to secure essential resources which has negative consequences on climate, biodiversity and ecosystem services. Biodiversity is often reduced by land use changes (Meyer and Turner 1994; Pimm and Raven 2000; Sala et al. 2000; DeFries et al. 2004; Rindfuss et al. 2004; Foley et al. 2005; Millennium Ecosystem Assessment 2005; Satake and Iwasa 2006; Quetier et al. 2007; Mitsuda and Ito 2011). When land form of native forest is converted to an agricultural land, the loss of native species of forest of that area may be immediate and complete. The areas exposed by land use change, especially near to human settlements are very susceptible to species invasions by non-native plants, animals and diseases. Habitat fragmentation, resulted due to land use changes like highway construction severely disturb the habitat suitability of forest and other ecosystems and exposed it to external influences and reduce the core habitat area. Smaller habitat areas generally support fewer species, and for species requiring undisturbed core habitat, fragmentation can cause local and even general extinction.

Land use change not only affects the biodiversity but also the climate at global, regional and local scales (Dixon et al. 1994; Ojima et al. 1994; Meyer and Turner 1994; Dale et al. 2000; DeFries et al. 2004; Foley et al. 2005; Yamaura et al. 2009). Global warming is the consequence of land use change at global scale, by releasing greenhouse gases to the atmosphere. Land use change can increase the concentration of atmospheric carbon dioxide by disturbance of soils and vegetation and release of carbon stored in soil by agricultural practices. Recently Singh et al. (2015) studied the impact of land use change on soil carbon storage and reported around 60% loss of stored carbon in soil through the change of land use type from natural forest to agroecosystem in about three decade. In the same study they (Singh and Ghoshal 2011) reported 75% loss of soil organic carbon in the land use change of natural forest to agroecosystem. The stored carbon exposed to the environment as carbon dioxide and play role in global warming in one hand and reduction in soil fertility as well as sustainability on the other. Singh et al. (2015) reported more than double rate of emission of CO_2-C from agricultural land than natural forest. Land use change also promotes the terrestrial emission of other greenhouse gases, like methane by changed surface hydrology, wetland drainage, rice paddies and cattle grazing, and nitrogen oxides through the extensive use of chemical fertilizers in agroecosystems, irrigation, cropping of nitrogen fixing plants and biomass combustion. Biomass combustion in croplands, land clearing and human settlements also contributes to climate change by releasing sulfur dioxide and particulates. These particulates reflect the sunlight and causes regional and global cooling by their effects on cloud cover. Land use change affects not only the greenhouse gases balance and stored carbon in soil but it also affects soil microbes. Singh and Ghoshal (2014) worked out the impact of land use change on soil microbial biomass carbon and nitrogen and reported 40 and 47% loss respectively through the conversion of natural forest to agroecosystem in about three decade.

Land use change play an important role in global climate change through the reflection of sunlight from land surfaces (i.e., albedo) which change surface heat balance by changing surface albedo (Brovkin et al. 1999; Betts 2000; Bala et al. 2007; IPCC 2007; Schwaiger and Bird 2010), as well as by altering evapo-transpiration ratio from vegetation and by changes in surface roughness.

Land use change is an important factor of soil, water and air pollution too (Foley et al. 2005). Removal of vegetation promotes soil vulnerability to soil erosion by wind and water, particularly on steep terrain, and when exposed to fire, releases pollutants to the atmosphere. This degrade soil fertility over time, and not only make soil unsuitable for future farming, but also releases enormous phosphorus, nitrogen, and sediments aquatic ecosystems, causing negative impacts on aquatic ecosystem such as sedimentation, turbidity, eutrophication and coastal hypoxia (Foley et al. 2005). Mining produces toxic metals in the process of extraction and negatively affect the ecosystem.

Modern agricultural practices like excessive use of fertilizers, herbicides, pesticides, growing livestock and their manures within small areas, have increased a lot the surface as well as groundwater pollution (Foley et al. 2005). The release of nitrogen oxides from agricultural lands and other altered regional and local hydrology like dam construction, irrigation projects, wetland drainage, increased impervious surface in urban areas, play an important role in destruction of stratospheric ozone. All the drivers of land use change and their consequences are observable simultaneously around the world today, and have resulted in land degradation.

Table 1. Land use classification and land use change in past 60 years in India. Million hectares

S. No	Land use Classification	1950-51	1990-91	2000-01	2009-10 (P)	2010-11 (P)	2011-12 (P)	2012-13	2013-14	2014-15
I	Geographical Area	328.73	328.73	328.73	328.73	328.73	328.73	328.73	328.73	328.73
II	Reporting Area for Land Utilization Statistics (1 to 5)	284.32	304.86	305.19	307.41	307.48	307.39	307.49	307.80	307.82
	1. Forest	40.48	67.81	69.84	71.59	71.60	71.60	71.57	71.83	71.79
	2. Not Available for Cultivation (A+B)	47.52	40.48	41.23	43.33	43.58	43.53	43.58	43.86	43.88
	(A) Area under Non-agricultural Uses	9.36	21.09	23.75	26.16	26.40	26.31	26.50	26.91	26.88
	(B) Barren and Un-cultivated Land	38.16	19.36	17.48	17.18	17.18	17.22	17.07	16.95	17.00
	3. Other Uncultivated Land excluding Fallow Land(A+B+C)	49.45	30.22	27.74	26.50	26.15	26.11	26.08	25.83	25.83
	(A)Permanent Pasture and other Grazing Land	6.68	11.40	10.66	10.34	10.30	10.31	10.26	10.26	10.26
	(B) Land under Miscellaneous Tree Crops and Groves not included in Net Area Sown	19.83	3.82	3.44	3.21	3.20	3.16	3.18	3.19	3.10
	(C) Cultivable Waste Land	22.94	15.00	13.63	12.95	12.65	12.64	12.64	12.39	12.47
	4. Fallow Lands (A+B)	28.12	23.37	25.04	26.85	24.60	25.18	26.32	24.85	26.18
	(A) Fallow Lands other than Current Fallows	17.45	9.66	10.27	10.84	10.32	10.67	11.04	10.69	11.09
	(B) Current Fallows	10.68	13.70	14.78	16.01	14.28	14.51	15.29	14.15	15.09
	5. Net Area Sown(6-7)	118.75	143.00	141.34	139.17	141.56	140.98	139.94	141.43	140.13
	6. Total Cropped Area (Gross Cropped Area)	131.89	185.74	185.34	188.19	197.68	195.80	194.25	200.95	199.36
	7. Area Sown more than once*	13.15	42.74	44.00	50.02	56.12	54.82	54.31	54.52	54.23

* Shows intensification of agricultural land over the past 60 years; (P) Provisional

Source: Pocket book of Agricultural Statistics 2017, Ministry of Agriculture, Govt. of India.

LAND DEGRADATION

Land degradation has been defined differently in literature (Dixon et al. 1994; Houghton et al. 1999; DeFries et al. 2002; Friedlingstein et al. 2010; Leite et al. 2012). Most accepted definition of land degradation was given by UN Convention to Combat Desertification (UNCCD) i.e., as some measurable loss of the "biological or economic productivity and complexity of rain fed cropland, irrigated cropland, or range, pasture, forest and woodlands, etc. arising from human activities and habitation patterns".

Degradation of natural and semi-natural lands has become very important from ecological as well as from economical prospective (Smith et al. 2008; Barbier and Hochard 2018). Land degradation is the composite term indicating the aggregate diminution of the productive potential of the land, including its major uses (rain-fed arable, irrigated, rangeland, and forestry), the farming systems (e.g., smallholder subsistence), and its value as an economic resource. This term refers to the decline of the biological productive potential of land, namely the entire geo-ecological system that includes soils, waters, climate, vegetation, topography, and land use. Land degradation may be an inherent property of the natural system e.g., eroded ravine lands, etc. In other cases, climatic change may also be implicated. More usually, land degradation is caused by a mismatch between the land's self-sustainable biological potential, its quality in human terms, and the way the land is used. Simply, the way the land is used causes more damage to the land than its restorative systems can compensate. Globally, land degradation affects about 15% of the world's dry land surface and 22% of its agricultural land area. Human induced chemical soil degradation, mainly due to salinization and pollution, affects more than 12% of the total degraded land area (Table 2).

Land Use Change and Carbon Dynamics

The major carbon stocks are the terrestrial ecosystems and it could play an important role in carbon emissions through anthropogenic activity (Pacala et al. 2001; Pan et al. 2011). Carbon storage capacity varied from one ecosystem to another. Decrease in vegetation cover results in the emission of carbon into the atmosphere. The land-use change not only directly reduces carbon storage in vegetation but also affects the amount of vegetation residues returned to the soil, which, in turn, are the main source of soil carbon storage (Post and Kwon 2000). Carbon storage as well as release can be affected due to different land use change (Apps and Price 2013). Distribution and quantity of global terrestrial carbon efflux is mainly affected by land use changes (Houghton 2010; Van der Werf et al. 2010). Land use change has been estimated to contribute about one-third of all human generated carbon emission since the industrial revolution (Houghton et al. 1999) and approximately13% of total emissions in the last decade (Friedlingstein et al. 2010). The impact of land use changes within terrestrial ecosystems on carbon balance thus has become very important from global change research (Leite et al. 2012). Many studies also reported that the disturbance of carbon pools by human activities using bookkeeping models, which track changes in the areas of different land-use types and use standard growth and decomposition curves to calculate changes in carbon pools (Houghton et al. 1999). The effects of land use change have been estimated by using process models that calculate the carbon concentration of vegetation and soils in different ecosystems internally based on climate and other factors used within the models (Shevliakova et al. 2009). It has been widely reported that the carbon emissions caused by different land use changes like deforestation and cultivation (Houghton 2003). Compared to biomass carbon pools, soil organic carbon stocks have been shown to undergo much larger changes due to land use change (Liu et al. 2004; Pan et al. 2004).

Land use change has a profound impact on C dynamics. Soil organic matter is considered as the largest terrestrial active carbon pool which stores ~1500 Gt carbon (in 1 meter soil depth) (Eswaran et al. 2000). The amount of carbon stored in soil depends on the balance between the photosynthetic input of carbon in soil and output through its decomposition. About 136 ± 55 Gt carbon was estimated to be lost from soil organic matter stock due to land use change, thus can subsequently alter soil organic matter dynamics. Monitoring and mediating the negative consequences of landuse change while sustaining the production of essential resources has therefore, become the major challenge.

Increasing C stock in the soil popularly known as soil carbon sequestration, has now been considered as one of the effective mechanisms for mitigating the loss of carbon from the soil by slowing or reversing the trend of increasing concentration of carbon dioxide in the atmosphere (Asante et al. 2011; Singh and Ghoshal 2011). Carbon sequestration in soil i.e., accumulation of carbon in soil, refers to taking carbon dioxide from the atmosphere through plants and storing the carbon in the soil in form of soil organic matter.

Table 2. Estimated extent of land degradation

Land use category	**Total area (million hectare)**	**Area affected by land degradation (per cent)**
Total land area	13,000	15
Total agricultural area	8,735	22
Total area of degraded land	1,965	100% from which:
• Degraded by water erosion	1,094	56
• Degraded by wind erosion	549	28
• Degraded by soil structural damage	83.3	4
• Degraded soil fertility- chemical degradation	135.3	7
• Salinization	76.3	4
• Soil pollution (inclusive of industrial)	21.8	1
• Acidification	5.7	0

Source: Haigh 2009, Land Rehabilitation. In Land Use, Land Cover and Soil Science- Vol. IV.

Soil is considered as ‘source’ when it stores C, and the ‘sink’ when it loses C (Powlson et al. 2011), and changes in the role of soil from ‘source’ to ‘sink’ or *vice- versa* depends mainly on the land uses, among various other factors. Any land use management strategy to enable the soil act as ‘sink’ also facilitate, storage of carbon in soil, and thus may eventually help in mitigation of global climate change. The assessment of soil CO_2 flux attracted interest worldwide as it is the major factor of carbon dynamics in the terrestrial ecosystem. It can be taken as the indicator of carbon availability in soil for decomposition (Al-Kaisi and Yin 2005), and also as the transformation of soil organic carbon to atmospheric CO_2 (Ryan and Law 2005). Many studies evaluated the effect of climatic factors on soil CO_2 flux (Raich and Potter 1995; Frank et al. 2002; Grover et al. 2012; Sugihara et al. 2012: Yilmaz 2012), biotic and abiotic factors (Litton et al. 2003). However, study regarding impact of land-use changes on soil CO_2-C flux is virtually lacking with particular reference to the role of soil structure in the dry tropics. The evaluation of seasonal variations in soil CO_2-C flux and C storage in soil owing to land-use changes in the dry tropics will be of immense significance for the better understanding of the carbon dynamics in this region.

Changes in the land-use pattern lead to the change in the level of atmospheric greenhouse gases especially CO_2 originating from the terrestrial ecosystem (Grover et al. 2012). Currently, the contribution of land-use change in emission of C to atmosphere is 1.4 Pg C yr^{-1} globally; the extent which is next to the fossil fuel consumptions (Le Quere 2010). There is global concern to contain the increasing, CO_2 concentrations either by reducing anthropogenic emissions or removing the atmospheric CO_2 through sequestration in the biosphere (Ingram and Fernandes 2001). Carbon storage in soil represents the balance between the two contrasting processes i.e., accumulation and loss of carbon from soil.

Soil CO_2-C flux or soil respiration in any terrestrial ecosystem, is the product of autotrophic (root) and heterotrophic (soil microbes) respiration (Raich and Schlesinger 1992), and is considered the major contributor to loss of C from soil. The contribution of plant roots and/or microbes in soil respiration is very difficult to estimate in isolation, however, estimations of composite soil surface CO_2-Cmay serve as the index of C output from soil. Since a substantial amount of carbon is allocated in the belowground biomass component of plants, its net productivity, may give an idea about the extent of root respiration (Raich and Nadelhoffer 1986; Palmroth et al. 2005). Therefore, quantification and the study of the role of belowground net productivity in regulation of soil CO_2-Cflux and C pool seems imperative. Small changes in the soil organic carbon pool could have dramatic impact on the concentration of CO_2 in the atmosphere. Therefore, dynamics of soil C pool may offer the insight about their role in controlling CO_2-Cflux.

LAND USE CHANGE AND SOIL MICROBIAL BIOMASS DYNAMICS

Soil microbial biomass is the active fraction of soil organic matter. It is considered as the 'eye' of the needle through which all the organic materials must have pass while entering the soil, often more than once as they are decomposed to the inorganic compounds (Jenkinson 1977). It is defined as the living part of soil organic matter, except the live root fractions and soil organisms larger than 5×10^{-15} m^3 (Jenkinson and Ladd 1981). Though soil microbial biomass comprises only 23 per cent of the total soil organic carbon, yet it is large enough to act as the source and sink for plant nutrients through decomposition of plant and animal residue and consequently the release of nutrient available to plants (Smith and Paul 1990; Gonzalez-Quinones et al. 2011). Soil microbial biomass plays dual role in soil. Firstly, it may act as the agent of transformation, through which, all the natural organic materials pass and enter the soil. Secondly, it functions as a dynamic pool containing appreciable reserves of nutrients (Jenkinson and Ladd 1981). Soil microbial biomass, the labile fraction of soil organic matter, responds faster than the total organic matter to any changes in the organic inputs (Powlson et al. 1987). Thus microbial biomass can be considered as the ecological marker for analyzing and predicting the long-term effects of disturbance in soil (Smith and Paul 1990). Microbial biomass is a sensitive indicator of change in organic matter turnover and so, it can be used as the tool for understanding and predicting the long-term effect of changes in soil organic matter in any particular land use.

Soil microbial biomass, as the predictor, is used to access changes in soil quality especially related to the land use change, sustainability and the restoration of ecosystems (Kennedy and Papendick 1995). The soil microbial biomass varies widely with the influence of land use change, and is strongly dependent on the particular ecosystem. The study regarding variation in soil microbial biomass with land use change has been carried out widely but limited reports are available in the dry tropics particularly with reference to restoration of degraded lands. The reports are available regarding the driver of seasonal variation in soil microbial biomass i.e., soil moisture and soil temperature in different climatic regions and ecosystems (Devi and Yadava 2006; Iqbal et al. 2010). The impact of climatic factors on the seasonal variations in soil microbial biomass in dry tropics, is scarce.

Measurement of soil microbial biomass is useful in evaluating the degree of disturbance and the subsequent recovery (Smith and Paul 1990). It regulates the soil organic matter and nutrient dynamics in soil. Since the turnover of nutrients in any ecosystem is of great importance for its functioning, therefore, dynamics of microbial biomass and its role in plant nutrition in the nutrient-poor soils of dry tropics is of major significance (Azam et al. 2003). In the seasonal dry tropical region, plant growth regulates the temporal and spatial distribution of organic input in soils in the form of plant roots, leaves and plant residue which in turn, affect the dynamics of soil microbial biomass (Singh et al. 1989). The distribution of microorganisms in soil is not uniform, and preferred the region where soil organic matter is centered like rhizosphere and along the root channels, and sharply decreases with soil depth (Murphy et al. 1998). Most of the studies were limited to the upper soil layer (Sahani and Behera 2001; Patel et al. 2010) whereas, for the proper restoration of degraded lands, knowledge about the dynamics of microbial biomass in deeper soil layers, seems imperative. Therefore, estimation of soil microbial biomass across the soil profile may provide insight of the status of soil organic carbon distribution.

SOIL AGGREGATES AND LAND USE CHANGE

Soil aggregate size fractions play important role(s) in the distribution and stability of soil organic carbon. Aggregation of soil particles may be used as the soil quality index integrating the major processes involved in the formation of soil structure, physico-chemical and biological processes of soil (Parent et al. 2011). Soil aggregates are closely packed structure of sand, silt, clay and organic particles, bound with each other by the adhesive force more strongly than the surrounding particles (Cambardella 2006). Soil aggregates are the product of reorganization, flocculation and cementation of soil particles (Duiker et al. 2003). Soil Science Society of America (1997) coined the term 'soil aggregate' to a group of primary soil particles bound with each other more strongly than the neighboring particles. The accumulation and transformation of soil organic matter play a critical role in aggregate stabilization.

Soil aggregates protect soil organic matter from decomposition by trapping and minimizing the direct contact of soil organic matter with soil microorganisms (Tisdall and Oades 1982). In addition, aggregates also affect the microbial community structure (Hattori 1988), aeration rate (Sexstone et al. 1985), mobility rate of water (Prove et al. 1990), soil erosion (Barthes and Roose 2002), and stabilization of soil organic matter (Six et al. 2000).

Land use change is an important factor affecting soil organic carbon accumulation and storage in soils, which controls the magnitude of soil organic carbon (Six et al. 2002). Land use change not only affects the total amount of soil organic carbon, but also the distribution of organic carbon in physical fractions (aggregate size fractions) and the soil organic carbon protecting processes, The knowledge of soil organic carbon in the aggregate-size fractions may be effective in monitoring and prediction the effects of land use change on soil organic carbon storage. Increase in soil aggregation results in more accumulation of soil organic carbon into the aggregates as soil organic carbon in aggregates is out of reach of biological decomposition by microorganisms (Guggenberger et al. 1999; Wilson et al. 2009).

One of the major challenges for this century is to sustain long term production of food and other resources especially when large areas are subjected to land use change from productive land to nonproductive uses, such as the conversion of agricultural land to residential use, degradation of rangeland by overgrazing, etc. An important solution to overcome the negative impact of land use change may be the 'sustainable land management'.

Sustainable Land Management: A Strategy to Mitigate Impact of Land Use Change

Sustainable land management implies the management of land degraded due to land use change in a way that it should ensure long term supply of food and other resources to the growing human populations in the eco-friendly way. In the present scenario, realization of sustainable land management is difficult, and may take decades to come. Land use changes even at local level have global impact, therefore it requires cooperation of the world's policymakers, land managers, and other stakeholders to overcome the problem at local, regional as well as global scales.

At the global scale, the Kyoto Protocol directed to reduce greenhouse gas emission which is a factor of climate change. It offers to encourage such land use practices which promote the storage of carbon in soil like plantation of trees, perennial crops, return of crop residues to soils, and no-till agricultural practices. The protocol also has the suggestions for reduction of emission of methane and nitrogen oxides.

Management of land to promote biodiversity, there is a need of wide range of policies and practices. The basic practice is to set-aside biodiverse habitats as conservation reserves from which humans are excluded. Protected areas like parks and sanctuaries should establish on biodiversity rich lands. More recently, several efforts are made to enhance the resilience of the managed landscapes as stated by studies that intensively managed ecosystems are more susceptible to climatic extremes, diseases, invasive species, etc. (Costanza 2000; Scheffer et al. 2001; Turner II et al. 2003; Foley et al. 2005). There should be establishment of corridors of biodiversity between existing patches of habitat, which connect smaller patches, distributed across landscapes, together and promote species migrations.

Dense populations in industrial era need land for industry and residential use which causes transformation of productive agricultural land to the unproductive land. Policy is running to avoid this loss of food production but their effectiveness in the economic view is often limited. Another problem is the wide adoption of automobile transportation, which transformed large areas of agricultural land to low density residential areas around cities and along highways (urban sprawl). "Smart growth" and other programs have been developed in such areas to promote more efficient and desirable land use and to protect agricultural land, which include increasing green spaces in urban areas, practicing agroforestry, etc. For example in India "smart city mission" has been launched with the objective to promote cities that provide core infrastructure and give a decent quality of life to its citizens, a clean and sustainable environment (Manjula et al. 2016). Further Ministry of Environment, Forest and Climate Change (MoEFCC), Government of India, has used Environmental Impact Assessment (EIA) Notification, 2006 as a major tool to regulate industrial development of the country for minimizing the adverse impact on environment and reversing the trends which may lead to land use change immediately and climate change in long run. The notification provides for screening, scoping, public consultation and appraisal of project proposals.

These suggested solutions and regional and global efforts have to prove their effectiveness to save the environment affected by land use changes. Therefore monitoring and mediating the negative consequences of land use change while sustaining the production of essential resources become a major challenge. There is a need to develop eco-friendly restoration/rehabilitation strategies that restore/rehabilitate the land degraded owing land use change in urban areas of dry tropics.

LAND REHABILITATION

Land rehabilitation is an intervention designed to make the geo-ecological improvement. This involves the mitigation or reversal of land degradation caused by poor land husbandry practices, especially the agricultural practices. There are two aspects of land rehabilitation, (a) the land be restored to nature, and (b) rehabilitation of land to human utilization.

If the land is to be restored to nature, the focus of the project becomes the establishment of biological control of the environment. This means the generation of an autonomous, self-sustaining geo-ecological system. Here, the key indicators of success include the vitality, productivity and resilience of the system. Often measured are the system's capacity to recycle organic and mineral components, and its capacity to affect self-catalyzing environmental improvement, which is often manifested through the normal patterns of ecological succession.

The key criteria for rehabilitation related to human use is long term sustainability and that is often measured in terms of the cost and benefits of maintaining environmental quality and the capacity of the local land management to preserve or enhance the productive quality of the land, howsoever defined.

The driving force for land rehabilitation is land degradation. Soil degradation can be mitigated if land is left for sufficiently longer periods "in fallow" which enables biological restoration of soil characteristics that have become degraded, notably its porosity and the nutrient contents. During this recuperation phase, biological processes can return the soil to optimal conditions for the growth of plant roots. However, land shortages are now reducing the time that the land can be allowed to remain in fallow, with the result that the rate of degradation and loss of underlying productive potential during the cropping period increases.

The soil is a key component in land rehabilitation. It is a living resource which is complex, dynamic, evolving and biologically modulated open system. The soil controls what grows on the land, and it determines what proportion of the rainfall that falls on the land infiltrates into the soil, runs off as surface or near surface flow, remains stored at the surface in ponds or in the soil attached to the soil particles, and how much is returned to the atmosphere through evaporation/evapo-transpiration. Directly or indirectly, through the vegetative cover, the soil also regulates the rate at which soil is removed by water or wind erosion.

Concerns have been raised for the restoration, rehabilitation and/or reclamation of degraded lands and that posses major challenge. In India, nearly ~ 200 Mha is under degraded land, which is a serious environmental problem. Society for Ecological restoration stated that restoration is the process which provides an avenue to ecosystem recovery that has been degraded, damaged or destroyed earlier. Re-establishment of structure, productivity and species diversity of the originally present forest is the restoration of degraded forest in which ecological function and processes match to the original one in due course of time. Restoration of degraded lands can occur either naturally (passive restoration) or through anthropogenic manipulation (active restoration). Since through passive restoration, the recovery of vegetation and soil quality may be slow by natural successional processes without human interventions (Ohsowski et al. 2012) therefore, active restoration is acceptable for quick recovery which enhanced the rate of recovery of ecosystem functioning by switching on the biogeochemical cycles of plant nutrients and carbon sequestration. Passive restoration is regulated by natural processes with minimum anthropogenic interference. Passive restoration takes place where environmental disturbances like cattle grazing and agricultural practices are excluded, and secondary succession takes place naturally by colonization of shrubs and trees whereas, active restoration is achieved through the management of land by planting vegetation, weeding to gain a desired structure (Rey et al. 2008).

SOCIAL FORESTRY

Social forestry is a land use for active restoration/rehabilitation of degraded lands. Social forestry is also a strategy to deal with the ecological balance and socio-economic crisis as the poorer people of the globe depend for its sustenance needs of fuel, fodder and small timber on the forest and other common lands which become increasingly denuded and degraded over the years due to land use change.

The term social forestry is difficult to define precisely, but is generally understood to mean tree-growing for the purpose of rural development. As social forestry has a rural development focus and is heavily dependent on the active participation of people, it is also known as "forestry for local community development" or "participatory forestry".

Social forestry for wastelands rehabilitation also forms a major component in India's strategy to deal with ecological balance along with socio-economic benefits. More than half of India - the poorer half - depends on the forests and other form of lands for fulfilling their basic requirements. In seventies, the Prime Minister of India has given a call to the nation to afforest 5 million hectares a year, primarily for fuel wood and fodder, and develop it as the people's movement. Government of India is trying to increase forest areas close to human settlements which have been degraded over the years due to human activities through afforestation. Plantation of trees has been suggested around agricultural fields, along railway lines and roadsides, river and canal banks, in village common land, government wasteland, and panchayat land. Social forestry scheme was initiated in India for making a difference in forest cover and benefiting rural and urban communities. The main objective of such schemes is to raise the standard of living and quality of life of the rural and the urban people by including the following goals:

1. Improvement of the environment for protecting agriculture from adverse climatic factors,
2. Enhancement in the supply of wood fuel for domestic use, small timber for rural housing, fodder for livestock, and minor forest produce for local industries,
3. Increment in the natural beauty of the landscape and also creation of the recreational forests for the benefit of rural and urban population,
4. Provision of jobs for unskilled workers,
5. Land rehabilitation, and

Although a wide range of activities are included in social forestry, five following main components can be distinguished in India:

1. Farm forestry (tree growing on private land),
2. Farmer leasehold or tree patta,
3. Village woodlots or community forestry,
4. Strip plantations alongside roads, canals, railways, etc. and
5. Reforestation or rehabilitation of degraded forest areas and degraded lands in rural and urban areas.

Farm forestry is tree planting undertaken by individual households on their own land or the land they have rented from others. Tree seedlings may be planted in blocks (small plantations), on field boundaries or around homesteads. They may be intermixed with agricultural crops in several forms of agroforestry, or may be planted alone on either agricultural land or uncultivable wastelands.

Farmer leasehold or tree patta denotes a kind of farm forestry in which poor farmers or landless laborers are given leases to tracts of public land on which, with varying degrees of public support, they are constrained to grow trees.

Village woodlots are small plantations on communal or government lands, operated by or on behalf of the village, for the benefit of the village as a whole, although there may be special arrangements which provide preferential treatment to the under-privileged.

Strip plantations are relatively narrow areas along the sides of roads, canals, railways, and rivers, established by the Government (usually the Forest Department) with the intention of providing the benefits of forest products to local people and to serve as demonstration areas.

The reforestation or rehabilitation of degraded forests and degraded lands refers to large plantations on public lands which have been severely degraded, and which are often in environmentally critical areas. Such plantations may or may not be considered a form of social forestry depending on whether or not there is significant involvement of local communities.

Common species used under social forestry schemes are *Acacia catechu*, *Acacia nilotica*, *Albizia procera*, *Albizia lebbeck*, *Azadirachta indica*, *Cassia siamea*, *Casuarina equisetifolia*, *Dalbergia sissoo*, *Dendrocalamus strictus*, *Eucalyptus sp*,. *Gmelina arborea*, *Jatropha curcas*, *Leucaena leucocephala*, *Madhuca indica*, *Melia azedarach*, *Phyllanthus emblica*, *Pongamia pinnata*, *Prosopis cineraria*, *Shorea robusta*, *Tamarindus indica*. *Tectona grandis*, *Terminalia arjuna*.

TREES OUTSIDE FORESTS (TOF)

TOF can play important roles in national economies, and international efforts for sustainability -and in many places, they already do. At the local level, people have long relied on TOF in various land-use settings for food security, income, and biological diversity. Forest professionals in many countries support local use of trees for these purposes, outside forests as well as in forest settings. More recently, international programmes are set on trees' roles in providing essential environmental services to encourage sustainable land management, carbon sequestration to mitigate climate change, and local economic development.

The concept of Trees Outside Forests (TOF) has emerged in 1995 to designate trees growing outside the forest and not belonging to Forest or Other Wooded Land (Bellefontain et al. 2002; FAO 2013). The term represents an effort to concentrate attention that had been spread out on components of this rather diffuse resource: agroforestry, silvo-pastoralism, urban and rural forestry, and other related disciplines. In policy and public discourse, these important resources were overlooked.

Trees Outside Forests (TOF) refers to trees, bamboos, palms, shrubs and bushes found in other lands (FAO 2013). This means that TOF are trees and shrubs on land not defined as forest and other wooded land. In other words, trees and shrubs on land spanning more than 0.05 ha with trees higher than 5 m and a canopy cover above 5 per cent, or with trees able to reach these thresholds *in situ*; or with a combined cover of shrubs and trees above 10 per cent or for linear structures a width of less than 20 m but more than 3 m. TOF are found on agricultural land (including meadows and pasture), on built-up land (viz.

industrial area, urban area, transportation, communication, and energy utilities) and on barren land (including sand dunes and rocky outcroppings). Trees and shrubs in windbreaks, shelterbelts, river galleries and corridors of trees of less than 20 m in width and below 0.5 ha in area are also TOF. *Cassia siamea*, *Tectona grandis* and *Jatropha curcas* are widely recommended for social forestry and also may be used as TOF especially in urban, peri-urban and quasi-urban areas of dry tropics.

Studies on land use change in dry tropics revealed that the soil under ten years of plantation on a cleared patch of degraded forest as compared to degraded forest and/or cropland has higher soil organic carbon, improved soil aggregate structure and stability in the whole profile and aggregate associated OC (organic carbon) stock and increased soil microbial biomass (Kumar and Ghoshal 2017, Singh et al. 2017, Singh et al. 2015 and Singh and Ghoshal 2014). In the dry tropics, among others plantation of *Cassia siamea*, *Tectona grandis*, and *Jatropha curcas* may be recommended to be included in rehabilitation strategies of the degraded lands, as it can increase C storage in soil and reduce soil CO_2-C flux, thereby help in mitigation of global climate change.

Conclusion

An important solution to overcome the impacts of land use change may be the "sustainable land management". Sustainable land management of the land degraded due to land use change may be achieved by developing eco-friendly rehabilitation strategies involving Social forestry, plantation of tree outside forest (TOF) in rural and urban lands so as to ensure long term supply of food and other resources to the growing population. Such strategies will help in rehabilitating degraded lands not only in terms of economical sustainability but also in terms of environmental sustainability. It may be suggested that for designing effective rehabilitation strategies for degraded lands, resulting from the population growth in the urban areas &/or the conversion of agricultural lands to build up structures and the remaining lands as grass fallows, especially in the dry tropics, should be based on holistic, site specific studies that include not only the upper soil layer but the entire soil profile.

Acknowledgments

We thank the Head and the Coordinator, Centre of Advanced Study in Botany, Department of Botany, for providing laboratory facilities. University Grants Commission, New Delhi, India provided financial support in form of Teacher Fellowship under FIP Scheme to Dr. Sunil Singh, University Research Fellowship to Dr. Mahesh Kumar Singh and CAS fellowship to Dr. Chandra Mohan Kumar.

REFERENCES

Al-Kaisi, M. & Yin, X. (2005). Tillage and crop residue effects on soil carbon and carbon dioxide emission in corn-soybean rotations. *J Environ Qual*, *34*, 437-445.

Alkama, R. & Cescatte, A. (2016). Biophysical climate impacts of recent changes in global forest cover. *Science*, *351*, 600-604.

Apps, M. J. & Price, D. T. (2013). Forest ecosystems, forest management and the global carbon cycle. Springer Science & Business Media.

Asante, P., Armstrong, G. W. & Adamowicz, W. L. (2011). Carbon sequestration and the optimal forest harvest decision: A dynamic programming approach considering biomass and dead organic matter. *J Forest Econ*, *17*, 3-17.

Azam, F. A., Farooq, S. & Lodhi, A. (2003). Microbial biomass in the agricultural soils-determination, synthesis, dynamics and role in plant nutrition. *Pak Soil Sci*, *6*, 629-639.

Bala, G., Caldeira, K., Wickett, M., Phillips, T. J., Lobell, D. B., Delire, C. & Mirin, A. (2007). Combined climate and carbon-cycle effects of large-scale deforestation. *Proc Natl Acad Sci*, *104*, 6550-6555.

Barbier, E. B. & Hochar, J. P. (2018). Land degradation and poverty. *Nat Sustain*, *1*, 623.

Barthes, B. & Roose, E. (2002). Aggregate stability as an indicator of soil susceptibility to runoff and erosion; validation at several levels. *Catena*, *47*, 133-149.

Bellefontaine, R., Petit, S., Pain-Orcet, M., Deleporte, P. & Bertault, J. G. (2002). Trees outside forests. *FAO Conservation Guide*, *35*. FAO, Rome.

Betts, R. A. (2000). Offset of the potential carbon sink from boreal forestation by decreases in surface albedo. *Nature*, *408*, 187.

Brovkin, V., Ganopolski, A., Claussen, M., Kubatzki, C. & Petoukhov, V. (1999). Modelling climate response to historical land cover change. *Glob Ecol Biogeogr*, *8*, 509-517.

Cambardella, C. A. (2006). Aggregation and organic matter, In: *Encyclopedia of Soil Science* (Ed. Lal R), Taylor and Francis, Boca Raton FL, pp. 52-55.

Census. (2011) Census of India 2011: towards a bright future. Office of the Registrar General & Census Commissioner, India, Ministry of Home Affairs, Government of India.

Costanza, R. (2000). Social goals and the valuation of ecosystem services. *Ecosystems*, *3*, 4-10.

Crooks, K. R., Burdett, C. L., Theobald, D. M., Rondinini, C. & Boitani, L. (2011). Global patterns of fragmentation and connectivity of mammalian carnivore habitat. *Philos Trans R Soc B*, *366*, 2642-2651.

Dale, V. H., Brown, S., Haeuber, R. A., Hobbs, N. T., Huntly, N., Naiman, R. J., Riebsame, W. E., Turner, M. G. & Valone, T. J. (2000). Ecological principles and guidelines for managing the use of land. *Ecol Appl*, *10*, 639-670.

DeFries, R. S., Foley, J. A. & Asner, G. P. (2004). Land-use choices: balancing human needs and ecosystem function. *Front Ecol Environ*, *2*, 249-257.

DeFries, R. S., Houghton, R. A., Hansen, M. C., Field, C. B., Skole, D. & Townshend, J. (2002). Carbon emissions from tropical deforestation and regrowth based on satellite observations for the 1980s and 1990s. *Proc Natl Acad Sci*, *99*, 14256-14261.

Devi, N. B. & Yadava, P. S. (2006). Seasonal dynamics in soil microbial biomass C, N and P in a mixed - oak forest ecosystem of Manipur, North -east India. *Appl Soil Ecol*, *31*, 220-227.

Diamond, H. L. & Noonan, P. E. (1996). Land use in America: report of the sustainable use of land project. Island Press, Covelo California, USA.

Dixon, R. K., Solomon, A. M., Brown, S., Houghton, R. A., Trexier, M. C. & Wisniewski, J. (1994). Carbon pools and flux of global forest ecosystems. *Science*, *263*, 185-190.

Douglas, I. (1994). Human settlements, In: *Changes in Land Use and Land Cover: A Global Perspective* (Eds. Meyer WB, Turner BL), Cambridge University, Press Cambridge, pp. 149-169.

Duiker, S. W., Rhoton, F. E., Torrent, J., Smeck, N. E. & Lal, R. (2003). Iron hydroxide crystaliinity effects on soil aggregation. *Soil Sci Soc Am J*, *67*, 606-611.

Ellis, E. C. & Ramankutty, N. (2008). Putting people in the map: anthropogenic biomes of the world. *Front Ecol Environ*, *6*, 439-447.

Environmental Impact Assessment (EIA) Notification. (2006). Ministry of Environment, Forest and Climate Change, Government of India.

Eswaran, H., Reich, P. F., Kimble, J. M., Beinroth, F. H., Padamnabhan, E., Moncharoen, P. & Kimble, J. M. (2000). Global carbon stocks. In: *Global climate change and pedogenic carbonates* (Eds. Lal R., Eswaran H., Stewart B. A.), Lewis Publishers, Boca Raton, Florida, pp. 15-25.

FAO. (1993). The state of food and agriculture 1993. Food and Agriculture Organization, Rome.

FAO. (1996). The state of food and agriculture 1996. Food and Agriculture Organization, Rome.

FAO. (2013). Towards the Assessment of Trees Outside Forests. Food and Agriculture Organization, Rome.

Foley, J. A., DeFries, R., Asner, G. P., Barford, C., Bonan, G., Carpenter, S. R., Chapin, F. S., Coe, M. T., Daily, G. C., Gibbs, H. K. & Helkowski, J. H. (2005). Global consequences of land use. *Science*, *309*, 570-574.

Frank, A. B., Liebig, M. A. & Hanson, J. D. (2002). Soil carbon dioxide fluxes in northern semiarid grasslands. *Soil Biol Biochem*, *34*, 1235-1241.

Friedlingstein, P., Houghton, R. A., Marland, G., Hackler, J., Boden, T. A., Conway, T. J., Canadell, J. G., Raupach, M. R., Ciais, P. & Le Quéré, C. (2010). Update on CO_2 emissions. *Nat Geosci*, *3*, 811-812.

Gonzalez-Quiñones, V., Stockdale, E. A., Banning, N. C., Hoyle, F. C., Sawada, Y., Wherrett, A. D., Jones, D. L. & Murphy, D. V. (2011). Soil microbial biomass-Interpretation and consideration for soil monitoring. *Soil Res*, *49*, 287-304.

Grover, S. P. P., Livesley, S. J., Hutley, L. B., Jamali, H., Fest, B., Beringer, J., Bahl, K. B. & Arndt, S. K. (2012). Land use changc and the impact on greenhouse gas exchange in north Australian savanna soils. *Bio Geo Sci*, *9*, 423-437.

Guggenberger, G., Frey, S. D., Six, J., Paustian, K. & Elliott, E. T. (1999). Bacterial and fungal cell-wall residues in conventional and no-tillage agro ecosystems. *Soil Sci Soc Am J*, *63*, 1188-1198.

Hansen, J., Ruedy, R., Sato, M. & Lo, K. (2010). Global surface temperature change. *Rev Geophys*, *48*, RG4004.

Hattori, T. (1988). Soil aggregates in microhabitats of microorganisms. Report of the Institute of Agricultural Research Tohoku University, *37*, 23-36.

Hijmans, R. J., Cameron, S. E., Parra, J. L., Jones, P. G. & Jarvis, A. (2005). Very high resolution interpolated climate surfaces for global land areas. *Int J Climatol*, *25*, 1965-1978.

Houghton, R. A. (1991). Tropical deforestation and atmospheric carbon dioxide. *Clim Change*, *19*, 99-118.

Houghton, R. A. (1995). Land-use change and the carbon cycle. *Glob Chang Biol*, *1*, 275-287.

Houghton, R. A. (2003). Revised estimates of the annual net flux of carbon to the atmosphere from changes in land use and land management 1850-2000. *Tellus B*, *55*, 378-390.

Houghton, R. A. (2010). How well do we know the flux of CO2 from land-use change? Tellus B: *Chem Phys Meteorol*, *62*, 337-351.

Houghton, R. A., Hackler, J. L. & Lawrence, K. T. (1999). The US carbon budget: contributions from land-use change. *Science*, *285*, 574-578.

Ingram, J. S. I. & Fernandes, E. C. M. (2001). Managing carbon sequestration in soils, concepts and terminology. *Agric Ecosyst Environ*, *87*, 111-117.

IPCC, (1996). Climate change 1995, Cambridge University Press, Cambridge, UK and New York, USA.

IPCC. (2007). Climate change 2007, Cambridge University Press, Cambridge, UK and New York, USA.

Iqbal, J., Hu, R., Feng, M., Lin, S., Malghani, S. & Ali, I. M. (2010). Microbial biomass and dissolved organic carbon and nitrogen strongly affect soil respiration in different land uses, a case study at three Gorges Reservoir Area, South China. *Agric Ecosyst Environ*, *137*, 294-307.

Jenkinson, D. S. (1977). The soil biomass. *NZ Soil News*, *25*, 213-218.

Jenkinson, D. S. & Ladd, J. N. (1981). Microbial biomass in soil, measurement and turnover. In: *Soil Biochemistry* (Eds. Paul EA, Ladd JN), Marcel Dekker, New York, Basel, pp. 415-471.

Kalnay, E. & Cai, M. (2003). Impact of urbanization and land on climate change. *Nature*, *423*, 528-531.

Kates, R. W., Clark, W. C., Norberg-Bohm, V. & Turner, III B. L. (1990). Human sources of global change: a report on priority research initiatives for 1990-1995, Discussion paper G-90-08. Global Environmental Policy Project, John E Kennedy School of Government, Harvard University, Cambridge, Massachusetts, USA.

Kennedy, A. C. & Papendick, R. I. (1995). Microbial characteristics of soil quality. *J Soil Water Conserv*, *50*, 243-248.

Kettunen, M., Terry, A., Tucker, G. & Jones, A. (2007). Guidance on the maintenance of landscape connectivity features of major importance for wild flora and fauna, In Guidance on the implementation of Article 3 of the Birds Directive (79/409/EEC) and Article 10 of the Habitats Directive (92/43/EEC). Institute for European Environmental Policy (IEEP), Brussels, Belgium.

Kumar, C. M. & Ghoshal, N. (2017). Impact of Land-use Change on Soil Microbial Community Composition and Organic Carbon Content in the Dry Tropics of India. *Pedosphere*, *27*, 974-977.

Lai, L., Huang, X., Yang, H., Chuai, X., Zhang, M., Zhong, T., Chen, Z., Chen, Y., Wang, X. & Thompson, J. R. (2006). Carbon emissions from land-use change and management in China between 1990 and 2010. *Sci adv*, *2*, e1601063.

Le Quere, C. (2010). Trends in the land and ocean carbon uptake. *Curropin Environ Sustain*, *2*, 219-224.

Leite, C. C., Costa, M. H., Soares-Filho, B. S. & Barros VianaHissa, L. (2012). Historical land use change and associated carbon emissions in Brazil from 1940 to 1995. *Global Biogeochem Cycles*, *26*.

Litton, C. M., Ryan, M. G., Knight, D. H. & Stahl, P. D. (2003). Soil-surface carbon dioxide efflux and microbial biomass in relation to tree density 13 years after a stand replacing fire in a lodgepole pine ecosystem. *Glob Chang Biol*, *9*, 680-696.

Liu, D. S., Iverson, L. R. & Brown, S. (1993). Rates and patterns of deforestation in the Philippines: application of geographic information system analysis. *For Ecol Manage*, *57*, 1-16.

Liu, J. Y., Wang, S. Q., Chen, J. M., Liu, M. L. & Zhuang, D. F. (2004). Storages of soil organic carbon and nitrogen and land use changes in China: 1990-2000. *Acta Geogr Sin*, *59*, 483-496.

Lugo, A. E. (1991). Cities in the sustainable development of tropical landscapes. *Nat Resour*, *27*, 27-35.

Mahlman, J. D. (1997). Uncertainties in projections of human-caused climate warming. *Science*, *278*, 1416-1417.

Manjula, S., Srivastava, A. K., Chatterjee, R., Ponneinselvan, M., Shinde, K. V., Chandra, V. & Sharma, A. (2016). India 2016 A Reference Annual. Publications Division, Ministry of Information and Broadcasting, Government of India, New Delhi, India.

McGarigal, K. & Cushman, S. A. (2002). Comparative evaluation of experimental approaches to the study of habitat fragmentation. *Ecol Appl*, *12*, 335-345.

Meyer, W. B. & Turner, II B. L. (1994). Changes in land use and land cover: a global perspective. Cambridge University Press, Cambridge, UK.

Millennium Ecosystem Assessment. (2005). Ecosystems and human well-being: Biodiversity synthesis. World Resource Institute, Washington, D.C.

Mitsuda, Y. & Ito, S. (2011). A review of spatial-explicit factors determining spatial distribution of land use/land-use change. *Landsc Ecol Eng*, *7*, 117-125.

Murphy, D. V., Sparling, G. P. & Fillery, I. R. P. (1998). Stratification of microbial biomass C and N and gross N mineralization with soil depth in two contrasting Western Australian agricultural soil. *Aus J Soil Res*, *36*, 45-55.

Ohsowski, B. M., Klironomos, J. N., Dunfield, K. E. & Hart, M. M. (2012). The potential of soil amendments for restoring severely disturbed grasslands. *Appl Soil Ecol*, *60*, 77-83.

Ojima, D. S., Galvin, K. A. & Turner, B. L. I. (1994). The global impact of land-use change. *Bio Sci*, *44*, 300-304.

Pacala, S. W., Hurtt, G. C., Baker, D., Peylin, P., Houghton, R. A., Birdsey, R. A., Heath, L., Sundquist, E. T., Stallard, R. F., Ciais, P. & Moorcroft, P. (2001). Consistent land- and atmosphere-based US carbon sink estimates. *Science*, *292*, 2316-2320.

Palmroth, S., Maier, C., McCarthy, H., Oishi, C., Kim, H. S., Johnsen, K., Katul, G. & Oren, R. (2005). Contrasting responses to drought of forest floor CO_2 efflux in a loblolly pine plantation and a nearby oak-hickory forest. *Glob Chang Biol*, *11*, 421-434.

Pan, G., Li, L., Wu, L. & Zhang, X. (2004). Storage and sequestration potential of topsoil organic carbon in China's paddy soils. *Glob Chang Bio*, *l10*, 79-92.

Pan, Y., Birdsey, R. A., Fang, J., Houghton, R., Kauppi, P. E., Kurz, W. A., Phillips, O. L., Shvidenko, A., Lewis, S. L., Canadell, J. G. & Ciais, P. (2011). A large and persistent carbon sink in the world's forests. *Science*, 1201609.

Parent, L. E., Almeida, C. X. d., Hernandes, A., Egozcue, J. J., Gülser, C., Bolinder, M. A., Kätterer, T., Andrén, O., Parent, S. E., Anctil, F., Centurion, J. F. & Natale, W. (2012). Compositional analysis for an unbiased measure of soil aggregation. *Geoderma*, *179*, 123-131.

Patel, K., Kumar, J. I. N., Kumar, R. N. & Bhoi, R. K. (2010). Seasonal and temporal variation in soil microbial biomass C, N and P in different types land uses of dry deciduous forest ecosystem of Udaipur, Rajasthan, Western India. *Appl Ecol Environ Res*, *8*, 377-390.

Pathirana, A., Denekew, H. B., Veerbeek, W., Zevenbergen, C. & Banda, A. T. (2014). Impact of urban growth-driven landuse change on microclimate and extreme precipitation-A sensitivity study. *Atmos Res*, *138*, 59-72.

Perlin, J. (1989). A forest journey: the role of wood in the development of civilization. Harvard University Press, Cambridge, Massachusetts, USA.

Pimm, S. L. & Raven, P. (2000). Extinction by numbers. *Nature*, *403*, 843-845.

Pocket book of Agricultural Statistics. (2014). Government of India, Ministry of Agriculture, Department of Agriculture & Cooperation, Directorate of Economics & Statistics, New Delhi, India.

Powlson, D. S., Brookes, P. C. & Cristenson, B. T. (1987). Measurement of soil microbial biomass provides as an early indicator of changes in total soil organic matter due to straw incorporation. *Soil Biol Biochem*, *19*, 159-164.

Powlson, D. S., Whitmore, A. P. & Goulding, K. W. T. (2011). Soil carbon sequestration to mitigate climate change, a critical re-examination to identify the true and the false. *Eur J Soil Sci*, *62*, 42-55.

Prove, B. G., Loch, R. J., Foley, J. L., Anderson, V. J. & Younger, D. R. (1990). Improvements in aggregation and infiltration characteristics of a krasnozem under maize with direct drill and stubble retention. *Aust J Soil Res*, *28*, 577-590.

Quéré, C. L., Andrew, R. M., Canadell, J. G., Sitch, S., Korsbakken, J. I., Peters, G. P., Manning, A. C., Boden, T. A., Tans, P. P., Houghton, R. A. & Keeling, R. F. (2016). Global carbon budget 2016. *Earth System Science Data*, *8*, 605-649.

Quetier, F., Lavorel, S., Thuiller, W. & Davies, I. (2007). Plant-trait-based modeling assessment of ecosystem-service sensitivity to land-use change. *Ecol Appl*, *17*, 2377-2386.

Raich, J. W. & Nadelhoffer, K. J. (1986). Belowground carbon allocation in forest ecosystems, global trends. *Ecology*, *70*, 1346-1354.

Raich, J. W. & Potter, S. C. (1995). Global patterns of carbon dioxide emissions from soil. *Global Biogeochem Cycles*, *9*, 23-36.

Raich, J. W. & Schlesinger, W. H. (1992). The global carbon dioxide flux in soil respiration and its relationship to vegetation and climate. *Tellus B*, *44*, 81-99.

Rey, B., Bullock, J. M. & Newton, A. C. (2008). Creating woodland islets to reconcile ecological restoration, conservation, and agricultural land use. *Front Ecol Environ*, *6*, 329-336.

Richards, J. F. (1990). Land transformations. In: *The earth as transformed by human action: global and regional changes in the biosphere over the past 300 years* (Eds. Turner II, B. L., Clark, W. C., Kates, R. W., Richards, J. E., Matthews, J. T., Meyer, W. B.), Cambridge University Press, New York, USA, pp. 163-178.

Richards, J. F. & Flint, E. P. (1994). A century of land-use change in south and southeast Asia, In: *Effects of land-use change on atmospheric CO_2 concentrations* (Ed. Dale VH), Springer-Verlag, New York, USA, pp. 15-66.

Riebsame, W. E., Parton, W. J., Galvin, K. A., Burke, I. C., Bohren, L., Young, R. & Knop, E. (1994). Integrated modeling of land use and cover change. *Biosci*, *44*, 350-356.

Rindfuss, R. R., Walsh, S. J., Turner, B. L., Fox, J. & Mishra, V. (2004). Developing a science of land change: challenges and methodological issues. *Proc Natl Acad Sci USA*, *101*, 13976-13981.

Ryan, M. G. & Law, B. E. (2005). Interpreting, measuring, and modeling soil respiration. *Biogeochemistry*, *73*, 3-27.

Sahani, U. & Behera, N. (2001). Impact of deforestation on soil physicochemical characteristics, microbial biomass and microbial activity of tropical soil. *Land Degrad Dev*, *12*, 93-105.

Sala, O., Chapin, F. S. & Armesto, J. (2000). Global biodiversity scenarios for the year 2100. *Science*, *287*, 1770-1774.

Satake, A. & Iwasa, Y. (2006). Coupled ecological and social dynamics in a forested landscape: the deviation of individual decisions from the social optimum. *Ecol Res*, *21*, 370-379.

Saunders, D. A., Hobbs, R. J. & Margules, C. R. (1991). Biological Consequences of Ecosystem Fragmentation: A Review. *Conserv Biol*, *5*, 18-32.

Scheffer, M., Carpenter, S., Foley, J. A., Folke, C. & Walkerk, B. (2001). Catastrophic shifts in ecosystems. *Nature*, *413*, 591-596.

Schwaiger, H. P. & Bird, D. N. (2010). Integration of albedo effects caused by land use change into the climate balance: Should we still account in greenhouse gas units? *For Ecol Manage*, *260*, 278-286.

Sexstone, A. J., Revsbech, N. P., Parkin, T. B. & Tiedje, J. M. (1985). Direct measurement of oxygen profiles and denitrification rates in soil aggregates. *Soil Sci Soc Am J*, *49*, 645-651.

Singh, J. S., Raghubanshi, A. S., Singh, R. S. & Srivastava, S. C. (1989). Microbial biomass acts as source of plant nutrients in dry tropical forest and savanna. *Nature*, *338*, 499-500.

Singh, M. K., Astley, H., Smith, P. & Ghoshal, N. (2015). Soil CO_2-C flux and carbon storage in the dry tropics: Impact of land-use change involving bioenergy crop plantation. *Biomass Bioenergy*, *83*, 123-130.

Singh, M. K. & Ghoshal, N. (2011). Impact of land use change on soil organic carbon content in dry tropics. *Plant Arch*, *11*, 903-906.

Singh, M. K. & Ghoshal, N. (2014). Variation in soil microbial biomass in the dry tropics: impact of land-use change. *Soil Res*, *52*, 299-306.

Singh, M. K., Singh, S. & Ghoshal, N. (2017). Impact of land use change on soil aggregate dynamics in the dry tropics. *Restor Ecol*, *25*, 962-971.

Six, J., Elliott, E. T. & Paustian, K. (2000). Soil macroaggregate turnover and microaggregate formation, a mechanism for C sequestration under no-tillage agriculture. *Soil Biol Biochem*, *32*, 2099-2103.

Six, J., Feller, C., Denef, K., Ogle, S. M., Sa, J. C. D. & Albrecht, A. (2002). Soil organic matter, biota and aggregation in temperate and tropical soils-effects of no-tillage. *Agronomie*, *22*, 755-775.

Smith, B., Knorr, W., Widlowski, J. L., Pinty, B. & Gobron, N. (2008). Combining remote sensing data with process modelling to monitor boreal conifer forest carbon balances. *For Ecol Manage*, *255*, 3985-3994.

Smith, J. L. & Paul, E. A. (1990). The significance of microbial biomass estimation. In: *Soil Biochemistry* (Eds. Bollag J. M., Stotsky G.), Marcel Dekker, New York, pp. 357-396.

Soil Science Society of America. (1997). Glossary of soil science terms. Soil Science Society of America, Madison, WI.

Song, X. P., Hansen, M. C., Stehman, S. V., Potapov, P. V., Tyukavina, A., Vermote, E. F. & Townshend, J. R. (2018). Global land change from 1982 to 2016. *Nature*, *560*, 639.

State of Environment Report India. (2009). Ministry of Environment & Forests, Government of India, India.

Sugihara, S., Funakawa, S., Kilasara, M. & Kosaki, T. (2012). Effect of land management on CO_2 flux and soil C stock in two Tanzanian croplands with contrasting soil texture. *Soil Biol Biochem*, *46*, 1-9.

Tisdall, J. M. & Oades, J. M. (1982). Organic matter and water-stable aggregates in soils. *J Soil Sci*, *62*, 141-163.

Turner, II B. L., Clark, W. C., Kates, R. W., Richards, J. F., Mathews, J. T. & Meyer, W. B. (1990). The Earth as Transformed by Human Action: Global and Regional Changes in the Biosphere Over the Past 300 Years. Cambridge University Press, New York, New York, USA.

Turner II, B. L., Matson, P. A., McCarthy, J. J., Corell, R. W., Christensen, L., Eckley, N., Hovelsrud-Broda, G. K., Kasperson, J. X., Kasperson, R. E., Luers, A., Martello, M. L., Mathiesen, S., Naylor, R., Polsky, C., Pulsipher, A., Schiller, A., Selin, H. & Tyle, N. (2003). Illustrating the coupled human-environment system for vulnerability analysis: three case studies. *Proc Natl Acad Sci*, *100*, 8080-8085.

Turner, II B. L., Moss, R. H. & Skole, D. L. (1993). Relating land use and global land-cover change: a proposal for an IGBP-HDP core project. HDP Report Number 5, International Geosphere-Biosphere Programme, Stockholm, Sweden.

Turner, B. L., Lambin, E. F. & Reenberg, A. (2007). The emergence of land change science for global environmental change and sustainability. *Proc Natl Acad Sci*, *104*, 20666-20671.

Turner, B. L. I. & Meyer, W. B. (1994). Global land-use and land-cover change: an overview. In: *Changes in land use and land cover: a global perspective* (Eds. Meyer WB, Turner BLI), Cambridge University Press, Cambridge, UK, pp. 3-10.

Verheye, W. H. (2009). Land use, land cover and soil sciences. Encyclopedia of Life Support Systems (EOLSS).

Vitousek, P. M., Mooney, H. A., Lubchenco, J. & Melillo, M. (1997). Human domination of Earth's ecosystems. *Science*, *277*, 494-504.

Wilson, G. W. T., Rice, C. W., Rillig, M. C., Springer, A. & Hartnett, D. C. (2009). Soil aggregation and carbon sequestration are tightly correlated with the abundance of arbuscular mycorrhizal fungi: results from long-term field experiments. *Ecol Lett*, *12*, 452-461.

WRI. (1997). World resources: 1996-1997. World Resources Institute, Washington, DC, USA.

Yamaura, Y., Amano, T., Koizumi, T., Mitsuda, Y., Taki, H. & Okabe, K. (2009). Does land-use change affect biodiversity dynamics at a macroecological scale? A case study of birds over the past 20 years in Japan. *Anim Conserv*, *12*, 110-119.

Yilmaz, G. (2012). Seasonal variation of cumulative CO_2 emission from a vertisol under apricot orchard in semi-arid southeast Yurkey. *Pedosphere*, *22*, 322-332.

In: Innovations in Life Science Research ISBN: 978-1-53615-868-7
Editors: R. P. Sinha, S. Pandey-Rai and N. Ghoshal ©2019 Nova Science Publishers, Inc.

Chapter 15

MULTIFUNCTIONAL TROPICAL DRY FOREST SYSTEMS OF INDIA: CURRENT NEED AND FUTURE DIRECTIONS

Apurva Rai[1,2,*], Ashutosh Kumar Singh[2], Shriya Mehrotra[1] and Nandita Singh[2]

[1]Centre of Advanced Study in Botany, Institute of Science,
Banaras Hindu University, Varanasi, India
[2]Plant Ecology and Environmental Science Division,
CSIR-National Botanical Research Institute, Lucknow, India

ABSTRACT

Tropical dry forests influence climate through physical, chemical and biological processes that affects the hydrologic cycle, composition of the atmosphere and the planetary energetic. The presence of tropical dry forests may potentially enhance the livelihood through providing food, fodder and firewood as well as in mitigation of the impact of climate change. Drawing on the representative literature, here, we critically review the contribution of tropical dry forestry systems in India to: (i) biodiversity conservation; (ii) supply of products and services; (iii) potential of carbon storage; (iv) improving the fertility of the soil (v) providing social and economic well being. Tropical dry forest contributes to various ecological, social and economic functions but they are only complementary and are still not considered as an alternative to natural ecosystem. Consequently, they are now considered among the world's most threatening system. To promote well being of society, management of tropical dry forests needs to be strengthened by innovations in its various systems. Products and services of tropical dry forests profoundly and disproportionally affect our lives from the impact of global weather, to the food we consume. Future research is required to eliminate many of the uncertainties that remain, and also carefully test the main functions certified to tropical dry forest system in

* Corresponding Author's Email: raiapurva@hotmail.com.

order to know as to what extent tropical dry forests served the purpose of sustainable development.

Keywords: tropical dry forests, biodiversity conservation, soil fertility, carbon sequestration, livelihood

INTRODUCTION

Land use options are necessary to increase the security of livelihood and for the susceptibility to climatic and environmental changes. Forests are the well-known natural resource management system that possesses the potential to improve the livelihoods through simultaneous production of firewood, fodder and food as well as mitigating the impact of climate change. Tropical forest system is one of the most diverse terrestrial ecosystems which shelters variety of natural resources and help to sustain the livelihood of local communities (Dash et al. 2009). They support the maximum diversity of plant species on the earth (Anonymous 1992). Tropical dry forests are socially, ecologically and economically highly valued all over the world (Mooney et al. 1995) and provide many goods and ecosystem services (Li et al. 2003; Wang 2003; Armentaras et al. 2009). High variability is found in tropical dry forests in the amount of precipitation, latitude, altitude and longitude and also with the duration of wet and dry period (Umashankar 2001). Tropical dry forests undergo changes in their structure and function. Tropical dry forests are rich in species diversity and in some cases are even comparable to tropical wet forests (Medina 1995). Characteristically tropical dry forests are classified as the 'Dry' forests, however in some cases they receive higher annual precipitation than other temperate forests (Chave et al. 2006).Where all the tropical forests have at least one period in the year with low water availability whereas in tropical dry forests the dry season is prolonged and severe.

Tropical dry forest system in India plays a complex and supporting role in agricultural systems helping the millions of farmers and providing them a wide range of environmental services (Chidumayo and Gumbo 2010). Apart from sequestering a huge amount of atmospheric carbon in the above ground (AGB) and below ground biomass (BGB), tropical dry forests supports the development of the social, cultural and economic conditions of the people living at the edge of it. Altogether having a fascinating characteristics, their management and conservation requires a different approach as the goods and services provided by them are significantly valuable and are different from the other types of forests (Wunder 2001; Makonda and Gillah 2007; Gumbo et al. 2010). However, they are not given much importance by the researchers (Dirzo et al. 2011; Sanchez-Azofeifa et al. 2005) and are exposed to a range of threats, mainly from human disturbances (Janzen 1998; Yosi et al. 2011; Baithalu et al. 2013) especially in India (Uma Shankar 2001). The large area of tropical dry forest has been converted into agriculture as the soil and climate which supports them are suitable for farming and ranching (Trejo and Dirzo 2000). In addition, they are impacted by several anthropogenic activities (Champion and Seth 1968; Singh et al. 1991) such as lopping, burning, overgrazing and clearing of land for cultivation (Jha and Singh 1990; Pattanayak et al. 2003; Sundarapandian and Pascal 2013) resulting in conversion of tropical dry forests into the dry deciduous scrubs savanna and grasslands which are progressively species-poor (Champion and Seth 1968; Sagar and Singh 2003). Furthermore, they are being disturbed due to, land use changes (Calvo Alvarado et al. 2009; Redo et al. 2012; Singh et al. 2015) and invasion of several exotic species (Miles et al. 2006; Gillespie et al. 2012). Tropical trees outside forestry are also greatly used in agro forestry systems or as community woodlots to provide a mixture of products and services to resident households, local communities, and regional cultures. Therefore, it can be a part of agro forestry, community forestry, farm forestry, social forestry, and in urban forestry (Long and Nair 1999). Owing to the role of tropical dry forests in ecological, social and environmental services to the society, it deserves attention particularly for the conservation and its development in India. The present assessment explores the characteristic features, carbon exchange capacity, socio economic and environmental benefits of tropical dry forest to Indian forestry system. Synthesis of the available literature also helps identify the remaining uncertainties and thus the future direction of research.

DISTRIBUTION OF FOREST AREA

In 2015 forest area covers 3999 M ha globally which is equivalent to approximately 31% of total global land area. In addition about 10% of the total global land is covered by wooded land. Forty-four per cent of global forest area is found in countries classified as tropical and another 8% is in sub-tropical countries. From 1990 to 2015 forest area and other wooded land area in tropical countries was reduced from 47.6% to 44% and other wooded land area reduced from 56% to 42.9% (Table 1).

In India, the total forest and tree cover is 24.39% of total geographical area. Champion and Seth (1968) recognized sixteen types of forest which are listed below (Figure 1, 2):

The predominant forest types of India are tropical dry deciduous and tropical moist deciduous. Of the total forest cover, tropical dry deciduous forests occupy an area of 2,17,713 km^2 (34.80%) followed in decreasing area by tropical moist deciduous forests, 2,07,649 km^2 (33.19%), tropical semi-evergreen forests, 48,295 km^2 (7.72%) and tropical wet evergreen forests, 47,192 km^2 (7.54%).

Table 1. Change in the area (K ha) of the forest and other wooded land from 1990 to 2015 in different global climatic domains (K ha)

Forest area (K ha)								other wooded land area (Kha)		
Domain	**1990**	**2000**	**2005**	**2010**	**2015**	**1990**	**2000**	**2005**	**2010**	**2015**
Boreal	1,219,309	1,219,820	1,218,856	1,224,873	1,224,452	121,212	117,735	119,590	121,999	121,187
Temperate	617,997	640,892	659,176	673,429	684,468	157,582	154,534	159,568	163,737	167,255
Sub-tropical	325,421	324,777	323,912	319,613	320,057	150,132	149,090	151,391	150,602	399,094
Tropical	1,965,542	1,870,112	1,830,799	1,797,757	1,770,156	549,529	533,090	523,143	537,825	516,935
Grand total	4,128,269	4,055,602	4,032,743	4,015,673	3,999,134	978,454	954,448	953,692	974,163	1,204,471

Source: FAO 2015.

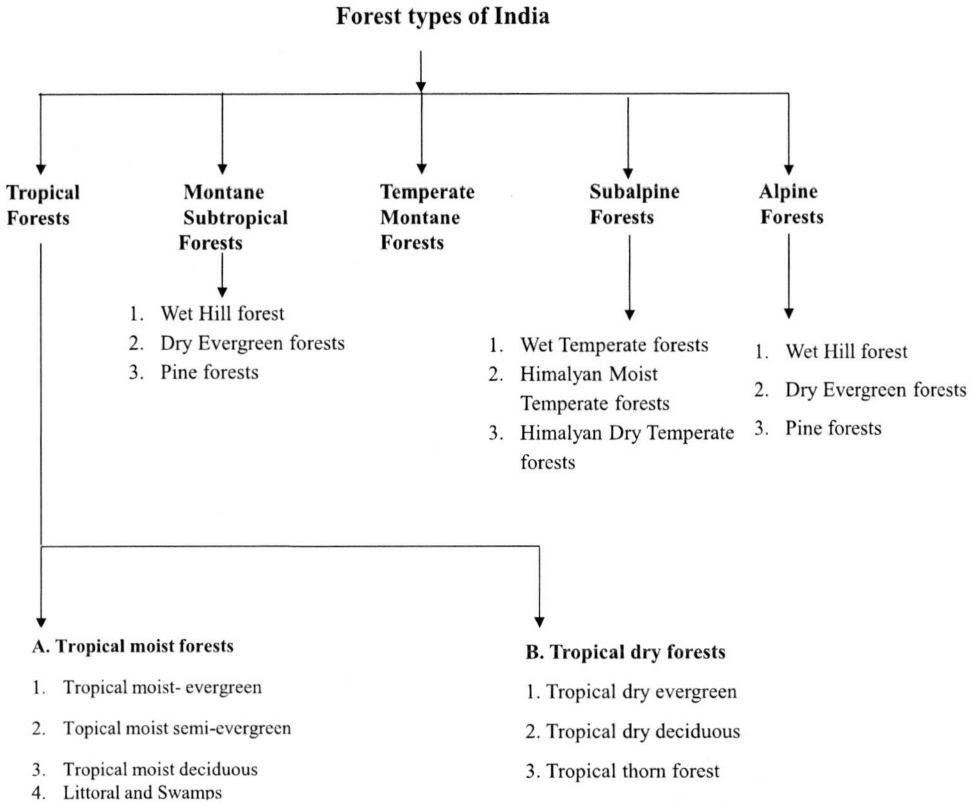

Figure 1. Classification of Forest types of India (Adapted from Champion and Seth 1968).

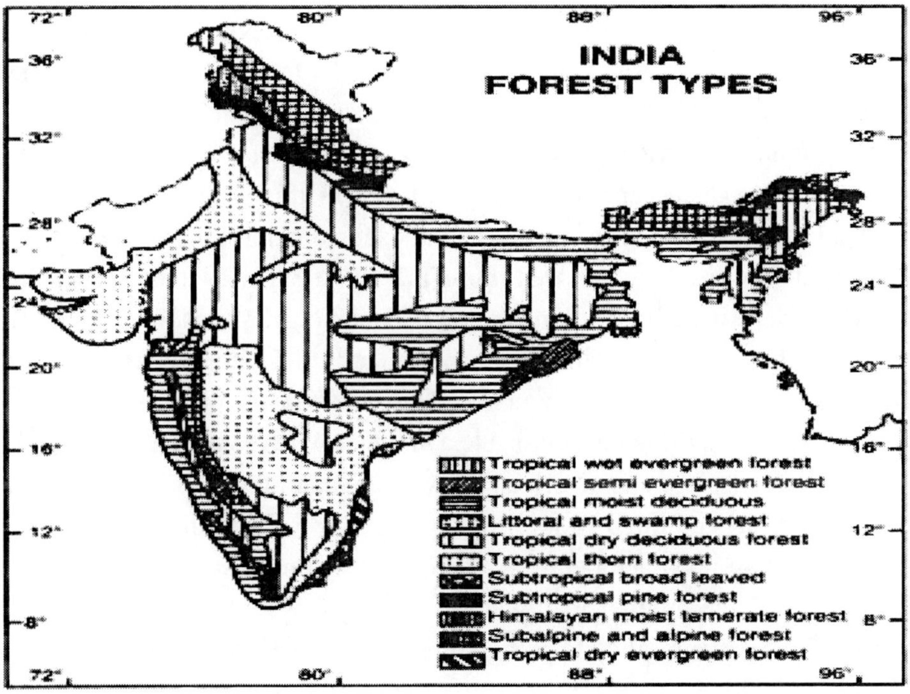

Figure 2. Distribution of different Forest types in India (Adapted from Champion and Seth 1968).

Tropical dry forests are classified into the following types:

(i) Tropical dry evergreen forests, which are hard leaved evergreen trees, predominate with some deciduous emergents, often dense, but usually under 20 m high.
(ii) Tropical dry deciduous forests, which are entirely deciduous or nearly so. Top canopy are relatively light and rarely over 25 m high and usually of 8-20 m.
(iii) Tropical thorn forests, which are deciduous with low thorny tress and xerophytes predominating with canopy more or less broken and under 10 m in height.

Characteristic Features of the Tropical Dry Forest

Tropical dry deciduous forests (TDFs) can be found in severe and extremely variable climates characterized by low annual rainfall, 5-6 dry months within the annual cycle, and nutrient-poor soil (Figure 3). Several terms have been used for this vegetation type such as seasonally dry tropical forest (SDTF), tropical dry deciduous forest, monsoon forest, caatinga, cuabal, etc. More than any other factor, the lack of precipitation during a prolonged portion of the year is what produces true dry forest, an ecosystem type characterized by plants and animals with specific adaptations to survive the long dry season. Deciduousness is the single most important adaptation among plants to the extended droughts. Most of the trees drop their leaves after the rains end, and essentially halt photosynthesis, as they would otherwise be unable to survive the water loss during the dry season. Tropical dry forests are subject to intensive anthropogenic disturbances and are among the most at-risk ecosystems in the world. In order to assess the conservation status of this forest type, information is required on its distribution pattern, climate, the structure and functional traits of its vegetation, phenology, strategies for coping with drought and nutrient poverty, and disturbances and their effects. In this book, we review important studies on TDFs around the globe, particularly those in the northern dry deciduous forests of India. We put forward the claim that those TDFs that experience drought and arise on nutrient-poor sites feature adaptations such as deciduousness, as well as a variety of nutrient conservation strategies. They also experience biotic disturbances, which can result in fragmentation and ecosystem conversion, and therefore exhibit changes in biomass, productivity, and soil microbial biomass, etc. (Singh and Chaturvedi 2017).

Tropical dry forests are characterized by warm temperatures and a dry period of 3 or more months (Murphy and Lugo 1986; Mooney et al. 1995, Jaramillo et al. 2011; Becknell et al. 2012) The mean annual temperature >25°C, and are dominated by 50% of the drought deciduous trees and the annual precipitation ranging from 700 to 2000 mm (Sánchez-Azofeifa et al. 2005). Evaporation to precpitation ratio of tropical dry forests is greater than one (Olivares and Medina 1992). Length of dry period is the controlling factor of vegetation structure and patterns in tropical dry forests (Gritti et al. 2010) as the plant species are subjected to water stress during the dry season (Eamus 1999). These forests are composed of mosaic of plant functional traits (PFTs) due to the difference and intensity of seasonal drought (Borchert 2000). During the process of succession tropical dry forests are more vulnerable to stress due to the severe and less predictable environment (Murphy and Lugo 1986). The very specific phenological attribute of tropical dry forests is its deciduousness which is affected by the pattern of rainfall, duration of solar radiation and variability in its temperature which in turn affects the balance of carbon, water and energy (Singh and Kushwaha 2005; Bohlman 2010).

Figure 3. Aerial photography of Tropical dry forest (canopy view) (A), Landscape of tropical dry forest in Indogangetic plains (Sonbhadra), India (B).

TROPICAL DRY FOREST SYSTEM AS A CARBON SINK AND SOURCE

The biomass storage of carbon in tropical dry forests was reported to be higher than the other forest ecosystems (Figure 4) (Kishwan et al. 2009). Carbon management through afforestation and reforestation in natural forests is a useful option and the presence of forests enhances the uptake of CO_2 from the atmosphere. About 82% of total vegetation carbon density is stored by the tree species of tropical dry forests in its perennial aerial structure, 4% in the foliage while the rest 14% is stored in its coarse roots (Singh and Singh 1991; Singh and Singh 1993). Above ground biomass (AGB) is considered as an important measure of C sequestration as it regulates the atmospheric carbon concentration and constitutes about 60% of the total phytomass (Ketterings et al. 2001). Saatchi et al. (2011) stated that globally, 8.7 Pg C is stored in the above ground biomass of tropical dry forests. In India, the carbon storage is 27-113 t C ha^{-1}in the AGB (Singh and Singh 1991; Chaturvedi et al. 2011). The maximum storage of carbon in tropical dry forests is its AGB (Gibbs et al. 2007). Storage of carbon in the AGB is important as it is directly linked with carbon stocks in other pool. Total litter fall of tropical dry forests which is a part of AGB contributes to 6 $tha^{-1}yr^{-1}$ (Chabra and Dhadwal 2004). The C stocks in litter or dead woods (down trees, standing dead, broken branches, leaves, etc.) is equivalent to 10-20% of AGB while 20% of it is stored in root (Chabra et al. 2004; Gibbs et al. 2007). The overall C pool in tropical dry forests of India is shown in (Figure 5). The C flux estimated from the total litter of Tropical Dry Forests during 80's was 53Tg C yr^{-1}(Figure 6) (Chabra and Dhadwal 2004). Evidence is now emerging that the tropical dry forest systems are proving to be a promising management practices for the mitigation of greenhouse gas by storing it in the above ground and as soil carbon stocks. Soil in forest system has the ability to act as a long term storage reservoir of carbon. Soil in large area within forest is involved in global scale and thus holds an important position in C budget (Jackson 2000). The large size along with the potential ability of sequestering a huge amount of carbon makes the forest soil an important part of carbon cycling.

The role of forest in carbon cycle is most considerable as it stores 62-78% of global carbon. The below ground storage of carbon as soil organic carbon (SOC) pool is 70% (Dixon et al. 1994; Schimel 1995; Houghton 1996). Estimation of below ground biomass (BGB) of tropical dry forests is important, as it acts as a biggest repository of carbon. The carbon stored in BGB indicates the quality as well as productivity of soil (Bauer and Black 1994; Krishan et al. 2009). Addition of most of the carbon in the soil is done by the decomposition of the leaf litter on the surface of soil and through the release of root exudates (Lynch and Whipps 1990; Nguyen 2003). The soil in tropical dry forests stores approximately 120 tha^{-1} C at 100 cm of its depth (Chabra et al. 2004). SOC pool for India has been about 4.13Pg for 50 cm, 35 Mt (for 30 cm) (FSI 2005; FSI 2008) (Figure 7), 2.2% (6.18Pg for top soil) for 1m and 2.6% for 2m depth. (Chhabra et al. 2002; Eswaran et al. 2000; Ravindranath et al. 1997; Dadhwal et al. 1998).SOC is the imperative sink of atmospheric C but sometimes it may act as a source if it is not managed appropriately. The C stored in the soil is lost by several natural and anthropogenic activities. The loss of C from the soils of tropical dry forests is also of great concern as it adds C to the atmosphere through respiration, leaching and soil erosion (Lal 2003). The loss of carbon from the terrestrial ecosystem accounts for 55-136Pg in which the release from soil contributes to 12-78Pg (Lal 2007). Accordingly, any change in the SOC will effect on the global carbon budget. At the present time the biggest concern is the rapid loss of forests.

A major uncertainty and therefore an issue for future research is that still our understanding of carbon cycling in tropical dry forests lags behind that of other ecosystems (Sanchez-Azofeifa et al. 2005). In order to exploit the unrealized potential of carbon sequestration through tropical dry forest requires innovative policies, based on rigorous research results, are required. Tropical dry forests act as both sink and source of organic carbon.

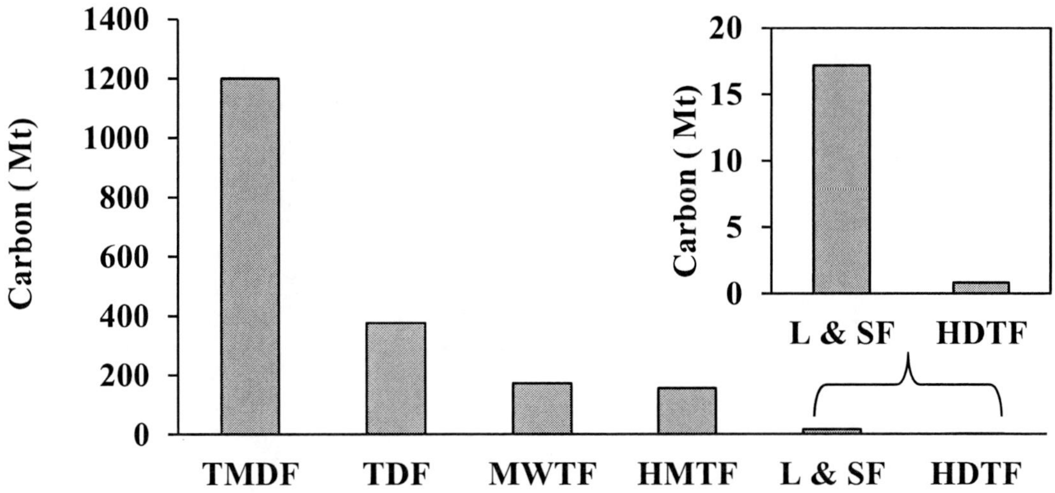

TDF-Tropical dry forest, TMDF-Tropical moist deciduous forest, HDTF-Himalyan dry temperate forest, L& SF-Littoral and swamp forest, MWTF-Montane wet temperate forest

Figure 4. Biomass carbon (Mt) in different forest type in India.

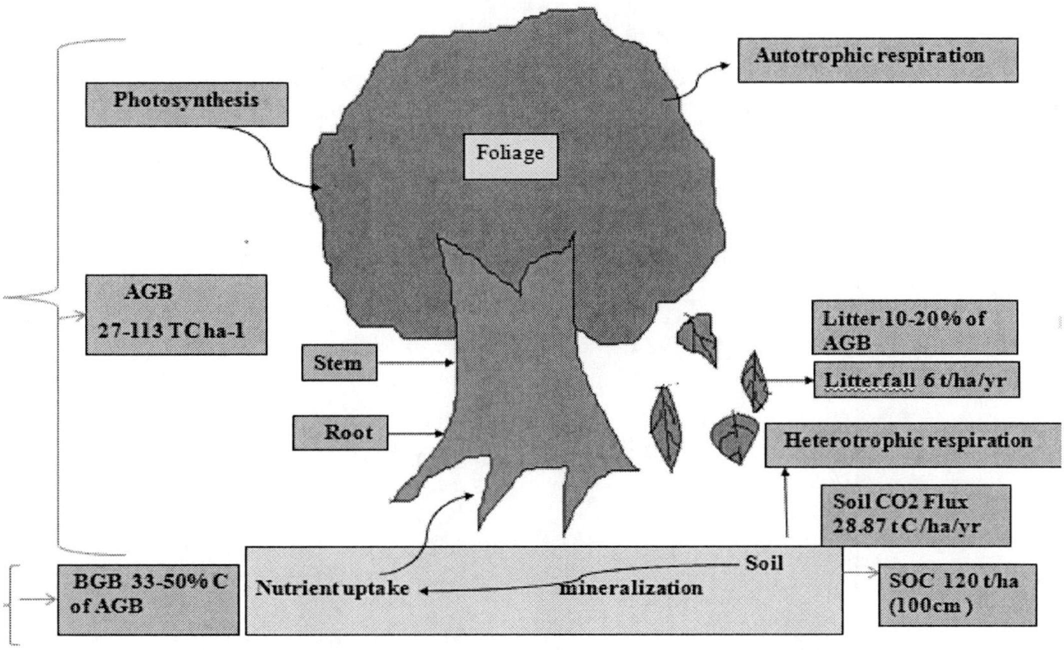

Figure 5. The overall carbon pool in India (Adapted from Chabra and Dhadwal 2004; Chaturvedi et al. 2011).

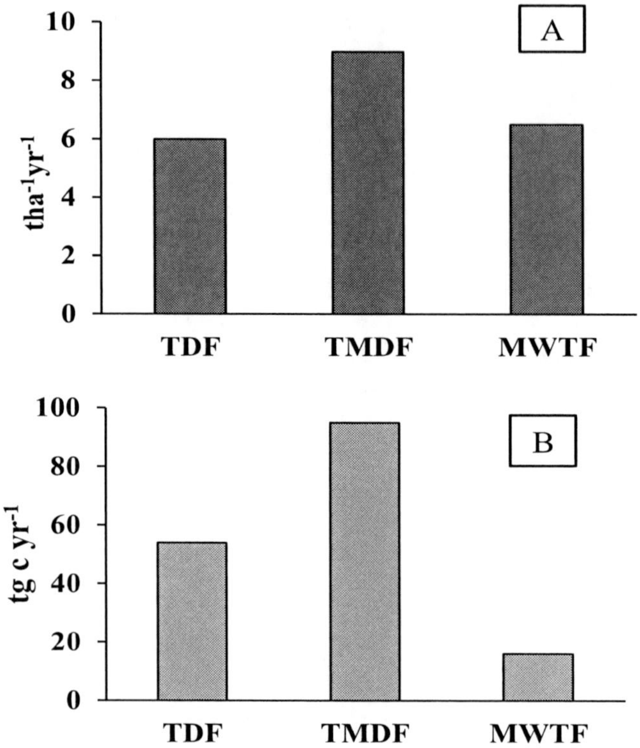

Tg C= 10^{12} g

CTDF = Tropical dry forest, TMDF = Tropical moist deciduous forest, HDTF = Himalyan dry temperate forest, L & SF = Littoral and swamp forest, MWTF = Montane wet temperate forest

Figure 6. Mean Total litter fall ($tha^{-1}yr^{-1}$) (A), C flux (tg Cyr^{-1}) in Indian forest (B).

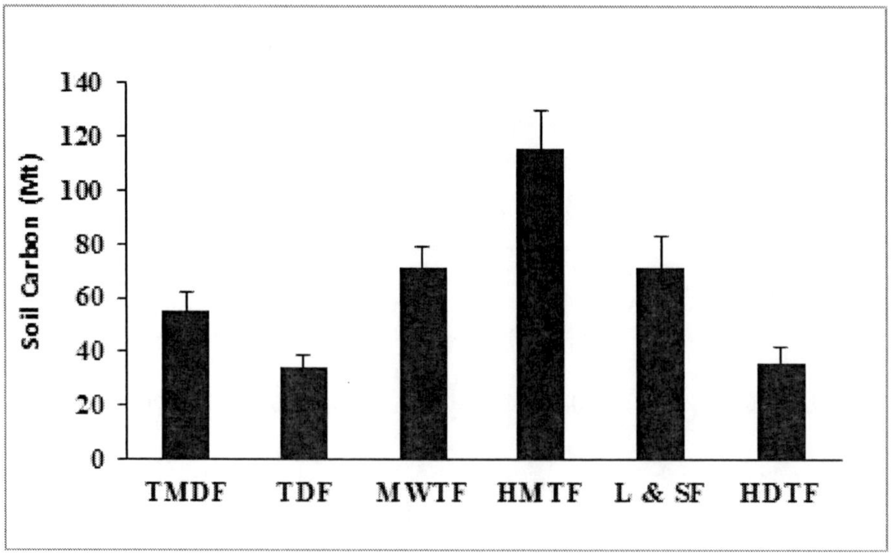

TDF-Tropical dry forest, TMDF-Tropical moist deciduous forest, HDTF-Himalyan dry temperate forest, L& SF-Littoral and swamp forest, MWTF-Montane wet temperate forest

Figure 7. Mean soil C (Mt C) of various forest types of India at 0-30 cm depth.

TROPICAL DRY FORESTS INCREASE SOIL FERTILITY AND WATER USE EFFICIENCY

Tropical dry forests plays a major role in soil erosion, flooding and controlling landslides due to deforestation the problem of soil erosion has increased, which causes landslides (Table 2). Tropical dry forests are reported to increase water infiltration which in turn results into less soil erosion and landslides (Calder and Aylward 2006). Although, the soil of tropical dry forests is nutrient poor (Sanchez 1976; Singh et al. 1989) yet the presence of moisture plays a crucial role in survival and growth of plant diversity (Lugo et al. 1978). Additionally, tropical dry forests helps in nutrient cycling by transfer of manure and leaf litter to the soil and thus improves its quality. Higher biodiversity leads to more efficient ecosystem services. Tropical dry forests improve the quality of water through transpiration.

Water consists of chemicals, nutrients, sediments, salts, and even pathogens (Brauman et al. 2007). The quality of water is improved with the hit of first raindrop. The water is treated by the processes of infiltration with vegetation; leaf litter, microbes and soils all removing and biochemically transforming the contaminants present (Brauman et al. 2007; Hall et al. 2011; Acreman et al. 2012).

THREATS TO TROPICAL DRY FORESTS

A need is required to design a synergy in the society among sustainable livelihoods, Convention on Biological Diversity, the Kyoto Protocol and the other international treaties. The biodiversity of the tropical forests is the highest in the world. India is known for its high biodiversity. The historic evidence states that forests with higher diversity have a higher number of species usages. Out of the 17,000 plants in India, 7,500 have been used for traditional medicines (Senthilkumar and Murugesan et al. 2012). Continued deforestation is a major challenge for forests and livelihoods. In addition, there is an emerging challenge of decreasing biological diversity through species reduction in forestry system. Tropical dry forests are known for their endemic characteristics. Highly specialized, rare, and endemic plants and animals are most vulnerable to anthropogenic pressure, and/or climate change is present in this biome. Extinction of species is fast, accelerating, and pervasive across plant and animal species, and is closely linked to tropical dry forests destruction and fragmentation and to direct overexploitation which is going to cause a crisis to the global biodiversity.

Deforestation and degradation is considered as the biggest source of GHG's emission in tropical countries. The deforestation and degradation of tropical dry forests has a direct impact on the surrounding atmosphere as the aboveground living biomass of trees has the largest pool of carbon stored in it (Gibbs et al. 2007). More than 70% of the total emission is reported by deforestation activities in some parts of the world (FAO 2005). A huge amount of carbon is being released into the atmosphere by the disturbance. The carbon released by deforestation of tropical forest was about 1-2 Btha^{-1}, which was approximately 15-20% of annual greenhouse gas emission which was even higher than the transport sector during 90's (Gibbs et al. 2007). In India, the cumulative net carbon flux to the atmosphere has been about 3.2Pg by the clearance of tropical dry forests (Chhabra and Dadhwal 2004). The loss in the forest cover will not only effect the above ground storage of carbon but it will also influence the below ground services. The conversion into grassland or cropland by various natural and anthropogenic activities brings alteration in structure and function of soil (Grünzweig et al. 2003; Sorenson 1974) by affecting the nutrient cycling and soil organic matter (SOM) dynamics (Glaser et al. 2001).

Table 2. Regional examples of soil fertility enhancement in multifunctional tropical dry forestry in India

Region	Challenge	Changes observed due to forestry	References
Indo gangetic plains (U.P)	Biomass production and nutrient dynamics in nutrient-deficient and toxic soils	Biomass production (49 t ha^1/decade)	Singh (1998)
Central India (M.P)	Restoration of mine spoil soil	Increase in soil nutrient (N and P)	Singh et al. 1998
Purulia district (W.B)	Enhancing litter production and soil nutrient dynamics	Increase in the flow of nutrient	Biswas and khan 2010
North India (Haryana)	Improvement Sodic soil	Increase in soil moisture ratio, organic carbon, Total nitrogen, Extractable phosphorous	Bhojvaid and Timmer, 1998
Singrauli (U.P)	Soil improvement	Increase in soil organic matter	Singh and Singh 1998
Western India	Improvement in soil	Increases the fertility of soil	Kumar et al. 2010
Assam	Restoration of degraded lands	Increase in the microbial biomass, enhanced P and N availability	Mishra et al. (2014)
Himalayas (Meghalaya)	Enhancing tree survival and crop yield	Crop yield did not decrease in proximity to *Albizzia* trees	Dhyani and Tripathi (1998)
Indo gangetic plains (U.P)	Optimality of fertilizer use	Redevelopment of mine soil by use of fertilizer, increase soil carbon	Singh and Singh (2006)
Eastern India	Reducing soil and water loss in agro-ecosystems	*Butea monosperma* increases the water holding capacity	Kumari et al. (2005)
Eastern India	Afforestation in tropical laterite wastelands	Improvement in nutritional status and microbial activity	Mukhopadhyay and Joy (2010)
Central India (Raipur)	Biomass production in N and P stressed soils	*Azadirachtaindica* trees were found to produce biomass in depleted soils.	Puri and Swamy (2001)

ROLE OF TROPICAL DRY FORESTS IN ENHANCING LIVELIHOOD

In India, where people have a long history and accumulated knowledge, forests contribute to a huge scope for the livelihood improvement. Ethno forestry practices along with the indigenous knowledge system on the growing tree are prominent in India. Forestry system in tropics offers a significant opportunity for the sustainable development by providing nutritional and economic security to the poor through the domestication of forest fruit tree and other species grown (Milne et al. 2006). Tropical dry forests support some of the world's poorest people (Murphy and Lugo 1986) and occupy half of the world's tropical and subtropical forests (Campbell 1996; Cunningham et al. 2008; Waeber et al. 2012). It also supports livestock by providing a source of grazing (Scoones and Toulmin 1993; Campbell 1996; Clarke et al. 1996; Cavendish 2000; Marunda and Bouda 2010; Dewees et al. 2011). Yet, there is complexity in the role and services of tropical dry forest ecosystem and the methods for assessing its contribution. Here, we discuss some of the key roles of tropical dry forest ecosystem. The direct role in providing food can be visualized particularly during scarcity and in extreme famine conditions. The nutritional intake of people residing near dry forests are mostly affected by the nutritional diversity and dietary of wild foods (Chidumayo and Marunda 2010; Shackleton et al. 2011).

Products from native tree species contribute significantly to rural livelihoods (Abebaw et al. 2012). But still research is required to explore the benefits to rural lively hoods by comparing studies of different regions. Over 500 million of people are living or are at the edge of tropical forests and tropical dry forests. Human health and ecosystems health are inextricably linked. Studies have started to show that tropical dry forests, influence human health by the numerous ways. Forest dwelling peoples make use of these chemicals and compounds traditionally. It was estimated that about 70-95% of people living in developing countries rely on the use of traditional medicines for their primary health care needs (Robinson and Zhang 2011). This forest dependent population relies not only for economic and environmental benefits but also form an integral part of their culture and spiritual traditions. Globally, industrial wood products account for about US$ 400 billion worth of global production (WCFSD 1998) in which 25% of this production is contributed by woods of tropical forests. Several species of trees viz (*Shorea robusta*, *Tectona grandis*, *Hardwickia binnata)* which are found extensively in these regions are used extensively as timber in industries as well as in household products. The demand in the form of wood and charcoal as a fuel is increasing (Arnold et al. 2003). Several studies have been conducted in India and abroad but the focus is rare on tropical dry forests (Sagar and Singh 2004; Top et al. 2004 a, b, 2006). Fallen twigs and litter are used as a fuel for cooking. There are numerous non timber forest products collected for wilderness for subsistence and cash income. Some of trees in tropical dry forests *(Butea monosperma*) produce lac which has its economic values. Several non-timber forests products such as fibers, resins, latexes, fruits, and traditional medicines are also provided. Traditional practices and sacred elements in cultural landscapes of India also have a substantial livelihood and conservation value (Bawa et al. 2004). Agro forestry system is a common practice in India. It gives us the solution to the query of hike in prices of fuel wood resulted due to higher demand and low supply generated as to forest degradation in India (Bowonder et al. 1988). It includes trees from community forests, farms and a variety of local forest management and ethno forestry practices (Pandey 1998). Trees from tropical dry forests are a very important part of agro forestry system in India. This system encompasses a variety of practices such as trees growing in close association to form a boundary in farms, to collect rain water in the collection ponds in villages, in silvo pastoral system, and to grow trees within settlements. Indian forests are mostly comprised of dry forests and cover a wide range of area in different states of India (FSI 1995).But, the destructive strategies driven by the market forces and the lack of knowledge about the growing species often directs to the unsustainable harvesting. The pressure on the natural ecosystem can be reduced through domestication of such growing species which aims at commercialization and production of valued products (Belcher et al. 2005).

THE FUTURE OF TROPICAL DRY FORESTS

Table 3. Unresolved challenges for tropical dry forest system research and innovations in India

Nutrients: additional supply or redistribution?	Matured trees are associated with the improved soil nutrient supply in traditional forestry system but there is still a lack in knowledge is the addition of nutrient is the total quantum or towards its redistributed i.e., horizontally and vertically
Carbon sequestration in above ground and below ground biomass	Holistic sights are required for the carbon sequestration in tropical dry forestry
Carbon sequestration in above ground and below ground biomass	Holistic sights are required for the carbon sequestration in tropical dry forestry
Land Amelioration and conservation	Mature trees capable of yielding enough litter are known to conserve soil but their attributes useful for agro climatic problems are required in India.
Domestication of useful species	Many wild populations of species that yield commercial products are being depleted. Research efforts are required to domesticate these species.
Value added innovations	Non timber forests products have the potential to improve the livelihoods, but vigorous knowledge is required for the value added innovations.
Genetically modified trees	Genetically improved trees may provide more biomass and the products, but presently research result is mostly remains in the laboratory. An approach is required to ensure the suitable stock to the society.

Even though various options are involved in the livelihood improvement, forestry system is considered as one option with multifunctional value. In India, proper implementation of strategies and a decentralized planning to promote the local biomass production in forestry system could be a path to the sustainable development. Along with the mitigation of climate change, the forestry system can meet the energy needs of about one billion people in India through the options of bio energy by the practical use of residues and biomass generated in forestry systems. Availability of the biomass energy can create rural wealth and the employment necessary for the improvement of livelihood and can further sequester a large amount of carbon in a decentralized manner. These strategies will also ensure social, economical and ecological well-being of the society. There are further some unresolved challenges which are required for further innovation and research in India (Table 3).

In order to use tropical dry forestry systems as in important option for livelihood improvement, mitigating climate change and for sustainable development in India, the research, policy and practices will have to progress towards:

- Documentation of studies on the dry forests in the provision of food and nutrition is only restricted to few studies.
- For the effective management of ecosystem it is important to understand the dry forests particularly in the face of deforestation.
- Further studies are required on the use and users of forests and towards its adaptation to climate change.
- Indian research should focus on the use of forests for other purposes rather than clearing it for agricultural practices.
- Energy efficiency of dry forests and its link to urbanization is still inadequate.
- A focused study should be done on direct provisioning and subsistence usage on non-timber forests products.
- Alternative institutions for conservation and management of tropical dry forests should be explored.
- Last but not the least, bringing researchers, planners, policy analysts, and practitioners together to learn, decide, and implement adaptive management. As the society has been greatly lacking in this regard. It is now imperative to integrate the relatively independent activities of research, planning, monitoring, assessment, and decision support into systems for societal learning and adaptive management of tropical dry forests.

Conclusion

Prevalence of tropical dry forest system in India offers multifunctional opportunities, worth reconsidering for carbon sequestration, biodiversity conservation, soil fertility enhancement, livelihood improvement and poverty reduction. The aim of this study is to figure out the importance of tropical dry forests by discussing its natural property to act as a repository of atmospheric carbon and also as a valuable system for the socio-economic development in India. Practically there is a threat to the tropical dry forests by both the anthropogenic (looping, grazing, clearing of forests) as well as natural calamities such as climate change which is stressing forests in India. With the decline of forests size and health, their services are affected to provide the sustainable livelihood. Appropriate policies should be made to protect the deforestation and degradation of tropical dry forests for the conservation of biodiversity. This is very imperative if we really want to ensure a lasting flow of this productive ecosystem. Thus some explicit practices for ensuring the everlasting services of this ecosystem is required. This may be concluded that the values of tropical dry forests are immense, distributed throughout and are of great importance to the economic as well as to the environmental sectors. Thus a priority should be made for the protection and conservation of tropical dry forests.

Acknowledgments

The authors are thankful to The Director, CSIR-National Botanical Research Institute, Lucknow and also the Head and the Coordinator, Centre of Advanced Study in Botany, Department of Botany for providing needful facilities. Apurva Rai and Ashutosh Kumar Singh are also thankful to Council of Scientific and Industrial Research under project BSC0109 and Department of Science & Technology, India for financial support.

References

Abebaw, D., Kassa, H., Kassie, G. T., Lemenih, M., Campbell, B. & Teka, W. (2012). Dry forest based livelihoods in resettlement areas of Northwestern Ethiopia. *Forest Policy Econ*, *20*, 72-77.

Acreman, M., Albertengo, J., Amado, T., Amis, M., Anderson, A., Bacchur, I. & Basch, G. (2012). Report of the work of the expert group on maintaining the ability of biodiversity to continue to support the water cycle.

Anonymous. (1992). Space Station wins again.

Armentaras, D., Rodriguez, N. &Retana, J. (2009). Are conservation strategies effective in avoiding the deforestation of the Colombian Guyana Shield? *Biol Conserv*, *42*, 1411-1419.

Arnold, M., Köhlin, G., Persson, R. & Shepherd, G. (2003). Fuelwood Revisited: What has Changed.

Baithalu, M. S., Anbarashan, M. & Parthasarathy, N. (2013). Two-decadal changes in forest structure and tree diversity in a tropical dry evergreen forest on the Coromandel Coast of India *Trop Ecol*, *54*, 397-403.

Bala, G. (2007).*Proc Natl Acad Sci USA*, *104*, 6550.

Baldocchi, D. D., Vogel, C. A. & Hall, B. (1997). *J Geophys Res*, *102*, 28939.

Baldocchi, D. D. & Xu, L. (2007).*Water Resour*, *30*, 2113.

Bauer, A. & Black, A. L. (1994). Quantification of the effect of soil organic matter content on soil productivity. *Soil Sci Soc Am J*, *58*, 185-193.

Bawa, K. S. (2004). Tropical ecosystems into the 21st century. *Science*, *306*, 227-228.

Becknell, J. M., Kucek, L. K. & Powers, J. S. (2012). Above ground biomass immature and secondary seasonally dry tropical forests: A literature review and global synthesis. *For Ecol Manage*, *276*, 88-95.

Bhojvaid, P. P. & Timmer, V. R. (1998). Soil dynamics in an age sequence of *Prosopisjuli-flora* planted for sodic soil restoration in India. *For ecol manage*, *106*, 181-193.

Blackie, R., Baldauf, C., Gautier, D., Gumbo, D., Kassa, H., Parthasarathy, N., Paum-garten, F., Sola, P., Pulla, S., Waeber, P. & Sunderland, T. (2014). Tropical dry forests: The state of global knowledge and recommendations for future research. CIFOR, Bogor, Indonesia.

Bohlman, S. A. (2010). Landscape patterns and environmental controls of deciduousness in forests of central Panama. *Global Ecol Bio Geogr*, *19*, 376-385.

Bonan, G. B. (2008). Forests and climate change: forcings, feedbacks and climate benefits of forests. *Science*, *320*, 1444-1449.

Borchert, R. (2000). Organismic and environmental controls of bud growth in tropical trees. *In: Dormancy in plants: From whole plant behavior to cellular control* (Eds. Viemont, J. D., Crabbe, J.), Wallingford: CAB International, pp. 87-107.

Bowonder, B., Prasad, S. S. R. & Unni, N. V. M. (1998). Dynamics of fuel wood prices in India: Policy implications. *Word De*, *16*, 1213-1229.

Brauman, K., Daily, A., Duarte, G. C., Harold, T. K. & Moon, A. (2007). "The Nature and Value of Ecosystem Services: An Overview Highlighting Hydrologic Services." *Annu Rev Environ Resour*, *32*, (November), 67-98.

Brown, S., Gillespie, A. J. & Lugo, A. E. (1989). Biomass estimation methods for tropical forests with applications to forest inventory data. *Forest Sci*, *35*, 881-902.

Brown, S. & Lugo, A. E. (1982). The storage and production of organic matter in tropical forests and their role in the global C cycle. *Biotropica*, *14*, 161-187.

Calder, I. R. & Aylward, B. (2006). "Forest and Floods: Moving to an Evidence-Based Approach to Watershed and Integrated Flood Management. *Water Int*, *31*, 87-99.

Calvo-Alvarado, J., McLennan, B., Sanchez-Azofeifa, A. & Garvin, T. (2009). Deforestation and forest restoration in Guanacaste, Costa Rica: Putting conservation policies in context. *For Ecol Manage*, *258*, 931-940.

Campbell, B. (1996). The Miombo in Transition: Woodlands and Welfare in Africa. Bogor, Indonesia: Center for International Forestry Research.

Canadell, J. G. & Raupach, M. R. (2008). Managing forests for climate change mitigation. *Science*, *320*, 1456-1457.

Cardinale, B. J. (2011). Biodiversity Improves Water Quality through Niche Partitioning. *Nature*,*472*, 86-90.

Cavendish, W. (2000). Empirical Regularities in the Poverty-Environment Relationship of Rural Households: Evidence from Zimbabwe. *World Deve*, *28*, 1979-2003.

Chadwick, O. A., Kelly, E. F., Merritts, D. M. & Amundson, R. G. (1994), Carbon dioxide Consumption during soil development. *Biogeochemistry*, *24*, 115-127.

Champion, H. G. & Seth, S. K. (1968).A revised survey of the forest types of India, Government of India Publication, New Delhi, IN.

Champion, H. G. & Seth, S. K. (1968). General silviculture for India (Delhi: Publication Division, Government of India).

Chaturvedi, R. K., Raghubanshi, A. S. & Singh, J. S. (2011). Leaf attributes and tree growth in a tropical dry forest. *J Veg Sci*, *22*, 917-931.

Chave, J., Muller-Landau, H. C., Baker, T., Easdale, T., Steege, H. & Webb, C. (2006). Regional and phylogenetic variation of wood density across 2456 neotropical tree species. *Ecol Appl*, *16*, 2356-2367.

Chhabra, A. & Dadhwal, V. K. (2004). Assessment of major pools and fluxes of carbon in Indian forests. *Clim Change*, *64*, 341-360.

Chhabra, A., Palria, S. & Dadhwal, V. K. (2002). Soil organic carbon pool in Indian forests. *For Ecol Manage*, *5877*, 1-13.

Chidumayo, E. & Gumbo, D. (2010).The Dry Forests and Woodlands of Africa: Managing for Products and Services. London: Earths can.

Clark, D. A. (2007). Detecting tropical forests' responses to global climatic and atmospheric change: current challenges and a way forward. *Biotropica*, *39*, 4-19.

Clarke, J., Cavendish, W. &Coote, C. (1996). Rural households and miombo woodlands: Use, value and management. *In: The Miombo in Transition: Woodlands and Welfare in Africa* (ed. Campbell B), Bogor, Indonesia: Center for International Forestry Research., 101-36.

Corlett, R. T. &Primack, R. B. (2011). Tropical Rain Forests: An Ecological and Biogeographical Comparison, 2nd edn. Wiley, Oxford, UK.

Cunningham, A. B., German, L., Paumgarten, F., Chikakula, M., Barr, C., Obidzinski, K., van Noordwijk, M., de Koning, R., Purnomo, H., Yatich, T., Svensson, L., Gaafar, A.

& Puntodewo, A. (2008). Sustainable trade and management of forest products and services in the COMES Aregion: an issue paper. Center for International Forestry Research, Bogor, Indonesia.

D'Amato, A. W., Bradford, J. B., Fraver, S. & Palik, B. J. (2011). Forest management for mitigation and adaptation to climate change: insights from long-term silviculture experiments. *For Ecol Manage*, *262*, 803-816.

Dash, P. K., Mohapatra, P. P. & Giri Rao, Y. (2009). Diversity and distribution pattern of tree species Niyamgiri hill Ranges, Orissa, India. *Ind For*, *135*, No.7.

Dewees, P. A., Campbell, B. M., Katerere, Y., Sitoe, A., Cunningham, A. B., Angelsen, A. & Wunder, S. (2010). Managing the Miombo woodlands of southern Africa: policies, incentives and options for the rural poor. *J Nat Resour Pol Res*,*2*, 57-73.

Dhadhwal, V. K., Pandya, N. & Vora, A. B. (1998). Carbon cycle for Indian forest ecosystem: a preliminary estimate. *In: Global change studies: Scientific Results ISRO-GBP*, (Eds. Subbaraya, B. H., Rao, D., Desai, P. S., Manikram Rajratnam, P.), ISRO, Bangalore, pp. 411-430.

Dhyani, S. K. &Tripathi, R. S. (1998). Tree growth and crop yield under agrisilvicultural practices in northeast India. *Agro for Syst*,*44*, 1-12.

Dirzo, R., Young, H. S., Mooney, H. A. & Ceballos, G. (2011). Introduction. *In: Seasonally Dry Tropical Forests: Ecology and Conservation.* (Eds. Dirzo, R., Young H. S., Mooney H., Ceballos, G.), Island Press, Washington, DC, USA.

Dixon, R., Winjum, J. K., Andrasko, J., Lee, J. & Schroeder, P. (1994). Integrated land-use systems: assessment of promising agroforest and alternative land-use practices to enhance carbon conservation and sequestration. *Clim Change*, *27*, 71-92.

Dixon, R. K., Brown, S., Houghton, R. A., Solomon, A. M., Trexler, M. C. & Wisniewski, J. (1994). Carbon pools and flux of global forest ecosystems. *Science*, *263*, 185-190.

Eamus, D. (1999). Ecophysiological traits of deciduous and evergreen woody species in the seasonally dry tropics. *Trends Ecol Evol*, *14*, 11-16.

Eswaran, H., Reich, P. F., Kimble, J. M., Beinroth, F. H., Padmanabhan, E. & Moncharoen, P. (2000). Global carbon stocks. *In*: *Global Change and Pedogenic Carbonate.* (Eds. Lal, R, Kimble, J. M., Eswaran H., Stewart, B. A.), CRC Press, Boca Raton, FL, pp. 15-25.

Eswaran, H., Van Den Berg, E. & Reich, P. (1993). Organic carbon in soils of the world. *Soil Sci Soc Am J*, *57*, 192-194

FAO. (2005). Global forest resources assessment. Global synthesis FAO forestry paper no. 124. Food and Agriculture Organisation of the United Nations, Rome.

FAO. (2006). Global forest resources assessment 2005 Food and Agriculture Organisation of the United Nations, Rome.

FAO. (2007). State of the World's Forests 2007; FAO of United Nations, Rome (Italy), p. 200

Field, C. B., Behrenfeld, M. J., Randerson, J. T. & Falkowski, P. (1998). Primary production of the biosphere: integrating terrestrial and oceanic components. *Science*, *281*, 237-240.

Foley, J. A. (1995). An equilibrium model of the terrestrial carbon budget. *Tellus*, *47*, 310-319.

Foley, J. A., DeFries, R., Asner, G. P., Barford, C., Bonan, G., Carpenter, S. R., Chapin, F. S., Coe, M. T., Daily, G. C. & Gibbs, H. K. (2005). Global consequences of land use. *Science*, *309*, 570-574.

FSI. (1995). Extent, composition, density, growing stock and annual increment of India's forests. Forest Survey of India, Ministry of Environment & Forests, Dehradun.

FSI. (2005) State of the forest report 2003 Forest Survey of India, Ministry of Environment and Forest, Government of India, Dehra Dun, India

FSI. (2008). State of Forest Report 2005 Forest Survey of India, Ministry of Environment and Forests Government of India, Dehra Dun, India.

Gentry, A. H., Bullock, S. H., Mooney, H. A. & Medina, E. (1995). Diversity and Floristic Composition of Neotropical Dry Forests. *In: Seasonally dry tropical forests* (Eds. Bullock, S. H., Mooney, H. A., Medina, E.) Cambridge University Press, Cambridge, pp. 146-194.

Gerhardt, K. (1993). Tree seedling development in tropical dry abandoned pasture and secondary forest in Costa Rica. *J Veg Sci*, *4*, 95-102.

Gibbs, H. K., Brown, S., Niles, J. O. & Foley, J. A. (2007). Monitoring and estimating tropical forest carbon stocks: making REDD a reality. *Environ Res Lett*,*2*, 045023.

Gillespie, T., Lipkin, B., Sullivan, L., Benowitz, D., Pau, S. & Keppel, G. (2012). The rarest and lprotected forests in biodiversity hotspots. *Biodivers Conserv*, *21*, 3597-611.

Glaser, B., Johannes, L., Führböter, M., Solomon, D. & Zech, W. (2001). Carbon and nitrogen mineralization in cultivated and natural savanna soil of Northern Tanzania. *Biol Fertil Soils*, *33*, 301-309.

Gritti, E. S., Cassignat, C., Flores, O., Bonnefille, R., Chalié, F., Guiot, J. & Jolly, D. (2010). Simulated effects of a seasonal precipitation change on the vegetation in tropical Africa. *Clim Past*, *6*, 169-178.

Grünzweig, J. M., Sparrow, S. D. & Chapin, F. S. (2003). Impact of forest conversion to agriculture on carbon and nitrogen mineralization in subarctic Alaska. *Bio geo chemistry*, *64*, 271-29.

Gumbo, D., Cunningham, T. & Shackleton, C. (2010). Opportunities for dry forests research within CIFOR. Unpublished report, Center for International Forestry Research, Bogor, Indonesia.

Hall, J. S., Ashton, M. S., Garen, E. J. & Jose, S. (2011). The Ecology and Ecosystem Services of Native Trees: Implications for Reforestation and Land Restoration in Mesoamerica. *For Ecol Manage*, *261*, 1553-1557.

Hansen, M. C., Potapov, P. V., Moore, R., Hancher, M., Turubanova, S. A., Tyukavina, A., Thau, D., Stehman, S. V., Goetz, S. J., Loveland, T. R., Kommareddy, A., Egorov, A., Chini, L., Justice, C. O. & Townshend, J. R. G. (2013). High-Resolution Global Maps of 21st-Century Forest Cover Change and Supplementary Materials for High-Resolution Global Maps of 21st-Century Forest Cover Change. *Science*, *342*, 850-853.

Hassan, R., Scholes, R. & Ash, N. (2005). Ecosystems and Human Well-being: Current State and Trends, Volume 1, Island Press, Washington, DC.

Hegde, R. & Enters, T. (2000). Forest products and household economy: a case study from Mudumalai Wildlife Sanctuary, Southern India. *Environ Conserv*, *27*, 250-259.

Holdridge, L. R. (1967). Life zone ecology (San Jose, Costa Rica: Tropical Science Centre).

Houghton, R. A. (1996). Converting terrestrial ecosystems from sources to sinks of carbon. *Ambio*, *25*, 267-272.

Houghton, R. A. (1996). Land-use change and terrestrial carbon: the temporal record. *In: Forest ecosystems*, *Forest management and the global carbon cycle* (Eds. Apps MJ, Price DT), Springer-Verlag Berlin Heidelberg, New York, NY.

Houghton, R. A. (1999). The annual net flux of carbon to the atmosphere from changes in land use 1850-1990. *Tellus*, *50*, 298-313.

Houghton, R. A. (2008). Carbon flux to the atmosphere from land-use changes: 1850-2005. In: Trends: A compendium of data on global Change, Carbon di oxide Information Analysis Center, oak Ridge National Laboratory, US Department of Energy, Oak ridge, TN, USA.

Houghton, R. A., Skole, D. L., Nobre, C. A., Hackler, J. L., Lawrence, K. T. &Chomentowski, W. H. (2000). Annual fluxes of carbon from deforestation and regrowth in the Brazilian Amazon. *Nature*, *403*, 301-304.

Hudson, R. J. M., Gherini, S. A. & Goldstein, R. A. (1994). Modelling the Global Carbon Cycle: Nitrogen Fertilization of the Terrestrial Biosphere and the "Missing" CO_2 Sink. *Global Bio geo chem Cycles*, *8*, 307-333.

Jackson, R. B., Schenk, H. J., Jobbagy, E. G., Canadell, J. &Colello, G. D. (2000). Below-ground consequences of vegetation change and their treatment in models. *Ecol Appl*, *10*, 470-483.

Janzen, D. H. (1998). *In: Tropical dry forests: the most endangered major tropical ecosystem* (ed. Wilson EO), Biodiversity National. Academy Press, Washington DC., Sundarapandian SM, Pascal J,Karoor (2013) Edge effects on plant diversity in tropical forest ecosystems at Periyar Wildlife sanctuary in the Western Ghats of India. *J For Res*, *24*, 403-418.

Jaramillo, V., Martínez-Yrízar, A. & Sanford, R. (2011). Primary Productivity and Biogeochemistry of Seasonally Dry Tropical Forests. *In: Seasonally Dry Tropical Forests* (eds.R Dirzo, H Young, H. A. Mooney and G Ceballos), Island Press, pp. 109-128.

Jha, C. S. & Singh, J. S. (1990). Compositions and dynamics of dry tropical forest in relation to soil texture. *J Veg Sci*, *1*, 609-614.

Jordan, C. F. (1983). Productivity of tropical rain forest ecosystems and the implications for their use as future wood and energy sources. *In: Tropical rain forest ecosystems* (ed. Golley FB), Ecosystems of the world. Elsevier, Amsterdam, N. L., pp. 117-136.

Jumbe, C. B., Bwalya, S. M. &Husselman, M. (2008).Contribution of Dry Forests to Rural Livelihoods and the National Economy in Zambia. Bogor, Indonesia: Center for International Forestry Research.

Ketterings, Q. M., Coe, R., van Noordwijk, M., Ambagau, Y. & Palm, C. A. (2001). Reducing uncertainty in the use of allometric biomass equations for predicting aboveground tree biomass in mixed secondary forests. *For Ecol Manag*, *146*, 199-209.

Kishwan, J., Pandey, R. & Dhadwal, V. K. (2009). *India's Forest and Tree Cover: Contribution as a carbon Sink.*, *130*, 1-12.

Krishan, G., Srivastav, S. K., Kumar, S., Saha, S. K. &Dadhwal, V. K. (2009). Quantifying the underestimation of soil organic carbon by the Walkley and Black technique-examples from Himalayan and Central Indian soils. *CurrSci*, *96*, 1133-1136.

Kumari, N., Sharma, V., Mikosch, M., Unfried, C., Gessler, A., Fischer Schliebs, E. &luettg, E. U. (2005). Seasonal photosynthetic performance and nutrient relations of *Butea monosperma* TAUB: in comparison to two other woody species of seasonal deciduous forests in SE-Rajasthan and to planted trees in the area. *Ind J For*, *28*, 116-126.

Lal, R. (2003). Global potential of soil C sequestration to mitigate the greenhouse effect. *Crit Rev Plant Sci*, *22*, 151-184.

Lal, R. (2007). Soil carbon stocks under present and future climate with specific reference to European eco regions, Nutr. Cycl. *Agro-ecosyst*, *81*, 113-127.

Lal, R., Follett, R. F. & Kimble, J. M. (2003). Achieving soil carbon sequestration in the United States: a challenge to the policy makers. *Soil Sci*, *168*, 827-845.

Li, Y. D., Zhou, G. Y., Zeng, Q. B., Wu, Z. M. & Luo, T. S. (2003). The values for ecological service function of tropical natural forest in Hainan Island, China. *For Res*, *16*, 146-152.

Long, A. J. & Ramachandra Nair, P. K. (1999). Trees outside forests: agro, community, and urban forestry. *New forests*, *17*, 145-174.

Lugo, A. E., Gonzalez-Liboy, J. A., Cintron, B. & Dugger, K. (1978). Structure, productivity and transpiration of a subtropical dry forest in Puerto Rico. *Biotropica*, *10*, 278-91.

Lynch, J. M. & Whipps, J. M. Substrate flow in the rhizosphere. *Plant soil*, *129*, 1-10.

Makonda, F. & Gillah, P. (2007). Balancing wood and non-wood products in miombo woodlands. *Working Papers of the Finnish Forest Research Institute*, *50*, 64-70.

Malhi, Y., Baldocchi, D. D. & Jarvis, P. G. (1999). The carbon balance of tropical, temperate and boreal forests. *Plant Cell Environ*, *22*, 715-740.

Marunda, C. & Bouda, H, (2010). Environmental services from the dry forests and woodlands of sub-Saharan Africa. *In: The Dry Forests and Woodlands of Africa: Managing for Products and Services* (Eds. Chidumayo E, Gumbo D), London, Earthscan, pp. 231-260.

Melillo, J. M., McGuire, A. D., Kicklighter, D. W., Moore, B., Vorosmarty, C. J. & Schloss, A. L. (1993). Global climate change and terrestrial net primary production. *Nature*, *363*, 234-240.

Miles, L., Newton, A. C., Defries, R. S., Ravilious, C., May, I., Blyth, S., Kapos, V. & Gordon, J. E. (2006). A global overview of the conservation status of tropical dry forests. *Biogeogr J*, *33*, 491-505.

Mo, E. F. (1999). National Forestry Action Plan, Ministry of Environment and Forests, Government of India, New Delhi.

Mukhopadhyay, S. & Joy, V. C. (2010). Influence of leaf litter types on microbial functions and nutrient status of soil: ecological suitability of forest trees for afforestation in tropical laterite wastelands. *Soil Biol Biochem*, *42*, 2306-2315.

Murphy, P. G. & Lugo, A. E. (1986). Ecology of tropical dry forest. *Annu Rev Ecol Syst*, *17*, 67-88.

Nath, S., Das, P. K., Gangopadhyay, S. K. & Banerjee, S. K. (1991). Suitability of different forest species for social forestry programme under different soil conditions. *Indian Forester*, *117*, 625-631.

National Forestry Action Programme. (1999)/ Ministry of Environment and Forests, Government of India, New Delhi, vols 1&2.

Nguyen, C. (2003). Rhizodeposition of organic C by plants: mechanisms and controls. *Agron*, *23*, 375-396.

Nirmal Kumar, J. I., Kumar, R. N., Bhoi, R. K. & Patel, K. (2010). Seasonal changes of bioelements in litter and their potential return to green leaves in five species of tropical dry deciduous forest., western India. *J For Res*, *21*, 33-38.

NoAA. (2014). National Oceanic and Atmospheric Administration.

Olivares. E. & Medina, E. (1992). Water and nutrient relations of woody perennials from tropical dry forests. *Veg Sci*, *3*, 383-392.

Olivier. J. G., Janssens-Maenhout, G., Muntean, M. & Peters, J. A. (2013). Trends in global CO_2 emissions 2013 Report, PBL Netherlands Environmental Assessment Agency The Hague, 2013.

Pacala, S. & Socolow, R. (2004). Stabilization wedges: solving the climate problem for the next 50 years with current technologies. *Science*, *305*, 968-972.

Pandey, D. N. (1998). Ethnoforestry: Local Knowledge for Sustainable Forestry and Livelihood Security, Himanshu/AFN. New Delhi.

Parry, M. L., Canziani, O., Palutikof, J., Van der Linden, P., Hanson, C., editors. Climate change 2007-impacts, adaptation and vulnerability: Working group II contribution to the fourth assessment report of the IPCC. Cambridge University Press, 2007.

Pattanayak, S., Sills, E. O., Mehta, A. D. & Kramer, R. A. (2003). Local uses of parks: uncovering patterns of household production from forests of Siberut, Indonesia. *Conserv Soc*, *1*, 209-222.

Phillips, O. L., Malhi, Y., Higuchi, N., Laurance, W. F., Nunez, P. V., Vasquez, R. M., Laurance, S. G., Ferreira, L. V., Stern, M., Brown, S. & Grace, J. (1998). Changes in the carbon balance of tropical forests: evidence from long-term plots. *Science*, 282, 439-442.

Powers, J. S. & Tiffin, P. (2010). Plant functional type classifications in tropical dry forests in Costa Rica: leaf habit versus taxonomic approaches. *Funct Ecol*, *24*, 927-936.

Prasad, R., Pandey, D. N. & Kotwal, P. C. (2000). Trees outside forests in India: a national assessment. Indian Institute of Forest Management, Bhopal, India., 2000.

Puri, S. & Swamy, S. L. (2001). Growth and biomass production in *Azadirachtaindica* seedlings in response to nutrients (N and P) and moisture stress. *Agroforest Syst*, *51*, 57-68.

Rai, S. N. & Chakarbarti, S. K. (2001). Demand and supply of fuelwood and timber in India. *Indian Forester*, *127*, 263-279.

Ramachandran, A. S., Jayakumar, R. M.., Haroon, A. B. & Arockiasamy, D. I. (2007). Carbon Sequestration: estimation of carbon stock in natural forests using geospatial technology in the eastern ghats of Tamil Nadu, India. *Curr Sci*, *92*, 323-331.

Ratter, J. A. & Pennington, R. T. (2006). Neotropical savannas and seasonally dry forests: plant diversity, biogeography, and conservation. CRC Press, Florida.

Ravindranath, N. H. & Murthy, I. K. (2003). Clean Development Mechanism and Forestry Projects: Strategy for operationalization in India. *Indian Forester*, *129*, 691-706.

Ravindranath, N. H., Somashekhar, B. S. & Gadgil, M. (1997). Carbon Flows in Indian Forests. *Clim Change*, *35*, 297-320.

Rawat, V. & Rawat, L. (2003). Climate Change Mitigation approach through plantation forestry in India. *Indian Forester*, *129*, 895-904.

Rawat, V. R. S. (2010). Reducing Emissions from deforestation in developing countries (REDD) and REDD Plus under the UNFCC negotiations Research Note. *Indian Forester*, *136*, 129-133.

Rawat, V. R. S. & Kishwan, J. (2008). Forest conservation-based, climate change mitigation approach for India. *Int Forest rev*, *10*, 269-280.

Redo, D., Aide, T. M. & Clark, M. L. (2012). Vegetation change in Brazil's dryland ecoregions and the relationship to crop production and environmental factors: Cerrado, Caatinga, and Mato Grosso, 2001-2009. *J Land Use Sci*, *8*, 1-31.

Robinon, M. R. & Zhang, X. (2011). The World Medicine Situation (Traditional Medicines: Global Situation, Issues and Challenges). Geneva. World Health Organization, Geneva, Switzerland.

Saatchi, S. S., Harris, N. L., Brown, S., Lefsky, M., Mitchard, E. T. A., Salas, W., Zutta, B. R., Buermannb, W., Lewisg, S. L., Hagen, S., Petrova, S., Whiteh, L., Silmani, M.

& Morel, A. (2011). Benchmark map of forest carbon stocks in tropical regions across three continents. *Proc Natl Acad Sci USA*, *108*, 9899-9904.

Sabine, C. L., Field, C. B. & Raupach, M. R. (2004). Integrating Humans, Climate, and the Natural World. *In: The Global Carbon Cycle* (Eds. Field CB, Raupach MR), Island Press, Washington, DC, pp. 1-16.

Sagar, R., Raghubanshi, A. S. & Singh, J. S. (2003). Tree species composition, dispersion and diversity along a disturbance gradient in a dry tropical forest region of India. *For ecol Manag*, *186*, 61-71.

Sagar, R. & Singh, J. S. (2003). Predominant phenotypic traits of disturbed tropical dry deciduous forests of northern India. *Comm Eco*, *4*, 63-71.

Sagar, R. & Singh, J. S. (2004). Local plant species depletion in a tropical dry deciduous forest of northern India. *Environ Conserv*, *31*, 55-62.

Sanchez, P. A. (1976).Properties and Management of Soils in the Tropics. New York, USA: John Wiley.

Sanchez-Azofeifa, G. A., Quesada, M., Rodriguez, J. P., Nassar, J. M., Stoner, K. E., Castillo, A., Garvin, T., Zent, E. L., Calvo-Alvarado, J. C., Kalacska, M. E. R., Fajardo, L., Gamon, J. A. & Cuevas-Reyes, P. (2005). Research priorities for neotropical dry forests. *Biotrop*, *37*, 477-485.

Schimel, D. S. (1995). Terrestrial ecosystems and the carbon cycle. *Global Change Biol*,*1*, 77-91.

Scoones, I. &Toulmin, C. (1993). Socio-economic dimensions of nutrient cycling in agropastoral systems of dryland Africa. *In Livestock and Sustainable Nutrient Cycling in Mixed farming Systems of sub-Saharan Africa: Addis Ababa* (eds. Powell J, Fernandez-Rivera S, Williams T, Renard C), Technical Papers. ILCA.

Senthilkumar, N. & Murugesan, S. (2012). Bioprospecting the renewable forest resources, an overview. *Curr Biotica*, *5*, 522-540.

Shackleton, S., Delang, C. O. & Angelsen, A. (2011). From subsistence to safety nets and cash income: exploring the diverse values of non-timber forest products for livelihoods and poverty alleviation. *In: Non-timber forest products in the global context* (Eds. Shackleton S, Shackleton C, Shanley P), Springer, Berlin, Heidelberg, pp. 55-81.

Shankar, U. (2001). A case of high tree diversity in a Sal (*Shorearobusta*)-dominated lowland forest of Eastern Himalaya: Floristic composition, regeneration and conservation. *Curr Sci*,*81*, 776-786.

Sharma, S., Bhattacharya, S. & Garg, A. (2006). Greenhouse gas emissions from India: a perspective. *Curr Sci*, *90*, 326-333.

Sheil, D. (2014). How plants water our planet: advances and imperatives. *Trends Plant Sci*, *19*, 209-211.

Sheil, D. & Murdiyarso, D. (2009). How forests attract rain: an examination of a new hypothesis. *Bioscience*, *59*, 341-347.

Shukla, J., Nobre, C. & Sellers, P. (1990). Amazon deforestation and climate change. *Science.*, *247*, 1322-1325.

Shukla, P. R. (2006). India's GHG emission scenarios: Aligning development and stabilization paths, *Curr Sci*, *90*, 384-395.

Silver, W., Ostertag, R. & Lugo, A. (2000). The potential for carbon sequestration through reforestation of abandoned tropical agricultural and pasture lands. *Restor Ecol*, *8*, 394-407.

Singh, A. N. & Singh, J. S. (1999). Biomass, net primary production and impact of bamboo plantation on soil redevelopment in a dry tropical region. *For Ecol Manag*, *119*, 195-207.

Singh, A. N. & Singh, J. S. (2000). Experiments on ecological restoration of coal mine spoil using native trees in a dry tropical environment, India: a synthesis. *New forests*, *31*, 25-39.

Singh, B. (1998). Biomass production and nutrient dynamics in three clones of *Populusdeltoides* planted on indo gangetic plains. *Plant Soil*, *203*, 15-26.

Singh, J. S., Raghubanshi, A. S., Singh, R. S. & Srivastava, S. (1989). Microbial biomass acts as a source of plant nutrient in dry tropical forest and savanna. *Nature*, *399*, 499-500.

Singh, J. S., Singh, K. P. & Agrawal, M. (1991). Environmental degradation of the Obra-Renukoot-Singrauli area, India and its impact on natural and derived ecosystems. *Environmentalist*, *11*, 171-180.

Singh, J. S. & Singh, V. K. (1992). Phenology of seasonally dry tropical forest. *Curr Sci*, *63*, 684-689.

Singh, K. & Kushwaha, C. (2005). Emerging paradigms of tree phenology in dry tropics. *CurrSci*, *89*, 964-975.

Singh, K. P. & Singh, J. S. (1988). Certain structural and functional aspects of dry tropical forests and savanna. *Int J Ecol Environ Sci*, *14*, 31-34.

Singh, L. & Singh, J. S. (1991). Species structure, dry matter dynamics and carbon flux of a dry tropical forest in India. *Ann Bot*, *68*, 263-273.

Singh, L. & Singh, J. S. (1993). Importance of short lived components of a dry tropical forest for biomass production and nutrient cycling. *J Veg Sci*, *4* , 681-686.

Singh, O. P., Datta, B. & Rao, C. N. (1991). Podochemical characterization and genesis of soils in relation to altitude in Mizoram. *J Indsoc Soil Sci*, *39*, 739-750.

Singh, S. P., Rawat, Y. S. & Garkoti, S. C. (1997). Failure of brown oak (*Quercussemi-carpifolia*) to regenerate in the Central Himalaya: a case of environmental semi-surprise. *Curr Sci*, *73*, 371-374.

Soni, P. (2003). Climate change and restoration of tropical forests. *Indian Forester*, *129*, 865-873.

Sorenson, L. J. (1974). Rate of decomposition of organic matter in soil as influenced by repeated air drying-rewetting and repeated additions of organic matter. *Soil Biol Biochem*, *6*, 287-292.

Srivastava, S. C. & Singh, J. S. (1991). Microbial C, N and P in dry tropical forest soil: effects of alternate land uses and nutrient flux. *Soil Biol Biochem*, *23*, 117-124.

Tamgadge, D. B., Gaikwad, S. T., Gajbhiye, K. S. & Gaikwad, M. S. (2000). Soil landform relationship of granite/gneissic terrain in Deccan plateau satpura range, Madhya Pradesh. *J Ind Soc Soil Sci*, *48*, 567-571.

Top, N., Mizoue, N., Ito, S. & Kai, S. (2004b). Spatial analysis of woodfuel supply and demand in Kampong Thom Province., Cambodia. *Forest Ecol Manag*, *194*, 369-378.

Top, N., Mizoue, N., Ito, S., Kai, S., Nakao, T. & Ty, S. (2006). Re-assessment of wood fuel supply and demand relationships in Kampong Thom Province, Cambodia. *Biomass Bioenerg*, 30, 134-143.

Top, N., Mizoue, N. & Kai, S. (2004a). Estimating forest biomass increment based on permanent sample plots in relation to wood fuel consumption: A case study in Kampong Thom Province, Cambodia. *J Forest Res*, *9*, 117-123.

Tosi, Jr. J. A. & Voertman, R. F. (1964). Some environmental factors in the economic development of the tropics. *Econ Geogr*, *40*, 189-205.

Trejo, I. & Dirzo, R. (2000). Deforestation of seasonally dry tropical forest: a national and local analysis in Mexico. *Biol Conserv*, 94, 133-142.

Van Schaik, C. P., Terborgh, J. & Dugelby, B. (1997). The silent crisis: the state of rain forest nature preserves. *In: Last stand: protected areas and the defense of tropical biodiversity* (Eds. Kramer R, van Schaik C, Johnson J), Oxford University Press, New York, NY, US, pp. 64-89.

Waeber, P., Ramesh, B., Parthasarathy, N., Pulla, S. & Garcia, C. (2012). Seasonally dry tropical forests in South Asia: A research agenda. A research agenda to contribute to the discussions on "Key Issues for the Global Dry Forests" workshop organized by CIFOR/For Dev in Zurich, 28-30th October, 2012.

Wang, X. C. (2003). Problems and relevant strategies on natural forest protection in change bai Mountain forest area. *J For Res*, *14*, 259-262.

World Commission on Forests and Sustainable Development. (1998) Our Forests. Our Future, March report, WCFSD Secretariat, Winnipeg, pp.126.

Wunder, S. (2001). Poverty alleviation and tropical forests: What scope for synergies? *World Dev*, *29*, 1817-33.

Yosi, C., Keenan, R. & Fox, J. (2011). Forest dynamics after selective timber harvesting in Papua New Guinea. *For Ecol Manage*, *262*, 895-902.

About the Editors

Prof. Dr. Rajeshwar P. Sinha

Dr. Rajeshwar P. Sinha, DAAD (German Academic Exchange Service) Fellowship Awardee and Fellow, Society for Applied Biotechnology, India, is a Professor of Molecular Biology and Coordinator of Applied Microbiology, Centre of Advanced Study in Botany, Banaras Hindu University (BHU), Varanasi, India. He received his Ph. D. (Biotechnology) form BHU, Varanasi and visited countries such as Argentina, Austria, Belgium, Canada, China, Germany, Greece, France, Italy, Japan, Luxembourg, Norway, Poland, Republic of Korea, Spain, Switzerland, The Netherlands, UK and USA, in the field of academics/research. He is a Lifetime Member of several national and international scientific societies and an Editorial Board Member of various national and international journals. He is working on UV radiation effects on aquatic ecosystems (DNA damage and repair, phycobiliproteins, mycosporine-like amino acids and scytonemin). He has published over 300 research papers/reviews/book chapters and edited/authored six books. He has over 8725 citations with an h-index of 46.

Prof. Dr. Shashi Pandey-Rai

Dr. Shashi Pandey-Rai is a professor in the Department of Botany, Institute of Science, Banaras Hindu University, Varanasi, India. She has made an outstanding contribution in the area of epigenetic/genetics/genomics of regulation of stress resistance/terpenoid biosynthetic pathway expression in *Artemissia annua*, *Catharanthus roseus* and *Withania coagulans*. The main focus of her research work is to identifying key elements involved in these processes which will allow generating novel tools for metabolic engineering. She has published several book chapters and around 68 papers in international and national journals. One US Patent FPO 1826, NF28/04 granted and published online in 2006, App. No. 10/838531, One technology transferred to Almet India Pvt. ltd. (Mahyco).

Prof. Dr. Nandita Ghoshal

Dr. Nandita Ghoshal is a Professor at Department of Botany, Institute of Science, Banaras Hindu University, Varanasi, India. Since February 2017 she is Course Coordinator, M Sc Environmental Science, Department of Botany, BHU. She did her Ph. D. in Botany from Banaras Hindu University. She has a teaching experience of about twenty four years. Her academic visit abroad includes several countries, Denmark, England, Scotland and USA. She has published more than 35 research papers in reputed international journals and book chapters. She has attended about forty international/ national conferences, seminars and symposium. Her area of specialization is Plant Ecology and Environmental Science. Her current research interest includes eecological analysis of organic matter dynamics of terrestrial ecosystems with special reference to agroecosystems, carbon and nitrogen balance in soil, global climate change, sustainable biological productivity, land use change and biodiversity.

INDEX

#

A

B

C

N

O

P

Q

R

S

T

U

W

Z

Related Nova Publications

TRENDS IN LIFE SCIENCE RESEARCH

EDITORS: Rajeshwar P. Sinha and Umesh P. Shrivastava

SERIES: Biotechnology in Agriculture, Industry and Medicine

BOOK DESCRIPTION: This book is primarily written with the objective of providing standard information on recent topics in the fields of stress biology, molecular biology, ecology, agriculture, bioremediation, human diseases and over all, several biotechnological approaches towards the studies ranging from green photosynthesizing plants to human pathogens.

HARDCOVER ISBN: 978-1-53613-241-0
RETAIL PRICE: $230

ADVANCES IN MATERIALS SCIENCE RESEARCH. VOLUME 38

EDITOR: Maryann C. Wythers

SERIES: Advances in Materials Science Research

BOOK DESCRIPTION: Chapter one of *Advances in Materials Science Research. Volume 38*, critically scrutinizes various recycling techniques implemented in the current scenario for the polyvinyl chloride based products.

HARDCOVER ISBN: 978-1-53615-597-6
RETAIL PRICE: $250

To see a complete list of Nova publications, please visit our website at www.novapublishers.com

Related Nova Publications

Mechanisms of Evolution

Author: Börje Ekstig

Series: Insights into Biological and Cultural Evolution

Book Description: In this book, we embark on an innovative, exploratory and interdisciplinary adventure, step by step following the author towards his quest of investigating evolution, its mechanisms, its direction and progress and, not least, the place of ourselves in it.

Hardcover ISBN: 978-1-53615-795-6
Retail Price: $160